Developing and Monitoring Smart Environments for Intelligent Cities

Zaigham Mahmood
University of Northampton, UK & Shijiazhuang Tiedao University, China

IGI Global
PUBLISHER of TIMELY KNOWLEDGE

A volume in the Advances in Civil and Industrial Engineering (ACIE) Book Series

Published in the United States of America by
IGI Global
Engineering Science Reference (an imprint of IGI Global)
701 E. Chocolate Avenue
Hershey PA, USA 17033
Tel: 717-533-8845
Fax: 717-533-8661
E-mail: cust@igi-global.com
Web site: http://www.igi-global.com

Library of Congress Cataloging-in-Publication Data

Names: Mahmood, Zaigham, editor.
Title: Developing and monitoring smart environments for intelligent cities / Zaigham Mahmood, editor.
Description: Hershey, PA : Engineering Science Reference, 2020. | Includes bibliographical references and index. | Summary: "This book explores development, management, and monitoring approaches to building digital cities and intelligent environments"-- Provided by publisher.
Identifiers: LCCN 2020014110 (print) | LCCN 2020014111 (ebook) | ISBN 9781799850625 (h/c) | ISBN 9781799857228 (s/c) | ISBN 9781799850632 (eISBN)
Subjects: LCSH: Smart cities. | Ambient intelligence. | City planning--Technological innovations.
Classification: LCC TD159.4 .D48 2020 (print) | LCC TD159.4 (ebook) | DDC 307.1/160285--dc23
LC record available at https://lccn.loc.gov/2020014110
LC ebook record available at https://lccn.loc.gov/2020014111

This book is published in the IGI Global book series Advances in Civil and Industrial Engineering (ACIE) (ISSN: 2326-6139; eISSN: 2326-6155)

British Cataloguing in Publication Data
A Cataloguing in Publication record for this book is available from the British Library.

All work contributed to this book is new, previously-unpublished material. The views expressed in this book are those of the authors, but not necessarily of the publisher.

For electronic access to this publication, please contact: eresources@igi-global.com.

Advances in Civil and Industrial Engineering (ACIE) Book Series

Ioan Constantin Dima
University Valahia of Târgovişte, Romania

ISSN:2326-6139
EISSN:2326-6155

Mission

Private and public sector infrastructures begin to age, or require change in the face of developing technologies, the fields of civil and industrial engineering have become increasingly important as a method to mitigate and manage these changes. As governments and the public at large begin to grapple with climate change and growing populations, civil engineering has become more interdisciplinary and the need for publications that discuss the rapid changes and advancements in the field have become more in-demand. Additionally, private corporations and companies are facing similar changes and challenges, with the pressure for new and innovative methods being placed on those involved in industrial engineering.

The **Advances in Civil and Industrial Engineering (ACIE) Book Series** aims to present research and methodology that will provide solutions and discussions to meet such needs. The latest methodologies, applications, tools, and analysis will be published through the books included in **ACIE** in order to keep the available research in civil and industrial engineering as current and timely as possible.

Coverage

- Hydraulic Engineering
- Optimization Techniques
- Engineering Economics
- Productivity
- Urban Engineering
- Transportation Engineering
- Materials Management
- Structural Engineering
- Coastal Engineering
- Operations Research

IGI Global is currently accepting manuscripts for publication within this series. To submit a proposal for a volume in this series, please contact our Acquisition Editors at Acquisitions@igi-global.com or visit: http://www.igi-global.com/publish/.

The Advances in Civil and Industrial Engineering (ACIE) Book Series (ISSN 2326-6139) is published by IGI Global, 701 E. Chocolate Avenue, Hershey, PA 17033-1240, USA, www.igi-global.com. This series is composed of titles available for purchase individually; each title is edited to be contextually exclusive from any other title within the series. For pricing and ordering information please visit http://www.igi-global.com/book-series/advances-civil-industrial-engineering/73673. Postmaster: Send all address changes to above address.

Titles in this Series

For a list of additional titles in this series, please visit:
http://www.igi-global.com/book-series/advances-civil-industrial-engineering/73673

Applications and Challenges of Maintenance and Safety Engineering in Industry 4.0
Alberto Martinetti (University of Twente, The Netherlands) Micaela Demichela (Politecnico di Torino, Italy) and Sarbjeet Singh (Luleå University of Technology, Sweden)
Engineering Science Reference • © 2020 • 321pp • H/C (ISBN: 9781799839040) • US $265.00

Reconstructing Urban Ambiance in Smart Public Places
Hisham Abusaada (Housing and Building National Research Center (HBRC), Egypt) Ashraf M. Salama (University of Strathclyde, Glasgow, UK) and Abeer Elshater (Ain Shams University, Egypt)
Engineering Science Reference • © 2020 • 280pp • H/C (ISBN: 9781799838562) • US $225.00

Additive Manufacturing Applications for Metals and Composites
K.R. Balasubramanian (National Institute of Technology, Tiruchirappalli, India) and V. Senthilkumar (National Institute of Technology, Trichy, India)
Engineering Science Reference • © 2020 • 348pp • H/C (ISBN: 9781799840541) • US $205.00

Multi-Objective Optimization of Industrial Power Generation Systems Emerging Research and Opportunities
Timothy Ganesan (Royal Bank of Canada, Canada)
Engineering Science Reference • © 2020 • 233pp • H/C (ISBN: 9781799817109) • US $195.00

Impact of Industry 4.0 on Architecture and Cultural Heritage
Cecilia Maria Bolognesi (Politecnico of Milano, Italy) and Cettina Santagati (Università di Catania, Italy)
Engineering Science Reference • © 2020 • 422pp • H/C (ISBN: 9781799812340) • US $225.00

Handbook of Research on Urban-Rural Synergy Development Through Housing, Landscape, and Tourism
Aleksandra Krstić-Furundžić (University of Belgrade, Serbia) and Aleksandra Djukić (University of Belgrade, Serbia)
Engineering Science Reference • © 2020 • 437pp • H/C (ISBN: 9781522599326) • US $275.00

Re-Coding Homes Through Flexible Interiors Emerging Research and Opportunities
Nilüfer Saglar Onay (Politecnico di Torino, Italy) S. Banu Garip (Istanbul Technical University, Turkey) and Ervin Garip (Istanbul Technical University, Turkey)
Engineering Science Reference • © 2020 • 165pp • H/C (ISBN: 9781522589587) • US $175.00

Green Building Management and Smart Automation
Arun Solanki (Gautam Buddha University, India) and Anand Nayyar (Duy Tan University, Vietnam)
Engineering Science Reference • © 2020 • 312pp • H/C (ISBN: 9781522597544) • US $215.00

IGI Global
PUBLISHER of TIMELY KNOWLEDGE

701 East Chocolate Avenue, Hershey, PA 17033, USA
Tel: 717-533-8845 x100 • Fax: 717-533-8661
E-Mail: cust@igi-global.com • www.igi-global.com

To Rehana Zaigham Mahmood

Editorial Advisory Board

Table of Contents

Detailed Table of Contents

In this chapter, the authors present the German city of Marburg as an alternative model for the conception of a smart city. Marburg's historical relation with its visually impaired citizens has shaped not only its infrastructure but also its human framework. Generally, smart cities are equated with world-class major metropolitan areas, with international airports, use of high band internet, internet of things, and other IT infrastructures. However, Marburg might be considered a smart city according to other criteria. This case study articulates the uniqueness of what Marburg has been able to achieve using a diverse approach of cultural acceptance of the blind, becoming a world reference for other small cities. The authors suggest that Marburg can serve as a model for other cities. This hypothesis was reached through critical investigations into concepts of smartness and disability, intersected with the insights obtained in a focus group interview.

Cities in developing countries are increasingly under stress through urbanization, which leads to the expansion of slum areas or informal settlements due to demand for low-cost housing. This chapter presents the social, environmental, and economic realities facing slum dwellers and discusses their redevelopment into intelligent cities. The concept of 'function accompanying intelligent' is introduced for the transformation of slums to functional intelligent cities. In this context, a city is intelligent if it serves both the functional and social needs of its entire population. The chapter overviews an approach to integrated data collection, data analytics, and user access to information. Geospatial analysis of demographic, economic, social, and environmental data is introduced to help delineate slums, and to monitor the outcomes of urban planning initiatives and the progress of social wellbeing. The city of Accra in Ghana is discussed as a potential slum city to functional intelligent city transformation.

Through the use of internet of things-based sensors in air quality monitoring stations, concentration of different pollutants and meteorological parameters can be regularly measured. In case of unusual conditions (e.g., increased levels of dangerous pollutants), a smart assessment system can produce warning so that appropriate air quality management process can be initiated. In this context, the objective of this study is to discover relationships and patterns among air pollution features and characteristics. In this case, determination of frequently observed association rules can trigger an appropriate background smart environment system when a critical situation is detected. In the experimental studies in the current project, traditional association rule mining and weighted association rule mining methods have been employed using real-world datasets collected from 21 monitoring stations in Turkey. In consequence, useful and outstanding association rules exceeding the user-defined support and confidence levels were obtained that can form basis for further research.

An intelligent city, also known as smart city or cognitive city, is a new vision for urban development to integrate the various sectors and functions of the society through the use of internet of things and sensor-based technologies. The main objective is to seamlessly integrate and efficiently manage a city's resources such as transportation, healthcare, commerce, manufacturing, education, utilities, law enforcement, as well as city's various e-government information systems. The aim is to improve the quality of life of citizens and efficiency of city's services and processes, through the use of latest distributed technologies. In this chapter, the authors discuss the connected vehicles vision, as a core element of the internet of vehicles paradigm, as part of the intelligent transportation system (ITS). In doing so, they discuss the underlying concepts, technologies, types of vehicular connectivity, and mobile and vehicular ad hoc networks (i.e., MANET and VANET).

This chapter focuses on a systematic and integrated approach to managing cities using multidisciplinary and technology-based approach and drawing on integrated knowledge from different fields. The chapter aims to discuss and analyse issues in an organic and holistic approach. It explores theory and combines praxis, applications, and futuristic conjectures. Emphasis is placed on issues of ethics, human rights, environmental friendliness, sustainability, and compliance with the 17 UN Sustainable Development Goals (SDGs). While the chapter has considerable input from existing scholarship, it adopts a narrative that is innovative and creative and engages with novel ideas and propositions for future progress. It also charts the breakthrough in transportation, commerce, medicine, education, the world of work, and contemporary health, with particular emphasis on issues relating to safety, confidentiality, privacy, facial recognition, city policing, crime detection, monitoring and evaluation, forecasting, and future of AI.

The rapid growth of the world's population is placing a huge strain on the existing infrastructures. As a quest for accommodating this growth, interest is turned to the internet of things (IoT). In fact, the IoT is significantly improving today's quality of life by innovating the provided services and enhancing communication and interaction. Furthermore, it has also empowered real-time decision making by introducing dynamic services for innovative traffic handling, energy-efficient infrastructure saving, and public safety ensuring. However, IoT applications for smart cities is still a major issue as it lacks assuring privacy and security within provided services. In this chapter, the authors pinpoint IoT's security risk assessment challenges and examine its critical influence on smart cities. Additionally, they highlight the key aspects characterizing a smart city which also represent the critical assets requiring security risk assessment. Moreover, they discuss the resulting issues and their related countermeasures.

In 2017, a Price Waterhouse report on intelligent cities advised that technology would be a key enabler for efficient management of resources for overpopulated cities. The increased reliance on technology to drive daily lives of people is the main reason why many believe that smart technology is a major part of intelligent cities. By design, intelligent cities are heavily reliant on connected devices. Consequently, reliance on technology poses the risk that any disruption of technology-based services may put human lives in danger. In this context, attacks on technological frameworks can mean attacks on the entire city. Unfortunately, there is no existing technology that is able to successfully defend against such cyberattacks. In this research, the authors review the concepts, principles, and technologies, and propose an approach for a cyber defence system, based on artificial intelligence and machine learning, to detect and prevent an integrated cyberattack.

Smart homes, which incorporate IoT technologies to provide home security, efficient environmental services, conveniences, and improved living standards, are becoming the centre of smart urban developments. With the increased inter-connectivity of smart objects and sensors, there is now, also, an increased level of cyber threats, which can compromise privacy and security. These threats either modify packets of information or inject modified packets into the networks. This chapter examines current intrusion detection systems (IDSs) and presents a unique solution to overcome intrusion detection challenges. It discusses the implementation of smart home IDS (SHIDS), using a machine learning based signature and anomaly intrusion detection scheme to detect network intrusions in the smart home. Suggested mechanism is based on naïve Bayes technique to improve the detection performance. The performance of SHIDS has been tested with network intrusions resulting from DoS, probe, remote-to-local (R2L), and user-to-root (U2R) attacks.

Preface

OVERVIEW

An Intelligent City is a new vision for urban development that brings together the various sectors of the society through the deployment of Distributed Computing technologies. The vision extends the Internet of Everything (IoE) paradigm, to efficiently manage city resources, optimise city operations, and connect with citizens and other sectors of the society. Relevant underlying technologies include the Internet of Things (IoT) paradigm, ubiquitous computing frameworks, cyber-physical cloud/edge computing, sensor-based networks, mobile crowdsourcing methods, device connectivity approaches, and mobile and telecommunication technologies. Relevant city services based on interconnection between smart objects include intelligent transportation system, smart healthcare, efficient provision of utilities, automated working environments, and, in general, smarter approaches to more comfortable living for citizens. There is no universally clear definition of what an Intelligent City (also known as Smart City) is; but its 'smartness' depends on the level of digitisation of its services, that the city provides. Intelligent Cities, often known as Digital Cities or Cognitive Cities, also imply a more interactive, open and responsive city administration, safer public spaces and meeting the needs of its citizens with improved quality of life (European Commission, 2020).

In this new vision of urban development, the Intelligent City vision aims to improve urban life through sustainable and technology based integrated solutions to living in a digital city. Modern cities around the world are growing larger; nearly 55% of the world's population now lives in urban areas. Since, these numbers are expected to rise to around 70% by the year 2050 (Apm.org, 2020), the ultimate objective is to develop smart living environments, improve working environments, automate city services, promote safety and security, improve citizens' mobility requirements, develop open and transparent systems, and, in general, build *smarter* and *connected* communities. Goal is to ensure that citizens' needs are met and they are technologically empowered to affect the city's functioning following a technologically-led and citizen-centered-government approach.

In this context, Internet of Things (IoT) can be regarded as the most essential underlying technology where 'things' refer to not just computers and processer embedded smart devices, but also everyday objects and systems such as mobile phones, smart body wearables, autonomous cars, home security systems, vehicle navigation systems, etc. Here, smart sensors are at the core of any IoT based system. These collect and transmit data to other connected objects for further automated actions. Big Data Analytics help to find trends in the many zeta bytes of data generated each day, for prompt decision making. 5G technologies, through the use of faster broadband, provide strong connectivity between millions of sensor-embedded devices that pervade within the Intelligent Cities to enable the IoT to work more efficiently. Geospatial technologies, such as GPS, help city planner to provide efficient management of transportation and connected vehicles. Robotics, Artificial Intelligence (IA) and Machine Learning allow automation of city operations and taking over of many hazardous processes from human operators (HereMobility, 2020).

All these technologies and related applications become essential elements of Smart Infrastructure of Intelligent Cities, that generally consist of three layers of "smartness", viz:

- Sensor based networks for seamless connectivity
- Smart applications for data processing and decision making in real time
- Adoptability protocols for human interaction, as necessary.

In this scenario, afore mentioned technologies, social media offerings, communication protocols, and device connectivity frameworks that are well established are already helping to do the following:

- Empower citizens to provision the digital services provided by the city managers
- Allow city officials to interact directly with the community at large and to monitor digital environments of Intelligent Cities
- Develop smart communities where citizens can enjoy improved quality of life living in smart environments in digital cities.

Intelligent Cities, and smart environments within, are developed and monitored mainly to deliver better quality of life for citizens. According to Mckinsey & Co. (2020), and there is a general agreement that, good quality of life includes a number of related aspects of a city e.g. public safety and security, faster multimodal commutes, healthier living, economic prosperity that also attracts venture and investment capital, and greener environments. There is no doubt that development and management of smarter environments is proving to be a highly successful model for the new style urbanisation. There are currently over 50 smart cities around the world, including Amsterdam, Hong Kong, Toronto, Seoul, Singapore, Reykjavik, Tokyo, Paris, London, and New York (Gigabit, 2019). The existing model will, no doubt, continue to evolve as the technology industry develops newer frameworks, smarter applications, better standards, and more efficient solutions for building the digital cities of the future.

Apart from the desire to use technology for the sake of speed, openness, and effectiveness, there are currently a number of inherent challenges that modern cities face. These include factors such as the following:

- Rise in unemployment due to global economic downturn and increased automation
- Pressures on housing sector due to growing demand for security due to growth in population
- Concerns about climate change, privacy and safety
- Paradigm shift towards online entertainment and internet-based services
- Pressures on public finances and city resources due to increased demands of the general public
- Need to educate and empower citizens to affect the functioning of city governments.

Fortunately, the said concerns are well understood and, therefore, being addressed as the development and innovation continues.

In the context as articulated above, this book aims to investigate the development, management, governance, and monitoring approaches to building Intelligent Cities, and smart environments therein. Emphasis is on the current and future use of IoT based frameworks, related methodologies, device connectivity approaches, communication protocols as well as inherent issues and limitations. Thirty researchers and practitioners of international repute have presented latest research on frameworks, current trends, case studies and suggestions for further enhancement of the IoT vision for the development of Intelligent Cities. Hopefully, the current text, *Developing and Monitoring Smart Environments for Intelligent Cities,* will fill a gap with respect to smart living in smarter environments; and further extend the existing body of knowledge in this field.

BOOK OBJECTIVE

This book, *Developing and Monitoring Smart Environments for Intelligent Cities*, aims to serve as a reference text that presents the latest research focusing on the use of distributed and mobile technologies, especially the applications of the IoT vision, to further develop smart environments and intelligent services, as necessary components of Intelligent Cities. Such systems and services being the Industrial IoT, smart healthcare, Intelligent traffic system, Internet of Vehicles, Internet based commerce, etc. The objective is to understand the use and effectiveness of such technologies and best practices that provide successful strategies towards new ways of automation of urbanisation. The objectives are as follows:

- To discuss frameworks, strategies. and case studies illustrating approaches, best practices and practical solutions
- To present state-of-the-art research and practice with respect to the usefulness, as well as limitations, of Intelligent City environments
- To develop a complete reference for students, researchers and practitioners of distributed computing systems and smart environments
- To present future research directions and technological innovations in the context of distributed smarter environments and the IoT.

TARGET AUDIENCE

This volume is a reference text aimed at supporting a number of potential audiences, including the following:

- *Communication Engineers, Software Developers,* and *Network Security Specialists* who wish to adopt the newer approaches to developing information systems, and to ensure security of data and implement seamless connectivity of smart objects.
- *Students* and *Lecturers* who have an interest in further enhancing the knowledge of latest technologies, mechanisms and frameworks relevant to smarter living environments from a distributed computing perspective.

- *Researchers* and *Practitioners* in this field who require up to date knowledge of the current practices, mechanisms, frameworks and limitations relevant to the development of Intelligent Cities and extension of the Internet of Things vision.

BOOK ORGANISATION

This book, *Developing and Monitoring Smart Environments for Intelligent Cities*, is organised in three sections with a total of 12 chapters, authored by 30 well known academics and practitioners from around the world, as follows:

Section 1: Concepts, Principles, and Reference Architectures

There are three contributions in this section. The first chapter discusses the concepts, principles and emerging technologies that help to transform the city infrastructures of an intelligent city, in terms of development, monitoring, and management. The second chapter presents a reference model for building intelligent cities taking into account the various dimensions relevant to intelligent cities; and the next chapter presents developing trends with respect to human-machine interaction in relation to the Industrial Internet of Things sector of an intelligent city.

Section 2: Development Approaches, Monitoring, and Case Studies

This section has five chapters. The first of these looks at smart cities of London, Amsterdam, Vienna and Stockholm as examples of modern digital cities. The second contribution discusses the inclusion of disability in a modern city using case study of city of Marburg in Germany. The next chapter discusses transformation of urban slums into functional intelligent cities and uses city of Accara in Ghana as an example. The 4th chapter presents an air pollution monitoring system for intelligent cities using Weighted Association Rule Mining approach. The final chapter in the section discusses the connected vehicles vision as a vital component of intelligent transportation for an intelligent digital city.

Section 3: Security Challenges and Cyber Defence Systems

This part of the book comprises four chapters. The first contribution presents a multidisciplinary approach to intelligent cities considering different fields of study including transportation, commerce, medicine, education, employment, and health care; whereas the second chapter focuses on the security risk assessment challenges inherent in the IoT based applications for an intelligent city. The 3rd contribution suggests a machine learning based cyber defence system for a modern digital City. The final contribution in the section presents a hybrid intrusion detection system for smart home applications with an intelligent city.

CHAPTER DESCRIPTIONS

Detailed abstracts of book chapters appear in the *Detailed Table of Contents* section in this book. Here, very brief summaries of chapters' content are presented.

Chapter 1 is entitled 'Building Intelligent Cities: Concepts, Principles, and Technologies'. Authored by Vijayaraghavan and Rajendra, it looks into the essential concepts, principles and technologies that help to transform the city infrastructures of an intelligent city. The aim is to enhance the provision and management of electronic services within the city. The chapter provides a comprehensive introduction to intelligent cities (also known as smart cities), and the associated requirements. It also articulates the evolution of a typical city to a truly integrated, responsive, open and connected intelligent city in terms of the underlying technologies. In this respect, the chapter serves as the basis for other related contributions, that appear later in the book.

Chapter 2 is jointly developed by Soares, Monteiro, Duarte, and Machado. Entitled 'Reference Models for Intelligent Cities: An Aligned Template', this contribution discusses the modelling approaches and reference architectures, that currently exist, for the development of intelligent cities. Based on a comprehensive literature review, and recognising the lack of a universally accepted reference model, the authors discuss the functional and non-functional requirements, present the various dimensions relevant to intelligent cities, and propose an aligned reference model in terms of the said dimensions, taking into consideration the degree of acceptance of each dimension. Future directions are also presented.

Chapter 3, co-authored by Demir and Turan, is entitled 'Developing Trends in Power and Networking Technologies for Intelligent Cities'. This contribution has a focus on a number of power and networking technologies that help to realise the vision of intelligent cities. The chapter starts with a discussion on Industry 4.0 and Industry 5.0. Then, outlining the prevailing robotics and ambient intelligent systems, the contribution presents developing trends with respect to human-machine interaction in relation to Industrial Internet of Things (IIoT) sector of an intelligent city. In this context, the chapter looks into the advancement in Artificial Intelligence and Robotics as means of man-machine integration to further the vision of digital intelligent cities.

Chapter 4 is contributed by Vijayaraghavan and Leevinson. It is titled as 'Next Generation of Intelligent Cities: Case Studies From Europe' and presents a discussion on unique characteristics and requirements of next generation of intelligent cities that will set them apart from the current level of digitisation within todays' modern cities. This contribution looks at smart cities of London, Amsterdam, Vienna and Stockholm as examples of innovation, ingenuity, technology, public policy making and sustainability; and examines the use of different emerging technologies including quantum computing, autonomous vehicles, Artificial Intelligence, Machine Learning, etc., that could potentially positively impact the next generation of digital cities.

Chapter 5 is entitled 'A Smart City for the Blind: Marburg as a Case Study'. In this contribution, the authors, Schelin, Gomes and Quandt present a case study of the city of Marburg (known as the city of the blind) in Germany, as an alternative model for the conception of a smart city. The study articulates the uniqueness of what Marburg has been able to achieve, using a diverse approach of cultural acceptance of the blind and visually impaired, becoming a world reference for other small cities. The authors suggest that Marburg can serve as a model for other cities. The hypotheses presented in the chapter were reached through critical investigation into the concepts of smartness and disability, intersected with the insights obtained through focus group interviews.

Chapter 6 is jointly developed by Cordes, Wongthongtham, and Morrison. Entitled 'Transforming Urban Slums: Pathway to Functional Intelligent Cities in Developing Countries', this contribution presents social, environmental and economic realities facing urban slum dwellers; and discusses their redevelopment into functional intelligent cities. The concept of 'function accompanying intelligence' is introduced for this transformation. In the chapter, a novel approach to integrated data collection, data analytics and user access to information is developed and further discussed. The city of Accra in Ghana is presented as a case study, and discussed as a potential slum city transformed into a functional intelligent city.

Chapter 7 is developed by Tuysuzoglu and Birant. Entitled 'Air Pollution Monitoring in Intelligent Cities Using Weighted Association Rule Mining', this contribution presents an air pollution monitoring system for intelligent cities. The chapter submits results of experimental studies, that look into the relationship and patterns among air pollution features and characteristics, considering the concentrations of air pollutants in the air. The study uses traditional Association Rule Mining and Weighted Association Rule Mining methods on real-world datasets collected from 21 monitoring stations situated at different locations in Turkey. The aim is to determine the useful and outstanding association rules that can hopefully form the basis for further research.

Chapter 8 is entitled 'Connected Vehicles: A Vital Component of Smart Transportation in an Intelligent City'. Authored by Zaigham Mahmood, this contribution discusses the connected vehicles vision, as part of the Internet of Vehicles (IoV) paradigm, as a core element of the Intelligent Transportation System (ITS), which is an essential mobility related component of an intelligent city. The author presents the concepts, architectures, and enabling technologies for developing an efficient transportation system, especially for the network of connected vehicles. Discussing the types of vehicle connectivity (including V2V, V2I, V2P, and V2X), this contribution outlines the technologies relevant to Mobile and Vehicular Ad Hoc Networks i.e. MANET and VANET.

Chapter 9 is titled 'Intelligent Cities: A Compendious and Multidisciplinary Approach – Issues and Opportunities'. Developed by Kwesi Sakyi, it focuses on an integrated approach to managing cities using multidisciplinary and technology-based methods, drawing on integrated knowledge from different fields of study including transportation, commerce, medicine, education, employment, and health care. The aim is to discuss and analyse issues in a holistic manner to explores theory; and combine praxis, applications and futuristic conjectures. Emphasis is on ethics, human rights, environmental friendliness, sustainability, and compliance with the seventeen goals as suggested by the UN Sustainable Development programme.

Chapter 10 is developed by Abbas, Baina, and Bellafkih. Entitled 'IoT Applications for Intelligent Cities: Security Risk Assessment Challenges', this contribution places a focus on the security and privacy concerns relating to Internet of Things based applications within an intelligent cities infrastructure, and the smart environments therein. The chapter examines the impact and implications of inherent security challenges. It highlights the key factors that characterize the smart environments, and determines the critical assets that require more careful security risk assessment and resolution. Research is based on a case study of the city of Casablanca in Morocco. The contribution also presents the countermeasures to the identified issues and risks.

Chapter 11 is a collaborative effort jointly authored by Piconi, Maruatona, Alex Ng, Kayes and Watters. Titled as 'A Machine Learning Based Cyber Defence System for an Intelligent City', this chapter states that too much reliance on technology, as is the case in technology based intelligent cities, poses huge risks that can cause disruption and denial of city services. Such disruptions can cause, in the worst-case scenario, the entire city to put human lives in danger. Within this context, this contribution reviews the principles and technologies relevant to the topic, and proposes a framework for a secure cyber defence system. The emphasis, is on the development of a system based on artificial intelligence and machine learning approaches.

Chapter 12 is entitled 'Hybrid Intrusion Detection System for Smart Home Applications'. Developed by Hussain, Induruwa and Man Qi, this chapter examines existing Intrusion Detection Systems (IDS) and presents a novel solution to overcome intrusion detection challenges to the Internet of Things based applications for smart homes. It discusses the implementation of Smart Homes IDS (SHIDS), using a machine learning based signature and anomaly intrusion detection scheme to detect network intrusions. The suggested mechanism is based on Naïve Bayes technique that has been tested in case of network intrusions resulting from Denial of Service, Probe, Remote-to-Local and User-to-Root attacks.

REFERENCES

Apm.org. (2020). *Smart Cities, Urbanisation, and Connectivity*. Available at: https://www.apm.org.uk/projecting-the-future/smart-cities-urbanisation-and-connectivity/

European Commission. (2020). *Smart Cities*. Available at: https://ec.europa.eu/info/eu-regional-and-urban-development/topics/cities-and-urban-development/city-initiatives/smart-cities_en

HereMobility. (2020). *Smart City Mobility*. Available at: https://mobility.here.com/smart-city-technologies-role-and-applications-big-data-and-iot

McKinsey & Co. (2018, June). *Smart Cities: Digital Solutions for a More Livable Future*. Available at: https://www.mckinsey.com/~/media/mckinsey/industries/capital%20projects%20and%20infrastructure/our%20insights/smart%20cities%20digital%20solutions%20for%20a%20more%20livable%20future/mgi-smart-cities-full-report.ashx

Mullan, L. (2019, June). *Top Ten Smart Cities in the World, Published in Gigabit*. Available at: https://www.gigabitmagazine.com/big-data/top-10-smart-cities-world

APPENDIX: OTHER BOOKS BY ZAIGHAM MAHMOOD

Software Engineering in the Era of Cloud Computing

This book has a focus on developing scalable complex software for distributed computing applications. It presents and discusses state of the art of software engineering (SE) in terms of methodologies, trends and future direction for cloud SE. Domain modelling, SE testing in the cloud, SE analytics, and software process improvement as a service rare also discussed. ISBN: 978-3030-24891-8

The Internet of Things in the Industrial Sector: Security and Device Connectivity, Smart Environments, and Industry 4.0

This reference text has a focus on the development and deployment of the Industrial Internet of Things (IIoT) paradigm, discussing frameworks, methodologies, benefits and inherent limitations of connected smart environments, as well as providing case studies of employing the IoT vision in the industrial domain. ISBN: 978-3030-24891-8

Security, Privacy and Trust in the IoT Environment

This book has a focus on security and privacy in the Internet of Things environments. It also discusses the aspects of user trust with respect to device connectivity. Main topics covered include: principles, underlying technologies, security issues, mechanisms for trust and authentication as well as success indicators, performance metrics and future directions. ISBN: 978-3030-18074-4

Guide to Ambient Intelligence in the IoT Environment: Principles, Technologies and Applications

This text discusses the AmI element of the IoT paradigm and reviews the current developments, underlying technologies, and case scenarios relating to AmI-based IoT environments. The book presents cutting-edge research, frameworks, and methodologies on device connectivity, communication protocols, and other aspects relating to the AmI-IoT vision. ISBN: 978-3-030-04172-4

Fog Computing: Concepts, Frameworks and Technologies

This reference text describes the state of the art of Fog and Edge computing with a particular focus on development approaches, architectural mechanisms, related technologies, and measurement metrics for building smart adaptable environments. The coverage also includes topics such as device connectivity, security, interoperability, and communication methods. ISBN: 978-3319-94889-8

Smart Cities: Development and Governance Frameworks

This text/reference investigates the state of the art in approaches to building, monitoring, managing and governing smart city environments. A particular focus is placed on the distributed computing environments within the infrastructure of smart cities and smarter living, including issues of device connectivity, communication, security and interoperability. ISBN: 978-3-319-76668-3

Data Science and Big Data Computing: Frameworks and Methodologies

This book has a focus on data science, and provides practical guidance on big data analytics. Expert perspective is provided by an authoritative collection of 36 researchers and practitioners, discussing latest developments and emerging trends; presenting frameworks and innovative methodologies; and suggesting best practices for efficient data analytics. ISBN: 978-3-319-31859-2

Connected Environments for the IoT: Challenges and Solutions

This comprehensive reference presents a broad-ranging overview of device connectivity in distributed computing environments, supporting the vision of IoT. Expert perspectives are provided, covering issues of communication, security, interoperability, networking, access control and authentication. Corporate analysis is also offered via several case studies. ISBN: 978-3-319-70102-8

Connectivity Frameworks for Smart Devices: The Internet of Things from a Distributed Computing Perspective

This is an authoritative reference that focuses on the latest developments on the Internet of Things. It presents state of the art on the current advances in the connectivity of diverse devices; and focuses on the communication, security, privacy, access control and authentication aspects of the device connectivity in distributed environments. ISBN: 978-3-319-33122-5

Cloud Computing: Methods and Practical Approaches

The benefits associated with cloud computing are enormous; yet the dynamic, virtualized and multi-tenant nature of the cloud environment presents many challenges. To help tackle these, this volume provides illuminating viewpoints and case studies to present current research and best practices on approaches and technologies for the emerging cloud paradigm. ISBN: 978-1-4471-5106-7

Cloud Computing: Challenges, Limitations and R&D Solutions

This text reviews the challenging issues that present barriers to greater implementation of the Cloud Computing paradigm, together with the latest research into developing potential solutions. This book presents case studies, and analysis of the implications of the cloud paradigm, from a diverse selection of researchers and practitioners of international repute. ISBN: 978-3-319-10529-1

Continued Rise of the Cloud: Advances and Trends in Cloud Computing

This reference volume presents latest research and trends in cloud related technologies, infrastructure, and architecture. Contributed by expert researchers and practitioners in the field, this book presents discussions on current advances and practical approaches including guidance and case studies on the provision of cloud-based services and frameworks. ISBN: 978-1-4471-6451-7

Software Engineering Frameworks for the Cloud Computing Paradigm

This is an authoritative reference that presents the latest research on software development approaches suitable for distributed computing environments. Contributed by researchers and practitioners of international repute, the book offers practical guidance on enterprise-wide software deployment in the cloud environment. Case studies are also presented. ISBN: 978-1-4471-5030-5

Cloud Computing for Enterprise Architectures

This text, aimed at system architects and business managers, examines the cloud paradigm from the perspective of enterprise architectures. It introduces fundamental concepts, discusses principles, and explores frameworks for the adoption of cloud computing. The book explores the inherent challenges and presents future directions for further research. ISBN: 978-1-4471-2235-7

Cloud Computing: Concepts, Technology & Architecture

This is a text book (in English but also translated in Chinese and Korean) highly recommended for adoption for university level courses in distributed computing. It offers a detailed explanation of cloud computing concepts, architectures, frameworks, models, mechanisms, and technologies - highly suitable for both newcomers and experts. ISBN: 978-0133387520

Software Project Management for Distributed Computing: Life-Cycle Methods for Developing Scalable and Reliable Tools

This unique volume explores cutting-edge management approaches to developing complex software that is efficient, scalable, sustainable, and suitable for distributed environments. Emphasis is on the use of the latest software technologies and frameworks for life-cycle methods, including design, implementation and testing stages of software development. ISBN: 978-3319-543246

Requirements Engineering for Service and Cloud Computing

This text aims to present and discuss the state-of-the-art in terms of methodologies, trends and future directions for requirements engineering for the service and cloud computing paradigm. Majority of the contributions in the book focus on requirements elicitation; requirements specifications; requirements classification and requirements validation and evaluation. ISBN: 978-3319513096

User Centric E-Government: Challenges & Opportunities

This text presents a citizens-focused approach to the development and implementation of electronic government. The focus is twofold: discussion on challenges of service availability, e-service operability on diverse smart devices; as well as on opportunities for the provision of open, responsive and transparent functioning of world governments. ISBN: 978-3319594415

Cloud Computing Technologies for Connected Government

This text reports the latest research on Electronic Government for enhancing the transparency of public institutions. It covers a broad scope of topics including citizen empowerment, collaborative public services, communication through social media, cost benefits of the Cloud paradigm, electronic voting systems, identity management, and legal issues. ISBN: 9781466686298

Human Factors in Software Development and Design

This reference text brings together high-quality research on the influence and impact of ordinary people on the software industry. With the goal of improving the quality and usability of computer technologies, topics include global software development, multi-agent systems, public administration Platforms, socio-economic factors, and user-centric design. ISBN: 9781466664852

IT in the Public Sphere: Applications in Administration, Government, Politics, and Planning

This nook evaluates current research and best practices in the adoption of e-government technologies in developed and developing countries, enabling governments to keep in touch with citizens and corporations in modern societies. Topics covered include citizen participation, digital technologies, globalisation, strategic management, and urban development. ISBN: 9781466647190

Emerging Mobile and Web 2.0 Technologies for Connected E-Government

This reference discusses the emerging mobile and communication technologies including social media, for use by governments and citizens. It presents a reference source for researchers, practitioners, students, and managers interested in the application of recent technological innovations to develop open, transparent and more effective e-government environment. ISBN: 9781466660823

E-Government Implementation and Practice in Developing Countries

This volume presents research on current undertakings by developing countries towards the design, development, and implementation of e-government policies. It proposes frameworks and strategies for the benefits of project managers, government officials, researchers, and practitioners involved in the development and implementation of e-government planning. ISBN: 9781466640900

Developing E-Government Projects: Frameworks and Methodologies

This text presents frameworks and methodologies for strategies for the design, implementation of e-government projects. It illustrates the best practices for successful adoption of e-government and thus becomes essential for policy makers, practitioners and researchers for the successful deployment of e-government planning and projects. ISBN: 9781466642454

Acknowledgment

The editor acknowledges the support and efforts of a number of colleagues. First and foremost, I would like to thank the contributors to this book, 30 authors and co-authors from academia as well as industry from around the world who collectively developed and submitted a total of 12 chapters. Without their efforts in developing quality chapters conforming to the required guidelines and meeting often the strict deadlines, this text would not have been possible. Their names and brief biographical notes are listed in a separate section in this book.

Secondly, my grateful thanks are due to the members of the advisory and editorial board of this book who willingly volunteered their time in reviewing the book chapters and providing further advisory and editorial support. Their names and affiliations also appear in a separate section in this book.

Finally, I would like to thank members of my immediate family—Rehana, Zoya, Imran, Hanya, Arif and Ozair—for their continued support and encouragement. Every good wish, also, for the youngest and the most delightful in our family: Eyaad Imran Rashid Khan and Zayb-un-Nisa Khan.

Thank you all.

Zaigham Mahmood
University of Northampton, UK & Shijiazhuang Tiedao University, China
30 April 2020

Section 1

Concepts, Principles, and Reference Architectures

Chapter 1
Building Intelligent Cities:
Concepts, Principles, and Technologies

Vijayaraghavan Varadharajan
Infosys Ltd, Bangalore, India

Akanksha Rajendra Singh
Infosys Ltd, Bangalore, India

ABSTRACT

A city may be regarded as an intelligent city when its services to citizens are connected and it is able to obtain data from every aspect of its technology infrastructure to leverage it in real time for resource allocation, monitoring, management, and decision making. Cities around the globe are ambitiously leveraging the latest technologies to transform their infrastructures to better provision and manage the e-services. Although they are setting goals for focusing on the appropriate financing, long-term planning, developing technology stack, and advancing data management, governments need to further encompass all relevant guidelines towards right technology frameworks before commencing their intelligent city projects. This chapter provides a comprehensive introduction to intelligent cities, also known as smart cities, and the associated requirements. It also articulates the evolution of a typical city to a truly integrated, responsive, open, and connected intelligent city and the required underlying technologies.

INTRODUCTION

In this first section, we attempt to define an Intelligent City (IC), also known as Smart City or Digital City, and what are the essential elements of such a city. However, before that, we briefly look at the evolution of cities, over the ages.

Cities Over the Ages

Cities can be simply defined as areas that serve as epicenters for commerce, trade routes, politics and culture, that have large, concentrated human settlements. Right from city states of Greece to modern

DOI: 10.4018/978-1-7998-5062-5.ch001

metropolises of today like New York and London; Cities have been a major catalyst in shaping up civilizations. The first cities were formed when mankind discovered agriculture for their livelihood during Neolithic revolution naturally leading to denser human settlements in specific geographies, usually alongside a river with fertile soil.

Depending on the requirements and the extent of development of human race, cities have fulfilled different needs at different times. In ancient times, Egyptian cities on the banks of Nile, Mesopotamian cities on banks of Tigris-Euphrates and Harappan cities close to the Indus river served as cradles of civilization with their agriculture-based economy. At the start of this millennium, city states of Greece laid the foundation for the modern democratic system. In Medieval times, trade and commerce flourished in city states of Venice, Genoa, Milan, and Florence. Furthermore, these city states aided the Renaissance revolution paving the way for great works in the arts and sciences. The industrial revolution led to an explosion of cities across Europe followed by rest of the world. Invention of steam engine, cars and electricity abetted the rapid urbanization across the globe.

In today's world, currently, the urban areas are home to about 55% of the world population and this is expected to grow to 68% by 2050 (United Nations Department of Economic and Social Affairs, 2018). In that case, urban spaces will not be able to support healthy and sustainable living for their people. This necessarily mandates a need for cities to transform into sustainable, connected and intelligent living spaces. As of 2019, there are 33 Megacities (Ellie Donelly, 2019) across the world with a population of more than 10 million people. catering to numerous industries, necessities and a cosmopolitan crowd. The current megacities are struggling with issues of climate change, rapid urbanization, sustainability, and successfully managing large populations, among others. A connected, engaged, intelligent city collects huge amounts of data, generated by all manner of smart objects and information systems, which then needs to be stored and analyzed in real time for resource allocation, management, and decision making. This, in turns requires mechanisms and data analytics frameworks.

The chapter is organized as follows: The immediate section discusses the definition of, and the need for an Intelligent City. The next section explains, in some depth, technology requirements for an Intelligent City and how the combination of technologies help to create unique value propositions for businesses as well as residents of a Smart City. The subsequent section will discuss Intelligent City frameworks followed by a conclusion.

In this contribution, we will use the Intelligent city and smart city terms interchangeably.

What is an Intelligent City?

A city is intelligent, or smart, when it is able to obtain data from every aspect of its infrastructure and leverage it for resource allocation, management and decision making in real time. The information revolution in the digital era has resulted in numerous technologies, mechanisms and frameworks, namely, Artificial Intelligence (AI), Machine Learning (ML), Ambient Intelligence (AmI), Big Data, Internet of Things (IoT), and next generation software platforms which can be leveraged across urban populaces to develop smart spaces (Teena Maddox, 2018). The microcosmic representation of a Smart City can be best seen in smart campuses across the world. One of the best example is Microsoft smart campus initiative. Here, Microsoft is transforming its Redmond Campus using "Big Data meets Internet of Things" approach to smart space management and increased operational efficiency, among other benefits (Jennifer Warnick, 2013). Smart campuses manage a sizable amount of population and use technologies to generate new experiences and services while driving transformation to achieve efficiency and sustainability. Intelligent Cities can also be considered to be massive scaling and combination of many Smart Spaces together. Although the definition of Intelligent City might differ depending on the context, there is broad agreement that Intelligent Cities are cities that use technologies to collect, analyze and derive insights from data collected from various aspects of its urban life and infrastructure to drive reforms and decisions, and make life of its citizens more comfortable and enjoyable.

Need for Intelligent Cities

A relevant question is, why do we need to develop our cities into Intelligent Cities? The answer is straight forward. It is the growing population, constrained resources, sustainability goals, need for improved quality of life, citizen participation, etc that are some of the major factors driving the emergence of digitally based Intelligent Cities. The current situation with respect to cities across the globe and the technology revolutions will necessarily drive the Intelligent Cities implementation across the world. In addition to making life better, Smart Cities also offer viable business and development opportunities. It is estimated that Smart Cities would drive a 34 billion dollars market just in the Information Technology sector. Figure 1 showcases the disruptive potential of Smart Cities.

Figure 1. Intelligent Cities in Numbers

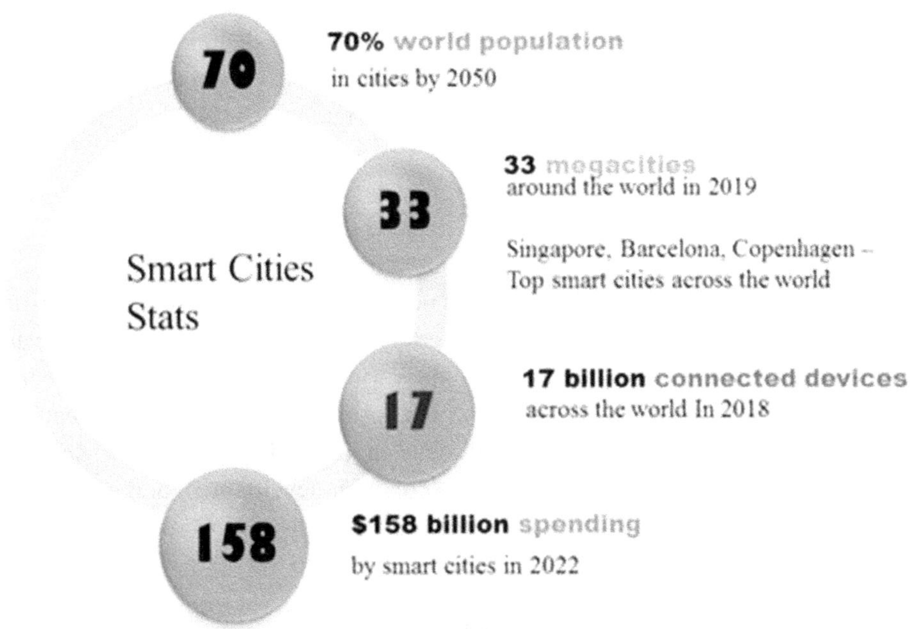

Urban population of the world has grown to 4.2 billion in 2018 and two-thirds of the population is predicted to reside in urban areas by 2050. Managing massive populations would be an important imperative for cities across the globe and accomplishing it would require leveraging all the modern technologies at disposal. Rapid urbanization across the world requires managing large population in city spaces, proper management and allocation of constrained resources in terms of connectivity, transport, healthcare, safety and quality living, to mention a few. Major cities across the world understand this imperative and have started transforming themselves, through the use of latest technology, to super-efficient urban living spaces.

Despite advancements in the energy sector, only 18% of the total energy consumed across the globe is renewable (World Bank, 2019). A large concentrated population will need a colossal amount of energy, the demand for which will only keep growing. Currently cities consume two-thirds of world's energy output and also contribute to 70% of the Carbon Dioxide emissions (C40.org, 2019). The current urban infrastructure without proper technology will not be able to handle the energy demands. An efficient system to manage energy allocation and consumption which is positively contributing to climate change requires a Smart City.

Most cities are currently in survival mode and are unable to offer quality standards of living to majority of its members. Hong Kong, for instance, has bed space apartments also known as cage homes where people survive in subdivided residential units of, sometimes only 5 square meters, leaves much to be desired in terms of living standards (Bryant Cheng, 2016). A world where everyone has access to optimum lifestyle is what current cities should strive for. And a Smart City can definitely create a sustainable and optimum lifestyle for all of its residents.

Inclusiveness is another emerging factor in the management of cities. A city where all citizens participate seamlessly in the decision-making process in all aspects of its governance is very important. All City precincts want engagement from their populace and Intelligent Cities with their platform technologies can help to increase citizen participation.

Taking up a Smart City initiative requires alignment of all the stakeholders on the governing concepts and principles, and technologies. The next sections discuss the concepts, principles, components, technologies and frameworks involved in a Smart City undertaking.

SMART CITIES: CONCEPTS AND PRINCIPLES

The concept of an Intelligent City differs across cities on the basis of goals, metrics, priorities, and governance system. But certain common parallels and generally accepted concepts can be derived, across the board, by looking at the Smart City initiatives across the world. One of the most apt conceptualization of a Smart City is provided by Cargaliu et al. (2009) which is stated as follows: "A city can be defined as *intelligent* or *smart* when investments in human and social capital and traditional (transport) and modern (ICT) communication infrastructure fuel sustainable economic development and a high quality of life, with a wise management of natural resources, through participatory action and engagement". Figure 2 depicts the goals of an Intelligent City as elicited from this definition.

Key Goals of an Intelligent City

Using the above definition of an Intelligent City, and studying various Smart City initiatives across continents, the key goals of a Smart City refer to the improvement in the following aspects of a city life. These are briefly discussed below.

Figure 2. Generic Goals of an Intelligent City

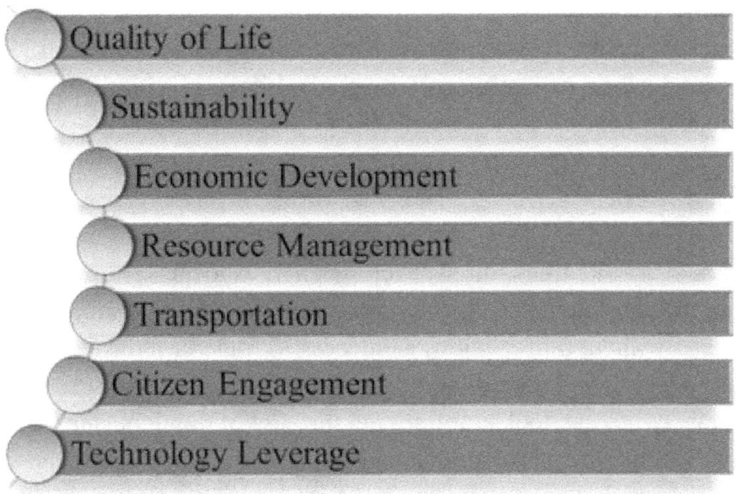

Quality of Life

A Smart City can enlist goals pertaining to improving the living standards of its citizens across political, economic, environmental and social arenas; and undertaking a variety of programs to achieve that. A case in point would be improving air quality of a city. Hamburg in Germany undertook to clean up its air by reducing the usage of marine fuels, while cargo ships docked in its sea port. Using Siharbor from Siemens, Hamburg enabled ships to use power from the mainland instead of burning marine fuels while stationed at the harbor, resulting in the improvement of air quality (new.siemens.com, 2019). Google cars used sensors in West Oakland city in the USA, to map out the polluted areas. The data collected via this exercise helped West Oakland to identify heavily polluted areas across its neighborhoods. Other areas of focus in improving quality of life for its citizens in a Smart City could be providing clean water, traceability of sources of consumer goods and food items, advanced education and healthcare. Such ventures help improve the quality of life of city residents.

Sustainability

According to United Nations, creating a sustainable city entails building strong and green ecosystems in all facets of City life. For example, cities could focus on reducing energy consumption and in order to reduce the CO2 emissions. In this regard, Copenhagen City would be a perfect example. The city is working towards achieving the status of the first carbon neutral capital of the world and wants to do so without affecting its economic growth. Accomplishing this goal requires tracking and managing resources efficiently, which they have managed to do well, across the city. Cisco estimates that cities that use information to manage its energy sources, can improve their energy efficiency by 30% in the next 20 years. Copenhagen utilizes sensors and gathered information in its decision making process. Copenhagen has programs for monitoring traffic in real time to track CO2 emissions, providing free Wi-Fi for tourists, embedding sensors to monitor sewer system, obtaining real time feedback from energy systems of city buildings, and suggesting to drivers, free parking spaces in real time (Niras.dk, 2019). Copenhagen is well on track to accomplishing its Intelligent City goals. Likewise, cities across the world can emulate the Copenhagen model to improve their sustainability scores.

Economic Development

Intelligent Cities should prioritize economic development next to quality of life and sustainability. Cities should realize that a thriving Intelligent City can exist only if there is commerce and trade. Otherwise, the city would be dead without a fiscal flow. A successful trade or industry attracts a large number of human populations to a city; and, therefore, city governments should ensure that cash and commerce are always flourishing. Cities can look at improving the ease of doing business and encourage a startup ecosystem to aid in the economic development and growth. Moreover, cities can identify some primary economic activities determined from their geo-political locations, natural wealth, and human wealth, and promote those capabilities. A case in point would be Hong Kong. This city is exploring options for promoting itself, both as a financial center and as important transit route for maritime trade.

A city should look at its indigenous advantages and develop itself in those areas before branching out to other methods for economic development.

Resource Management

Resource management, especially of energy, waste, water, sanitation, and recreation, is essential for a Smart City to subsist and flourish. Smart resource management especially in terms of energy is a pertinent mandate for Intelligent Cities. An apt example is smart street lights in the city which can bring immediate benefits by enabling street lights to be dimmed, switched on/off depending on the needs of individuals or localities; thereby preventing unnecessary wastage of energy. A remotely controlled city-wide lighting systems can, also help identify and fix faulty street lights rapidly. Los Angeles in the US was an early adopter to LED based energy efficient smart street lights connected to a central management system which led to realizing energy savings of sixty-five percent (Cree, 2014). Smart street poles which integrate LED lights with Wi-Fi, and also connected to the central system, has furthermore enabled the city government to install various sensors on the street poles to monitor pollution, air quality traffic or noise across the city of Los Angeles. The data thus generated was utilized to garner various insights and make decisions.

To ensure sustainability and efficient resource management, Intelligent Cities should watch out for novel sources of energy and resources. Cities around the world consume a major chunk of global energy and Intelligent Cities, by design, will be highly resource intensive. However, Smart Cities can manage their energy by consistently exploring possibilities of new sources of energy. Zero carbon natural gas power plants, hot solar cells are some of the options a Smart City can explore in this context. Germany has already launched the world's first hydrogen powered train. Intelligent Cities have an imperative to be sustainable and this can be accomplished through doing the following:

- Efficient management of resources and energy.
- Employing new sources of energy.
- Adopting renewable sources of energy.

For example, consider the problem of water management in an Intelligent City. Water is becoming a scarce resource with every passing year. There have been altercations over water resources and many cities are grappling with their own water crisis. Cape Town in South Africa came close to level 7 water restrictions in mid-2018s before the crisis was averted. Level 7 restriction refers to when the City government switches off water supply and rations it (Robbie Parks, Megan Mclaren, et al., 2019). Other than constantly monitoring the water levels and usage with help of sensors, Intelligent Cities can also look out for new sources of water, for instance, mega scale desalination. Desalination of seawater at large scale using reverse osmosis is possible but has not been a viable option until recently due to heavy costs involved. Many enterprises, notably, Sorek Desalination (Water Technology, 2019) plant in Israel has been able to bring down desalination costs. Similarly, Intelligent Cities can consider other viable resources and zero-in on alternate management strategies and sources for energy.

Transportation

Among the various important initiatives, transportation garners an important place when designing Smart Cities. Commute is a major pain point in cities across the world. People on average waste 54 hours in traffic every year and citizens of Los Angeles spend an average of 119 hours yearly stuck in traffic (Nicole Lyn Pesce, 2019). Intelligent Cities, therefore, have turned to various pilot studies to mitigate transportation problems across their expanse. This could be paving bike lanes, piloting autonomous road vehicles, creating efficient and integrated public private transportation systems, designing smart parking applications, and monitoring and navigating traffic flows in real time. An example is the city of Copenhagen which has proposed a host of novel solutions targeting mobility related issues clubbed together as Copenhagen Connecting project. Copenhagen Connecting deals with optimizing traffic, finding parking spaces, promoting bike lanes and reducing transportation costs.

Intelligent Cities can implement any of the mobility related schemes to provide equitable, accessible and fast transportation for its citizens.

Citizen Engagement

Inclusion of participation across all strata of citizens is important for a thriving Smart City. From drafting policies to making civic decisions to measuring impact of its numerous programs, if the city has to get it right, then the citizens will have to participate and also be well informed. Ancient city state democracies of Greece thrived because the citizens gathered in town squares regularly and took an active interest on various political and philosophical issues related to governance and politics.

Modern megalopolises have to deal with populations much larger than the city states across far flung areas. However, this does not reduce the need to ensure the everyone participates and contributes to the functioning of the city, in some way. Citizen engagement is crucial to Intelligent City success. Cities should leverage virtual interactions, electronic voting, dialogues and campaigns through platforms and online social media forums.

Citizen engagement would also help to drive innovations, co-creations and policies at a grassroots level rather than the typical top-down approach.

Technology Leverage

Making the most of the available Information and Communication Technologies (ICT) and the latest distributed computing paradigms, is the cornerstone of any Intelligent City undertaking. Utilizing the power of cheap sensors to collect data, super-fast internet connectivity to transmit data, cloud based data stores to process the large amounts of data generated, using artificial Intelligence data analytics to derive insights and take smart decisions is the crux of Smart City technologies. All these technologies can be used individually for a plethora of undertakings, but the greatest benefit is realized only when used together as an interoperable, interconnected and unified technology stack on a common platform, instead of storing data in silos. Open APIs and transparency with respect to data usage is pertinent in assuaging the citizens to proactively share data.

The large amounts of data generated will also bring on the onus of ensuring privacy, security, safety, and proper storage in data lakes. City governments should have a proper framework in place for managing data and its ensuing issues.

Key Principles of an Intelligent City

Most Intelligent Cities pick one or more or all of the above goals when deciding on Smart City development. Improvement in quality of life, technology leverage i.e. technology based services, and resource management are the omnipresent themes in most of them.

There are seven principles that serve as a guide to governments when designing an ideal Intelligent City. These principles are centered around accomplishing the goals discussed above. Some generic principles are given below.

1. Analyze the current state of a city, rethink the city systems, prioritize important action items, define personalized, suitable to you, perfect, sustainable Intelligent City model with focus on long term solutions (Vicente Guallart, 2015).
2. Aim to build a livable city where the well-being of the living population takes precedence before everything else.
3. Encourage experimentation and pilots in various arenas but do not work on creating solutions for transient and temporary problems (Smart Circle, 2014).
4. Develop the economy of the city by focusing and promoting the primary trade and commerce activities related to that city (Smart Cities Mission, 2017).
5. Data privacy and security frameworks should be in place for any existing and new data collection initiatives (Christina Kakderi, 2013).
6. Interoperability between systems, transparency and Open APIs should be the cornerstones for any technology implementation in a Smart City.
7. Intelligent City technologies and infrastructure should be solutions oriented, intuitive, easy to access, oriented, fully digital, completely connected and experience oriented (Brian Zanghi, 2017).

When designing a Smart City and while developing its components and architecture, it is imperative that these principles are the focus as the underlying principles would lead to quality execution.

Components of an Intelligent City

Taking into account the goals and founding principles, a Smart City can be subdivided into a number of components as shown in Figure 3:

Figure 3. Components of an Intelligent City

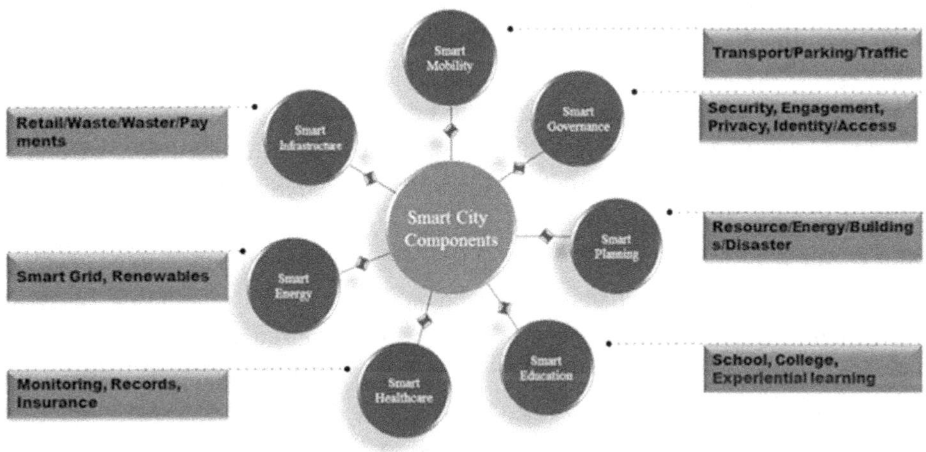

Each aspect of an Intelligent City (or Smart City as it is also known as), as illustrated in Figure 3, is briefly explained below:

1. Smart Governance: that indicates all aspects regarding administration and policies of a Smart City, that is, citizen engagement, identity and access management, transparent and accountable practices and transactions, citizen co-creation, interoperable integrated and open source technology stack with strong data privacy and cybersecurity.
2. Smart Planning: which refers to aspects of Intelligent Cities pertaining to urban planning of cityscape, smart buildings, disaster management and security.
3. Smart Education: that connotes to providing the best skills and access to knowledge resources for its workforce using interactive, immersive and data driven classroom tech.
4. Smart Healthcare: which refers to providing accessible and state-of-the-art healthcare to all of its citizens. Technologies, namely, AI, Internet of Medical Things, wearables, robotics, precision drugs, practical Blockchain, tele-medicine can be explored to provision Smart Healthcare. The presence of stringent regulations in the healthcare space which might potentially slow down the implementation of smart healthcare initiatives.
5. Smart Energy: that aims at monitoring and analyzing energy sources and their consumption in the city and also identifying new and clean sources of energy generation.

6. Smart Infrastructure: that refers to handling of waste, water, consumer goods, food items, air quality, sanitation, recreation, retail, security across an Intelligent City.
7. Smart Mobility: which is a theme that is common and very important to majority of Smart City initiatives. This connotes to improving three areas in the residents life in a Smart City: transport, parking, and traffic. Intelligent Cities should strive at improving both traditional form of transportations as well as designing the city to be future ready for next-gen transport systems, that is, drones, flying taxis and autonomous vehicles. Cites can gear themselves towards implementing electric vehicles, ride hailing and mobility regulation at scale to further enhance themselves in the smart mobility compass.

Most of the current frameworks in Smart Cities limit themselves to certain chosen aspects of urban living, depending on the priority of the citizens and the desired preferences of the government. But a truly smart/intelligent City should be integrated with all aspects of its inhabitant's life as well as every infrastructure. A framework for Intelligent City should consist of digital and emerging technologies architecture geared towards improving the following: healthcare, education, commerce, governance, water, food, Internet, energy, traffic, mobility, communication, environment, waste, transport, parking, entertainment, safety, and other necessary public services.

The various components of the Intelligent Cities can be integrated together to go above and beyond, in fulfilling the goals of a Smart City, that is, improved quality of life, sustainability, transport, economic development, resource management, citizen participation and cutting edge technology. These components can be further sub-divided into relevant areas for appropriate technologies to be implemented.

The ultimate aim is to design a fully optimized, efficient and sustainable urban living space that may be considered 'smart'. Less than 31% of cities across the world have Intelligent City strategy to develop an integrated, intelligent infrastructure (Sue Weeks, 2019). The focus of any Smart City initiative should be investing in core infrastructure and technologies, promoting innovation and open participation, implementing transparency, privacy, and security. The seven goals with seven guiding principles and seven components of a Smart City provide a good guide to city stakeholders for deciding and forming the technology architecture of a Smart City. A quote from Tormod Larsen, CTO of ExteNet which exemplifies the ambition of cities today: *Cities basically go out and want to do everything. Free Wi-Fi, meter reading, environmental sensors, information screens, they want them all and they want them for free. Key is to begin by developing a road map of desired services and then prioritize them.*

SMART CITY TECHNOLOGIES

Sensors based Internet of Things, Big Data, Artificial Intelligence, Machine Learning and citizen participation, at all levels, is what makes a city 'smart'. This section discusses the technology triumvirate of IoT, Big Data and AI/ML that is leading the Intelligent City revolution, followed by potential technologies that can further support the development of Intelligent Cities. The last part of this section discusses a few use cases for these technologies, relevant to the Smart City ecosystem.

IoT in Sensing Cities

Internet of Things (IoT) has ushered a new generation where machines, devices, objects, and people are equipped with sensors that are collecting and transmitting data all the time. Sensor based Internet of Things usually works in the following steps: Firstly, Sensors are placed in objects and devices, and kept with humans as smart gadgets (e.g. smart phones); these collect data based on the required parameters. Connectivity and data storage and processing happens in the cloud, that forms the next step. The collected data is transmitted using a gateway to cloud environment for data processing; and sensors communicate to other smart devices and machines without an intermediary computer or human; medium of transmission is the Internet. Finally, the processed data is made available to the end users via their smart phones and computing capable devices.

The era of the IoT is enabled by low cost sensors to allow sensors embedded inter-connected machines and other smart objects to communicate with each other. It is predicted that there are going to be approximately 20.4 billion connected devices across the world by 2020 (Gartner, 2017). Cities can utilize this to their advantage.

IoT is the chief enabler of Intelligent Cities technology architecture as cities can deploy sensors across every nook and corner of their infrastructure to collect, process and transmit data. Intelligent Cities can combine IoT with location of such smart things. Intelligent Cities should ensure, when installing sensors to use positioning systems to map the sensors to their locations in real time, while gathering data. IoT with location tracking would make it easier to track fast moving consumer goods, food supplies, medicines, transportation in real time. While GPS with IoT sensors would give real time location information and is ideal, there is a concern that it might become bandwidth intensive so alternate location systems should be investigated and utilized.

To design a city-wide network of sensors, an IoT based architecture, right from the hardware level up to the end user presentation, should be in place (Ivan Ganchev, Zhanlin Ji, et al., 2014). The city government should decide on the type of sensors that should be deployed across wide range of sectors. IoT sensors can be placed right from rivers to plumbing systems to forests to light poles to goods to home. A multitude of sensors will help to accumulate varied data parameters, and keep up with anytime-anyhow-anywhere concept. The cities should ensure that all these sensors are able to communicate seamlessly with each other and that their data sets are not stored in silos. In a city, data should be collected from all the sources (sewage, energy, traffic, streets, and other connected devices) but this data will hopefully be most useful when put to work in a unified fashion. Only sensors that comply to open and seamless communication should be embedded in the city networks. The cities should provide ubiquitous and fast connectivity to ensure that the data from sensors are relayed to the cloud environment in near real time if not real time. A proper security protocol should be in place to mask, encrypt and protect the data during transmissions. There is a great risk of data hacks during transmission and Intelligent Cities should implement secure mechanisms in place, to prevent that. The transmitted data can be processed in big data lakes and sent to the end user.

Internet of City Things (IoCT), that refers to a city-wide system of IoT based sensors and other networks, can be applied to any sector of a Smart City's existence. Internet of City Things can be utilized in large scale city initiatives such as smart healthcare, intelligent wearables, traffic management systems, etc. It can also be leveraged for smaller enterprises for scenarios such as controlling an entire home through IoT automation to monitoring energy consumption in a building.

The vast amounts of datasets, collected by the sensors spread across the city, can be considered useless if they are not utilized properly. The value does not lie in the data but on the analytics and insights derived from the data, that might help the governing bodies of the city to make smarter decisions. For example, city bodies can look at the pollution levels in various neighborhoods over a period of time and can introduce mobility regulations, or ban further industrialization of highly affected areas. Another example would be using data collected from traffic systems to map out the flows and divert the traffic to less congested areas, in real time. IoT and location systems together would be a potent tool, but only when employed efficaciously.

Big Data in Informed Cities

A large megacity with millions of people and vast technological infrastructure means that a large amount of data will be generated every second which the city will have to transmit, store, curate and analyze. As an example, a connected car generates 4TB (terabytes) of data every day and for a city with thousands of vehicles, this could amount to handling millions of terabytes of vehicular data every day. By 2020, wearable devices would be generating around 28PB (petabytes) of data on a daily basis, where 1PB=1000TB (visualcapitalist.com, 2019). A fully functional Intelligent City will face an onslaught of data from every sector, that is, governance, mobility, healthcare, education, infrastructure, energy. The amount of data collected across the city for various components will bring on a data explosion that requires a well-planned information and data architecture for performing analytics for further decision making in real time.

To implement the concept of Intelligent Cities, municipalities need a basic infrastructure with reliable network connections. Intelligent Cities strive to collect data from its devices, people and sensors to gain an edge in optimizing availability and efficiency of various functions. To accomplish that, cities are required to provision its datasets easily and effectively to its citizens and civic bodies in real time via high speed internet and powerful Cloud systems. Each data point at the fundamental hardware level needs to be transmitted to the public cloud via a wireless gateway. Having an industrial grade Internet through fiber optic cables providing fast access to the sensors will aid in rapid data transmission and processing. A heterogeneous set of networks including the emerging 5G wireless networks should also be further explored for Intelligent Cities development.

Cloud based services suited to specific Intelligent City functions should be deployed. The data in the Cloud environment should be stored in data lakes, which would have the raw data from various systems, that can then be accessed by administrators, citizens and smarty city businesses. The data lakes can further use Cloud services to process and curate the data for further analysis, modeling, forecasting and reporting. Considering the data regulations, and also the safety of data, it is advisable that a Smart City has its server farms for storing information within the city premises. Cloud infrastructures give a city a plethora of choices in the form of Software as a Service (SaaS), Infrastructure as a Service (IaaS) and Platforms as a Service (PaaS). Any city can go for an apt hybrid cloud model to suit its needs. Moreover, Cloud is elastic in terms of provisioning rapid scaling options that cities can use when the data load is high. Moreover, the pay as you go model where only usage is charged can help save large amounts of fiscal capital for provision of cloud infrastructure, software, and platforms in the cities.

Robust connectivity, effective communication protocols, elastic cloud environment and efficient data lakes are the necessities for an effective big data ecosystem in a Smart City.

Data in Intelligent Cities opens up three areas of immediate concern: Data security, Data Privacy and Data Transparency. Data security and privacy must be tackled head on as. Otherwise, hackers can hold cities hostage across the world. Recently, many cities across the USA were attacked by hackers and cities had to pay millions to avoid the ransomware infections (Patrick Howell O'Neill, 2019). Commencing from the sensors to transmission and to data storage, all standards and guidelines must be stringently followed. Cyber Security should be integral part of every Intelligent City program right from the start.

Data transparency connotes to ensuring that the data collected citywide is not monopolized by only a set of stakeholders, rather, it should be available for all citizen's and enterprises in the city. Many a times, under the guise of data ownership, very few entities have access to the data and thus gaining them an unfair advantage. Intelligent Cities should establish Data Trusts who would act as custodians of the data collected and would appropriately mask individual information before opening it up to the city. Open data should be the foundation for all Intelligent City data sets. Data trusts should also provide Open APIs using which any individual/enterprise should be able to easily and fully access the data. Additionally, having an Open Data policy, enables faster innovation and citizen co-creation in cities.

As Intelligent Cities store large tracts of information about their citizens, there should be a clear policy regarding data privacy and security issues. In many Intelligent Cities, Data privacy has been a show stopper for many operations as the citizens did not feel safe sharing large amounts of data and are ignorant of how their data is or will be used. While data is the underlying driving force, Intelligent Cities also need to draw lines on when collecting data becomes too much. A fine balance between personalization and respect for privacy of its citizen's life needs to be struck. At the same time, we cannot ignore the fact that Intelligent Cities need the large sets of data to gain real insights in order to optimize processes. To mitigate this differential privacy, zero knowledge proof based systems should be explored. Data sets should be made available to the wider audience only after masking individual information. To sum up, citizens should be made well aware of how and where their data are utilized.

With rapid technological and software innovation, as well as a regulatory environment that cannot persevere with the speed of change, transparency, neutrality and objectivity must be an integral part of the data and corresponding processes. Intelligent City data and services must be neutral and objective for coverage of the urban environment. They should cover the entire population and comply with data licensing rules, regulations and data protection rules.

AI and ML in Intelligent Cities

Two of the technologies that can turn a city into a truly intelligent and responsive city, is Artificial Intelligence (AI) and Machine Learning (ML). As discussed earlier, cheap sensors, data explosion and low cost data processing has enabled AI and ML in applications that support every aspect of life. AI revolution has always been preceded by data revolution where large quantities of data are available. Intelligent Cities with their massive data lakes thus serve as the biggest playground for piloting AI and ML use cases. Furthermore, the evolution of cloud computing and neural network algorithms has also allowed cities to scale analytics and AI. Due to the ease of applying AI and ML, there has been a big increase in Intelligent City initiatives utilizing these technologies.

Artificial Intelligence algorithms are used increasingly in Intelligent City applications, especially in terms of informed and timely decision making. The use of Artificial Intelligence in tandem with Internet of Things and data analysis has been critical to devising impactful Intelligent City solutions. There are many industries in the advanced urban services sector that are seeking to exploit Artificial Intelligence:

industries such as healthcare, manufacturing, critical infrastructures and transportation. AI and ML can provide insights, and thus helping cities improve their existing infrastructure, improve sustainability, prevent crime andf undesirable situations, and improve quality of living, overall. A robust AI based system could provide not only autonomous monitoring, but also make decisions about how each of the elements in a system should behave, how the behavior changes over the course of the day or time, and how elements respond to city systems. Artificial Intelligence can also be used for training autonomous vehicles and intelligent transport applications and deciding on the next generation systems.

AI in Smart Cities can find patterns within a set of data and across disparate sets of data while identifying correlation and causes for certain trends. Reinforcement Learning algorithms can be used to train autonomous vehicles and drones in a Smart City. Deep Learning algorithms can be used for facial recognition and surveillance. If a child is lost in a Smart City, AI based facial recognition systems should be easily able to locate him or her. If the city wants to see if there is any correlation between two completely disconnected entities, for instance holidays and education, it can use the appropriate data sets to run Machine Learning algorithms and discern interesting insights. Intelligent Cities of the future can train their robot workforce using reinforcement learning. AI opens up a plethora of opportunity for Smart Cities. Smart Cities can transform themselves from sensing to responsive cities with AI.

Using Big Data, IoT, Artificial Intelligence (AI), to accomplish the required goals, is the crux of a Smart City vision. Following is a case in point which demonstrates the part played by these technologies. A Smart City will leverage cheap sensors to monitor the state of affairs of its infrastructure in real time. Geographical Positioning System embedded along with IoT sensors would feed in the data about the Intelligent City components. Relaying information from several sources would mandate a need for big data infrastructure that can handle massive real time data sets. Using these data sets, Machine Learning algorithms can be implemented for both predictive as well as prescriptive analytics.

Beyond the afore mentioned technologies, a number of other relevant technologies are discussed below which can have a huge impact on further developing Smart Cities.

Other Emerging Technologies for Intelligent Cities

Discussed below are other technologies beyond IoT, AI, ML, and Big Data that are playing a significant role in the design and development of Intelligent Cities.

Next Generation Software Platforms

The digital layer of the City should be converging towards an integrated, unified platform. All the data and operations, instead of being fragmented, should operate from a single source. Additionally, all the functions related to the Intelligent City governance, engagement, open APIs and activities should be available to citizens from a Single Platform. This should translate into one, and at the most three, apps on a Smart Citizen's smartphone that will provide all the services. The platform for a Smart City should be designed in a manner that, in addition to the services provided by the Government, anyone else can create new services/applications in the same ecosystem. Citizen co-creation should be one of the top most priorities of a Smart City and its digital platform should give its citizens and small enterprises full freedom to be creative and innovative, so they can reach a much wider audience. The software platform should have an interactive, intuitive and ever evolving user interface so the Citizens are engaged with the platform on an ongoing basis.

Identity and Access Management

If 5G technology is the most important development to enable a real time Intelligent City, a robust and integrated Identity and Access Management system (IAM) of inhabitants should be the gatekeeper for any real Intelligent City. In large enterprises, an employee's electronic ID card, enabled with biometrics, opens up access to both digital and physical systems within the enterprise, with the ascribed privileges. Similarly, in a Smart City environment, a unique ID for each of its citizens provides access to public services and related facilities. The local government in China's Chan Cheng district has already developed a Blockchain based identity system to provide access to its public services.

Cities can go a step further and integrate biometrics (finger print scanning, retinal imprint and facial identification) to securely tie a person to their identity, and thereby preventing potential misuse of their identity. Furthermore, biometrics based identity can render the need of carrying an identity card useless. A city can use facial recognition and biometrics in enabling its citizens to make payments, board airplanes, travel on public transportation, access healthcare, track earnings and taxes, and even to unlock their homes and cars without the use of any keys. Intelligent Cities should strive to implement a seamless identity management system which would enable its citizens to access public services via smartphones with robust set of login and authentication protocols and also by leveraging its citizen's biometrics.

In a Smart City, Identity management does not end with people. Small Cities should go further to ensure identity management of all the sensors, devices and other connected objects under its purview, to reduce incidents of fraudulent use of these objects. Identity of Things (IDOT) refers to an initiative which requires assigning a Unique Digital ID (UDID) to all IoT based devices, sensors and objects, so that they can effectively communicate with each other over the Internet. IDOT's addressability helps objects/goods/things to be targeted and located, aids secure communication between objects and data transfer over the networks, helps to perform authentications, and thus prevent breaches on a City's IoT network. IDOT can also help track the arrival and delivery of consumer goods and determine location of assets in a city, as and when required. Blockchain based IDOT can even trace back the origin of goods and food items in the city to ensure quality. Identity of Things will be pertinent in a city's Internet of Things vision for authentication, identification, access management, security of data, and monitoring. A unified approach to the identity and access management of users and objects based on IAM and IDOT will serve as the answer to unlocking a Smart City's services and privileges.

Computer Vision

Currently, around the world, a gargantuan amount of data is produced every minute and majority of this data is in the form of images and videos. Around 4 million videos on YouTube are watched at any given time and approximately 47 thousand photos are posted on Instagram every minute, as domo.com (2019) have suggested. On similar lines, a Smart City will also be generating a huge number of images and videos in a day via its applications and through its citizens, be it for social media, street cams, work, or entertainment. All pervasive, low cost cameras in Smartphones and network of street cams generating nonstop stream of images has created a persistent need to analyze images and videos across the city. Analyzing texts is easy, however analyzing images and videos without any human intervention is highly problematic. Computer Vision addresses the need of video/image analysis and enables intelligent systems to sense and respond to images/videos. Computer Vision technologies utilize AI/ML algorithms to process, analyze, understand and derive insights from visual content.

Computer Vision can be, and is being, used for object detection, object tracking, traffic control, surveillance, autonomous vehicle navigation in Intelligent Cities.

Practical Blockchain

Most Blockchain projects at enterprise level are not successful due to scalability, interoperability and resource demands. Despite the naysayers, major cities across the world are utilizing Blockchain for transactions, tracking and access.

China has recently implemented a Blockchain based identity system for its countrywide system of Intelligent Cities by assigning a unique, global ID to its citizens (Yogita Khatri, 2019). Intelligent Cities can benefit from aspects of Blockchain such as data transparency, smart contracts, and secure ledgers. Intelligent Cities can either wait another 5 to 10 years when full-fledged Blockchain technology becomes fully functional and scalable (Laurence Goasduff, 2019) or start using Distributed Ledger Technologies for initiatives pertaining to identity management, goods tracking and transactions. Blockchain is becoming a popular and useful technology in industry and many other sectors of human life.

5G Technologies

Ubiquitous connectivity is the need of the hour for Intelligent Cities. But the reality is quite far off from the ideal. Intelligent Cities use a combination of wired and wireless technologies to achieve high speed connections. Yet, there is latency in data transmissions between the source and destination. In mission critical applications which are sensitive to time, latency can cause massive failures. For example, latency can cause collisions in autonomous vehicles.

5G or fifth generation technologies have already been rolled out in some cities across the world. 5G offers four specific spectrum bands: Low, medium, high and ultra-high. The network speed varies between 100 Mbps to 10 Gbps. It is worth noting that 5G is 100 times faster than current technologies. 5G is an enabling technology for Device-to-Device communication (D2D): both for massive device to device communication as well as for critical device to device communication due to its almost instantaneous speed feature.

5G technologies can be used in real-time crime prevention, self-driving vehicles and the IoT framework in a Smart City.

Edge Computing

Edge computing is predicted to be a 9 billion USD market by 2024 (Markets and Markets, 2019) and Intelligent Cities would be one of the mainstream adopters of Edge technologies. Edge Computing enables processing at the edge of the network and on the hardware devices without the need to process data at a distant (e.g. cloud) data lake. Edge enables IoT devices and sensors to perform analytics, and make decisions at device level instead of transmitting data to and from an over centralized cloud. Another aspect of edge computing is the availability of frequently used services and critical services on the edge reducing latency and achieving near real time speed. Edge Computing also fortifies security by keeping the data at or near the device. Edge Computing in Intelligent Cities can be applied in almost any area of a city's life e.g. Smart Monitoring, Smart parking, Autonomous vehicles, Adaptive systems, identification systems. Edge computing coupled with 5G can truly help a Smart City in becoming a Real Time City.

Immersive Technologies

These technologies include Augmented Reality (AR), Virtual Reality (VR), and Mixed Reality (MR). under the generic name of Extended Reality (XR), all these various technologies come under the gamut of Immersive Technologies. Smart Cities can use the various applications of these in entertainment, education, entertainment and retail.

Already, immersive technologies are making a foray in these areas. In the entertainment arena, immersive technologies are used in video gaming, interactive story telling. Pokemon Go, an AR based mobile game with 28.5 million peak users per day, shows the upward demand for immersive technology based entertainment. In the area of education, immersive technologies can be used for variety of training purposes. Google Expeditions, a VR app takes students to go on virtual field trips. Immersive technologies can be used to study battles and design simulation exercises in military trainings. AR/VR/XR can provide exemplary hands on training for any tool or manufacturing operations. In retail, immersive technologies can bring the world of retail to the citizen's door step by creating a virtual store or virtual shopping events. These technologies can easily measure a shopper's fit and size in real-time and design custom goods. They also show promise in the area of healthcare, construction and customer service.

Autonomous Technologies

Autonomous vehicles are going to change the way the world travels and lives. The effect of autonomous vehicles will be more pronounced in Intelligent Cities. Autonomous vehicles, though not yet in the mainstream, are predicted to be commercially viable in the next 5 to 10 years, but this has not stopped many cities around the world from piloting projects in this area. The world's first self-driving tram is already operating in the city of Potsdam, Germany (Charlotte Jee, 2018). Self-driving car experiments are being conducted on a large scale in every major city of the world and currently Cadillac and Tesla offer Level 3 Automation in their commercial car models. A truly self-driving car should achieve a level 5 automation, that is, it can operate in any environment without any geo fencing limits and with no human intervention.

In a Smart City, with the help of detailed city maps and 3D models, roadside sensors and separate lanes; self-driving car fleets can be deployed and operated with much more ease when compared to rest of the world without the aforementioned autonomous vehicle infrastructure. Self-driving cars in a Smart City can be used for public transportation and thereby could boost a sharing economy as well as a sustainable city. Self-driving trucks can be used for cargo transportation in a Smart City. Flying cars might become a reality in the forthcoming future. Dubai and Singapore are piloting flying taxis. Uber is working on air based ride sharing models and its flying taxi pilot, Uber Air will be debuted in 2020. Drones in a Smart City can be used for deliveries, surveying, inspection, agriculture and surveillance. Barcelona is using drones to inspect the sewer systems.

5G technologies combined with Internet of Things (IoT) and immersive technologies will be the key drivers in adoption of autonomous vehicles. Intelligent Cities should design an intelligent transit system that should be a mix of train/trams and cars which operate autonomously connecting the whole city throughout the day. There is a strong possibility that there will be millions of drones flying in a city sky in future. Self-driving vehicles will need separate lanes and Unmanned Aerial Vehicles (UAVs) will need revamped aviation control policies and systems. Flying taxis, delivery drones, recreational flying machines will crowd a city's airspace and will require air traffic navigations systems to prevent collisions and manage aerial movement of vehicles. Ensuring public safety will be a crucial priority in the age of autonomous vehicles both on land and in air. Airspace management in the future Intelligent City will require serious planning as the place will be teeming with aerial vehicles.

Intelligent Cities should future proof themselves by taking into account the deluge of Autonomous things in near future while planning their infrastructure.

Quantum Computing

Quantum Computers use quantum bits instead of binary bits and employ quantum mechanical rules of superposition and entanglement to achieve very high processing power and speeds when compared to a traditional computer. Google has recently claimed supremacy in Quantum Computing, where their quantum computer, code named Sycamore, has performed calculations in a few minutes which would have taken the most powerful super computer of today 10,000 years (Chelsea Whyte, 2019). Quantum computers are predicted to be in the mainstream in the next 5 to 10 years and the possibilities they offer can greatly benefit Smart Cities.

Quantum computers can be used for any scenarios which requires massive processing of datasets at super-fast rates or in real time. Smart Cities can use Quantum Computers to churn data throughout the city in real time to predict weather more accurately, build more accurate climate change models, develop personalized and effective drugs and vaccines, build autonomous vehicles, managing complex traffic snags and optimizing energy consumption.

3D Printing

3D printing or additive manufacturing refers to the process of building a solid object from its digital design by adding layer after layer of required materials. 3D printing has use cases ranging from manufacturing consumer goods, producing food, manufacturing personalized prosthetics, clothing and even constructing houses. Its applications for Intelligent Cities were brought to notice by the 3D canal project. The 3D print canal house project in Netherlands is a publicly accessible project where designers and architects all over the world are collaborating together to print a 3D canal house in Amsterdam. Cycling bridges have also been opened in Netherlands using 3D printers. World's largest 3D printed pedestrian bridge was opened in Shanghai in 2019 (Neil Patrick Walsh, 2019). This technology can also encourage citizens' co-creation, as anyone from anywhere can creatively conceptualize items and develop a prototype rapidly without undergoing the time consuming and cost intensive manufacturing process.

3D printing ensures that anyone can manufacture products and market it quickly. 3D printing can stimulate the startup ecosystem in the manufacturing domain. It enables the concept of fail fast in manufacturing industry, while saving cost, time and materials. Unlike the current mode of manufacturing, also known as subtractive manufacturing, where the excess materials are cut out to create an object, 3D printing uses additive technique where the materials are added progressively to accomplish a design. This aspect of 3D printing saves large quantity of materials from going waste and increases the sustainability quotient.

This technology will take another 10 to 15 years to go mainstream and whenever it goes mainstream it will be a game changing technology similar to Quantum Computing. Intelligent Cities should consider the imminent emergence of Quantum Computing and 3D printing in near future while laying down the architecture of their technology systems.

Zero Knowledge Proofs

As discussed earlier, the dominant problem plaguing the development of Intelligent Cities, is Data privacy, confidentiality and Data transparency. For an optimally functioning city, a large amount of data is necessary and if the citizens have suspicions with respect to their data, collecting such data becomes a hassle. Having a robust data policy and data trust mechanisms can assuage citizens. Similarly, developing proof of concepts in Zero Knowledge Proof and Differential Privacy systems can ensure Data Privacy in forthcoming years.

Differential privacy is a method where an individual's privacy is protected while at the same time extracting the aggregate information of the populace capturing all important trends and parameters. Zero Knowledge Proof is better utilized in areas where there are barriers to entry and the user has to provide data to get in. An example would be providing proof that confirms the age of a user before gaining access to a social media site. In this process, the user would have to reveal his/her age date, month and year of birth which is a potentially sensitive information. A zero-knowledge proof allows information to be validated without revealing or transmitting the information.

A zero-knowledge proof system has three attributes: Soundness, Completeness and Zero-knowledge. A system which can soundly verify the data, completely convince the assessing authority that it has verified the data and at the same time provides zero knowledge of the identifying information is the perfect Zero Knowledge Proof. It can be applied in any field that deals with sensitive data such as healthcare, insurance, financial transactions. Intelligent Cities can explore Zero knowledge proof and differential privacy techniques to resolve their data complications.

The current, upcoming and emerging technologies of Intelligent Cities are showcased in Figure 4. Digital transformation of many aspects of a city can turn it into a Smart City (Hafedh Chourabi, et al.,

Figure 4. Enabling Technology in Intelligent Cities

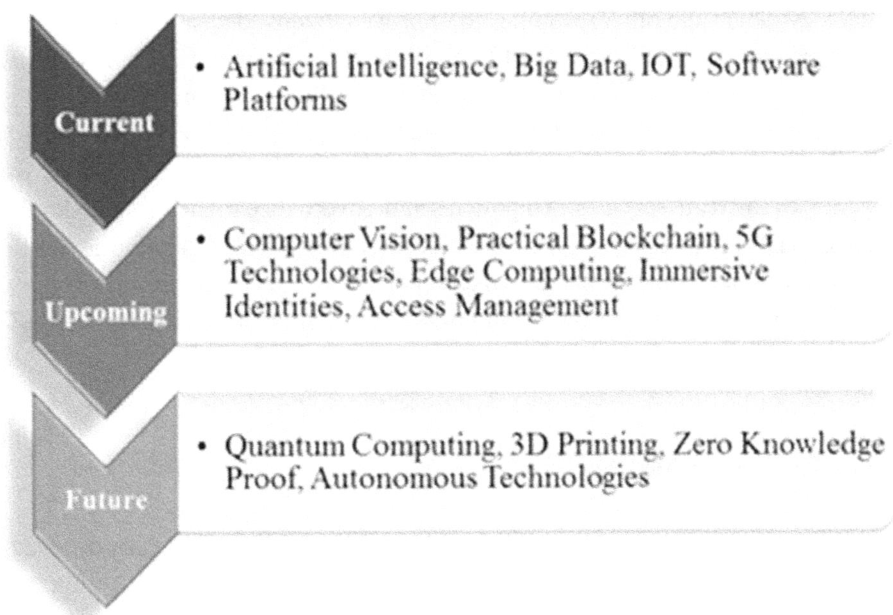

2012; J. Ramon Gil-Garcia, et al, 2012). The technologies discussed above can be integrated together to achieve the goals of Smart healthcare, Intelligent Manufacturing, Smart transportation, Smart payments, Smart governance, Smart grids and so on.

Integrated Technologies for Intelligent Cities - Use Cases

Intelligent Cities can use the technologies mentioned in the above sections to become a post industrialized information society. Below are the two relevant use cases of various technologies working in a synchronized manner:

Smart Payments

Smart Cities can form a completely cashless society where smart phones can trigger payments which can be activated every time a person shops, eats or conducts a financial transaction. Smart Cities can go a step further and use next generation of biometric techniques to enable seamless payments. Alipay, a Chinese unicorn has recently unveiled Facial Recognition based payments system "Smile to Pay" leveraging advanced likeness detection algorithms. Intelligent Cities can create payment systems which require multi factor authentication via citizens smart phones, and mandate cashless payments across businesses and transactions. Intelligent Cities can run statistical analytics on the payments and transactions of its citizens to assign a fair credit score or track the path of transactions. Smart payments can also bring down the occurrences of white collar financial crimes.

Smart Governance

The main motive behind Intelligent Cities development is to give its inhabitants a more comfortable life, through the use of latest technology. To begin with, smart governance should focus on three specific aspects of Smart City: digital government, citizen services, and public safety. Smart governance would be able to use data, evidence and other resources to improve citizens' decision-making and compliance. Smart governance requires the intelligent use of ICT in the IoT environment to improve decision-making through improved cooperation between all stakeholders, including government and citizens. Intelligent Cities can architect an open software platform with inclusive design to obtain citizen participation on policies and issues. The platform should be equipped with Open APIs for data use as well as enable citizens and businesses to create services. All policies and governance related issues should provide unbiased data to the citizens followed by Citizen opinions and polling. These capabilities and activities can happen in real time through the platform and citizen sentiments, and can be easily analyzed in real time.

The above two use cases highlight how technologies can completely transform a city. The next section discusses Smart City generations, frameworks, and evolution.

INTELLIGENT CITY - FRAMEWORKS

The main issues of concern with incumbent cities is that they come with baggage of legacy infrastructure and legacy technologies which, when trying to accommodate the Smart City imperative, creates difficulties, though, getting right is not impossible. A case in point would be comparing the developed cities of China with cutting edge technology to legacy infrastructure of New York subway. Building smart technologies from the ground up will be much easier, as compared to trying to merge them with the existing twentieth century old infrastructure. This section provides some guidelines on the change management of an incumbent old style city's transformation to a Smart City.

Intelligent Cities have been divided into three generations based on their fundamental characteristics as Smart City 1.0, Smart City 2.0 and Smart City 3.0 (George Martinidis, 2017):

- Smart City 1.0 is driven by technology, connotes to pushing of technologies in city territory and driven mostly by tech conglomerates.
- Smart City 2.0 refers to cities driven by technology but headed by City governments and focused on solving specific City problems or improving city metrics.
- Smart City 3.0 are cities where Citizen's co-creation and participation is the corner stone and city initiatives are driven by technology as well as complete Citizen involvement and inclusion.

The difference between the generation is subtle and refer to the level of connectivity of city-wide systems and the level of automation therein. Smart Cities usually oscillate between these three generations but sometimes directly jump to an evolved state which is beyond the three generations. Chinese Cities who do not have to struggle with the baggage of legacy infrastructure are able to develop into a highly advanced Smart City without having to transition between these generations. Smart Cities should have frameworks when transitioning between these generations.

A relevant question is, how does an incumbent city become a Smart City? Copenhagen in Denmark whose government has undertaken many Smart City initiatives to transform its urban landscape, is a perfect example. Copenhagen has implemented sustainability initiatives, asset tracking to ensure public safety, smart street lighting systems and other urban labs to experiment with multifold technologies and areas. Copenhagen has now become the innovation hub for Smart City enterprises and was awarded the "World Smart Cities" award in 2014.

Smart Cities roughly fall into two categories: A Smart City which is built from Scratch and a Smart City transitioning between the three generations. Whatever the vision, before embarking on a Smart City project, city managers should decide on the guiding principles, measurable goals, services priorities, technology stack, governance mechanisms, and required standards. Each City is unique in its own way and at a different stage towards becoming a Smart City. Therefore, a city, before embarking on its Smart City journey, decides on its guiding principles, first. This should be followed up by determining measurable goals and priorities. It could be reducing pollution by a certain percentage or improving water savings by certain million cubic feet. Intelligent City goals should be specific and measurable. The next step in a Smart City project would be the prioritization in terms of goals, features, and funding. Irrespective of the prioritization, a Smart City should always be oriented towards open stack, citizen co-creation and future proofing. Every Intelligent City should decide on a technology stack which supports data privacy, interoperability, ubiquitous connectivity and security. Intelligent Cities should have frameworks for data sharing, communication between all the infrastructures in place. Although a Smart City might begin with initiatives in a few areas, any such a city should have technology and protocols in place which would enable interoperability and data sharing between all infrastructure areas or initiatives. Intelligent City technologies should strive towards creating an ecosystem of services. Finally, Intelligent City's governance and standards should be well-defined based on its citizen's needs which is inclusive and fair.

There are multiple frameworks an Intelligent City can further consider as it deems fit, namely, KPIs bench marking, maturity assessment, components case study, and Smart City Framework suggested by British Standards Institute. All these frameworks have their unique strengths. A Smart City can develop its own hybrid framework based on the requirements and goals of its Smart City mandate.

Intelligent City - Risks

A Smart City enterprise may raise concerns in areas of cyber security, data privacy, funding, and policies. Smart Cities should build security protocols right from the start of any service to ensure data leakage and cyber security attacks are reduced to minimum. With respect to data privacy, Intelligent Cities need to strike the right balance between personalization and protection of their citizen's personal data. Another critical area of concern for a Smart City is the formulation of technology policies. Technologies on the lines of quantum computing, autonomous vehicles are continuously changing paradigms, with currently no clear mandates for guiding their use. For example, autonomous vehicles might completely change the landscape of vehicles insurance. Governing bodies might take years to come up with policy and frameworks defining the use of such technologies; and this may delay the cause of Smart Cities vision.

Intelligent Cities should constantly conduct deep studies about the impact of technologies across all areas much before the technology goes into the mainstream. Additionally, Intelligent Cities should be weary of using technology just for technology's sake. Cities conduct due diligence and only then allow the scaling of a technology to a higher level. Pilot studies in small pockets are strongly advised before city wide implementation of any initiative.

It should be noted that any Intelligent Cities related initiatives will have limited funding, So, maximum care should be taken to avoid investing in pointless and ineffective technologies. Purchasing equipment versus leasing it, should also be carefully considered, at least initially.

Sensing City to Connected City to Real Time City to Responsive City

The three generation of Smart Cities (Smart City 1.0, 2.0 and 3.0) that we discussed earlier, are called sensing cities whose technologies primarily function to sense, collect and analyze data. The next generation of Smart Cities, also known as City 4.0, is where every area of city infrastructure is interconnected and can communicate with each other seamlessly. For example, a city's water systems should be able to share data with the energy grid or waste management system, for more efficient functioning of these sectors. Ideally, all this data should be captured and kept in a single data lake so it can be easily leveraged for cross-functional analytics. City 4.0 is also known as a connected city. The Figure 5 illustrates the journey of a city through various stages in its evolution.

With the advent of 5G technologies and high-speed networking, Intelligent Cities would be able to transmit data at super-fast rates and would then become Real-time Cities. The move from connected to real-time city would be parallel in nature as ubiquitous connectivity and integration will drive the evolution of connected and real time cities. The combination of real time and connected city will lead to the formation of a new genre of intelligent city called Responsive City. These refer to cities which are real-time, can analyze data at high speeds and respond to change in the environment or make intelligent decisions based on the available parameters. Facets of Responsive city have been emerging in Chinese Smart Cities in the form of real time crime detection and surveillance. In the city of Shenzhen, cameras empowered with Artificial Intelligence can detect and instantly fine jaywalkers.

Figure 5. Smart City Evolution

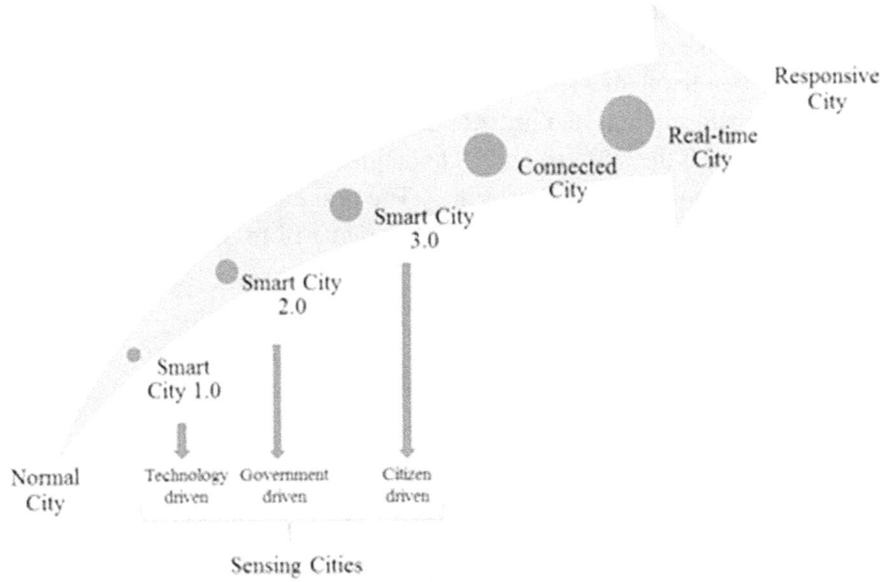

Another city which has the potential to truly go beyond a responsive city is the city of Neom in Saudi Arabia, which is currently in the construction phase. Neom is a cross-border city, being built from scratch, which, if successfully built, would open up a new era in the history of cities. Neom is aiming to leverage solar energy, desalinate water from red sea, mobilize 100% green transport systems to achieve sustainability. Neom is also planning to employ Building Information Modelling (BIM), cutting edge genetic technologies, next-generation farming techniques, 3D printing and immersive technologies etc. to create an example of a city of the future. A robotic police force, flying taxis, and advanced industry 4.0 manufacturing processes are on Neom's to do list towards building the world's first responsive city.

Building a responsive city should be the ultimate ambition of any city; and a city should not rest until it is fully responsive in every facet of its existence.

CONCLUSION

International Data Corporation's Intelligent Cities Spending Guide estimated that in 2018 over $81 billion was spent on Intelligent City initiatives and expect this to grow to $158 billion by 2022 (IDC, 2018). Technological progress will further affect humankind's ways of working, methods of traveling, and manners of living. The three trends with the greatest potential to disrupt or shape the future of mankind are: advances in the field of artificial intelligence, smart transport, and new Intelligent Cities. Big Data Analytics, Internet of Things (IoT) based applications, Artificial Intelligence (AI), and 5G networking will ensure the evolution of the current Smart City 3.0 generation. The convergence of the afore mentioned technologies will lead to innovations that will enable further advances in various industries and other technologies, such as robotics and virtual reality, and finally leading to the creation of a responsive city.

Currently, the vast majority of IoT applications take place in pockets of metropolitan areas and eventually become integrated into Smart City ecosystems. But to be a truly responsive city, Smart Cities should respond and make intelligent decisions without human intervention in all of its components. Smart Cities 3.0 is the beginning of a series of advanced city generations.

Amidst all these transformations and growth, cities should never lose sight of their most precious resource: its inhabitants. All the systems and technologies of a Smart City should be built around its citizens following their requirements and priorities. The best cities are designed with their people and sustainability goals at the core. To conclude, here is a quote from Shakespeare "*What is the City but the People?*"

REFERENCES

C40.org. (2019). *Why cities: Ending Climate change begins in the cities.* https://www.c40.org/ending-climate-change-begins-in-the-city,C40Cities

ArchDaily. (2019). *Shore connection for berthed ships: SIHARBOR.* https://new.siemens.com/global/en/products/energy/medium-voltage/solutions/siharbor.html,Siemens

Caragliu, A., Del Bo, C., & Nijkamp, P. (2009). Smart Cities in Europe 49-59. *Proceedings of the 3rd Central European Conference in Regional Science—CERS 2009.*

Cheng, B. (2016). *Number of 'cage home' residents rises to nearly 200,000.* https://www.hongkongfp.com/2016/03/20/hkfp-lens-number-of-cage-home-residents-rises-to-nearly-200000/,HKFPLens

Chourabi, H., Nam, T., Walker, S., Gil-Garcia, J. R., Mellouli, S., & Nahon, K. (2012). Understanding Smart Cities: An Integrative Framework. In *2012 45th Hawaii International Conference on System Sciences.* IEEE Explore.

Circle, S. (2014). *Five principles for a smarter city.* https://www.smart-circle.org/smartcity/blog/five-principles-smarter-city/

Cree.com. (2014). *Cree LED Street Lights Transform California City's Streets, Cut Energy Use by 65 Percent.* https://www.cree.com/news-media/news/article/cree-led-street-lights-transform-california-city-s-streets-cut-energy-use-by-65-percent

domo.com. (2019). *Data Never Sleeps 7.0.* https://www.domo.com/learn/data-never-sleeps-7

Donnelly, E. (2019). *Future success: megacities of more than 10 million set to provide big business opportunities.* https://www.independent.ie/business/world/future-success-megacities-of-more-than-10-million-set-to-provide-big-business-opportunities-37961796.html

Ganchev, I., & Ji, Z. (2014). *Generic IoT Architecture for Smart Cities.* Telecommunications Research Centre (TRC) University of Limerick.

Gartner. (2017). *Gartner Says 8.4 Billion Connected "Things" Will Be in Use in 2017, Up 31 Percent From 2016.* https://www.gartner.com/en/newsroom/press-releases/2017-02-07-gartner-says-8-billion-connected-things-will-be-in-use-in-2017-up-31-percent-from-2016,Gartner

Goasduff, L. (2019). *Top Trends on the Gartner Hype Cycle for Artificial Intelligence, 2019*. https://www.gartner.com/smarterwithgartner/top-trends-on-the-gartner-hype-cycle-for-artificial-intelligence-2019/

Guallart, V. (2015). *The 8 principles on How to Make a Smart City*. Barcelona: Architect Institute for Advanced Architecture of Catalonia; https://smartnet.niua.org/content/3e4c6fcf-f273-4722-a531-e9decc8bf5de

IDC. (2018). *IDC Forecasts Smart Cities Spending to Reach $158 Billion in 2022, with Singapore, Tokyo, and New York City Among Top Spenders*. https://www.idc.com/getdoc.jsp?containerId=prUS44159418

Jee, C. (2018). *The world's first autonomous tram just launched in Germany*. https://www.technologyreview.com/f/612186/the-worlds-first-autonomous-tram-just-launched-in-germany/

Kakderi, C. (2013). *Smart City Design Principles*. https://www.urenio.org/2013/10/09/smart-city-design-principles/

Khatri, Y. (2019). *China launches blockchain-based identification system for Smart Cities*. https://www.theblockcrypto.com/linked/45654/china-launches-blockchain-based-identification-system-for-smart-cities

Maddox, T. (2018). *Smart Cities: A cheat sheet*. https://www.techrepublic.com/article/smart-cities-the-smart-persons-guide/

Markets and Markets. (2019). *Edge Computing Market by Component (Hardware, Platform, and Services), Application (Smart Cities, IIoT, Content Delivery, Remote Monitoring, AR and VR), Organization Size (SMEs and Large Enterprises), Vertical, and Region - Global Forecast to 2024*. https://www.marketsandmarkets.com/Market-Reports/edge-computing-market-133384090.html

Martinidis, G. (2017). *Three Generations of Evolving Smart Cities*. https://www.urenio.org/2017/07/19/three-generations-evolving-smart-cities/

Mission, S. C. (2017). *Smart City Features*. http://smartcities.gov.in/content/innerpage/smart-city-features.php

Niras.dk. (2019). *Copenhagen Smart City*. https://www.niras.dk/media/1585/kimspiegelbergsteltzer.pdf

O'Neill, P. H. (2019). *Ransomware hackers hit nearly two dozen Texas cities*. https://www.technologyreview.com/f/614200/ransomware-hackers-hit-nearly-two-dozen-texas-cities/

Parks, Mclaren, Toumi, & Rivett. (2019). *Experiences and lessons in managing water from Cape Town*. Grantham Institute Briefing paper No 29.

Pesce, N. L. (2019). *Commuters in this city waste 119 hours sitting in traffic every year*. https://www.marketwatch.com/story/commuters-in-this-city-waste-119-hours-sitting-in-traffic-every-year-2019-08-22

Smartnet. (2019). *A Day in Data*. https://www.visualcapitalist.com/wp-content/uploads/2019/04/data-generated-each-day-wide.html

United Nations Department of Economic and Social Affairs. (2018). *68% of the world population projected to live in urban areas by 2050, says UN*. https://www.un.org/development/desa/en/news/population/2018-revision-of-world-urbanization-prospects.html

Walsh, N. P. (2019). *World's Largest 3D-Printed Concrete Pedestrian Bridge Completed in China*. https://www.archdaily.com/909534/worlds-largest-3d-printed-concrete-pedestrian-bridge-completed-in-china

Warnick, J. (2013). *88 Acres How Microsoft Quietly Built the City of the Future*. https://www.microsoft.com/en-us/stories/88acres/

Water Technology. (2019). *Sorek Desalination Plant*. https://www.water-technology.net/projects/sorek-desalination-plant/

Weeks, S. (2019). *Whitepaper: Smart Cities - Beyond the hype*. Smart Cities World in Association with Interact.

Whyte, C. (2019). *Google claims it has finally reached quantum supremacy*. https://www.newscientist.com/article/2217347-google-claims-it-has-finally-reached-quantum-supremacy/

World Bank. (2019). *Sustainable Energy for All (SE4ALL) database. Renewable energy consumption (% of total final energy consumption)*. https://data.worldbank.org/indicator/EG.FEC.RNEW.ZS?end=2015&start=1990,World

Zanghi, B. (2017). *4 Key Principles for Smart City Solutions*. https://www.masabi.com/2017/01/17/4-key-principles-for-smart-city-solutions/

ADDITIONAL READING

Berger, R. (2019). *Report: The Smart City Breakaway: How a small group of leading digital cities is outpacing the rest*. Roland Berger.

Bhalavat, K. (2019). *Smart City projects and technologies to keep an eye on*. https://pluto-men.com/insights/smart-city-projects-and-technologies-to-keep-an-eye-on/,Pluto-Men

KEY TERMS AND DEFINITIONS

Artificial Intelligence: Artificial intelligence (AI) refers to the ability of machines to have cognitive capabilities similar to humans using advanced algorithms and quality data.

Connected City: It refers to a city where all aspects of its infrastructure and different sectors of the society are inter-connected to each other through IoT based sensors via the internet.

Data Lakes: Data lakes are massive repositories for original, raw and unstructured data which is collected from various sources across a smart city. The data from data lakes can be cleansed and transformed for further analytics and modeling.

Deep Learning: This is also a subset of AI where unstructured data is processed using layers of neural networks to identify, predict and detect patterns. Deep learning is used when there is a large amount of unlabeled data and problem is too complex to be solved using machine learning algorithms. Deep learning algorithms are used in computer vision and facial recognition systems.

Machine Learning: This can be regarded as a subset of AI which refers to analyzing structured data and identifying trends (correlations) for specific outcomes and using that information to predict future values (causation).

Real-Time City: This is a smart city where the information highways operate in real-time and data is transmitted and processed instantaneously for further decision making and instant actions.

Reinforcement Learning: Reinforcement learning is also a subset of AI algorithms which creates independent, self-learning systems through trial and error. Any positive action is assigned a reward and any negative action would result in a punishment. Reinforcement learning can be used in training autonomous vehicles where the goal would be obtaining the maximum rewards.

Responsive City: A responsive city is a self-regulating and autonomous city space which uses advanced AI and quantum computing techniques to respond to changes and make decisions in real time without human intervention. Responsive cities are "always on" cities.

Smart City: A city using internet of things (IOT) sensors to collect data and garner insights gained from the data to manage all aspects encompassing the city efficiently and transparently.

Chapter 2
Reference Models for Intelligent Cities:
An Aligned Template

Nuno Soares
ALGORITMI Research Centre, University of Minho, Portugal

Paula Monteiro
CCG/ZGDV Institute, Guimaraes, Portugal

Francisco J. Duarte
https://orcid.org/0000-0001-9270-2226
ALGORITMI Research Centre, University of Minho, Potugal

Ricardo J. Machado
ALGORITMI Research Centre, University of Minho, Portugal

ABSTRACT

Cities are multifaceted and highly complex interconnected systems of people, objects, and machines, bringing endless design, construction, and operational challenges. To be intelligent, the parts of a city and the relationships and connectivity among them need to be smart. Much work has been done on models and reference architectures for intelligent cities to achieve the smart integration of city domains. Nevertheless, a universally complete smart city reference architecture does not exist. This chapter aims to discuss accepted dimensions of an intelligent city and the best arrangement for connectivity of them. Through literature review of relevant studies from academia, international organizations, and standards organizations that analysed contributions ranging from enunciations of smart cities' relevant dimensions, conceptual models, reference architectures, and urban platforms, this chapter aggregates the dimensions to align the reference model for intelligent cities that reflects the importance of the core relevant dimensions.

DOI: 10.4018/978-1-7998-5062-5.ch002

INTRODUCTION

Intelligent Cities, also known as Smart Cities, are multidimensional and highly complex interconnected systems of people, software, machines, smart objects, and data. Systems with this high level of complexity and connectivity have been studied from diverse perspectives and are referred to by numerous different designations e.g. system of systems (SoS) (Maier, 1998); cyber-physical systems (Lee, 2008), (Griffor, Greer, Wollman, & Burns, 2017); cyber-physical-social systems (Wang, 2010); socio-technical systems (Fischer & Herrmann, 2011); and multi-scale systems (Kevrekidis, Gear, & Hummer, 2004).

Besides intelligent cities have attracted extensive and emerging interest from both academia and industry with an increasing number of international examples and use cases emerging from all over the world. There is no unique and consensual definition for what constitutes a smart city, though the most common characteristic among all definitions is the strong recurrence to Information and Communications Technology (ICT), including latest tecknological paradigms such as Cloud Computing, Internet of Things (IoT), Indjustrial IoT, connectivity protocols, and Social Media offerings.

These emerging complex and interactive connected systems, based on latest technologies, that constitute different aspects and components of the cities ecosystem, bring new design, construction, and operational challenges. A significant challenge of these types of evolving projects, related to the improvement of the quality of life for citizens, is how to manage complexity, including interoperability, in an SoS with many continuously changing components.

If there is still no single definition of what an intelligent (or smart) city is, even less is there a unified reference model, a description of the basic arrangement and connectivity of the parts of the city. There still is missing a complete broader overview and skeleton of the interlinked items, which might constitute an all-encompassing architecture that provides guidance, design patterns, common vocabulary, and enables an end-to-end information flow, lowering barriers to interoperability; and thereby enabling stakeholders of smart city projects to perform more and faster implementations.

The infrastructures of a city were often developed and provided through stand-alone (silo-vertical) systems. Now, however, enhanced smart services interworking across domains, in combination with data analytics, is required to leverage the opportunities of digitalization. Current architectural standardization efforts have not yet converged, exhibiting a lack of consensus on a common language/taxonomy and architectural principles. Thus, the interworking across multiple urban infrastructures requires a coordinated approach avoiding that the different management systems are all dealing independently.

Dedicated domain-specific systems are still required to operate the related urban infrastructures, thus such urban systems need to follow an open design approach, e.g. with open interfaces, supporting open standards for exchanging urban data and supporting new urban services extending the scope of the Intelligent City services over time by increased use of digitalization. To provide this SoS like approach, requires interoperability of systems at all levels as a core principle.

A reference architecture provides the common framework around which more detailed discussions can centre. By staying at a higher level of abstraction, it enables the identification and comprehension of the most important issues and patterns across its applications in many different use cases. An intelligent city reference architecture will contribute to alleviate real-world deployment of intelligent cities, bringing advantages like interoperability assurance, integration facilitation, reuse, risk reduction, better quality, and knowledge transfer.

As said before, there is no single definition for smart cities, neither is there a single global and unified reference model, nor a reference architecture that permeates the entire operation of a smart city. However, several models for understanding and conceptualizing smart cities have been developed, which aim to define their definition, scope, objectives, benefits, and architectures.

The aim of this chapter is to analyse the existing smart city modelling approaches, contributing to the definition of components that must be part of such models. First, the benefits of a reference model for smart cities are presented. Next, we identify functional and non-functional requirements that form the basis for a detailed architecture design. It follows review of the literature research and analyses the gathered relevant dimensions for smart cities, grouping them according to a twofold classification for technological and non-technological dimensions. A segmentation for the overall of the ICT dimensions, given its disparate intentions is also stated. Lastly, the chapter proposes an aligned intelligent cities reference model based on the said classification and the degree of acceptance of each dimension. The chapter concludes with remarks about the undertaken work and some considerations for future work. In this comntribution, the terms - intelligent city and smartr city – are used interchangeably.

MODELLING CONCEPTS

Models are abstractions of reality representing views of reality and, thus, allowing its users to focus only on the needed characteristics of the modelled reality. Models can minimize the complexity of reality they represent by including representations understandable and more easily handled by humans or machines. Models are also used to transfer pieces of knowledge between reality domains, namely from one organization into another, sharing a set of characteristics. A conceptual model is the most generic, abstract and simple model. Nevertheless, it clarifies the fundamentals of the systems by asking for the overall set of entities and their relationships, and the key concepts involved, and the identity of the actors.

An architecture is a description of the basic arrangement and connectivity of parts of a system (either a physical or a conceptual object or entity) (ISO 15704, 2000). The software engineering community considers that architecture is the fundamental organization of a system embodied in its components, their relationships to one another and to the environment and the principles guiding its design and evolution (ISO/IEC/IEEE 42010, 2011). It is a set of conventions, rules, and standards employed in a system's technical framework, plus customer requirements and specifications, that the system's manufacturer or a system integrator follows in designing, or integrating, the system's various components (such as hardware, software and networks).

A reference architecture provides guidance for the development of system, solution and application architectures. It provides common and consistent definitions in the system of interest, its decompositions and design patterns, and a common vocabulary which allows discussing the specification of implementations so that options may be compared (Lin et al., 2017). A framework is a structure for the content or process that can be used as a tool to structure thinking, ensuring consistency and completeness (The Open Group, 2018).

The proposed architecture summed with additional consideration of entities and their interaction leads to a reference model. Such a model integrates aspects of the related entities such as human users, devices implementation, and server structures; and provides a more complete view of, or templates for, the overall setup and its domain implementation (Weyrich & Ebert, 2016). Although abstract, a step up from the conceptual model, the reference model is a framework for understanding significant relationships among the entities of an environment, and for the development of consistent standards or specifications supporting that environment (The Open Group, 2018). A reference model consists of a minimal set of unifying concepts, axioms and relationships within a particular problem domain, and is independent of specific standards, technologies, implementations, or other concrete details (OASIS RM-SOA TC, 2019).

The identification of proposed models for smart cities in the literature has led to consider all these terms and concepts which although conceptually can be distinguished; as these are closely intertwined and are together mentioned when it comes to the need for a definition of an all-encompassing high-level abstraction of architecture patterns and descriptions for the cities ecosystems.

A high-level abstraction of these architecture patterns and descriptions has obvious benefits since it provides common and consistent definitions, patterns, and a collective vocabulary, that facilitates easy sharing of experience and know-how in designing, implementing and operating the systems (Cloutier et al., 2010)(Muller & van de Laar, 2010), in a way that: 1) encourages reuse of common system building blocks (Cloutier et al., 2010)(Galster & Avgeriou, 2011), 2) reduces the risk through the use of proven and partly prequalified architectural elements (Cloutier et al., 2010), 3) provides better quality by facilitating the achievement of quality attributes (Nakagawa, Antonino, & Becker, 2011), and 4) contributes to the overall interoperability of the different systems (Cloutier et al., 2010)(Galster & Avgeriou, 2011).

REQUIREMENTS OF AN ARCHITECTURE FOR INTELLIGENT CITIES

A reference architecture can be understood as contexts provided with common features, vocabulary, and requirements, together with supporting artefacts to enable their use. A reference architecture handles the requirements and generates a superset of functionalities, information structures, and mechanisms (Weyrich & Ebert, 2016). At this point, we strive to elicit the requirements, the common features and the design principles that can define a reference architecture for smart cities.

Believing that the complete architecture for intelligent cities is yet to come, da Silva et al. (2013) state that there has not been an established standard for integrating the available technologies in order to generate a coherent, flexible, scalable, repeatable and effective system. Another author who shares this belief is Anthopoulos (2015a) that is aimed at defining a common smart city architecture which serves government purposes for innovation and sustainability, while it utilizes experiences from prestigious cases and corresponding theoretical context.

da Silva et al. (2013) discusses various architectures, highlighting the main requirements that they aim to fulfil. Additionally, based on those different architectures with the most varied purposes, the authors presented a set of requirements for the implementation of a smart city. To obtain this minimum set of requirements, they considered the sum of requirements able to compose and define the domains that compose a city, such as social, technological, cultural and infrastructure. Moreover, these requirements should be easily adapted to the peculiarities of each city, such as climate, size and personal issues of each citizen.

Seeking for the requirements which must be met for minimally implementing an architecture for smart cities, da Silva et al. (2013) conducted a survey, concluding that no architecture has fulfilled all requirements classified as essential in the context of their work and getting a final list of eleven requirements: Object interoperability, Sustainability, Real-time monitoring, Historical data, Mobility, Availability, Privacy, Distributed sensing and processing, Service composition and integrated urban management, Social aspects, and Flexibility/Extensibility.

Kakarontzas, Anthopoulos, Chatzakou, & Vakali (2014) also attempted to discover important quality properties relating to smart cities. More precisely, functional and non-functional requirements defined in ISO-25010[1] quality model. The authors conducted a survey based on a questionnaire of 34 questions, divided into six major smart cities domains, submitted to eighteen experts/engineers of fifteen real smart cities throughout the world. The authors analysed responses and concluded that the most prominent quality drivers that must be considered for a smart cities architecture are in the order of importance: 1) Interoperability and Compatibility; 2) Usability; 3) Authentication and Authorization (i.e. Security); 4) Availability; 5) Recoverability and Reliability; 6) Maintainability; and 7) Confidentiality (referring to Security).

It is clear that some of the requirements are similar to the ones proposed by da Silva et al. (2013) but they are far from an exact correspondence. Joining the authors above (particularly in the assumption that the lack of standardization causes smart city platforms to not support several key requirements and references architectures and thus failing to provide a solid foundation to develop city solutions which meet the requirements of stakeholders and systems), Bastidas, Helfert, & Bezbradica (2018) explored and defined a set of requirements to mitigate the lack of standardization and to guide the design of reference architectures for smart cities, so that researchers and practitioners can use these requirements as a comprehensive foundation to make more effective design decisions.

They conducted a literature review to determine the requirements, which would be suitable for, and aligned with, the reference architectures, but they did not include any of the works mentioned above. They suggested a set of twelve functional requirements and sixteen non-functional requirements, as shown in Table 1.

Table 1. Requirements for the design of smart cities reference architecture

Non-functional Requirements		**Functional Requirements**	
Scalability	Context-aware	Resource discovery	External data access
Security	Flexibility	Resource management	Software engineering tools
Trust	Heterogeneity	Data management	City model
Privacy	Integrity	Event management	Business models
Availability	Adaptability	Code management	City Oriented
Reliability	Extensibility	Application run-time	Cost minimization
Real-time	Configurability		
Interoperability	Service-based		

A large number of specific domains of applications under the umbrella topic of smart cities, most of the time prefixed with 'Smart' in order to emphasise the fact they embed intelligence and global awareness, are quintessential examples of the Internet of Things (IoT) application areas. The core of smart city implementation is the vision of the IoT. In other words, the IoT is the technical backbone of intelligent cities. Such cities need to have three key features that the IoT can provide: intelligence, interconnection, and instrumentation. Use of the IoT can, therefore, make intelligent cities feasible. (Mohanty, Choppali, & Kougianos, 2016).

Applications for intelligent cities exploit the full potential of IoT related technologies, however, unfortunately, they appear only as vertical silos, meaning specific applications with very specific architectures, deprived of inter-communication and inter-operation. The smartness of those new applications can only reach its pinnacle whenever full collaboration between those vertical silos can eventually be achieved. Unsurprisingly, research, studies and efforts to achieve full collaboration between vertical silos for IoT applications are quite pertinent to the task of defining the requirements and the design of reference models for smart cities. Therefore, the common requirements of the Internet of Things (ITU-T, 2012) (ITU-T, 2014) can be seen as a significant contribution to the definition of the requirements for smart cities architecture.

The common requirements of the IoT specified in this recommendation are classified into the categories of non-functional requirements, application support requirements, service requirements, communication requirements, device requirements, data management requirements, and security and privacy protection requirements. For instance, non-functional requirements include Interoperability, High Availability, Scalability, Adaptability, Reliability, and Manageability.

If IoT can be brought to the discussion in relation to intelligent cities, the same can be done concerning Cyber-Physical Systems (CPS). It is difficult to distinguish between CPS and the IoT based on the review of literature. CPS is a much larger entity than the IoT; i.e., the IoT is a network/communication subset of CPS. It is the implementation of the IoT in a physical system that leads to CPS (Mohanty et al., 2016). IoT has four main components: 1) the Things such as interconnected smart ojects, 2) the local area network, 3) the Internet, and 4) the cloud. So, the IoT along with these physical entities, such as the buildings of a city, can make CPS.

The National Institute of Standards and Technology (NIST), established the CPS Public Working Group (CPS PWG) to bring together a broad range of CPS experts in an open public forum to help define and shape key characteristics of CPS, The major deliverable of the CPS PWG is a document that presents a set of high-level concepts about CPS, their relationships, and a vocabulary for clear communication among stakeholders. The NIST CPS Framework (Griffor et al., 2017) is an abstract framework, or meta-model, for understanding and deriving application domain-specific CPS architectures. The NIST CPS Framework consists of three facets viz: conceptualization, realization, and assurance. Each facet is presented and understood from its set of activities and artefacts. The activities in turn address aspects and concerns throughout the CPS development cycle.

Aspects identify a wide perspective of analysis by which an application (e.g. for smart cities) can be evaluated and this perspective is further specified by concerns. CPS Framework established nine aspects. These groups of relevant interests in a system include: Functional, Business, Human, Trustworthiness, Timing, Data, Boundaries, Composition, and Lifecycle. The concerns were identified and further analysed, producing a set of cross-cutting concern groupings called aspects. Concerns and aspects are not orthogonal. That is, within the analysis of a given concern, consideration must also be given to related concerns.

The validity of these aspects and concerns to the requirements elicitation for smart cities architecture is not complete. Among aspects and concerns, we can derive truly architectural requirements for an architecture for smart cities, but they are not explicitly identified. Close targeting of the common requirements of the Internet of Things and the sets of requirements can and has been identified from aspects and concerns as defined in the NIST Framework for CPS, into a true set of requirements of an architecture for smart cities. The International Technical Working Group on IoT-Enabled Smart City Framework (IES-City PWG, 2018) have produced a valuable mapping and harmonization, Refer to Table 2 that shows an abbreviated and adapted synthesis of these.

IoT-Enabled Smart City Framework, led by the NIST was created to collect consensus and expertise from research and academia on the best ways to draft a specific framework for smart cities. The first public draft was released for public comments. The mapping presented in Table 2, on the one hand, expurgate some of the common IoT requirements that are not so relevant to smart cities geo-domains[2], on the other hand, substantiates the concerns into abstract requirements. For instance, the concern 'actuation', belonging to the aspect 'functional' and described in the NIST CPS Framework as 'Concerns related to the ability of the CPS to effect change in the physical world' gains an instantiation to the abstract requirement 'Device control and configuration (support of remote monitoring, control, and configuration of devices)' after being mapped and harmonized with the (ITU-T, 2014) Common IoT Requirements, namely requirement #D2, from category 'Device', with description 'Support of remote monitoring, control and configuration of devices is required so that device manageability in IoT is increased', and summary 'Remote monitoring, control and configuration of devices are required'.

As is clear from Table 2, concerns produce a set of abstract requirements. But these abstract requirements must be enhanced to achieve the implementation requirements for intelligent cities applications. More work needs to be done transforming the abstract requirements into implementation requirements, just like in the example proposed by Griffor et al. (2017) to illustrate the use of the NIST CPS Framework concepts, applying it to an Emergency Response use case.

REFERENCE MODELS FOR INTELLIGENT CITIES

Proposals of elements and their relationships to integrate on an overarching reference model for intelligent cities could come from diverse propositions, though obviously based on existing models for smart cities, but also from urban platforms; either working on practice or only conceptually projected, and even smart cities definition attempts, and enunciations of their relevant dimensions. Altogether, definitions of smart cities, models, and platforms indicate the scope, the concepts, and the dimensions involved in smart cities.

We have surveyed literature not only about reference models, architectures, reference architectures, and frameworks for smart cities, but also to determine definitions, core concepts, attributes, indicators, key performance indicators (KPI), and requirements for smart cities. Some of the identified contributions already propose what can be considered as an architecture, a reference architecture, a framework, or even a reference model for smart cities. Other studies describe what can be considered a suitable platform for cities, particularly open urban platforms (OUP), and a third depict dimensions to take into account in a city's ecosystem.

Besides recognizing that models and platforms can further contribute to the present study, simple conceptual models or enunciations of dimensions were also included in the evaluation, through a selection process subject to the inherent subjectivity due to the particular appreciation we made about the validity and originality of the proposals. A total of more than forty contributions from the literature were selected for proposing a reference architecture, a reference model, an urban platform, a set of KPI for cities, or at least a set of city dimensions.

Dimensions of an Intelligent City

Many authors dedicated their work to point-out dimensions, domains, planes, process fields, viewpoints, and underlying drivers for intelligent cities, in which, for standardization and simplification of concepts

Table 2. Mapping of ITU-T Common IoT Requirements in the NIST CPS Framework – adapted from IES-City PWG (2018)

Aspect	Concern	Abstract Requirement (harmonized with ITU)
Functional	Actuation	Device control and configuration
	Communication	Communication modes
		Communication control
		Intelligent communication
		Heterogeneous communication support and support for heterogeneous device-related communication technologies
	Monitorability	Device control and configuration
		User management
	Functionality	Monitoring of things
		Service management
	Controllability	Device control and configuration
	Physical context	Device Mobility
		Mobility services
	Sensing	Device Mobility
		Autonomic Services
Human	Usability	Programmable interfaces
Trustworthiness	Security	Communication security
		Data management security
		Service provision security
		Integration of security policies and techniques
		Mutual authentication and authorization
		Reliability
Timing	Synchronization	Time synchronization
Boundaries	Behavioural	Capacity to interact with the system from other domains
Data	Data semantics	Semantic annotation and access to data of things
		Semantic storage, transfer, and aggregation of data of things
	Operations on data	Data storage
		Data processing
		Data query
		Data access control
		Data exchange
		Data validation
	Relationship between data	Semantic storage, transfer, and aggregation of data of things

will be identified as the diverse relevant dimensions. Available contributions can be understood as simple conceptual models, providing a common structure and definitions for describing the concepts of smart cities, and relationships among the entities involved. The works presented in Table 3 also identify stakeholders and point out concerns i.e. interests in the cities ecosystem relevant to one or more of its stakeholders (ISO/IEC/IEEE 42010, 2011). Concerns may pertain to any aspect of the system's functioning, development, or operation. TOGAF (The Open Group, 2018) describes the identification of stakeholders, concerns, and business requirements as one of the steps to create the architecture vision.

Anthopoulos (2015b) have performed a literature review and identified seven major smart city dimensions. Giffinger et al. (Giffinger et al., 2007) identified six such characteristics. Glebova, Yasnitskaya, & Maklakova (2014) defend that the smart city concept can be divided into five constituent parts. Hancke et al. (Hancke, Hancke Jr, & others, 2012) advocate seven sensor areas in a smart city. Mahizhnan (1999) identified four key dimensions of a smart city, defined for the Singapore Intelligent Island.

Eger (2009) defends four key dimensions of a smart community. Thuzar (2011) advocates a different set of four key dimensions for a smart city. Barrionuevo, Berrone, & Ricart (2012) identified five types of capital that contribute to a city's intelligence. Kourtit and Nijkamp (2012) argue that smart cities are based on a promising mix of four types of capital. Fromhold-Eisebith (2017) has based on the categorization of process fields that constitute a smart city made by several scholars, and further added her own urban arenas proposing eleven fields that have the potential for CPS-driven processes.

Neirotti et al. (2014) advocate six smart city domains, which in turn include some subdomains. (Albino, Berardi, & Dangelico (2015) sought the dimensions advanced by various scholars of the phenomenon and identified the four most common characteristics emerging from them. The International Telecommunication Union (ITU) Telecommunication Standardization Sector (ITU-T) Focus Group on Smart Sustainable Cities (ITU-T FG-SCC, 2014a) also identified another group of four core themes related to smart cities.

De Oliveira Fernandes et al. (2011) advocate three levels of city smartness for city authorities' actions. Lee et al. (2014) conducted a research study that identified six key conceptual dimensions and 17 sub-dimensions of smart city practices that form a theoretical foundation for classification of practices. Finally, the United Nations (UN) Habitat (UN-Habitat, 2013) identified five dimensions of city prosperity.

Models for Intelligent Cities

More interlinked approaches to the intelligent cities' core concepts and requirements can be seen in what can be considered a framework or equivalent. In this section, we present the works approximated to a reference model or reference architecture. Models presented intended and detailed, with clear relationships between the entities.

Desouza & Flanery (2013) presented a model with five city components divided between physical and social spheres. Nam & Pardo (2011) introduced a representation of smart cities with the three key factors: human, institutional and technological, applied to theirs more than six, key dimensions. Chourabi et al. (2011) presented a framework with five outer and three inner factors that explain the integration of technology, organization and policy. CISCO Systems Inc. (Falconer & Mitchell, 2012) envisioned a smart city framework with four layers, a decision methodology that enables both the public and private sectors to plan and implement smart city initiatives more effectively.

International Business Machines (IBM) (Söderström, Paasche, & Klauser, 2014) introduced the technological functionalities of a smart city and highlighted the role of the ICT infrastructure and the information services stating a theory founded on two main assumptions: 1) the city is based on nine main pillars organized in three major categories, with each category pillar being an individual system, and 2)

Table 3. List of reviewed studies presenting a conceptual model for cities

Designation	Dimensions
Smart City dimensions (Anthopoulos, 2015b)	resource, transportation, urban infrastructure, living, government, economy, coherency
Smart City characteristics (Giffinger et al., 2007)	smart economy, smart governance, smart people, smart mobility, smart living, smart environment
Smart City conceptual parts (Glebova et al., 2014)	intellectual transport system, public security, energy consumption management and control, environmental protection, ICT
Sensor areas in Smart City (Hancke et al., 2012)	smart infrastructure, smart surveillance, smart electricity & water distribution, smart buildings, smart healthcare, smart services, smart transportation
Key dimensions of a Smart City (Mahizhnan, 1999)	IT education, IT infrastructure, IT economy, quality of life
Key dimensions of a Smart Community (Eger, 2009)	technology, economic development, job growth, increased quality of life
Key dimensions of a Smart City (Thuzar, 2011)	quality of life, sustainable economic development, management of natural resources through participatory policies, convergence of economic, social, and environmental goals
Types of capital for city's intelligence (Barrionuevo et al., 2012)	economic (GDP, sector strength, international transactions, foreign investment); human (talent, innovation, creativity, education); social (tradition, habits, religion, families); environment (energy policies, waste and water management, landscape); institutional (civic engagement, administrative authority, elections)
Smart Cities' types of capital (Kourtit & Nijkamp, 2012)	human capital (skilled labour force); infrastructural capital (high-tech communication facilities); social capital (intense and open network linkages); entrepreneurial capital (creative and risk-taking business activities)
Potential CPS-driven process fields (Fromhold-Eisebith, 2017)	smart ICT infrastructure, smart energy, smart mobility, smart construction, smart security, smart metabolism (reuse, recycling, water and waste management), smart industries, smart economy, smart education, smart consumption, smart governance (e-government, e-democracy)
Smart City domains (Neirotti et al., 2014)	natural resources and energy, transport and mobility, buildings, living, government, economy and people, (each one divided in several sub-domains)
Smart City Dimensions (Albino et al., 2015)	city's networked infrastructure, social and cultural development, business-led urban development, natural environment.
Levels-of City Smartness (De Oliveira Fernandes et al., 2011)	levels of city smartness: 1st lead by example, 2nd level of city smartness: govern the private urban actors, 3rd level of city smartness: integrated approach (hi/medium/no resilience)
Core themes (ITU-T FG-SCC, 2014a)	society, economy, environment, governance
Framework for SC analysis (J. H. Lee et al., 2014)	urban openness, service innovation, partnerships formation, urban proactiveness, smart city infrastructure integration, smart city governance
Dimensions of City Prosperity (UN-Habitat, 2013)	productivity and the prosperity of cities, urban infrastructure: bedrock of prosperity, quality of life and urban prosperity, equity and the prosperity of cities, environmental sustainability and the prosperity of cities

the city seen as an SoS and IBM's '3-Is equation' where a smarter city is the result of instrumentation, interconnection of data, and intelligence brought by software.

Naphade, Banavar et al. (2011) also have an SoS perspective with a set of seven interdependent public and private systems that the city can integrate and optimize. Leydesdorff & Deakin (2011) proposed a neo-evolutionary perspective of the Triple-Helix model for smart cities and Lombardi, Giordano, Farouh, & Yousef, 2012) have come up with also a revised version of the triple helix model, for smart cities, as always, based on networks of universities, industry, and government, that proposes a novel framework for analysis and performance measurement.

Liu, Wei et al. (2014) advocate a smart city value chain model with primary and supportive activities. Edvinsson (2006) envisioned the city as a knowledge tool model. Yovanof & Hazapis (2009) conceptualized the digital city architectural framework for smart service provision with three main dimensions. Zygiaris (2012) introduced his smart city reference model in a form of a multi-tier model with several components and entities that smart city planners could use to define the conceptual layout of a smart city and describe the smart innovation characteristics.

The Spanish standard (UNE, 2017) has five smart cities layers in a structure devoted to assuring Open Urban Platforms (OUP) interoperability requirements. Abu-Matar and Mizouni (2018) multiple-view smart city reference architecture meta-model has nine views interrelated with each other, where the cyber views run and control the physical views. Batty et al. (2012) defend the structure of FuturICTs Knowledge Accelerator smart city program that organizes smart cities research directions into seven distinct but overlapping areas. Anthopoulos et al. (2016) propose a unified smart city model that identifies six major dimensions in literature and discovered eight conceptual models of viewpoints that address smart cities. Elhoseny, Elhoseny, Abdelrazek, Bakry, & Riad (2016) have performed an elicitation of requirements and capabilities for the ideal smart city service-oriented architecture (SOA) framework, proposing one of their authorship.

Clement at al. (2017) propose a reference architecture for smart cities incorporating the viewpoints of the traditional city infrastructure, services, and business roles. These authors augment the traditional city architecture layers of environment, infrastructure, logistics, and human elements, with a computing infrastructure based around SOA principals, traversing these computing elements and business roles across several levels of city architecture from the utilities through to the service provisions. The smart cities model advocate by Clement et al. (2017) is expected to utilize simulation so that the system can predict future outcomes and react accordingly.

Refer to Figure 1. Here, based on McKee, Clement, Almutairi, & Xu, (2017), the reference architecture integrates IoT, Cloud Computing and Edge technologies with existing city infrastructure. A service-oriented approach has been proposed for the development of new systems that can augment the existing systems in the city at all levels. The benefits of the service-oriented approach are the ability for services to be continually evolved and added organically to the smart city. These services do not necessarily need to be only based on infrastructure and could be provided by for-profit autonomous systems as another revenue stream.

ISO 37120 (ISO, 2014) defines and establishes methodologies for a set of indicators (a quantitative, qualitative or descriptive measure) to steer and measure the performance of city services and quality of life. The indicators are structured around seventeen themes. Recognizing the differences in resources and capabilities of cities worldwide, the overall set of indicators for city performance has been divided into core indicators (mandatory) and supporting indicators (recommended).

The ITU-T performed a components analysis for a smart sustainable city and derived a modular Smart and Sustainable Cities (SSC) ICT architecture approach (ITU-T FG-SCC, 2016) with six sub-systems organised in modules. For almost all the modules, a hypothetical assignment of standards is presented.

The International Electrotechnical Commission (IEC) city model presented in ISO/IEC JTC 1 (2014) synthesizes a set of city characteristics where smartness is applied. City subsystems surrounded by environmental context, city history and characteristics, societal context, and city governance.

Figure 1. Reference architecture for smart cities, incorporating viewpoints of the traditional city infrastructure, services and business roles - based on McKee, Clement, Almutairi, & Xu, (2017)

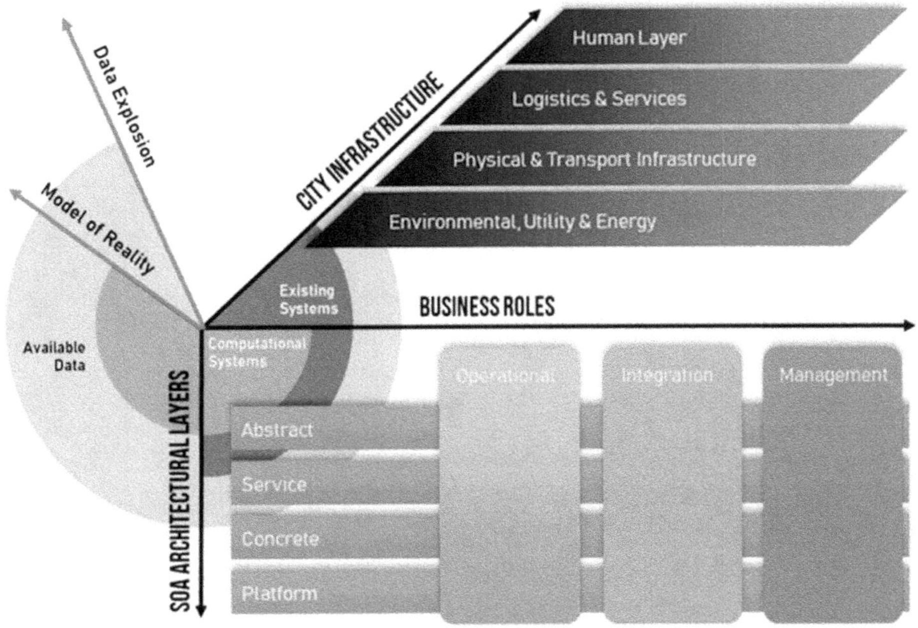

National Institute of Standards and Technology (NIST) and some international partners, convened an international public working group to compare and distil a consensus language, taxonomy, and framework of common architectural features for smart cities. The first version of the IES-City Framework (Internet-of-Things-Enabled Smart City Framework) was recently released (IES-City PWG, 2018) introducing a reference framework for the development of architectures for incremental and modular smart cities. The framework provides the tools to evaluate the breadth of functional requirements, assess the readiness of the municipality infrastructure, and measure the benefits to the citizenry.

The ESPRESSO H2020 project (Cox et al., 2016) produced a smart cities reference architecture with three vertical layers formed by groups of services, and six horizontals layers topped by two services consumer layers for respectively, sectorial specific businesses and city stakeholders. The ESPRESSO reference architecture joined with the results of the European Innovation Partnership on Smart Cities and Communities (EIP SCC) urban platform initiative (EIP_SCC Urban Platform, 2016), were on the basis of the DIN (the German Standardisation Institute) standard (DIN, 2017) for a reference system architecture of an open urban platform (OUP), with its multi-layered approach where each capability cluster is represented by a single layer, and three priority vertical areas. Refer to Figure 2.

A common smart city architecture was the desideratum of Anthopoulos (2015a). The author jointly with the ITU-T Study Group 20 (ITU-T FG-SCC, 2016) conducted an analysis with a literature review within the period 1997 (appearance of smart city concepts in literature) to early 2015[3]. Findings suggest that a common architecture must be multi-tiered, consisting of five layers, as follows:

- Natural Environment: it concerns all the environmental features where the city is located (landscape, rivers, lakes, sea, forests, etc.).

Figure 2. Reference System Architecture of an Open Urban Platform based on DIN (2017)

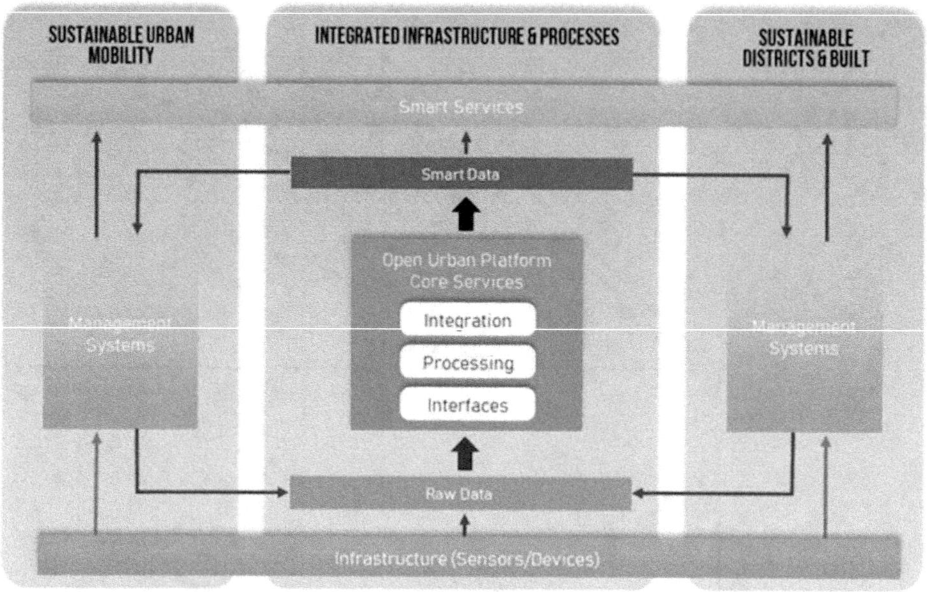

- Hard Infrastructure (Non-ICT-based): it contains all the urban features, which have been installed by human activities and are necessary for city operation (buildings, roads, bridges, energy-water-waste utilities, etc.)
- Hard Infrastructure (ICT-based): it concerns all smart hardware, with which SSC services are offered (i.e., data centres, supercomputers and servers, networks, IoT, sensors, etc.)
- Services: this is all types of smart city services, grouped in the smart city six dimensions and organized according to international urban key-performance indicators.
- Soft Infrastructure: it refers to individuals and groups of people living in the city, as well as applications, databases, software, and data, with which the SSC services are realized.

These layers address all potential smart city solutions, while it leaves space to incorporate services that are delivered across the urban space. This architecture, named Generic Multi-tier ICT Meta-Architecture for Smart and Sustainable Cities (SSC) meets the existing ISO-2014 standardization, and more specifically, the UN-Habitat key-performance indicators (UN-Habitat, 2013) (ITU-T FG-SCC, 2014b), which define a model for urban measurement at an international level.

Yin et al. (2015) suggest a smart city architecture proposal with four general layers and two planes based on what they observed in the literature, a data-centric and multidisciplinary smart city. Also, Silva et al. (2018) synthesize the commonalities of most of the works they have analyzed in a four layers architecture for a generic smart city, with a key concern on sensitive data protection.

Santana, Chaves, Gerosa, Kon, & Milojicic (2017) have produced extensive work about Smart Cities architectures, platforms, and related concepts, deriving a novel reference architecture for software platforms for smart cities with four technological levels and an additional all-encompassing security vertical tier. The authors have based their analysis on several urban platforms and architectures alfreday suggested by other researchers. They have questioned the literature about the enabling technologies for state-of-the-art software platforms for Smart Cities requirements (that a software platform for Smart Cities should meet) and determined the main challenges and open research problems in the development of the next-generation software platforms for Smart Cities. The extracted main requirements, components and features from smart city platforms then formed the basis to derive a reference architecture that facilitates the development of smart city applications. Each level in the reference architecture proposed by Santana et al. (2017) supports some functional and non-functional requirements. For instance, the cloud and networking level is responsible for the management and communication of the city network nodes. This level ensures some fundamental non-functional requirements, including scalability and extensibility.

The resulting reference architecture originated from an extensive survey with 23 platforms analysed, and primarily grounded on two other particular architectures: City Data and Analytics Platform (CiDAP) described by Cheng, Longo, et al. (2015), and OpenIoT[4],[5]. Santana et al. (2017) define the software platform for Smart Cities as an integrated middleware environment that supports software developers in designing, implementing, deploying, and managing applications for Smart Cities. By using open data and open services provided by a city, common citizens and users may also execute, or even develop, novel applications to run on top of the city's infrastructure. The applications will use the services and data from the platform, and also generate new data to be stored on it. The platform should provide a Software Development Kit (SDK) to facilitate the development of applications, including tools such as an Integrated Development Environment (IDE), libraries, and frameworks for commonly used programming languages, and a simulator for debugging and experimenting with applications before real deployment.

Figure 3. Reference Architecture for Smart City Platforms based on Santana et al. (2017)

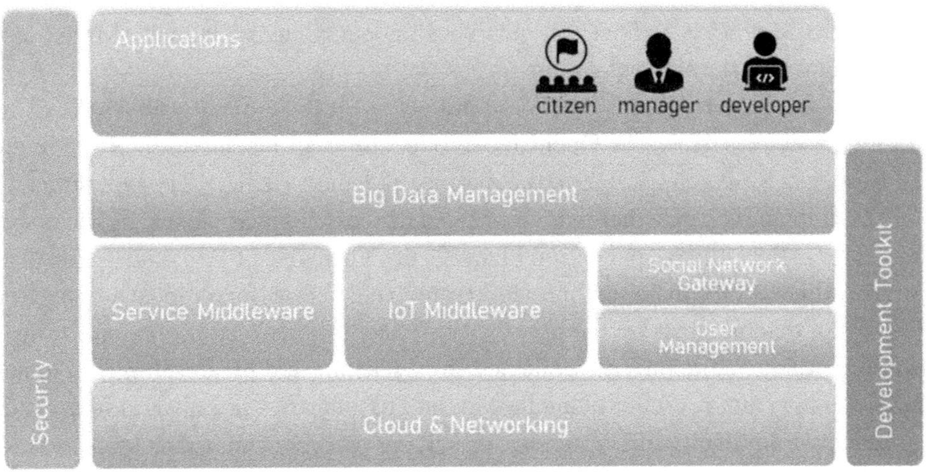

As stated before, the reference architecture by Santana et al. (2017), as presented in Figure 3 combines aspects from other architectures presented in the literature. The big data module is similar to the one in CiDAP considering batch and real-time processing and big data storage components. However, the idea of an application repository to store data and meta-data associated with applications so that we can better manage and reflect on the applications executing in the city was added, as well as a model repository to store different types of models associated with various city-related phenomena.

Similar to OpenIoT, a Cloud and Networking layer to manage the devices was also fomed part of the architecture. Included, also, are the service middleware to support many service-related operations, such as deployment, management, composition (via orchestrations and choreographies), and enactment. OpenIoT also provides a Service Delivery component with more limited support for some of these operations.

Urban Platforms

As seen in the work of Santana et al. (2017), existing OUPs are an inestimable source of architectural patterns to synthesise the most recurrent components used in reference models for smart cities. The analysis continues identifying literature testimonials of urban platforms.

Although we cannot directly extract a reference architecture from the OUPs, the application architecture that we can derive from them, generally through their views or viewpoints, is still valuable. An application architecture is a description of the structure and interaction of the applications as groups of capabilities that provide key business functions and manage the data assets (The Open Group, 2018). Such models could not be technology agnostic and provide specific solutions, which is not what we pretend for a reference architecture, but are useful to abstract the specific urban platform concerns. If a specific or concrete architecture, abstracted from a platform or application, is resulted from the adaptation for use of a reference architecture, then in the presence of a specific architecture, we can undertake a generalization process and infer the main components of the reference architecture that could have guided the definition of that architecture and others, in the same domain.

City-Data and Analytics Platform (CiDAP) has been deployed and integrated with a running IoT experimental testbed SmartSantander, one of the largest smart city testbed in the world (Cheng et al., 2015). The system architecture of CiDAP has four main modules and three layers: modules to collect, store, process, intensive process and perform analytics to data, along with a City Model server responsible for interfacing with external applications.

OpenIoT is an IoT platform used by the Vital project, with three planes: physical, virtualized, and utility-app (Soldatos et al., 2015). The OpenIoT Architecture is an instantiation of the reference architecture of the European Research Cluster on the Internet of Things (IERC). The OpenIoT architecture is comprised of seven main elements that belong to three different logical planes, as illustrated in Figure 4. These planes are: 1) the Utility/Application Plane, 2) the Virtualised Plane, and 3) the Physical Plane. OpenIoT handles almost all the main requirements enunciated before, for a robust architecture for Smart Cities. The merits of this platform are the use of an IoT middleware to configure and collect data from devices, the middleware to store the data collected from sensors, and the development tools.

Sii-Mobility[6] architecture is presently in place in the Tuscany area, in Italy (Badii et al., 2017). With the aim of producing a smart city infrastructure for solving problems related to transportation systems, services for mobility, services and systems for goods transportation, ordinances and public services, Sii-Mobility workes on enabling technologies for smart city and mobility, to integrate and produce the support of integrated interoperability (SII). The SII aims to create a smart city big data framework for providing integrated data, supporting services as mobile App, decision support, integrated ticketing, personal assistants, participative portals, crowdsourcing, dashboarding, etc., via Smart City API. It also aims to support data analytics and data intelligence based on integrated data collected from public administrations open data, private data from operators, and personal data coming from social media and city users.

Figure 4. OpenIoT Architecture based on Soldatos et al. (2015)

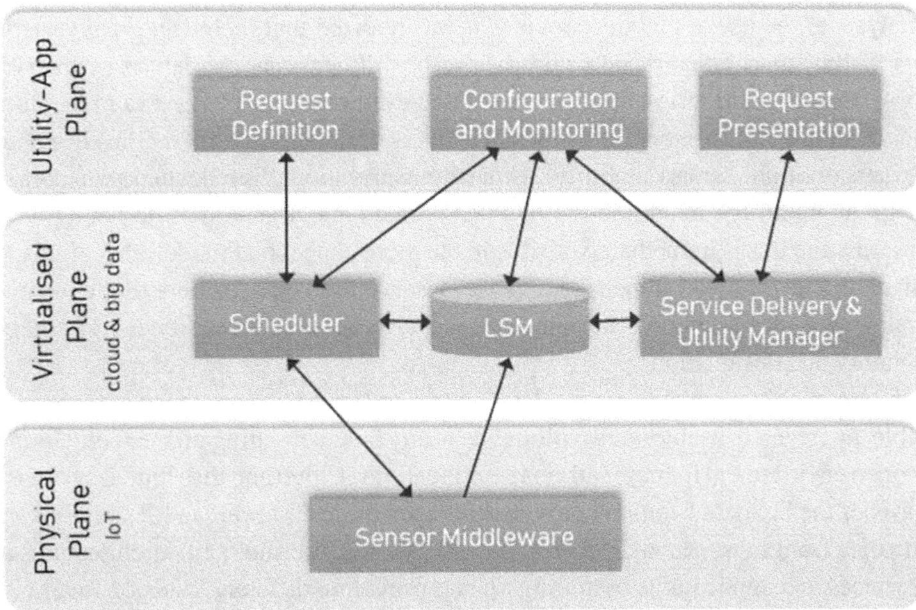

For the semantic aggregation of data and service, it has been decided to exploit and improve the Km4City Ontology[7] (Bellini, Benigni, et al., 2014), as the main ontological model. Km4City is modelling multiple domain aspects related to mobility.

The Open Machine Type Communications (Open-MTC) platform is an IoT/M2M middleware (Fraunhofer FOKUS OpenMTC, 2019) with a backend assuring several functions and several gateways for vertical domains. The Fiware platform (Fiware, 2019) provides a set of APIs that ease the development of smart applications in multiple vertical sectors, with its library of generic enablers.

Semantic Processing of Urban Data (SPUD) proposed by IBM (Kotoulas et al., 2014) exploits commercial non-open solution via a non-accessible ontology. An end-to-end semantic approach to extract business results in a combination of open datasets, proprietary datasets and social media: all pertaining to urban information. The authors assert that semantic technologies are indeed applicable to complex business problems, and can cope in smart city use cases, in scenarios such as ambulance response times in a big city, with acceptable performance overheads, and a focused domain. On the contrary of SPUD, envisaging a fully open upper-level ontology for smart cities is the recommendation comprised in (ISO/IEC, 2016) Information technology - Smart City ICT reference framework Part2 - Smart City Knowledge Management Framework.

Finally, CitySDK[8] has been developed in an EC project involving major cities and providing specific REST API. It is grounded on OASC[9] (Open & Agile Smart Cities) that adopted the FIWARE[10] NGSI API agnostic model.

REFERENCE MODELS COMPARISON AND ALIGNMENT

From the analysis of the above literature review, it was realized that there is a great variety in the way the elements of the smart cities models and the various dimensions, perceived as intervening in the complex ecosystem of smart cities, are designated and even enunciated. For example, all the thematic dealing with the mobility of persons and goods in a city is designated in smart cities literature in several ways, like 'transportation', 'smart mobility', 'smart transportation', 'intellectual transport system', and 'sustainable urban mobility'. In addition to this taxonomic topic, the models do not agree on the same set of dimensions, and this is immediately visible in the more conceptual models that al-most exclusively include human and institutional dimensions. If we observe the layers or tiers that constitute the more technological models, this scenario is even more evident, with a great diversity of ICT components with their functionality and designation.

Most of the smart cities architectures are data-centric and have been proposed with the aim of making data accessible, aggregated, usable, and exploitable. Many failed in setting up the foundation for creating a smart city open environment for new and smart applications. Cloud and distributed systems approaches are at the basis of the big data solutions provided for a smart city, as well as IoT solutions are the basis for collecting data from sensors and devices in the city. Most of the smart city architectures are enabling smart city services and applications by setting up and providing different kinds of smart city APIs.

Though there is abundant literature available on intelligent cities, there is no standardized and commonly accepted set of terminologies to aptly identify a set of univocal dimensions that an intelligent city must take into account. Given the lack of standardized terminologies, a comparative analysis was performed to map and compare the designations and definitions of the various smart city dimensions and organize the results to reveal commonalities and consensual dimensions.

This section presents a study of the dimensions associated with intelligent and sustainable digital cities. For this purpose, several research papers were collected from academic literature, as well as reports and articles from other databases. These sources were reviewed and analysed to help consolidate a wide range of dimensional perspectives of smart cities. These dimensions were obtained from a variety of sources comprising academic and research communities, government initiatives including EU, international organizations (like United Nations, ITU, European Commission, etc.), corporate/company profiles, and standards development organizations, briefly exposed in the previous section.

More than forty sources of proposals about the components or dimensions to be considered for a smart city were analysed. All sources were scrutinized and a spreadsheet was used to collect the dimensions indicated by each one, which facilitated the subsequent aggregation. The decomposition operated for the dimensions pertinent to the overall eco-system of smart cities, showed a total of one hundred and one (101) dimensions, even after some depuration obtained from eliminating evident synonyms and aggregating obviously similar dimensions.

To sort and group all dimensions found, dimension categories have been defined. These categories were established taking into consideration the classification of Nam and Pardo (2011) and Lea (2017), that arranged the components of smart cities in three major aspects: institutional, human, and technological. Some dimensions cross several boundaries just like Nam and Pardo (2011) and Lea (2017) conveyed in theirs overlapping classifications, so, two more categories were taken into account with the purpose of univocity and intended and casuistic specialization: city component-related aspects, and environmental aspects. So, all the viewpoints of what an intelligent city is, can be segmented into the following categories:

- **Human** viewpoint – this pertains to human infrastructure and social capital, involving education and governance, lifestyle, historical legacy, culture, recreation, economics, finance, social equity, engagement, laws, division of power, and politics.
- **Institutional** viewpoint – this relates to institutions, governance, government, policy, regulations/directives, planning, procurement, safety, emergency response, security, healthcare, city infrastructure.
- **Technological** viewpoint – this viewpoint concerns all topics related to ICT, namely software, mobile and virtual technologies, and digital infrastructure and networks.
- **City Component-related** viewpoint – this is used to denote cities' specific resources and processes like transportation, hard infrastructures, buildings, waste management, water, energy, and some city functions and services in general.
- **Environmental** viewpoint – this view indicates environmental management, protection and sustainable usage.

The above classification provided the first grouping level and more noticeable than the five categories with respect to the underlying distinction between technological dimension and all the others. Each non-technological dimension perceived in the literature was solely identified as belonging to one of the categories above. The technological dimensions have been also distinguished and grouped under subcategories or layers.

A second grouping level would need to detail this segmentation, particularly for the yet not segment technological dimensions. TOGAF (The Open Group, 2018), the Open Group standard for enterprise architecture includes four different architectural domains, which are accepted as sub-sets of an overarching enterprise architecture. A complete enterprise architecture should address all four architecture domains: data, application, and technology architectures that build on the business architecture.

An "enterprise" is any collection of organizations that have common goals. It can be a partnership or alliance of businesses working together, such as a consortium or supply chain. The purpose of enterprise architecture is to optimize across the enterprise the often fragmented legacy of processes (both manual and automated) into an integrated environment that is responsive to change and supportive of the delivery of the business strategy (The Open Group, 2018): an intent that is not far from the main objective of smart cities if we take strategy in a much broader sense than just the business.

This aggregation and classification of the technological intelligent cities dimensions identified in the overall literature used the TOGAF architecture domains. It is influenced from the OSI reference model (Zimmermann, 1980):

- **Type A** – Consumers (including any smart city stakeholder who wishes to interact with and consumes smart city services) and the business services themselves which are demanded from the non-technological smart cities dimensions. Represents the application and presentation levels of the OSI reference model and are contained in the TOGAF application architecture domain, but also to some extent, in the business architecture domain.
- **Type B** – Includes data services for data management, processing, exploitation and dissemination capabilities, and the software applications that manipulate that data, to support the Type A application and business services. Modelling & Analytics, Integration, Data Management, Reporting, Visualization, Systematization, Choreography and Orchestration, Technology and Supporting

Services (Transaction Management, Orchestration, BI, Collaboration). Represents the TOGAF data architecture domain.
- **Type V** – Supports the common services. Some layers can be 'vertical' as they cut across various other layers within the overall architecture. For example, sensitive data protection is a key concern, thus security modules shall be integrated into each layer.
- **Type C** – Includes communications, network & transport capabilities. Represents the TOGAF technology architecture domain and the session, transport, and network levels of the OSI reference model.
- **Type D** – Includes the operation and management of physical devices, sensors, actuators. Represents the TOGAF technology architecture domain and the data link and physical levels of the OSI reference model.

As mentioned above, the decomposition operated for the dimensions pertinent to the overall ecosystem of smart cities, has shown a total of one hundred and one (101) dimensions, even after some depuration obtained from eliminating evident synonyms and aggregating similar dimensions.

AN ALIGNED REFERENCE MODEL FOR INTELLIGENT CITIES

Following the review of the various research contributions, definitions and the 100+ dimensions or layers for an intelligent city, there was a need to perform some in-depth analysis to determine what would be a comprehensive and inclusive proposal of a smart city reference architecture. In the literature survey, there are numerous divergent visions of smart city architectures. Through these different expressions, it is possible to still find some common characteristics, evidenced by the way they were grouped in the categories described above.

The need to have a second level of grouping for non-tech dimensions, such as the classification presented in the previous section for the technological ones, has led to sort all these dimensions taking into account their similarity. Therefore, appriopriate groups has become apparent.

Excluding the technological components addressed below, we identified thirteen main dimensions from the literature (the most referred to) which, although mentioned by different designations as presented in Table 4, fairly represent the same principles and seem more than consensual, given the profusion of citations of information sources.

The number of times each dimension is referred to in the selected literature is presented in parenthesis and contributes to the perception of which dimensions are considered as most important and inescapable. The table shows the exact designations used by the selected scholars and professionals, to the right of the aggregating designation that represents them.

Outside of this unification attempt were placed dimensions that occurred seldom, such as innovation, only advanced by Zygiaris (2012) that dedicates to the cities' fertile innovation environment with new business models and opportunities pushed by the emerging technologies, an entire layer in the top of his smart city conceptual reference model. However, not itemized as an individual dimension, this aspect is not forgotten by the rest of the authors that prefer to spread it over the remaining dimensions. Almost forgotten seems to be the Industry subject. Only three works from the ones selected defend a smart industry dimension in the context of smart cities. Leydesdorff and Deakin (2011) and Lombardi et al. (2012) in the scope of their evolutionary perspectives for the triple-helix model for smart cities,

Table 4. Non-technological dimensions

Dimensions	Alternative designations and sub-dimensions
Government and Governance (17+2)	Government (Anthopoulos, 2015b) (Chourabi et al., 2011) (Neirotti et al., 2014) (Söderström et al., 2014) (Liu et al., 2014) (ISO, 2014) (ISO/IEC JTC 1, 2014) (Zygiaris, 2012); Smart governance (Giffinger et al., 2007) (Fromhold-Eisebith, 2017) (Anthopoulos et al., 2016)*(Cox et al., 2016); Government services (Naphade et al., 2011); e-Government (ITU-T FG-SCC, 2016); City governance for the smart city & decision support and participation (Batty et al., 2012)*; *sub-dimensions*: Urban planning (ISO, 2014); Smart procurement (Liu et al., 2014).
Economy and Business (17+3)	Economy (Anthopoulos, 2015b) Chourabi et al., 2011) (Neirotti et al., 2014) (ISO, 2014)*(Anthopoulos et al., 2016); Smart economy (Giffinger et al., 2007); Economic development (Eger, 2009)*; Sustainable economic development (Thuzar, 2011); Economic capital (Barrionuevo et al., 2012); Socio-economic development (IES-City PWG, 2018); Urban economics and commerce (DIN, 2017); e-Business (ITU-T FG-SCC, 2016), (Cox et al., 2016); Intelligent services layer (UNE, 2017); Domain application layer (Yin et al., 2015); *sub-dimensions*: Job growth (Eger, 2009); Finance (ISO, 2014); IT economy (Mahizhnan, 1999).
Transportation (16)	Transportation(Anthopoulos, 2015b) (Söderström et al., 2014) (Naphade et al., 2011) (Neirotti et al., 2014) (Clement et al., 2017) (ISO, 2014) (Anthopoulos et al., 2016) (Cox et al., 2016)(IES-City PWG, 2018); Smart mobility (Giffinger et al., 2007) (Fromhold-Eisebith, 2017); Smart transportation (Hancke et al., 2012) (Nam & Pardo, 2011) (ITU-T FG-SCC, 2016); Intellectual transport system (Glebova et al., 2014); Sustainable urban mobility(DIN, 2017);
Environment (15+10)	Environment (Chourabi et al., 2011) (Söderström et al., 2014) (Neirotti et al., 2014) (Lombardi et al., 2012) (Albino et al., 2015) (Clement et al., 2017) (ISO, 2014) (ISO/IEC JTC 1, 2014) (Anthopoulos et al., 2016) (Cox et al., 2016); Smart environment (Giffinger et al., 2007) (Nam & Pardo, 2011); Environmental capital (Barrionuevo et al., 2012); Smart metabolism (Fromhold-Eisebith, 2017); Green city layer (Zygiaris, 2012); *sub-dimensions*: Environmental protection (Glebova et al., 2014); Smart consumption (Fromhold-Eisebith, 2017); Solid waste (ISO, 2014) (IES-City PWG, 2018); Waste management (ITU-T FG-SCC, 2016); Water and sanitation (ISO, 2014) (ITU-T FG-SCC, 2016); Water and wastewater (Cox et al., 2016) (IES-City PWG, 2018).
People (13+3)	People (Chourabi et al., 2011) (Neirotti et al., 2014) (Albino et al., 2015) (Clement et al., 2017) (Anthopoulos et al., 2016) (Cox et al., 2016); People (social) (Desouza & Flanery, 2013); Smart people (Giffinger et al., 2007); People - human factors (learning) (Nam & Pardo, 2011); Human capital (Barrionuevo et al., 2012) (Kourtit & Nijkamp, 2012); Actors (ISO/IEC JTC 1, 2014); Civil society (Lombardi et al., 2012); *sub-dimensions*: Entrepreneurial capital (Kourtit & Nijkamp, 2012); Society and entrepreneurship (Edvinsson, 2006); Coherency (Anthopoulos, 2015b).
Urban infrastructure (11)	Urban infrastructure (Anthopoulos, 2015b) (Chourabi et al., 2011) (Söderström et al., 2014) (Zygiaris, 2012); Smart infrastructure (Hancke et al., 2012) (Albino et al., 2015) (Clement et al., 2017) (Liu et al., 2014); Infrastructural capital (Kourtit & Nijkamp, 2012); Infrastructure (ISO/IEC JTC 1, 2014); Integrated infrastructure & processes (DIN, 2017).
Safety and Security (11)	Smart safety (Nam & Pardo, 2011) (ISO, 2014) (ITU-T FG-SCC, 2016) (Cox et al., 2016); Public safety (Söderström et al., 2014) (Naphade et al., 2011); Public safety policing and emergency response (IES-City PWG, 2018); Public security (Glebova et al., 2014) (DIN, 2017); Smart security (Fromhold-Eisebith, 2017); Smart surveillance (Hancke et al., 2012).
Smart Buildings (10)	Smart buildings (Hancke et al., 2012) (Neirotti et al., 2014) (Söderström et al., 2014) (ITU-T FG-SCC, 2016) (Cox et al., 2016); Smart construction (Fromhold-Eisebith, 2017); Shelter (ISO, 2014); Facilities and buildings (ISO/IEC JTC 1, 2014); Built environment (IES-City PWG, 2018); Sustainable district and built environment (DIN, 2017).
Energy (10+3)	Energy (Neirotti et al., 2014) (Clement et al., 2017) (ISO, 2014) (ITU-T FG-SCC, 2016) (Anthopoulos et al., 2016) (Cox et al., 2016) (IES-City PWG, 2018); Smart energy (Nam & Pardo, 2011) (Fromhold-Eisebith, 2017); Energy consumption management and control (Glebova et al., 2014); *sub-dimensions: (paired with Water)* Smart electricity and water distribution (Hancke et al., 2012); Energy and water (Söderström et al., 2014) (Naphade et al., 2011).
Healthcare (9)	Healthcare (Naphade et al., 2011); Smart healthcare (Hancke et al., 2012) (Nam & Pardo, 2011) (ITU-T FG-SCC, 2016); Health (ISO, 2014) (Anthopoulos et al., 2016) (Cox et al., 2016) (IES-City PWG, 2018) Health and welfare (DIN, 2017).
Institutions (8)	Institutions (Desouza & Flanery, 2013); Community - institutional factors (Nam & Pardo, 2011); Institutional capital (Barrionuevo et al., 2012); Activities (ISO/IEC JTC 1, 2014); Social capital (Barrionuevo et al., 2012) (Kourtit & Nijkamp, 2012); Societal context (ISO/IEC JTC 1, 2014).
Education (8+3)	Education (Naphade et al., 2011) (ISO, 2014) (ITU-T FG-SCC, 2016) (Cox et al., 2016) (DIN, 2017) (IES-City PWG, 2018); Smart education (Nam & Pardo, 2011) (Fromhold-Eisebith, 2017); *sub-dimensions*: University (Lombardi et al., 2012) (Edvinsson, 2006) IT education (Mahizhnan, 1999).
Smart Living (8+3)	Smart living (Anthopoulos, 2015b) (Giffinger et al., 2007) (Neirotti et al., 2014) (Lombardi et al., 2012) (Anthopoulos et al., 2016); Quality of life (Mahizhnan, 1999) (Eger, 2009) (Thuzar, 2011); *sub-dimensions*: City history & characteristics (ISO/IEC JTC 1, 2014); Recreation (ISO, 2014); Knowledge cafes/cathedrals (Edvinsson, 2006).

and Fromhold-Eisebith (2017) that considers the industry an obvious process field for the potential use of CPS-driven processes.

Table 4 shows only the non-technological dimensions. ICT is one additional and inescapable dimension of smart cities. According to the particular classification of technological layers discussed above that incorporates influences of TOGAF (The Open Group, 2018) and the OSI reference model (Zimmermann, 1980), all layers proposed by the studied reference models, architectures, and platforms for smart cities, were identified and organized by type in Tables 5 to 9: one for each proposed technological layer, identifying the dimensions proposed in literature by its original designation and authors.

The Type A layers showed in Table 5, define the ICT Level A, characterized by the business services demanded from the non-technological smart cities dimensions presented in Table 4 and the interaction with the stakeholders involved. These layers are basically the application and presentation levels of the model.

Table 5. ICT Level A dimensions - Consumers

ICT Level A
Applications (Santana et al., 2017), Application Layer (Cox et al., 2016) (UNE, 2017) (Elhoseny et al., 2016) (Silva et al., 2018), Applications (ICT) (ITU-T FG-SCC, 2016), Services and Business (ICT) (ITU-T FG-SCC, 2016), External Applications (Cheng et al., 2015), Application View (Cyber) (Abu-Matar & Mizouni, 2018)
Business Process View (Abu-Matar & Mizouni, 2018), Business Roles Viewpoint (Management, Integration, Operational) (Clement et al., 2017)
Smart City API (Badii et al., 2017), Utility-App Plane (Request Definition, Config & Monitor, Request Presentation) (Soldatos et al., 2015)
Stakeholder Engagement & Collaboration Capabilities (DIN, 2017)
City / Community Specific Capabilities (DIN, 2017)
User Interface (UNE, 2017)
Presentation Layer (Elhoseny et al., 2016)
Gateways (Protocol Adapters) for Vertical Domains (Fraunhofer FOKUS OpenMTC, 2019)

The ICT Layer B in Table 6 represents Type B technological dimensions, which are mostly services, predominantly data-related, assuring data collection, management, storage, processing, and vitalisation. Similarly, includes management and integration capabilities. These deal with analytics, big data and semantics.

Type V layers define the ICT Level V, which is the vertical layer in the sense that it should transverse the other layers. It provides supporting services assuring management, and critical functions like security, safety and privacy. Refer to Table 7.

ICT Level C is composed of the Type C dimensions as presented in Table 8. These layers refer explicitly to the support capacity to interconnect people, smart nodes, workstations, and other embedded devices. Assures communication and interconnection facilities.

Lastly, Type D layers shape the ICT Level D. Refer to Table 9. This layer includes the operation and management of all the physical things in the cities ecosystem, that need monitoring and control. It is responsible for the capturing of all types of data from all types of sensors and physical devices.

Table 6. ICT Level B dimensions - services

ICT Level B
Service View (Abu-Matar & Mizouni, 2018), Middleware (Santana et al., 2017), Services Viewpoint (Clement et al., 2017), Common Data and Service Layer (Yin et al., 2015)
Open Integration Layer (Zygiaris, 2012), Integration Layer (Elhoseny et al., 2016), Integration, Choreography and Orchestration Capabilities (DIN, 2017), Generic City/Community Capabilities (DIN, 2017)
Integrated Databases Management (Batty et al., 2012) Management (ICT) (ITU-T FG-SCC, 2016) View Planning and Management (Anthopoulos et al., 2016)
Data (UNE, 2017) Data Management Layer (Silva et al., 2018) Data Services (Cox et al., 2016) View Data & Knowledge (Anthopoulos et al., 2016) Data View (Abu-Matar & Mizouni, 2018)
Data Management & Analytics Capabilities (DIN, 2017) Data Analysis and Modelling (Batty et al., 2012) Analytics View (Cyber) (Abu-Matar & Mizouni, 2018) Analytics and City Model Server (Cheng et al., 2015)
Data Collection (IoT Broker and IoT-agents) (Cheng et al., 2015) Semantic Aggregation & Big Data Processing (Badii et al., 2017) Data Vitalization Layer (Yin et al., 2015) Data Context Broker (Fiware, 2019)
Data Storage (Big Data Repository) (Cheng et al., 2015) Big Data Management (Santana et al., 2017) Data Processing (Intensive Big Data Processing) (Cheng et al., 2015)
Virtualized Plane Cloud and Big Data (Linked Stream Middleware, Scheduler, Service Deliver & Utility Manager) (Soldatos et al., 2015) Backend (Connectivity Management, Data Management, Authentication & Authorization) (Fraunhofer FOKUS OpenMTC, 2019)

Table 7. ICT Level V dimensions – supporting and common services

ICT Level V
Security (Santana et al., 2017) Privacy & Security Capabilities (DIN, 2017) Security Services (Cox et al., 2016) Security & Authentication (Yin et al., 2015)
Technology Services (Cox et al., 2016)
Supporting Services (Cox et al., 2016) (UNE, 2017) Common Services Capabilities (DIN, 2017)
Development Toolkit (Santana et al., 2017)

After this thorough analysis of multiple existing models and architectures, we derived the aligned and weighted reference architecture as illustrated in Figure 5, conjoint with a majority of proposed works. The figure illustrates the dimensions grouped in the categories human, institutional, city-component related and environmental, having a proportional dimension to their popularity in the studied literature, and supported by the ICT layers according to the classification justified above.

The non-technological dimensions appear at the top of the reference architecture as columns and in their role of consumers of the technological services provided below by the technological layers. These

Table 8. ICT Level C dimensions – communications capabilities

ICT Level C
Interconnection Layer (Zygiaris, 2012) Communications, Network & Transport Capabilities (DIN, 2017) Transmission Layer (Silva et al., 2018) Cloud and Networking (Santana et al., 2017)
Device Asset Management & Operational Services Capabilities (DIN, 2017) IoT Backend (Fiware, 2019)
Positioning Services (Cox et al., 2016)

Table 9. ICT Level D dimensions – physical operation

ICT Level D
Sensing layer (Silva et al., 2018) Field Equipment / Device Capabilities (DIN, 2017) Sensor/Actuator (UNE, 2017) Sensing Services (Cox et al., 2016) Physical Plane IoT (Sensor Middleware) (Soldatos et al., 2015) Instrumentation Layer (Zygiaris, 2012) IoT Edge (inc. IoT Gateways) (Fiware, 2019)
Data Acquisition Layer (Yin et al., 2015) (Elhoseny et al., 2016)
Infrastructure View (Physical) (Abu-Matar & Mizouni, 2018)
Place View (Physical) (Abu-Matar & Mizouni, 2018)
Participant View (Physical) (Abu-Matar & Mizouni, 2018)

dimensions are grouped according to the viewpoints previously presented and which were derived and extended from Nam and Pardo (2011) and Lea (2017). The chosen colour gradation denotes this categorization. The number of times each non-technological dimension is referred to in the selected literature, as computed in Table 4 arises from the height of the respective column in the figure and denotes the relative importance and inescapability of each dimension.

At the bottom of the reference architecture, appears the technological viewpoint. Each level or dimension provides services to the level immediately above. The exception and the reason for being a vertical and all-encompassing level in opposition to the others, is the ICT Level V, responsible for supporting and common services for all the remaining ICT levels.

Associated with each level is a list of the components derived from the dimensions aggregation as illustrated in Table 5 to Table 9. Particularly, a component designation that could represent and identify the set of designations used in the literature for the similar dimensions that were arranged by row below each type. For instance, dimensions whose authors in the literature have designated as Applications,

Figure 5. Aligned reference architecture for intelligent cities

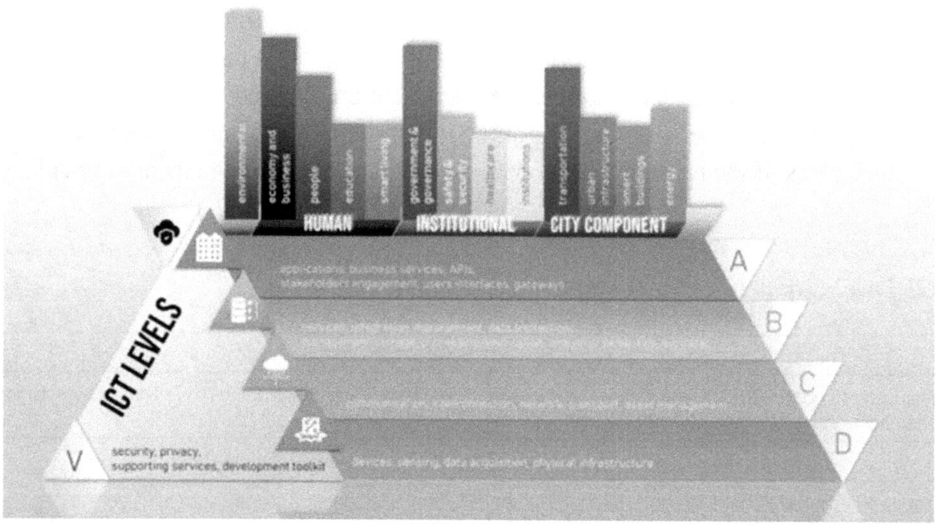

Application Layer, Applications (ICT), Services and Business (ICT), External Applications, and Application View (Cyber), are recognized as the component Applications inside ICT Level A.

Figure 5 shows that the non-technological dimensions are based directly on ICT Level A, consuming the services provided by this layer, and having here the point of interaction of the city stakeholders with the entire ICT component.

CONCLUSION

In this chapter, we reveal the mostly accepted aspects or dimensions of intelligent cities, supported by research on models and reference architectures for intelligent cities, in order to ease real-world deployment.

We stated the evident benefits of having such models sculpting the entire ecosystem of smart cities, with all the gains that such common and consistent definitions, patterns, collective vocabulary, and modelling artefacts represent to the design, implementation, and operation of the cities SoS. We reviewed existing smart city modelling proposals, ranging from the more conceptual to the more technological, from simples conceptual models to almost complete reference models and concrete architectures for urban platforms, and we found a huge variability of concepts and dimensions involved. However, because of the variability of concepts and the various arrangements of key components for cities, we could not find a universal smart city reference architecture or model. Lack of consensus on a common language/taxonomy that begins with a multiplicity of names chosen for the societal and physical dimensions involved, continues with the variety of the arrangements proposed for them and ends in heterogeneous supporting ICT building blocks.

The dimensions retrieved from all the modelling propositions were synthesized into an aligned smart city reference architecture that contributes to the definition of the components that must be part of such models. To perform the alignment of the non-technological dimension, categories have been defined. The same procedure was applied for the ICT dimension, resorting on influences from the TOGAF architecture domains and the OSI reference model. The result is a set of dimensions organized by the categories: human, institutional, city-component related and environmental, and supported by a services provider five layers ICT framework, which composes our aligned reference model for intelligent cities.

Our study shows that the overall vision of an intelligent city recognizes the growing importance of ICT and latest technologies as a core element and driver to foster smartness and the resulting sustainability and liveability of cities and citizens. Use of emerging technologies is inescapable in the smart cities ecosystem and as services enabler for the cities other aspects and dimensions. Although a variety of building blocks have been proposed to architect the overall technological infrastructure of a city, the proper model to present these components is multi-layered, where each layer uses services from the lower level and provides services to the upper level. As an exception, some of the technology based supporting functions must be arranged vertically as they cut across various other layers within the architecture. The upper technology layer must interface with the various domains of intelligent city stakeholders who consume these services. This current work presents these domains grouped in a number of categories, viz: human, institutional, city component-related and environmental. It testifies their relative popularity and recognition, also denoting the importance that satisfying the requirements of each domain may have to achieve the ultimate goal of an increasingly smart and sustainable city.

A critical remark and perplexity, when studying the contributions to this subject, a dimension that, despite its importance, appears much neglected and outside the cast of the most mentioned: industry.

Factories have their existence in cities, at least in the broader concept of cities adopted by smart cities, best embodied by regions or territory, and share many resources and restrictions with cities. ICT and emerging technologies have the exact same role in intelligent cities and initiatives aiming at digital industries, conducting factories to the fourth industrial revolution.

In the industrial domain, these initiatives are strongly supported on reference architectures, like Platform Industrie 4.0 (Kagermann, Wahlster, & Johannes, 2013), Industrial Internet Consortium (IIC) Industrial Internet (Lin et al., 2017), and NIST Smart Manufacturing (NIST, 2016). Those reference architectures for digital factories strongly rely on standards, which are the key to the effectiveness of information exchanging, sharing and integration. A universal reference architecture for intelligent cities should prescribe this sort of landscape of standards for cities, being based upon multi-dimension standards that can lend discipline and conformance, mainly to the most human and societal sectors of the cities ecosystem, providing the inestimable common and consistent definitions, patterns, and collective vocabulary.

HL7 FHIR (Lehne, Luijten, Vom Felde, & Thun, 2019) is an interoperability standard intended to facilitate the exchange of healthcare information, consisting of a content model in the form of resources that can be seen as an ontology, and a specification for the exchange of these resources. What HL7 FHIR makes for the healthcare dimension, other forthcoming standards should do for the remaining human and societal cities dimensions listed on our aligned reference architecture for smart cities.

The ultimate goal, yet somewhat far from reality but theoretically possible, is the achievement of a high-level overarching architectural framework for intelligent city concepts and relationships that can be used to describe and analyse data from any sector and thus mediate between different cities sector-specific ontologies.

ACKNOWLEDGMENT

This research was supported by national funds through FCT (Fundação para a Ciência e Tecnologia) within the Project UID/CEC/00319/2019 and by the Doctoral scholarship PDE/BDE/114567/2016 funded by FCT, the Portuguese Ministry of Science, Technology and Higher Education, through national funds, and co-financed by the European Social Fund (ESF) through the Operational Programme for Human Capital (POCH).

REFERENCES

Abu-Matar, M., & Mizouni, R. (2018). Variability Modeling for Smart City Reference Architectures. In *The Fourth IEEE Annual International Smart Cities Conference (ISC2 2018)*. Kansas City, MO: IEEE. 10.1109/ISC2.2018.8656967

Albino, V., Berardi, U., & Dangelico, R. M. (2015). Smart cities: Definitions, dimensions, performance, and initiatives. *Journal of Urban Technology*, *22*(1), 1–19. doi:10.1080/10630732.2014.942092

Anthopoulos, L. (2015a). Defining Smart City Architecture for Sustainability. *Proceedings of the 14th IFIP Electronic Government (EGOV) and 7th Electronic Participation (ePart) Conference 2015.*

Anthopoulos, L. (2015b). Understanding the smart city domain: A literature review. In *Transforming city governments for successful smart cities* (pp. 9–21). Springer. doi:10.1007/978-3-319-03167-5_2

Anthopoulos, L., Janssen, M., & Weerakkody, V. (2016). A unified smart city model (USCM) for smart city conceptualization and benchmarking. In Smart Cities and Smart Spaces: Concepts, Methodologies, Tools, and Applications (pp. 247–264). IGI Global.

Badii, C., Bellini, P., Cenni, D., Difino, A., Nesi, P., & Paolucci, M. (2017). Analysis and assessment of a knowledge-based smart city architecture providing service APIs. *Future Generation Computer Systems, 75*, 14–29. doi:10.1016/j.future.2017.05.001

Barrionuevo, J. M., Berrone, P., & Ricart, J. E. (2012). Smart cities, sustainable progress. *IESE Insight, 14*(14), 50–57. doi:10.15581/002.ART-2152

Bastidas, V., Helfert, M., & Bezbradica, M. (2018). A Requirements Framework for the Design of Smart City Reference Architectures. *Proceedings of the 51st Hawaii International Conference on System Sciences*. 10.24251/HICSS.2018.317

Batty, M., Axhausen, K. W., Giannotti, F., Pozdnoukhov, A., Bazzani, A., Wachowicz, M., ... Portugali, Y. (2012). Smart cities of the future. *The European Physical Journal. Special Topics, 214*(1), 481–518. doi:10.1140/epjst/e2012-01703-3

Bellini, P., Benigni, M., Billero, R., Nesi, P., & Rauch, N. (2014). Km4City ontology building vs data harvesting and cleaning for smart-city services. *Journal of Visual Languages and Computing, 25*(6), 827–839. doi:10.1016/j.jvlc.2014.10.023

Cheng, B., Longo, S., Cirillo, F., Bauer, M., & Kovacs, E. (2015). Building a big data platform for smart cities: Experience and lessons from santander. In *Big Data (BigData Congress), 2015 IEEE International Congress on* (pp. 592–599). IEEE.

Chourabi, H., Nam, T., Walker, S., Gil-Garcia, J. R., Mellouli, S., & Nahon, K., … Scholl, H. J. (2011). Understanding smart cities: An integrative framework. Proceedings of the Annual Hawaii International Conference on System Sciences, 2289–2297. 10.1109/HICSS.2012.615

Clement, S. J., McKee, D. W., & Xu, J. (2017). Service-oriented reference architecture for smart cities. In *Service-Oriented System Engineering (SOSE), 2017 IEEE Symposium on* (pp. 81–85). IEEE. 10.1109/SOSE.2017.29

Cloutier, R., Muller, G., Verma, D., Nilchiani, R., Hole, E., & Bone, M. (2010). The concept of reference architectures. *Systems Engineering, 13*(1), 14–27.

Cox, A., Parslow, P., De Lathouwer, B., Klien, E., Kempen, B., & Lonien, J. (2016). *D4.2 - Definition of Smart City Reference Architecture, ESPRESSO systEmic Standardisation apPRoach to Empower Smart citieS and cOmmunities*. Academic Press.

da Silva, W. M., Alvaro, A., Tomas, G. H. R. P., Afonso, R. a., Dias, K. L., & Garcia, V. C. (2013). Smart cities software architectures - a survey. *Proceedings of the 28th Annual ACM Symposium on Applied Computing - SAC '13*, 1722. 10.1145/2480362.2480688

De Oliveira Fernandes, E., Meeus, L., Leal, V., Azevedo, I., Delarue, E., & Glachant, J.-M. (2011). *Smart Cities Initiative: how to foster a quick transition towards local sustainable energy systems*. European University Institute. doi:10.2870/34539

Desouza, K. C., & Flanery, T. H. (2013). Designing, planning, and managing resilient cities: A conceptual framework. *Cities (London, England)*, *35*, 89–99. doi:10.1016/j.cities.2013.06.003

DIN. (2017). *DIN SPEC 91357, Reference Architecture Model Open Urban Platform (OUP)*. Berlin, Germany: Deutsches Institut für Normung.

Edvinsson, L. (2006). Aspects on the city as a knowledge tool. *Journal of Knowledge Management*, *10*(5), 6–13. doi:10.1108/13673270610691134

Eger, J. M. (2009). Smart growth, smart cities, and the crisis at the pump a worldwide phenomenon. *I-WAYS-The Journal of E-Government Policy and Regulation*, *32*(1), 47–53. doi:10.3233/IWA-2009-0164

EIP_SCC Urban Platform. (2016). *Requirements Specification For Urban Platforms v2.2 (consultation stage)* (Vol. 2). Academic Press.

Elhoseny, H., Elhoseny, M., Abdelrazek, S., Bakry, H., & Riad, A. (2016). Utilizing service oriented architecture (SOA) in smart cities. *International Journal of Advancements in Computing Technology*, *8*(3), 77–84.

Falconer, G., & Mitchell, S. (2012, Sept.). Smart City Framework: A Systematic Process for Enabling Smart+Connected Communities. *Point of View*.

Fischer, G., & Herrmann, T. (2011). Socio-technical systems: A meta-design perspective. *International Journal of Sociotechnology and Knowledge Development*, *3*(1), 1–33. doi:10.4018/jskd.2011010101

Fiware. (2019). *Fiware Wiki*. Retrieved from https://forge.fiware.org/plugins/mediawiki/wiki/fiware/index.php

Fraunhofer FOKUS OpenMTC. (2019). *OpenMTC*. Retrieved from https://www.open-mtc.org

Fromhold-Eisebith, M. (2017). Cyber-Physical Systems in Smart Cities - Mastering Technological, Economic, and Social Challenges. *Smart Cities: Foundations, Principles, and Applications*, 1–22.

Galster, M., & Avgeriou, P. (2011). Empirically-grounded reference architectures: a proposal. In *Proceedings of the joint ACM SIGSOFT conference--QoSA and ACM SIGSOFT symposium--ISARCS on Quality of software architectures--QoSA and architecting critical systems--ISARCS* (pp. 153–158). ACM. 10.1145/2000259.2000285

Giffinger, R., Fertner, C., Kramar, H., Kalasek, R., Pichler-Milanovic, N., Meijers, E., & Studies, M. (2007). Smart cities Ranking of European medium-sized cities. *October*, *16*, 13–18.

Glebova, I. S., Yasnitskaya, Y. S., & Maklakova, N. V. (2014). Assessment of Cities in Russia According to the Concept of" Smart City" in the Context of the Application of Information and Communication Technologies. *Mediterranean Journal of Social Sciences*, *5*(18), 55. doi:10.5901/mjss.2014.v5n18p55

Griffor, E. R., Greer, C., Wollman, D. A., & Burns, M. J. (2017). *Framework for Cyber-Physical Systems: Volume 1, Overview*. Academic Press.

Hancke, G. P., Hancke, G. P. Jr, & ... (2012). The role of advanced sensing in smart cities. *Sensors (Basel), 13*(1), 393–425. doi:10.3390130100393 PMID:23271603

IES-City PWG. (2018). *IES-City Framework - A Consensus Framework for Smart City Architectures v1.0*. Academic Press.

ISO. (2014). ISO 37120:2014, Sustainable development of communities - Indicators for city services and quality of life (Vol. 2014). Geneva, Switzerland: ISO.

ISO 15704. (2000). *ISO 15704, Industrial automation systems - Requirements for enterprise-reference architectures and methodologies*. Geneva, Switzerland: International Organization for Standardization.

ISO/IEC. (2016). *ISO/IEC AWI 30145-2, Information technology - Smart city ICT reference framework - Part2 - Smart City Knowledge Management Framework*. Geneva, Switzerland: International Organization for Standardization/International Electrotechnical Commission.

ISO/IEC JTC 1. (2014). *Smart Cities - Preliminary Report 2014*. Geneva, Switzerland: International Organization for Standardization/International Electrotechnical Commission.

ISO/IEC/IEEE 42010. (2011). ISO/IEC/IEEE 42010:2011 - Systems and software engineering - Architecture description (Vol. 2011). Geneva, Switzerland: International Organization for Standardization/ International Electrotechnical Commission. doi:10.1109/IEEESTD.2011.6129467

ITU-T. (2012). *Recommendation ITU-T Y.2060 - Overview of the Internet of things*. Geneva, Switzerland: International Telecommunication Union.

ITU-T. (2014). *ITU-T Y.2066 - Common requirements of the Internet of things*. International Telecommunication Union.

ITU-T FG-SCC. (2014a). *Smart sustainable cities: An analysis of definitions*. International Telecommunication Union.

ITU-T FG-SCC. (2014b). *Technical Report on "Overview of key performance indicators in smart sustainable cities."*. Geneva, Switzerland: International Telecommunication Union.

ITU-T FG-SCC. (2016). *ITU-T Y.4400 series – Smart sustainable cities – Setting the framework for an ICT architecture - Supplement 27*. Geneva, Switzerland: ITU.

Kagermann, H., Wahlster, W., & Johannes, H. (2013). *Recommendations for implementing the strategic initiative INDUSTRIE 4.0: Final report of the Industrie 4.0 working group*. Frankfurt: Acatech.

Kakarontzas, G., Anthopoulos, L., Chatzakou, D., & Vakali, A. (2014). A conceptual enterprise architecture framework for smart cities: A survey based approach. *11th International Conference on E-Business, ICE-B 2014 - Part of 11th International Joint Conference on e-Business and Telecommunications, ICETE 2014*, 47–54. 10.5220/0005021400470054

Kevrekidis, I. G., Gear, C. W., & Hummer, G. (2004). Equation-free: The computer-aided analysis of complex multiscale systems. *AIChE Journal. American Institute of Chemical Engineers, 50*(7), 1346–1355. doi:10.1002/aic.10106

Kotoulas, S., Lopez, V., Lloyd, R., Sbodio, M. L., Lecue, F., Stephenson, M., ... Mac Aonghusa, P. (2014). SPUD—Semantic processing of urban data. *Journal of Web Semantics*, *24*, 11–17. doi:10.1016/j.websem.2013.12.003

Kourtit, K., & Nijkamp, P. (2012). Smart cities in the innovation age. *Innovation (Abingdon)*, *25*(2), 93–95. doi:10.1080/13511610.2012.660331

Lea, R. J. (2017). *Smart cities: An Overview of the technology trends driving smart cities*. Academic Press.

Lee, E. A. (2008). Cyber physical systems: Design challenges. In *2008 11th IEEE International Symposium on Object and Component-Oriented Real-Time Distributed Computing (ISORC)* (pp. 363–369). IEEE.

Lee, J. H., Hancock, M. G., & Hu, M.-C. (2014). Towards an effective framework for building smart cities: Lessons from Seoul and San Francisco. *Technological Forecasting and Social Change*, *89*, 80–99. doi:10.1016/j.techfore.2013.08.033

Lehne, M., Luijten, S., Vom Felde, G. I. P., & Thun, S. (2019). The Use of FHIR in Digital Health-A Review of the Scientific Literature. *Studies in Health Technology and Informatics*, *267*, 52–58. PMID:31483254

Leydesdorff, L., & Deakin, M. (2011). The triple helix model of smart cities: A neo-evolutionary perspective. *Journal of Urban Technology*, *18*(2), 53–63. doi:10.1080/10630732.2011.601111

Lin, S.-W., Miller, B., Durand, J., Joshi, R., Didier, P., Chigani, A., … Witten, B. (2017). Industrial Internet Reference Architecture 1.8 - Volume G1: Reference Architecture. Industrial Internet Consortium.

Liu, Y., Wei, J., & Rodriguez, A. F. C. (2014). Development of a strategic value assessment model for smart city. *International Journal of Mobile Communications*, *12*(4), 346–359. doi:10.1504/IJMC.2014.063652

Lombardi, P., Giordano, S., Farouh, H., & Yousef, W. (2012). Modelling the smart city performance. *Innovation (Abingdon)*, *25*(2), 137–149. doi:10.1080/13511610.2012.660325

Mahizhnan, A. (1999). Smart cities: The Singapore case. *Cities (London, England)*, *16*(1), 13–18. doi:10.1016/S0264-2751(98)00050-X

Maier, M. W. (1998). Architecting principles for systems-of-systems. *Systems Engineering*, *1*(4), 267–284. doi:10.1002/(SICI)1520-6858(1998)1:4<267::AID-SYS3>3.0.CO;2-D

McKee, D. W., Clement, S. J., Almutairi, J., & Xu, J. (2017). Massive-scale automation in cyber-physical systems: Vision & challenges. In *Autonomous Decentralized System (ISADS), 2017 IEEE 13th International Symposium on* (pp. 5–11). IEEE.

Mohanty, S. P., Choppali, U., & Kougianos, E. (2016). Everything you wanted to know about smart cities: The internet of things is the backbone. *IEEE Consumer Electronics Magazine*, *5*(3), 60–70. doi:10.1109/MCE.2016.2556879

Muller, G., & van de Laar, P. (2010). Researching reference architectures. In *Views on Evolvability of Embedded Systems* (pp. 107–119). Springer. doi:10.1007/978-90-481-9849-8_7

Nakagawa, E. Y., Antonino, P. O., & Becker, M. (2011). Reference architecture and product line architecture: A subtle but critical difference. In *European Conference on Software Architecture* (pp. 207–211). Springer. 10.1007/978-3-642-23798-0_22

Nam, T., & Pardo, T. A. (2011). Conceptualizing smart city with dimensions of technology, people, and institutions. In *Proceedings of the 12th annual international digital government research conference: digital government innovation in challenging times* (pp. 282–291). 10.1145/2037556.2037602

Naphade, M., Banavar, G., Harrison, C., Paraszczak, J., & Morris, R. (2011). Smarter Cities and Their Innovation Challenges. *Computer*, *44*(6), 32–39. doi:10.1109/MC.2011.187

Neirotti, P., De Marco, A., Cagliano, A. C., Mangano, G., & Scorrano, F. (2014). Current trends in Smart City initiatives: Some stylised facts. *Cities (London, England)*, *38*, 25–36. doi:10.1016/j.cities.2013.12.010

NIST. (2016). *Current standards landscape for smart manufacturing systems. National Institute of Standards and Technology, NISTIR* (Vol. 8107). U.S. Department of Commerce, National Institute of Standards and Technology.

OASIS RM-SOA TC. (2019). *OASIS SOA Reference Model (SOA-RM) Technical Committee FAQ*. Retrieved October 17, 2019, from https://www.oasis-open.org/committees/soa-rm/faq.php

Santana, E. F. Z., Chaves, A. P., Gerosa, M. A., Kon, F., & Milojicic, D. S. (2017). Software platforms for smart cities: Concepts, requirements, challenges, and a unified reference architecture. *ACM Computing Surveys*, *50*(6), 78.

Silva, B. N., Khan, M., & Han, K. (2018). Towards sustainable smart cities: A review of trends, architectures, components, and open challenges in smart cities. *Sustainable Cities and Society*, *38*, 697–713. doi:10.1016/j.scs.2018.01.053

Söderström, O., Paasche, T., & Klauser, F. (2014). Smart cities as corporate storytelling. *City*, *18*(3), 307–320. doi:10.1080/13604813.2014.906716

Soldatos, J., Kefalakis, N., Hauswirth, M., Serrano, M., Calbimonte, J.-P., Riahi, M., & … . (2015). OpenIoT: Open source internet-of-things in the cloud. In *Interoperability and open-source solutions for the internet of things* (pp. 13–25). Springer. doi:10.1007/978-3-319-16546-2_3

The Open Group. (2018). *The TOGAF Standard, Version 9.2 - Reference: C182*. Author.

Thuzar, M. (2011). Urbanization in SouthEast Asia: developing smart cities for the future? *Regional Outlook*, 96.

UN-Habitat. (2013). *State of the World's cities 2012/2013 Prosperity of Cities.* UN-Habitat (United Nations Human Settlements Programme).

UNE. (2017). *Sistemas Integrales de Gestión de la Ciudad Inteligente. Requisitos de interoperabilidad para una Plataforma de Ciudad Inteligente*. Madrid: UNE.

Wang, F.-Y. (2010). The emergence of intelligent enterprises: From CPS to CPSS. *IEEE Intelligent Systems*, *25*(4), 85–88. doi:10.1109/MIS.2010.104

Weyrich, M., & Ebert, C. (2016). Reference Architectures for the Internet of Things. *IEEE Software*, *33*(1), 112–116. doi:10.1109/MS.2016.20

Yin, C., Xiong, Z., Chen, H., Wang, J., Cooper, D., & David, B. (2015). A literature survey on smart cities. *Science China. Information Sciences*, *58*(10), 1–18. doi:10.100711432-015-5397-4

Yovanof, G. S., & Hazapis, G. N. (2009). An architectural framework and enabling wireless technologies for digital cities & intelligent urban environments. *Wireless Personal Communications*, *49*(3), 445–463. doi:10.100711277-009-9693-4

Zimmermann, H. (1980). OSI reference model - the ISO model of architecture for open systems interconnection. *IEEE Transactions on Communications*, *28*(4), 425–432. doi:10.1109/TCOM.1980.1094702

Zygiaris, S. (2012). Smart city reference model: Assisting planners to conceptualize the building of smart city innovation ecosystems. *Journal of the Knowledge Economy*, *4*(2), 217–231. doi:10.100713132-012-0089-4

ADDITIONAL READING

ISO/IEC. (2016). *ISO/IEC CD 30141 - Information Technology – Internet of Things Reference Architecture (IoT RA)*. Geneva, Switzerland: International Organization for Standardization/International Electrotechnical Commission.

ITU-T FG-SCC. (2015). *Technical Report on "Setting the framework for an ICT architecture of a smart sustainable city." Focus Group Technical Specifications*. Geneva, Switzerland: International Telecommunication Union.

IVI. (2018). *IVRA Next - Strategic implementation framework of industrial value chain for connected industries*. Tokyo: Industrial Value Chain Initiative.

Komninos, N., Bratsas, C., Kakderi, C., & Tsarchopoulos, P. (2015). Smart City Ontologies: Improving the effectiveness of smart city applications. *Journal of Smart Cities*, *1*(1), 31–46. doi:10.18063/JSC.2015.01.001

MIIT, & SAC. (2015). *National Intelligent Manufacturing Standard System Construction Guidelines (Version 2015)*. Ministry of Industry and Information Technology of China.

Ostrom, E. (2008). *Doing institutional analysis: Digging deeper than markets and hierarchies* (pp. 819–848). Springer Berlin Heidelberg.

Piai, S., & Claps, M. (2013). *Enabling Sustainable and Healthier Living in the Urban Era: the Healthier Cities Maturity Model*. IDC Health Insights.

Scottish Cities Alliance, & Urban Tide. (2014). *Smart Cities Maturity Model and Self-Assessment Tool - Guidance Note for completion of Self-Assessment Tool.*

KEY TERMS AND DEFINITIONS

Architecture: The architecture of a system is a metaphor analogous to the architecture of a building, in the sense that it functions as a blueprint for the system and the development. This refers to a description of the arrangement and connectivity of the parts of a system, their relationships to one another and to the environment and the principles guiding its design and evolution.

Aspect: It refers to the high-level grouping of conceptually equivalent or related concerns that manifest interest in a system relevant to at least one of its stakeholders.

Conceptual Model: It refers to a generic, abstract and simple structure and definitions for describing the concepts of, and the relationships among, the entities within systems. In order to achieve this goal, it is important to clarify the fundamentals of the systems by asking for the overall model of entities and their relationships, the key concepts in a typical system, and who and where are the actors.

Concern: Concerns refer to an interest in a system relevant to one or more of its stakeholders. They may relate to the system's functioning, development, or operation, including considerations related to non-functional requirements.

Design Principle: Design principles provide a systemization of knowledge describing the constituents of a phenomenon. They support practitioners in developing appropriate solutions and from an academic perspective. These principles form the foundation of a design theory.

Framework: This refers to a structure for content or process that represents a broad overview, skeleton, or outline, of interlinked items which supports a particular approach to a specific objective, and serves as a guide that can be modified as required by adding or deleting items.

Reference Architecture: A reference architecture is an architecture where the structures and respective elements and relations provide templates for concrete architectures in a particular domain or family of systems. It provides a template solution for an architecture, delivering a common vocabulary with which to discuss implementations, and stressing commonality.

Reference Model: A reference model describes both types of entities or domains and their relationships, including a description of the problem that it solves, and the concerns of the stakeholders who need the problem to be solved. A reference model is intended to promote understanding of a class of problems, not to provide specific solutions for those problems, being technology agnostic. In addition, a reference model provides a more complete view of, or templates for, the overall setup and its domain implementation.

Requirement: In systems engineering, a requirement can be defined as something that needs to be conceived. It refers to the properties that the systems (under consideration) shall possess when designed and built, and express the users' necessities and restrictions that are placed on the system and that must be considered during the development.

Urban Platform: This refers to a system that facilitates the exploitation of city data and intelligent monitoring and control of the cities infrastructures and assets, and that enables new and improved services for public and private ends and stakeholders. Desirably should utilise open standards and interfaces to guarantee compatibility and interoperability with other systems.

ENDNOTES

1 ISO/IEC. 2011. ISO/IEC 25010: Systems and software engineering - Systems and software Quality Requirements and Evaluation (SQuaRE) — System and software quality models.

2 (IES-City PWG, 2018) defines eight geo-domains (home, building, campus/aggregation, district, city, land, country, mainland), that specifies the geographical domain on which the applications can act.

3 The literature review is fully described in another work of the author ("Understanding the smart city Domain: A Literature Review," in Bolivar M.P. (Ed) Transforming City Governments for successful Smart Cities, Public Administration and Information Technology Series, Vol. 3, Springer Science+Business Media New York, 2015.

4 http://www.openiot.eu/

5 https://github.com/OpenIotOrg/openiot/wiki/OpenIoT-Architecture

6 http://www.sii-mobility.org

7 http://www.km4city.org

8 http://www.citysdk.eu

9 https://oascities.org/

10 http://www.fiware.org

Chapter 3
Developing Trends in Power and Networking Technologies for Intelligent Cities

Kadir Alpaslan Demir
https://orcid.org/0000-0002-8304-6324
Turkish Naval Research Center Command, Turkey

Buğra Turan
Koc University, Turkey

ABSTRACT

The introduction of Industry 4.0 has increased the focus on a number of technologies. These technologies also help realize the vision for intelligent cities. Furthermore, there are already discussions of Industry 5.0. One emerging aspect of Industry 5.0 is human-robot co-working. With the help of artificial intelligence, the internet of things paradigm, Industry 4.0, and Industry 5.0 visions, there will be two predominant types of systems interfacing with people in intelligent cities. These are robotic and ambient intelligence systems. The increasing deployment of these will help make cities even smarter. However, we need to see advancements in a number of relevant key technologies, including power and networking technologies. In this chapter, first, the authors briefly discuss Industry 4.0, Industry 5.0, and intelligent cities paradigm, as well as robotic and ambient intelligence systems. Then, they focus on developing trends in power and networking technologies.

INTRODUCTION

Artificial Intelligence (AI) research has been on the rise in recent years. AI is one of the key technologies for the realization of Industry 4.0. There are also other key technologies such as the Internet of Things (IoT), big data, robotics, and cloud computing. 3D printing, virtual and augmented reality, smart logistics, ambient intelligence are also among other supporting technologies for Industry 4.0 (Demir, Turan, Onel, Ekin, & Demir, 2019). Note that all these technologies support the development of each other. The

DOI: 10.4018/978-1-7998-5062-5.ch003

motto of Industry 4.0 is "Smart Manufacturing for the Future". Its focus is industrial development and mainly "Smart Factories". Recently, various scholars and futurists started discussions on Industry 5.0.

There are various visions for Industry 5.0 (Demir and Cicibas, 2018). One emerging vision attracting more attention than others is "Human-Robot Co-working". Another aspect of Industry 5.0 is the bio-economy (Demir and Cicibas, 2018). While we do not know which vision of Industry 5.0 will be realized, both visions are relevant to intelligent cities, also known as smart cities or digital cities. Therefore, it is possible to argue that Industry 5.0 and its related technologies will support the realization and further development of intelligent cities. We will discuss these revolutions and their relevance to intelligent cities in the following sections.

An intelligent city comprises numerous smart devices and intelligent information services. We believe these services will be delivered by two predominant types of systems, viz: robotic (Samani and Zhu, 2016) and ambient intelligence systems (O'grady and O'hare, 2012; Streitz, 2011). These systems will interface with other smart systems and interact with people. Robotic systems will mainly be used in cases where we need physical assistance such as cleaning or transportation services. In other cases, ambient intelligence systems will likely be utilized to make cities even smarter. Imagine a smart cashier-less store, where shelves are stocked by robots and payments are handled by an ambient intelligence system. Amazon recently opened a smart store called Amazon Go (Amazon Go, 2019; Demir, et al., 2019), where, there are no cashiers. Thus, there are no queues and no physical checkout. Customers just "grab-and-go". Amazon Go is just an earlier example of a smart store that will be predominant in intelligent cities. In the next sections, we discuss robotic and ambient intelligence systems in some detail.

While robotic and ambient intelligence systems will be the technologies interfacing humans, there will be other technologies supporting intelligent cities in the background. Power and networking technologies are among the important ones for intelligent cities. Power is required for any type of information system. However, it is more important for mobile systems such as smartphones and robots. Power has currently been a limiting factor for advancement in both these types of systems. For most smartphones, a recharge is needed almost every day. Many drones are limited in operation time due to limited power supply. Internet of Things (IoT) will also be an important enabler for smart cities. However, the IoT also brings many challenges. One of the main challenges is the massive networking that the IoT paradigm brings. Networking demand will likely increase exponentially in a smart city. Therefore, efficient power and networking technologies are crucial for smart devices, systems, and services in such digital cities. We discuss the developing trends in power and networking technologies in the following sections.

The organization of this chapter is as follows: There are three main sections. In the first section, we discuss the industrial revolution, particularly the vision of Industry 4.0 and Industry 5.0. We naturally focus on their relevance to intelligent cities. In the second section, we discuss robotic and ambient intelligence systems for intelligent cities. The third section presents the developing trends in power and networking technologies. Finally, we conclude the chapter. In what follows, we will use the *smart city* and *intelligent city* terms, interchangeably.

INDUSTRIAL REVOLUTION

There were three notable industrial revolutions in the past. Today, according to many, we are in the fourth one. Figure 1 shows the previous industrial stages as a timeline. As the figure shows, the periods between industrial revolutions are shortening. The time between the first and second industrial revolutions is around a century. The same period exists between the second and the third. However, it took around 40 years to reach the fourth industrial revolution. If the trend continues, it will likely take less than 40 years to reach the fifth industrial revolution. Next, we will discuss Industry 4.0 and Industry 5.0.

Industry 4.0

The German government is attributed to have initiated the Industry 4.0 revolution. One of the main reasons behind this initiative is to react to the rising Chinese economy. In recent years, the Chinese economy was significantly boosted by low-cost labor. China almost became the production center of many products in the world. The German government wanted to counter low-cost labor by introducing smart technology into manufacturing. Thus, the theme of Industry 4.0 is "Smart Manufacturing for the Future". The goal is to utilize innovation and the latest technology to increase automation, productivity, and efficiency in manufacturing. This is the same goal for the previous industrial stages as well. While there are many types of technologies helping to realize Industry 4.0, the main ones are the Internet of Things, robotics, artificial intelligence (AI), big data, and the cloud computing paradigm. 3D printing, smart logistics, virtual reality, augmented reality, ambient intelligence are some of the other technologies for moving Industry 4.0 from vision to reality. These technologies are not specifically researched for Industry 4.0. In fact, the vision of Industry 4.0 is about bringing these technologies together to accomplish smart manufacturing and developing smart working spaces (Demir, Döven, and Sezen, 2019).

While there are many studies supporting the Industry 4.0 vision, there is also criticism regarding Industry 4.0 (Maier and Student, 2015; Demir and Cicibas, 2017). One particular criticism is that Industry 4.0 is a top-down government initiative with a predefined notion (Sachsenmeier, 2016) rather than a widely accepted industrial movement stemming from many types of organizations including SMEs (Demir and Cicibas, 2018).

Figure 1. Evolution of Industry 1.0 to Industry 5.0

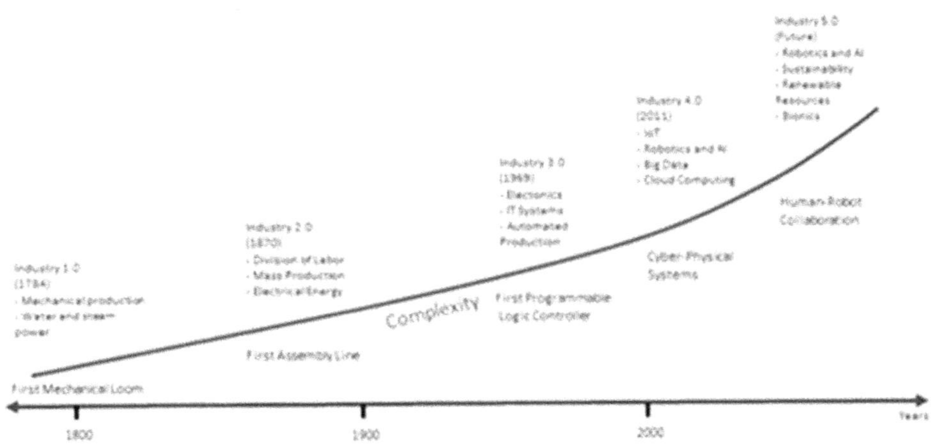

Industry 5.0

Although Industry 5.0 spans a much broader view (Demir, Döven, and Sezen, 2019), the focus of most research is on human-robot co-working. According to this aspect of Industry 5.0, humans and robots will work together and share many tasks based on their abilities. Some argue that Industry 5.0 will bring back the human touch to the factory floor. Here, humans will deal with tasks requiring creativity and higher-level thought, the robots will deal with the practical and routine work. While Industry 4.0 focuses on smart manufacturing, Industry 5.0 has a broad focus on all aspects of any work. At first, robots will primarily be responsible for Dull, Dirty, and Dangerous tasks (Ravikumar, 2019). Then in time, as robots evolve, they may be able to handle other types of tasks. The technologies that will support Industry 5.0 will not be much different from the technologies supporting Industry 4.0. Internet of Things, artificial intelligence, robotics, and other information and communication technologies will help advance the Industry 5.0 revolution. We believe that in this era, research areas such as human-robot interaction and human-robot collaboration will be quite important. We will advance current interaction methods or even develop new ways of interacting with robots.

Currently, it is too early to determine if human-robot co-working will actually become the vision of Industry 5.0. If this vision becomes a reality, then Industry 5.0 will deeply affect how we live and work. The advancements brought by human-robot co-working will naturally advance human-robot co-living. Robots will likely be a part of our lives away from work, as well. Therefore, the societal impact due to Industry 5.0 will not be limited to work only. Table 1 shows a comparison of Industry 4.0 (Smart Manufacturing) and Industry 5.0 (Human-Robot Co-working vision).

In the following sections, we discuss human-robot co-working and human-robot co-living. In the human-robot co-working section, we briefly discuss the current and future cases of human-robot co-working.

Table 1. A Comparison of Industry 4.0 vs Industry 5.0 Visions (adapted from Demir, Doven, & Sezen, 2019)

	Industry 4.0	**Industry 5.0**
Motto	Smart Manufacturing	Human-Robot Co-working
Motivation	Mass Production	Smart Society
Power Source	Electrical power Fossil-based fuels Renewable power sources	Electrical power Renewable power sources
Involved Technologies	Internet of Things (IoT) Cloud Computing Big Data Robotics and Artificial Intelligence (AI)	Internet of Things (IoT) Human-Robot Collaboration Renewable Resources
Involved Research Areas	Organizational Research Process Improvement and Innovation Business Administration	Smart Environments Organizational Research Process Improvement and Innovation Business Administration

INTELLIGENT CITIES

According to Silva, Khan, and Han (2018), an intelligent city is *an urban environment that utilizes ICT and other related technologies to enhance performance efficiency of regular city operations and quality of services (QoS) provided to urban citizens.* Another comprehensive definition provided by International Telecommunication Union (2014) Telecommunication Standardization Sector (ITU-T) Focus Group on Smart Sustainable Cities (ITU-T FG-SSC) is as follows: *A smart sustainable city is an innovative city that uses information and communication technologies (ICTs) and other means to improve quality of life, efficiency of urban operation and services, and competitiveness, while ensuring that it meets the needs of present and future generations with respect to economic, social, environmental as well as cultural aspects.* Note that the second definition has a sustainability focus. There are various other definitions focusing on the integration of sustainability and smartness for smart cities (Höjer and Wangel, 2014). Intelligent cities, also known as smart cities, will be an important trend for the future. In addition to smartness, we will also expect sustainability, and connectivity between smart devices, in intelligent cities.

There are generally two types of systems that will interface with humans in intelligent cities. These are robotic and ambient intelligence systems. In intelligent cities, there will be smart homes, smart buildings, smart offices, smart stores, smart schools, smart traffic systems, smart cars, smart utility services, etc. We will simply be adding intelligence into every aspect of living in the city. Having intelligent cities will not be an instantaneous event but it will be a transition to more automation and digitization. Smart things in smart cities will evolve and the transition will continue for a long time. In some aspects of smart cities, ambient intelligence systems will be dominant. In other aspects, we will interface with robotic systems more. We believe robotic systems will mainly be used in cases where we need physical assistance. For example, we will have cleaning robots at homes or in our offices. There will be smart taxies transporting people. There will be smart trucks or unmanned aerial vehicles (UAVs) transporting goods. In other cases, ambient intelligence systems will likely be utilized. For example, in smart cashier-less stores, there will be ambient intelligence systems recognizing people and charging them for the goods they buy. In smart offices, ambient intelligence systems will adjust the air conditioning depending on the number of people and their preference. In some cases, both types of systems will be used. For example, in a smart store, the shelves will be stocked by robots and payments will be handled by ambient intelligence systems.

Figure 2 presents the essential technologies for robotic systems and ambient intelligence systems. Note that many other technologies are supporting the development of these systems. However, in Figure 2, we focus on the most important ones necessary for interfacing humans.

A detailed discussion regarding the intelligent cities is outside the scope of this chapter. Several current information sources on intelligent systems and cities are provided in the section on Further Reading.

Ambient Intelligence (AMI)

Ambient Intelligence (AmI) refers to *electronic or digital environments that are sensitive and responsive to the presence of people* (IST Advisory Group, 2003; Aarts and Encarnação, 2005; Ramos, Augusto, & Shapiro, 2008). According to Augusto (2007), ambient intelligence is *a digital environment that supports people in their daily lives by assisting them in a sensible way.* Ambient intelligence (AmI) also refers to smart environments (SmE) or intelligent environments (IE) (Augusto and McCullagh, 2007). The idea of the invisible computer appears to be the basis for the concept of ambient intelligence (Norman,

Figure 2. Essential Technologies for Robots and Ambient Intelligence Systems

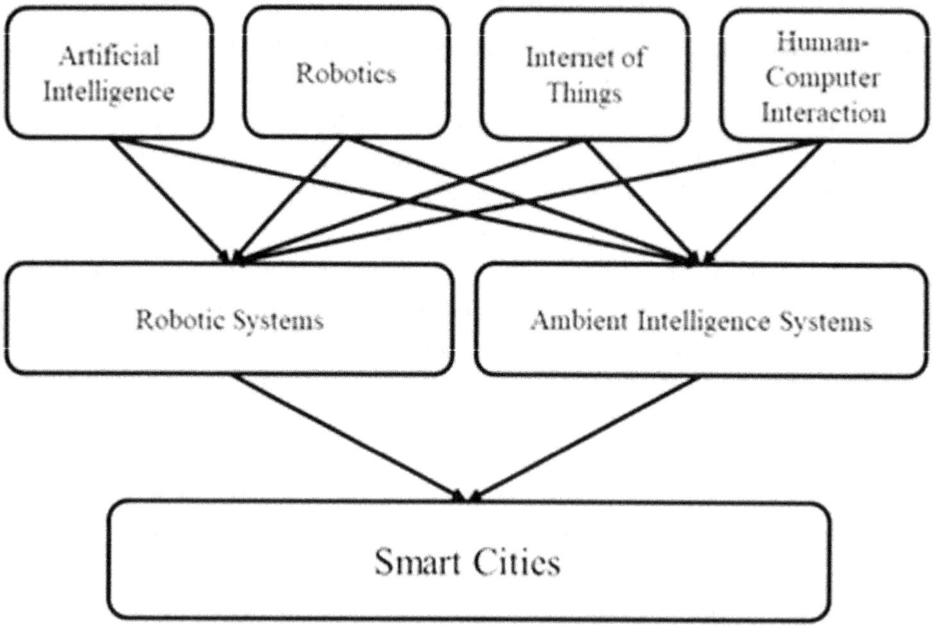

1998). Weiser (1991) elegantly described the idea of disappearing computers by stating that *the most profound technologies are those that disappear and they weave themselves into the fabric of everyday life until they are indistinguishable from it.*

In the ambient intelligence paradigm, small appliances interface with humans and powerful computers collectively handle the necessary computing in the background (Remagnino and Foresti, 2005). In an ambient intelligence system, there are sensors and actuators connected to each other and to a powerful central computer or to a distributed computer system. Sensors gather sensory data from the environment. This data is sent to the computer system for processing via the network. Then, the computer system sends commands to the necessary actuators to accomplish a specific task.

Amazon Go and Alibaba's Tao Café are recent examples of smart stores utilizing ambient intelligence technology (Demir et al., 2019). While they are in experimental stages, what they offer is quite promising for the future. It is only a matter of time that various other types of stores start utilizing this essential smart city technology. AmI is expected to help with various intelligent city applications such as smart parking, structural health, development of urban maps, smartphone detection, electromagnetic field leveling, traffic congestion management, smart lighting, waste management, smart roads, etc. (Libelium, 2019). O'grady and O'hare (2012) state that, in the long term, ambient intelligence may be experienced as part of the smart city e-services. Furthermore, Streitz (2011) believes that AmI will influence the future of our cities and further develop and improve the concepts of hybrid, smart, and humane cities. A detailed discussion of ambient intelligence vision may be found in the study by Markopoulos (2016). Discussions related to how AmI may help realize the concept of smart cities from various perspectives may be found in Gams, Gu, Härmä, Muñoz, & Tam (2019); Brown (2019); Mahmood, Butler, Sheng, Zhang, & Jennings (2019); and Buhalis (2019). Grand challenges of ambient intelligence for smart societies are discussed by Streitz, Charitos, Kaptein, & Böhlen (2019). In short, ambient intelligence provides plenty of opportunities for a smarter world. See a broad category of application areas in Libelium (2019).

Robotic Systems

Robots have been in use in factories for quite some time now (Demir et al., 2017). The use of robots is expanding (Ray, Mondada, & Siegwart, 2008) and robots are becoming a part of our lives (Lin, Abney, & Bekey, 2011; Demir, 2017). Robots and drones may be used in various areas and services such as intelligent transportation (Menouar, Guvenc, Akkaya, Uluagac, Kadri, & Tuncer, 2017), smart city maintenance services (Abbasi, Majidi, & Manzuri, 2018), smart medical services (Samani and Zhu, 2016; Grigorescu, Argatu, Paturca, Cepisca, Seritan, Adochiei, & Enache, 2019), elderly care (Skouby, Kivimäki, Haukiputo, Lynggaard, & Windekilde, 2014), service robotics (Alonso, 2011), and personal care (Lin et al., 2011), etc. There will be many other highly useful robotic application areas further advancing the notion of smart cities.

Currently, we do not have a widely accepted definition for a robot (Demir, 2017). When people are inquired about what a robot is, they will give many different definitions. Most of these views are affected by media coverage about robots. The discussions on the definition of robots will likely continue. One good definition is provided by Object Management Group (OMG). They define robotic systems as *systems that provide intelligent services and information by interacting with their environment, including human beings, via the use of various sensors, actuators and human interfaces* (Object Management Group, 2005). However, we can classify robots based on various aspects such as automation levels. Sheridan and Verplank (1978) developed such automation levels as presented in Table 2. We may also develop a taxonomy from an ethical perspective. Veruggio and Operto (2008) provide such a taxonomy based on a roboethics perspective. This taxonomy is presented in Table 3. The automation levels and the roboethics taxonomy are especially helpful in identifying types of robots that will be part of the smart city's vision. There may be other classifications from various perspectives. As a result, there will be many types of robotic systems in intelligent cities. Furthermore, these systems will interface with humans and other types of systems such as ambient intelligence systems. Roboethics is the discipline dealing with ethical

Table 2. Automation Levels for Robots

Automation Level	Automation Description
1	No computer assistance: the human must make all decisions and take actions.
2	The computer offers a complete set of decision/action alternatives.
3	The computer narrows the selection down to a few.
4	The computer suggests one alternative.
5	The computer executes a suggestion if the human approves.
6	The computer allows the human a restricted time to veto before the automatic execution.
7	The computer executes automatically, then informs humans when necessary.
8	The computer informs the human only if asked.
9	The computer informs the human only if it decides so.
10	The computer decides everything and acts autonomously, ignoring the humans.

Table 3. A Taxonomy of Roboethics

Humanoids
Outdoor Robotics
Military Robotics
Artificial Body
Surgical Robotics
Educational Robot Kits
Industrial Robotics
Biorobotics
Robot Toys
Adaptive Robot Servants
Biomechatronics
Entertainment Robotics
Distributed Robotic Systems
Health Care & Quality of Life
Robotic Art

issues surrounding robotics (Veruggio, 2007). To successfully utilize robots in intelligent cities, we need many research studies both on roboethics and robotic law (Demir, 2017). Table 4 presents several key research areas for roboethics (Demir, 2017).

Human-Robot Co-working

Human-robot co-working will be an important part of the intelligent city vision; although, the concept of human-robot co-working is still at an early stage. There are only a limited number of studies on the subject (Cormier, Newman, Nakane, Young, & Durocher, 2013; Banh, Rea, Young, & Sharlin, 2015;

Table 4. Key Research Issues in Roboethics

The Definition of a Robot
The Definition of a Cyborg
Designing Ethical Behaviour in Robots
Integrating Robots into Society
Integrating Robots into Organizations and Workplaces
Moral Responsibility
Roboethics Taxonomy
Ensuring Privacy
Automation Levels to be used in Robotics and Roboethics
Code of Ethics for Robotics Engineers
Killer Robots

Geiskkovitch, Cormier, Seo, & Young, 2016; Demir, Caymaz, & Elçi, 2017; Bagdasarov, Martin, & Buckley, 2018; Demir, Döven, & Sezen, 2019; Demir, 2019). We need many more research studies to advance human-robot co-working. Table 5 presents human-robot co-working issues that require investigation (Demir, 2019).

Table 5. Human-Robot Co-working Issues

Evolution in Organizational Behaviour
Acceptance of Robots in the Workplace
Evolution in Organizational Structures and Workflows
Evolution in Work Ethics
Discrimination against Robots or People
Privacy and Trust in a Human-Robot Collaborative Work Environment
Education and Training
Redesign of Workplaces for Robots
Legal and Regulatory Issues
Personal Preference toward Working with Robots
Psychological Issues Resulting from Human-Robot Co-working
Social Implications of Human-Robot Co-working
The Changing Role of Human Resources Departments
The Changing Role of Information Technology Departments and Emerging of Robotics Departments
Ethical Issues Resulting from Human-Robot Co-working – Ethical Status of Robots
Preference toward Types of Robots to Work with (Learning or Rule-based Robots)
Learning to Work with Robots
Negative Attitude Toward Robots due to Shrinking Human Workforce
Humans Competing with Robots or Robots Complementing Humans

Today, we still do not have a widely-accepted definition for a robot (Demir, 2017). Most people envision robots as machines with physical bodies. However, today, there are also software applications called robots without a physical body. These types of robots are already becoming a part of industry and institutions. Currently, various companies list robots as employees in their personnel databases. These robots perform tasks that are well-defined and somewhat mundane. For example, in a bank, a robot may be tasked with temporarily authorizing banking software functionalities to the bank employees requesting authorization for a banking operation. A robot is able to handle such tasks based on a certain well-defined set of rules. Previously, such tasks were handled by human employees.

A company called Softbanks Robotics develops various kinds of robots that are both affordable and functional enough to handle a variety of tasks. The robots called Nao and Pepper are among the most prominent ones. The current prices, in 2020, are just below twenty thousand US dollars. While Nao was initially designed for research and educational purposes, Pepper was designed for a number of industrial purposes. Pepper is a semi-humanoid robot designed with the ability to read emotions. This robot is envisioned to be used in retail, finance, government, healthcare, tourism, education, and research. According to the designers, Pepper may be used as a greeter, service provider, sales associate, loyalty management associate, receptionist, survey conductor, health assistant, communicator, entertainer, and teaching assistant. Pepper robot is already being tried in several stores, cafes, and banks. The Pepper robot still needs more development. First of all, Pepper shall converse in many languages. Furthermore, the robot should be able to understand cultural cues and act accordingly. Depending on the work environment, Pepper should be able to handle various unexpected situations. As the market grows, there will be other companies offering robots at affordable prices. Some of these robots may be customized for specific tasks with improved features.

Human-Robot Co-living

Human-robot co-living will be an integral part of the intelligent city's vision. Human-robot co-living is not only limited to homes. In all aspects of our lives except work, whenever we interact with robots, it is the subject of human-robot co-living. Some may even argue that human-robot co-working is actually a part of human-robot co-living. We will not further discuss this topic in this contribution. For now, we simply separate these two areas. If we interact with a robot in the workplace as part of our work, then it is human-robot co-working. All other interactions except for work is about human-robot co-living.

Various robotic products are being developed to assist and entertain people at homes, institutions, and in the workplace (Forlizzi, 2007). We can even argue that we already started living with robots considering the robotics toys or entertainment robots our children have (Han, Jo, Park, & Kim, 2005). Some people have cleaning robots at their homes (Forlizzi, & DiSalvo, 2006). Children perceive a robot dog as a pet (Bartlett, Estivill-Castro, & Seymon, 2004). Humanoid robots may be the tutors of our children (Serholt, Basedow, Barendregt, & Obaid, 2014). Robots may be companions to our elders (Kachouie, Sedighadeli, Khosla, & Chu, 2014). Robots may even become our friends (Sullins, 2008). A study shows that a human is affected by bad behavior toward a robot (Briggs and Scheutz, 2014).

There are several key research domains that need further study to realize the vision of smart cities. In Table 6, we briefly describe some of the relevant key technologies. The focus of this chapter is power and networking technologies. Therefore, in the following sections, we discuss the developing trends in these key areas.

DEVELOPING TRENDS IN POWER TECHNOLOGIES

Providing the necessary power source for intelligent city sensors will be an important research area. Intelligent city sensors will be deployed in our cities with almost no additional power supply and few maintenance requirements. Therefore, Energy Harvesting (EH) and Wireless Power Transfer (WPT) technologies will be key enablers for intelligent city sensor prevalence. Figure 3 presents the current trends in power technologies.

Solar Energy Harvesting

Solar energy harvesting depends on the photovoltaic solar cells to harvest energy mainly for outdoor usages. Solar cells installed at shadow-free locations can provide adequate power for smart sensors with

Table 6. Technologies and Research Areas for Intelligent Cities

Information and Enterprise Systems	Naturally, information systems will be at the heart of smart cities. While we may need to improve existing information system applications, we will also need to develop new types of systems. As other smart city technologies improve, the performance, efficiency, and capabilities of information systems will improve. Information systems research will be an ongoing effort for the future.
Communication Technologies	The need for communications is only increasing. Today, communications not only occur between humans but also between machines. As the internet of things becomes widespread, we will need faster and reliable connections with higher bandwidth capacity.
Urban City Planning	Intelligent cities may necessitate changes in our current city plans. Furthermore, future city developments may need to account for intelligent city requirements. City planners will learn about ICTs that will help them plan better smart cities.
Sensor Technologies	Intelligent cities will be equipped with many types of sensors supporting a wide range of applications. As the types and capabilities of these sensors increase, smart cities will likely become smarter.
Internet of Things (IoT)	The Internet of things will provide a conceptual base for integrating all kinds of systems and devices in smart cities. The IoT technology will help realize the idea of smart cities.
Big Data	The sensors in intelligent cities will generate huge amounts of data. This data has to be efficiently analyzed and effectively used. The big data research will help in this endeavor.
Information Security	In intelligent cities, huge amounts of data will be generated and a significant portion of this data will be private. Furthermore, some of this data may also be sensitive. Naturally, residents will expect a high level of information, application, and network security. As a result, information security will be an important issue in intelligent cities.
Artificial Intelligence (AI)	Smartness in cities is the result of artificial intelligence research. Simply, AI research will help make cities smarter.
Renewable Energies	The sensors and other types of devices deployed in smart cities will require an increased amount of energy. In addition, we are seeking to make our cities greener. As a result, this encourages research on renewable energies. This research area will help satisfy the energy requirements of smart cities.
Robotics	There will be many types of robotic systems in intelligent cities. There are already companies testing autonomous cars on roads. These cars will likely be used as taxis in intelligent cities. Amazon is testing package delivery system using autonomous unmanned aerial vehicles. Other types of services will be provided by robots. Robotics research will help advance the concept of intelligent cities.
Human-Computer Interaction (HCI)	Many types of applications interface with humans in smart cities. The seamless operation of smart city applications requires well-designed interfaces. Moreover, children and handicapped citizens are also part of cities. The human-interaction research may need to focus on these types of users as well.

increased efficiencies by up to 36%. (Essig, et al., 2017). However, they suffer from weather conditions, such as snow, fog, and dust accumulation. Furthermore, smart sensors equipped with solar cells need a battery to store energy for night time and cloudy days.

Recently, self-powered image sensors have also been proposed (Park, Lee, Song, & Yoon, 2018). These sensors are capable of harvesting energy and capturing images at the same time. They can capture 15 images per second powered only by the daylight (Moore, 2018). This kind of self-powered sensors may play an important role in all types of environment and other types of monitoring. For example, these devices may help monitor forests to prevent forest fires. They may be used for monitoring traffic and all other types of smart city services.

Figure 3. Trends in Power Technologies

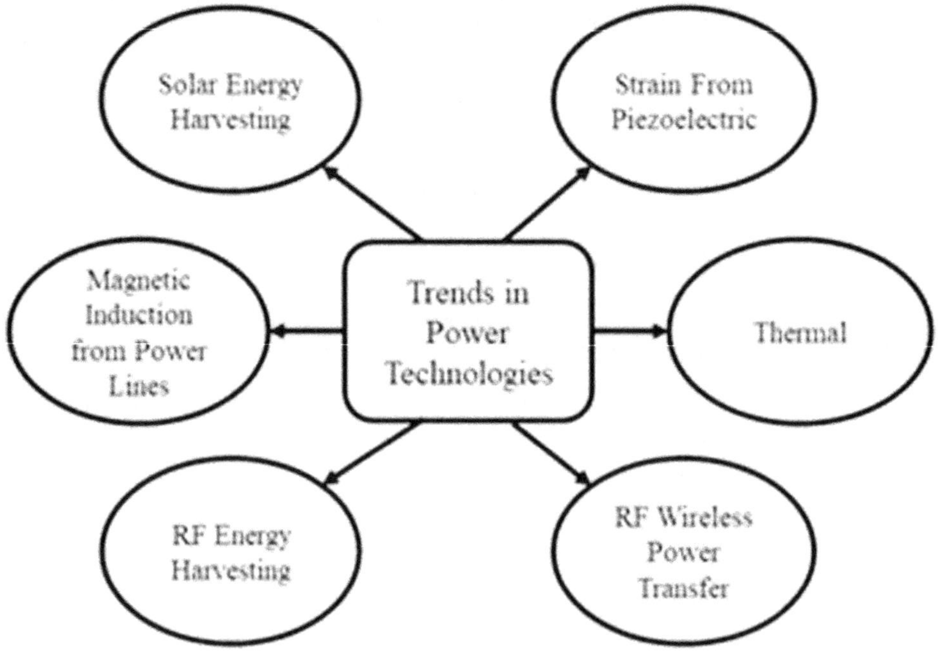

Strain from Piezoelectric

Piezoelectric elements are able to capture energy through vibration and compression. A human walking over a large piezoelectric plate can produce up to 34 W electric power (Nia, Zawawi, & Singh, 2017). This is an important energy source that can be utilized to power smart devices. Humans passing through a busy bus station or a train station may be able to power up tens or hundreds of smart city sensors. Furthermore, excess energy from the piezo plate can be paired with a small NiMH battery, enabling longer lifetimes with less dependency on the people.

Thermal Energy Harvesting

Photovoltaic solutions for solar cells are maturing fast. Solar cells are regarded either as a primary energy source or as a secondary source charging a battery to extend battery life. However, for indoor applications, limited available light hinders the usage of solar cells. The typical energy level obtained from a solar cell placed in an office is 1 μW/ 〚cm〛 ^2 @40 lux (Vuckovic, 2014). This level of power is inadequate for many types of applications. Thereby, alternatives such as thermal energy harvesters can be considered for indoor sensor nodes. Thermal energy harvesters use the temperature difference within a material to generate electricity as long as the ambient temperature is above absolute zero.

Thermal energy harvesting is the process of converting thermal gradients into electrical energy via thermoelectric generators (TEGs). These consist of thermocouple arrays sandwiched between ceramic plates. When TEGs are exposed to a temperature gradient, one side of the TEG stays warm, whereas the other side is cold. This process generates electrical energy. Basically, it converts the heat flux into electrical energy. Thus, TEG with one side facing the outer wall and the other side facing the inner wall of a building may provide a practical solution for thermal energy harvesting.

There are various issues associated with thermal energy harvesting; as temperature changes have dynamic behavior. There may be significant temperate differences between day and night and also between seasons. These differences may go up to a few degrees to tens of degrees. These conditions result in notable variances in energy generation. Therefore, energy outputs range from microwatts to kilowatts. Power management circuits should be able to handle these ranges and be a reliable power source. Sophisticated boost converter designs are required to provide reliable power.

Magnetic Induction from Power Lines

The time-variant magnetic field is capable of generating an induced electrical voltage. Therefore, smart city sensors placed close to power lines inside the walls can harvest energy through magnetic induction. Despite its simple structure, there are two main drawbacks of magnetic induction technology. First, energy harvested from the power line decreases with square law. Secondly, the power line supplies energy to the smart sensor node only when another device requires power from the line. When there is no current flow, (i.e. the device draws no power from the line), there will be no energy available for the other smart city node.

Radio Frequency (RF) Energy Harvesting

Energy harvested from natural resources such as sun and thermal may vary randomly over time. The power output depends on the location, environment, and weather conditions. Hence, RF/microwave energy sources are strong candidates to provide power during the lifetime of a sensor. RF sources are widely deployed and can be efficiently utilized via smart power beaming techniques (Liu, Xiong, Fan, & Zhong, 2017) either with RF energy harvesting (EH), or wireless power transmission (WPT). Omnipresent cellular and Wi-Fi signals are mainly foreseen for EH, as resonant antenna size of the smart sensors in the order of 10–50 cm can be effectively utilized. To obtain 100 μW of power from cellular technologies, 330-1000 〚cm〛 ^2 area is required (Visser and Vullers, 2013). The efficiency of the RF energy harvester mainly depends on the selected antenna gain and the accuracy of the impedance matching between the antenna and the voltage multiplier.

Radio Frequency (RF) Wireless Power Transfer (WPT)

Radio Frequency (RF) Wireless Power Transfer (WPT) relies on the fact that radio signals in the frequency range from 3 kHz to 300 GHz can be used to carry energy through the air. In RF-WPT networks, transmitters can proactively replenish relevant devices by transmitting RF signals. However, propagation loss and regulated RF transmit power are the main limitations of RF-WPT realization. With the introduction of fifth-generation (5G) networks, cellular base stations will be diversified into small cells such as the femtocell, microcell, and picocell base station technologies to increase spectral efficiency and provide better coverage. Hence, with closer distance to sensor nodes and RF beamforming capabilities, small cell architectures will simultaneously enable efficient wireless power and information transfer. On the other hand, considering the ultra-dense sensor node networks, fixed WPT transmitter utilization will increase the cost and hinder large-scale distributions of WPT systems due to interference issues. Hence,

Xu, Zeng, and Zhang (2018) proposed UAV enabled RF-WPT to recharge low-power distributed energy receivers. This will provide a ubiquitous wireless energy supply.

In addition to state of the art EH and WPT methods, effective utilization of ultra-low-power electronics, flex batteries, and supercapacitors will play an important role in practical realizations of smart sensor-based applications.

Power-sharing between nodes and access points may also be a key enabler for efficient resource utilization. Consider a traffic camera and weather monitoring station sharing a single photovoltaic battery. The camera can be on to monitor rush hour traffic for traffic routing, whereas the weather monitoring station will only be activated to inform drivers about real-time severe weather conditions.

DEVELOPING TRENDS IN NETWORKING TECHNOLOGIES

Intelligent city systems aim to support everyday routines, whether work-related or leisure, in a city. They will be sensitive to the presence of people, vehicles, and the environment. These systems are composed of intelligent sensors and devices with enhanced connectivity and advanced processing features. As the number of smart sensors and devices increases, the connectivity will exponentially increase. Developing networks capable of handling massive connectivity will be among the most important challenges of intelligent cities. Processing of huge amounts of big data, generated by interconnected smart devices, will also be a serious concern. In this section, we discuss the developing trends in networking technologies. Figure 4 presents the current trends in this field of research.

Massive Connectivity

According to Diamandis and Kotler (2012), 45 trillion networked sensors will be integrated into our daily lives within 20 years. There are many application areas and unique challenges brought by the widespread use of networked sensors (Alam, Tehranipoor, & Guin, 2017). Intelligent systems should be able to handle the increased network traffic and process vast amounts of data. Furthermore, smart systems will need to act fast and accurately in some applications such as emergency services. In some other cases, they need to execute decisions with little to no human intervention, and in real-time.

Currently, smart sensor connectivity features and challenges can be discussed under two perspectives, viz: Backhaul Supported Connectivity and Machine to Machine Type Networks.

Backhaul Supported Connectivity

Smart sensor technologies, such as Internet of Things (IoT) and Industrial Internet of Things (IIoT), are mainly studied for devices with Internet access through cellular backhaul. As ubiquitous internet reachability is required for most use cases, and sensor nodes have limited resources, they will also take advantage of backhaul supported connectivity.

For backhaul supported connectivity, increased sensor utilization will create RF spectrum scarcity issues. Thereby, efficient utilization of the RF spectrum is important. For example, cognitive radio utilizes RF spectrum in an opportunistic manner, where currently used frequencies are monitored to avoid network interference. Moreover, for backhaul supported environments, especially for urban environments, energy harvesting through RF signals will be feasible for seamless connectivity. Furthermore, handover

Figure 4. Trends in Networking Technologies

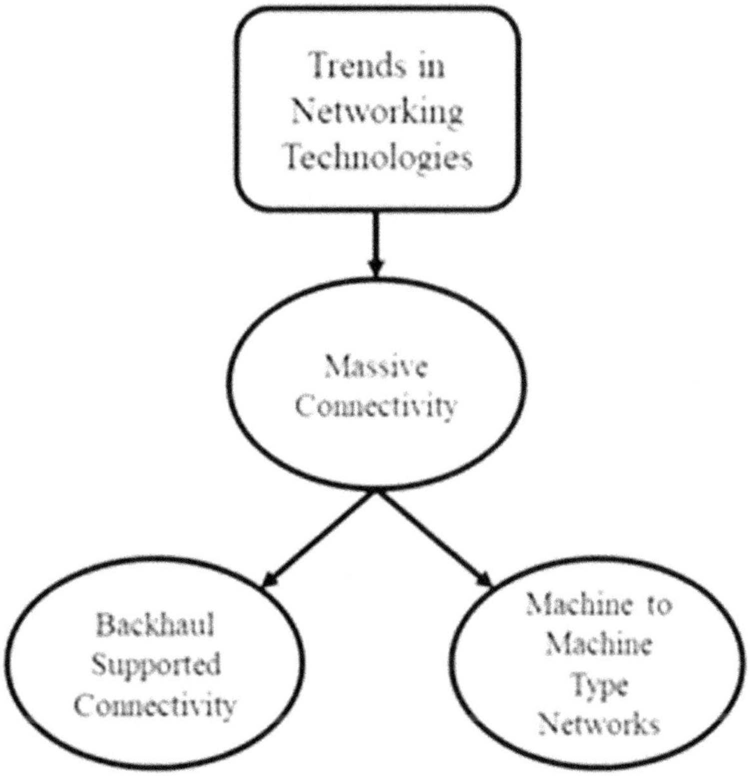

between sensor nodes is of critical importance for continuous support. On the other hand, there will be sensor nodes with limited battery and processing capabilities. For these sensors, computing, network control, and node interactions will be handled either in the cloud or at the edge. These networks are called "Enterprise Networks". Software-defined networking technologies may be used to manage these enterprise networks.

Cellular base stations will be equipped with sensor uplink and downlink communication technologies, e.g. collecting data from gas, water, and power meters via uplink (UL) transmissions. However, with the increasing number of nodes, reliability will become an important issue (Ding and Pérez, 2018). With the introduction of fifth-generation networks (5G) and advancements of IEEE 802.11 standards, new concepts such as network function virtualization (NFV), network slicing, and channel bonding will also play a major role in ultra-dense network sensor connectivity. NFV aims to pool signal processing resources in the cloud to avoid using dedicated processing units on each networked sensor/device. Resource pooling through NFV reduces sensor complexity, size, and power consumption while decreasing costs. Network slicing aims to provide an end-to-end virtual network with computation, storage, and networking functions. Network slicing will ultimately allow a cellular operator to slice a physical network into multiple isolated logical networks. As a result, these novel concepts are believed to contribute to smart city sensor networks via mitigating complexity, revenue generation, and scaling (Ericsson, 2019).

Backhaul connectivity embracing different connectivity technologies from mmWave communications to optical wireless communications will pave the way for high bandwidth applications. Thus, with higher bandwidths and bandwidth aggregation schemes (e.g. LTE-WLAN/ LWA), raw sensor data exchange with higher data rates will enable enhanced connectivity for intelligent city applications. Disruptive technologies, such as UAV based communications, wireless power transfer, energy harvesting, and optical communications, will enable massive amounts of data exchange.

Backhaul supported connectivity will enable remote intelligent city applications such as smart metering, smart parking, and public transportation planning. Moreover, high data rate communications will support real-time high definition video streaming for public safety and surveillance to better serve smart city residents.

Machine to Machine Based Networks

Machine to Machine (M2M) type networks will be prominent for backhaul support and network infrastructure lacking areas. As cellular technologies are not homogenously available throughout the world, devices connecting each other without any need to access points and control mechanisms will also play a major role in smart sensor proliferation. One of the key requirements for M2M type networks is simple protocol setups to avoid pairing. This simplifies the process of establishing links.

Redundant M2M type networks help to overcome temporary unavailability due to weather conditions and disasters. Networked smart sensors may play a key role to inform the environment about upcoming disasters. M2M type peer to peer networking technologies are promising especially for limited internet connection scenarios. Certain applications such as IBM Research Weather Channel App provide imminent disaster alerts. This smartphone application utilizes location assisted peer to peer Wi-Fi and Bluetooth to take preventative actions without relying on cellular communications. Thereby, M2M type network selection for smart city sensors is expected not only to ease the city life but also to provide vital information for health and survivability.

Moreover, UAVs will be utilized as dynamic access points (e.g. base stations). Hence, the lack of infrastructure to provide internet access, may not be a significant issue for smart sensor proliferation. UAVs are also provisioned to be utilized for data collection stored at smart sensor nodes and provide power through wireless power transfer (Zeng, Zhang, & Lim, 2016).

Ubiquitous dual-purpose sources such as LEDs and Laser Diodes, providing simultaneous illumination and communication, are other strong candidates for M2M type networks, due to low cost, low complexity, and low power consumption. Recently, Visible Light Communication supported Light Fidelity (Li-Fi) systems are demonstrated to provide secure and reliable connectivity. Moreover, solar panels are demonstrated to act as optical wireless communication receivers (Wang, Tsonev, Videv, & Haas, 2015). From street lights to vehicles, either visible or non-visible light radiating sources (i.e. LED and laser) can act as transmitters front-end to communicate with solar panel equipped smart sensor nodes.

In intelligent cities, dynamic sensor-based nodes such as vehicles should be taken into account in M2M type network design. For static nodes, database-dependent connectivity is assured. For mobile nodes, more complex topology-based wireless sensor networks should be considered. Sensors equipped with the M2M type connectivity will support intelligent city services such as waste management, pollution monitoring, smart lighting, and traffic re-routing. M2M should enable massive sensor deployment that requires low-cost architectures, low complexity, and low power consumption. Moreover, time-critical services such as traffic re-routing for better traffic management will require low latency M2M type communications.

Table 7 presents the intelligent city network types and a comparison between backhaul supported connectivity and machine to machine type connectivity in terms of core requirements and smart city applications.

Table 7. Smart City Network Types, Core Requirements, and Smart City Applications

	Core Requirements	Smart City Applications
Backhaul Supported Connectivity	Ultra-Reliable Communications High Data Rate Transmissions	Smart Metering Smart Parking Public Safety Public Transportation Planning
Machine to Machine (M2M) Type Connectivity	Low Power Consumption Low Complexity Sensors Low Latency Communications	Waste Management Pollution Monitoring Smart Lighting Traffic Re-routing

A BRIEF DISCUSSION ON INTEROPERABILITY

Interoperability among smart city sensor nodes can be built upon standard driven Internet of Things technologies. Here, IoT device requirements vary widely in terms of power, connectivity, size, price, processing power. Hence, there are numerous standards to achieve interoperability between IoT devices (IEEE Standards, 2019). An intelligent city infrastructure will be composed of many types of smart sensor nodes. A diligent product search will be necessary to achieve interoperability.

As the processing, power, and storage capabilities of smart sensor nodes will be limited, connectivity will be the determining factor in the selection of smart node products for smart city services. Moreover, these sensors will have the ability to learn from other types of sensors. This results in heterogeneous inter-connected sensor networks. The ability to learn from other sensors will be quite valuable in creating pattern recognition and optimization applications. Consider bus stops equipped with smart sensor nodes. These smart sensors will send the number of passengers at the bus stops to the city bus. Then the bus sends this information to public transport planners for optimizing the bus route while skipping empty bus stops. To ensure this kind of seamless integration and interoperability of smart city sensor nodes, heterogeneous connectivity features and network technology selection such as discovery and data protocols will be prominent (Lee, Bae, & Kim, 2017).

Due to various sensor device sizes, communications protocols. and power requirements, it may not be possible to use the same network technology at all times. Thereby, machine learning-based coexistence control mechanisms will be required for reliable wireless sensor connectivity mechanisms. Technologies, satisfying identical QoS requirements (e.g., range and data rate), such as IEEE 802.11ah and IEEE 802.15.4g may be considered for coexistence control, as they are likely to be used interchangeably.

To create value from smart sensor data, data should be accessible across agencies, departments, and third-party providers. Semantic web technologies are proposed to provide a standards-based approach on top of the content delivery. Semantic web integrates data as a knowledge graph with metadata labels, enabling easier access from disparate systems. Hence, new insights can be gained from metadata of smart city sensor data using artificial intelligence. It should be noted that even smart sensors of the same type may deliver data in various formats or different measurement units. Thus, semantics address heterogeneity issues for smart sensor nodes. Consider complex systems such as Connected Autonomous Vehicles within the Internet of Vehicles (IoV) paradigm integrated with smart city sensor networks. Semantic web providing machine-readable data structures with the ability to interconnect heterogeneous data sources into a homogeneous structure of shared knowledge can support efficient processing of shared information (Andročec, Novak, & Oreški, 2018) and better utilization of network bandwidth (Viktorović, M., Yang, D., de Vries, B., & Baken, N., 2019).

CONCLUSION

In recent years, the notion of Industry 4.0 dominated information technology-related research by providing an encapsulating context. While Industry 4.0 is being criticized due to various reasons, Industry 5.0 discussions have already started. One emerging theme for Industry 5.0 is human-robot co-working. These industrial revolutions provide the necessary conceptual and technological infrastructure to intelligent cities. The concept of intelligent cities will be based on many types of information technology developments. We believe two types of systems will be dominant in intelligent cities: these are robotic and ambient intelligence systems. Robotic systems will be used in tasks requiring physical activities. Other tasks may be handled by ambient intelligence systems derived from the idea of the invisible computer. Artificial intelligence, robotics, internet of things, and human-computer interaction research will significantly help develop necessary robotic and ambient intelligence systems.

Power and networking technologies will be crucial in advancing robotic and ambient intelligence systems capabilities. We need innovations in providing the necessary power source to these types of systems. One of the main limitations of ubiquitous and mobile computing is to do with power technologies. Developing trends in this area include solar energy harvesting, strain from piezoelectric, thermal energy harvesting, magnetic induction from power lines, radio frequency (RF) energy harvesting, and radio frequency (RF) wireless power transfer. Advancements in networking technologies will increase robotic and ambient intelligence systems capabilities. Handling massive connectivity in intelligent cities will be a challenge. Research related to backhaul supported connectivity and machine-to-machine (M2M) type networks will help overcome challenges brought by massive connectivity between smart objects.

In this chapter, we first discussed Industry 4.0 and Industry 5.0. We suggested that industrial revolutions will play a significant role in shaping the intelligent city concept. Furthermore, we briefly discussed robotic and ambient intelligence systems since these two types of systems will likely dominate systems interfacing humans. Human-robot co-living and co-working will be the concepts supporting the intelligent city concept. There will be some important technologies that will provide the necessary infrastructure for intelligent cities. Power and networking technologies are among these technologies. We provide brief discussions on the developing trends in these areas. We also briefly discussed interoperability due to its relevance. We believe this chapter provides an introduction to the interested readers and researchers willing to work in technology areas such as those described above.

ACKNOWLEDGMENT

The views and conclusions contained herein are those of the authors and should not be interpreted as necessarily representing the official policies or endorsements, either expressed or implied, of any affiliated organization or government.

REFERENCES

Aarts, E., & Encarnação, J. L. (2005). Into Ambient Intelligence. In True Visions: Tales on the Realization of Ambient Intelligence. Springer Verlag.

Abbasi, M. H., Majidi, B., & Manzuri, M. T. (2018). Deep cross altitude visual interpretation for service robotic agents in smart city. In *2018 6th Iranian Joint Congress on Fuzzy and Intelligent Systems (CFIS)* (pp. 79-82). IEEE.

Andročec, D., Novak, M., & Oreški, D. (2018). Using Semantic Web for Internet of Things Interoperability: A Systematic Review. *International Journal on Semantic Web and Information Systems*, *14*(4), 147–171.

Alam, M., Tehranipoor, M. M., & Guin, U. (2017). Tsensors vision, infrastructure and security challenges in trillion sensor era. *Journal of Hardware and Systems Security*, *1*(4), 311–327.

Alonso, I. G. (2011). Service robotics. In *Service robotics within the digital home* (pp. 89–114). Dordrecht: Springer.

AmazonGo. (2019). https://www.amazon.com/b?ie=UTF8&node=16008589011

Augusto, J. C. (2007). *Ambient Intelligence: The Confluence of Ubiquitous/Pervasive Computing and Artificial Intelligence. In Intelligent Computing Everywhere* (pp. 213–234). Springer.

Augusto, J. C., & McCullagh, P. (2007). Ambient intelligence: Concepts and Applications. *Computer Science and Information Systems*, *4*(1), 1–27.

Bagdasarov, Z., Martin, A. A., & Buckley, M. R. (2018). Working with robots: Organizational considerations. *Organizational Dynamics*. doi:10.1016/j.orgdyn.2018.09.002

Banh, A., Rea, D. J., Young, J. E., & Sharlin, E. (2015). Inspector baxter: The social aspects of integrating a robot as a quality inspector in an assembly line. In *Proceedings of the 3rd international conference on human-agent interaction* (pp. 19-26). ACM.

Bartlett, B., Estivill-Castro, V., & Seymon, S. (2004). Dogs or robots: why do children see them as robotic pets rather than canine machines? In *Proceedings of the fifth conference on Australasian user interface-Volume 28* (pp. 7-14). Australian Computer Society, Inc.

Briggs, G., & Scheutz, M. (2014). How robots can affect human behavior: Investigating the effects of robotic displays of protest and distress. *International Journal of Social Robotics*, *6*(3), 343–355.

Brown, T. L. (2019). Ambient Intelligence Changes the Office Environment. In Advances in the Technology of Managing People: Contemporary Issues in Business (pp. 149-160). Emerald Publishing Limited.

Buhalis, D. (2019). Technology in tourism-from information communication technologies to eTourism and smart tourism towards ambient intelligence tourism: A perspective article. *Tourism Review*.

Cormier, D., Newman, G., Nakane, M., Young, J. E., & Durocher, S. (2013). Would you do as a robot commands? An obedience study for human-robot interaction. *International Conference on Human-Agent Interaction*.

Demir, K. A. (2017). Research Questions in Roboethics. *Mugla Journal of Science and Technology*, *3*(2), 160–165.

Demir, K. A. (2019). *Human-Robot Co-working and Roboethics* (Master's Thesis). Gebze Technical University, Izmit, Turkey.

Demir, K. A., & Cicibas, H. (2017). *Industry 5.0 and a Critique of Industry 4.0*. In *4th International Management Information Systems Conference*, Istanbul, Turkey.

Demir, K. A., & Cicibaş, H. (2018). *The Next Industrial Revolution: Industry 5.0 and Discussions on Industry 4.0. In Industry 4.0 from the Management Information Systems Perspectives*. Peter Lang Publishing Group.

Demir, K. A., Caymaz, E., & Elçi, M. (2017). *Issues in integrating robots into organizations*. In The 12th International Scientific Conference "Defence Resources Management in the 21st Century", Brazov, Romania.

Demir, K. A., Döven, G., & Sezen, B. (2019). Industry 5.0 and Human-Robot Co-working. *Procedia Computer Science*, *158*, 688–695.

Demir, K. A., Turan, B., Onel, T., Ekin, T., & Demir, S. (2019). Ambient Intelligence in Business Environments and Internet of Things Transformation Guidelines. In *Guide to Ambient Intelligence in the IoT Environment* (pp. 39–67). Cham: Springer.

Diamandis, P. H., & Kotler, S. (2012). *Abundance: The future is better than you think*. Simon and Schuster.

Ding, M., & Pérez, D. L. (2018). Promises and caveats of uplink IoT ultra-dense networks. In *2018 IEEE Wireless Communications and Networking Conference (WCNC)* (pp. 1-6). IEEE.

Ericsson. (2019). https://www.ericsson.com/4a45a8/assets/local/digital-services/trending/scalable-network/executive_guide_network_slicing.pdf

Essig, S., Allebé, C., Remo, T., Geisz, J. F., Steiner, M. A., Horowitz, K., ... Young, D. L. (2017). Raising the one-sun conversion efficiency of III–V/Si solar cells to 32.8% for two junctions and 35.9% for three junctions. *Nature Energy*, *2*(9), 17144.

Forlizzi, J., & DiSalvo, C. (2006). Service robots in the domestic environment: a study of the roomba vacuum in the home. In *Proceedings of the 1st ACM SIGCHI/SIGART Conference on Human-Robot Interaction* (pp. 258-265). ACM.

Forlizzi, J. (2007). How robotic products become social products: an ethnographic study of cleaning in the home. In *Proceedings of the ACM/IEEE international conference on Human-robot interaction* (pp. 129-136). ACM.

Gams, M., Gu, I. Y. H., Härmä, A., Muñoz, A., & Tam, V. (2019). Artificial intelligence and ambient intelligence. *Journal of Ambient Intelligence and Smart Environments*, *11*(1), 71–86.

Geiskkovitch, D. Y., Cormier, D., Seo, S. H., & Young, J. E. (2016). Please continue, we need more data: An exploration of obedience to robots. *Journal of Human-Robot Interaction*, *5*(1), 82–99.

Grigorescu, S. D., Argatu, F. C., Paturca, S. V., Cepisca, C., Seritan, G. C., Adochiei, F. C., & Enache, B. (2019). Robotic Platform with Medical Applications in the Smart City Environment. In *2019 11th International Symposium on Advanced Topics in Electrical Engineering (ATEE)* (pp. 1-6). IEEE.

Han, J., Jo, M., Park, S., & Kim, S. (2005). The educational use of home robots for children. In *Robot and Human Interactive Communication, 2005. ROMAN 2005. IEEE International Workshop on* (pp. 378-383). IEEE.

Höjer, M., & Wangel, J. (2014). Smart Sustainable Cities: Definition and Challenges. In L. M. Hilty & B. Aebischer (Eds.), *ICT Innovations for Sustainability. Advances in Intelligent Systems and Computing 310*. Springer International Publishing.

IEEE Standards. (2019). https://standards.ieee.org/initiatives/iot/stds.html

International Telecommunication Union. (2014). https://www.itu.int/en/ITU-T/focusgroups/ssc/Pages/default.aspx

International Telecommunication Union Telecommunication Standardization Sector Focus Group on Smart Sustainable Cities (ITU-T FG-SSC). (2014). *Smart sustainable cities: An analysis of definitions. Technical report*. ITU.

IST Advisory Group. (2003). *Ambient Intelligence: From Vision to Reality*. European Commission.

Kachouie, R., Sedighadeli, S., Khosla, R., & Chu, M. T. (2014). Socially assistive robots in elderly care: A mixed-method systematic literature review. *International Journal of Human-Computer Interaction*, *30*(5), 369–393.

Lee, S., Bae, M., & Kim, H. (2017). Future of IoT networks: A survey. *Applied Sciences (Basel, Switzerland)*, *7*(10), 1072.

Libelium. (2019). *Top 50 Sensor Applications for A Smarter World*. http://www.libelium.com/resources/top_50_iot_sensor_applications_ranking

Lin, P., Abney, K., & Bekey, G. A. (2011). *Robot ethics: the ethical and social implications of robotics*. MIT Press.

Liu, J., Xiong, K., Fan, P., & Zhong, Z. (2017). RF energy harvesting wireless powered sensor networks for smart cities. *IEEE Access: Practical Innovations, Open Solutions*, *5*, 9348–9358.

Mahmood, A., Butler, B., Sheng, Q. Z., Zhang, W. E., & Jennings, B. (2019). Need of Ambient Intelligence for Next-Generation Connected and Autonomous Vehicles. In *Guide to Ambient Intelligence in the IoT Environment* (pp. 133–151). Cham: Springer.

Maier, A., & Student, D. (2015). Industrie 4.0 – Der Große Selbstbetrug. *Manager Magazine*. https://www.manager-magazin.de/magazin/artikel/digitale-revolutionindustrie-4-0-ueberfordert-deutschen-mittelstand-a-1015724.html

Markopoulos, P. (2016). Ambient intelligence: Vision, research, and life. *Journal of Ambient Intelligence and Smart Environments*, *8*(5), 491–499.

Menouar, H., Guvenc, I., Akkaya, K., Uluagac, A. S., Kadri, A., & Tuncer, A. (2017). UAV-enabled intelligent transportation systems for the smart city: Applications and challenges. *IEEE Communications Magazine*, *55*(3), 22–28.

Moore, S. K. (2018). Self-powered image sensor could watch you forever. *IEEE Spectrum*.

Nia, E. M., Zawawi, N. A. W. A., & Singh, B. S. M. (2017). A review of walking energy harvesting using piezoelectric materials. *IOP Conference Series. Materials Science and Engineering*, *291*(1), 012026.

Norman, D. (1998). *The Invisible Computer*. The MIT Press.

Object Management Group (OMG). (2005). https://www.omg.org/robotics/robotic_systems_rfi.htm

O'grady, M., & O'hare, G. (2012). How smart is your city? *Science*, *335*(6076), 1581–1582. PMID:22461597

Park, S. Y., Lee, K., Song, H., & Yoon, E. (2018). Simultaneous Imaging and Energy Harvesting in CMOS Image Sensor Pixels. *IEEE Electron Device Letters*, *39*(4), 532–535.

Ramos, C., Augusto, J. C., & Shapiro, D. (2008). Ambient intelligence—The next step for artificial intelligence. *IEEE Intelligent Systems*, *23*(2), 15–18.

Ravikumar, G. V. V. (2019). Advanced Robotics: Autonomous Solutions Take on Dull, Dirty, and Dangerous Work. *Infosys Insights*. https://www.infosys.com/insights/ai-automation/advanced-robotics.html

Ray, C., Mondada, F., & Siegwart, R. (2008). What do people expect from robots? In *Intelligent Robots and Systems, 2008. IROS 2008. IEEE/RSJ International Conference on* (pp. 3816-3821). IEEE.

Remagnino, P., & Foresti, G. L. (2005). Ambient Intelligence: A New Multidisciplinary Paradigm. *IEEE Transactions on Systems, Man, and Cybernetics. Part A, Systems and Humans*, *35*(1), 1–6.

Sachsenmeier, P. (2016). Industry 5.0—The Relevance and Implications of Bionics and Synthetic Biology. *Engineering*, *2*, 225–229.

Samani, H., & Zhu, R. (2016). Robotic automated external defibrillator ambulance for emergency medical service in smart cities. *IEEE Access: Practical Innovations, Open Solutions*, *4*, 268–283.

Serholt, S., Basedow, C. A., Barendregt, W., & Obaid, M. (2014). Comparing a humanoid tutor to a human tutor delivering an instructional task to children. In *2014 IEEE-RAS International Conference on Humanoid Robots* (pp. 1134-1141). IEEE.

Sheridan, T. B., & Verplank, W. L. (1978). *Human and computer control of undersea teleoperators*. Man-Machine Systems Laboratory Report, MIT. www.dtic.mil/cgi-bin/GetTRDoc?AD=ADA057655,

Silva, B. N., Khan, M., & Han, K. (2018). Towards Sustainable Smart Cities: A Review of Trends, Architectures, Components, and Open Challenges in Smart Cities. *Sustainable Cities and Society*, *38*, 697–713.

Skouby, K. E., Kivimäki, A., Haukiputo, L., Lynggaard, P., & Windekilde, I. M. (2014). Smart cities and the ageing population. *The 32nd Meeting of WWRF*.

Streitz, N. A. (2011). Smart cities, ambient intelligence and universal access. In *International Conference on Universal Access in Human-Computer Interaction* (pp. 425-432). Springer.

Streitz, N., Charitos, D., Kaptein, M., & Böhlen, M. (2019). Grand challenges for ambient intelligence and implications for design contexts and smart societies. *Journal of Ambient Intelligence and Smart Environments*, *11*(1), 87–107.

Sullins, J. P. (2008). Friends by design: A design philosophy for personal robotics technology. In P. E. Vermaas, P. Kroes, A. Light, & S. Moore (Eds.), *Philosophy and Design* (pp. 143–157). Springer Netherlands.

Turan, B., Demir, K. A., Soner, B., & Ergen, S. C. (2019). Visible Light Communications in Industrial Internet of Things (IIoT). In *The Internet of Things in the Industrial Sector* (pp. 163–191). Cham: Springer.

Veruggio, G. (2007). *EURON Roboethics Roadmap. Release 1.2. January 2007*. Retrieved from http://www.roboethics.org/index_file/Roboethics%20Roadmap%20Rel.1.2.pdf

Veruggio, G., & Operto, F. (2008). Roboethics: Social and Ethical Implications. In B. Siciliano & O. Khatib (Eds.), *Springer Handbook of Robotics* (pp. 1499–1524). Springer International Publishing.

Visser, H. J., & Vullers, R. J. (2013). RF energy harvesting and transport for wireless sensor network applications: Principles and requirements. *Proceedings of the IEEE*, *101*(6), 1410–1423.

Vuckovic, D. (2014). *Design and implementation of energy harvesting powered wireless sensor networks* (PhD Thesis). Technical University of Denmark (DTU).

Wang, Z., Tsonev, D., Videv, S., & Haas, H. (2015). On the design of a solar-panel receiver for optical wireless communications with simultaneous energy harvesting. *IEEE Journal on Selected Areas in Communications*, *33*(8), 1612–1623.

Weiser, M. (1991). The computer for the 21st century. *Scientific American*, *265*(3), 94–104. PMID:1675486

Xu, J., Zeng, Y., & Zhang, R. (2018). UAV-enabled wireless power transfer: Trajectory design and energy optimization. *IEEE Transactions on Wireless Communications*, *17*(8), 5092–5106.

Zeng, Y., Zhang, R., & Lim, T. J. (2016). Wireless communications with unmanned aerial vehicles: Opportunities and challenges. *IEEE Communications Magazine*, *54*(5), 36–42.

ADDITIONAL READING

Nakashima, H., Aghajan, H., & Augusto, J. C. (Eds.). (2009). *Handbook of ambient intelligence and smart environments*. Springer Science & Business Media.

Crawford, K., & Calo, R. (2016). There is a blind spot in AI research. *Nature*, *538*, 311–313. doi:10.1038/538311a PMID:27762391

Demir, K. A., Cicibas, H., & Arica, N. (2015). Unmanned Aerial Vehicle Domain: Areas of Research. *Defence Science Journal*, *65*(4), 319–329. doi:10.14429/dsj.65.8631

Lin, P. (2012). Introduction to Robot Ethics. In P. Lin, K. Abney, & G. A. Bekey (Eds.), *Robot Ethics – The Ethical and Social Implications of Robotics* (pp. 3–17). Cambridge, MA: The MIT Press.

Mehlman, M., Berg, J. W., & Ray, S. Robot Law (2017). Case Legal Studies Research Paper No. 2017-1. January 30, 2017. Retrieved from https://ssrn.com/abstract=2908488

Petit, N. (2017). Law and Regulation of Artificial Intelligence and Robots-Conceptual Framework and Normative Implications. 9 March 2017. (Working Paper).

Riek, L. D., & Howard, D. (2014). A code of ethics for the human-robot interaction profession. In Proceedings of We Robot 2014.

Shim, J., & Arkin, R. C. (2017). An intervening ethical governor for a robot mediator in patient-caregiver relationships. In M. I. A. Ferreira, J. S. Sequeira, M. O. Tokhi, E. E. Kadar, & G. S. Virk (Eds.), *A World with Robots* (pp. 77–91). Springer International Publishing.

Sullins, J. P. (2011). Introduction: Open questions in roboethics. *Philosophy & Technology*, *24*(3), 233–238. doi:10.100713347-011-0043-6

Sullins, J. P. (2015). Applied Professional Ethics for the Reluctant Roboticist. In Riek, L.D., Hartzog, W., Howard, D., Moon, A., Calo, R. (Ed.) In Proceedings of the 10th ACM/IEEE Conference on Human-Robot Interaction (HRI 2015): The Emerging Policy and Ethics of Human-Robot Interaction Workshop, Portland, OR, US.

Ucar, S., Turan, B., Ergen, S. C., Ozkasap, O., & Ergen, M. (2016). Dimming support for visible light communication in intelligent transportation and traffic system. In NOMS 2016-2016 IEEE/IFIP Network Operations and Management Symposium (pp. 1193-1196). IEEE.

Said, O., & Masud, M. (2013). Towards internet of things: Survey and future vision. *International Journal of Computer Networks*, *5*(1), 1–17.

Li, S., Da Xu, L., & Zhao, S. (2018). 5G Internet of Things: A survey. *Journal of Industrial Information Integration*, *10*, 1–9.

Singh, D., Tripathi, G., & Jara, A. J. (2014). *A survey of Internet-of-Things: Future vision, architecture, challenges and services. In 2014 IEEE world forum on Internet of Things (WF-IoT)* (pp. 287–292). IEEE.

Viktorović, M., Yang, D., de Vries, B., & Baken, N. (2019). Semantic web technologies as enablers for truly connected mobility within smart cities. *Procedia Computer Science*, *151*, 31–36.

Germany Trade and Invest. (2011). Industrie 4.0: Smart manufacturing for the future. www.inovasyon.org/pdf/gtai.industrie4.0_smart.manufact.for.future.july.2014.pdf. Accessed October 2019

KEY TERMS AND DEFINITIONS

Ambient Intelligence: Ambient intelligence refers to electronic or digital environments that are sensitive and responsive to the presence of people (IST Advisory Group, 2003; Aarts & Encarnação, 2005; Ramos, Augusto, & Shapiro, 2008). Furthermore, ambient intelligence is a digital environment that supports people in their daily lives by assisting them in a sensible way (Augusto, 2007).

Human-Robot Co-Working: Human-robot co-working may be defined as human persons working with robots in a collaborative manner in which robots handle part of a work task.

Industry 4.0: Industry 4.0 is the latest industrial revolution initiated by the German Government with the motto "Smart Manufacturing for the Future".

Industry 5.0: Industry 5.0 is the industrial revolution brought by the widespread use of robots in a wide range of industrial and business settings. The motto of Industry 5.0 is "human-robot co-working".

Robotic Systems: Robotic systems are defined as systems that provide intelligent services and information by interacting with their environment, including human beings, via the use of various sensors, actuators and human interfaces (Object Management Group, 2005).

Smart Sustainable City: A smart sustainable city is an innovative city that uses information and communication technologies (ICTs) and other means to improve quality of life, the efficiency of urban operation and services, and competitiveness, while ensuring that it meets the needs of present and future generations with respect to economic, social, environmental as well as cultural aspects (International Telecommunication Union Telecommunication Standardization Sector [ITU-T] Focus Group on Smart Sustainable Cities [ITU-T FG-SSC]).

Section 2

Development Approaches, Monitoring, and Case Studies

Chapter 4
Next Generation of Intelligent Cities:
Case Studies From Europe

Vijayaraghavan Varadharajan
Infosys Ltd, Bangalore, India

Rian Leevinson J.
Infosys Ltd, Bangalore, India

ABSTRACT

Over the past decade, intelligent cities have undergone rapid transformation. The definition of an intelligent city may vary based on the context and the purpose served. However, the next generation of intelligent cities will have unique characteristics that will set them apart from the existing intelligent cities. They will be more people-centered, and they will be formed through the fusion of technology, government, organizations, and people. This chapter explores four intelligent cities in Europe that are setting examples for innovation, ingenuity, technology, public policy making, and sustainable development: London, Amsterdam, Vienna, and Stockholm. With growing emphasis on people involvement in decision making, the intelligent city ecosystem is continuously evolving. However, technology continues to play a prominent role in shaping the intelligent city paradigm. In this contribution, the authors also examine different emerging technologies such as quantum computing, autonomous vehicles, AI, ML, etc. that could potentially impact the next generation of such cities.

INTRODUCTION

Throughout modern human history, cities have been the epicenter of the world for culture, governance, art, and technology, paving way for economic and socio-political growth. The urban population is testimonial to a country and region's development and growth. Cities act as the fusion of various sectors such as information technology, industry, governance, transportation, security, healthcare, economics and infrastructure. Phenomenal growth in urban population from 250 million at the beginning of the 20th

DOI: 10.4018/978-1-7998-5062-5.ch004

century, to 2.8 billion at the beginning of the 21st, and to about 9 billion by 2050 has made it imperative that cities plan efficient management systems and sustainable living initiatives. Such challenges render traditional approaches to city planning inefficient and inadequate. To address these challenges, governments and planners are propagating the development of intelligent cities, also referred to as Smart Cities.

Advanced integrated public transportation systems, sustainable and clean private vehicles, high level of connectivity, reliability, safety and ease of commute are an integral part of mobility in intelligent cities. They have excellent connectivity within the city and with external areas. Intelligent traffic management systems aim to regulate the flow of traffic through the city and control transportation. The environmental aspects of intelligent cities include sustainable development, green spaces, micro renewable power generation, waste management, pollution control, resource management and efficiency. Smart living covers the life of the city's inhabitants and their interactions with the city policies, infrastructure and decisions. It includes the aspects of smart homes, safety and security, healthcare, art and culture.

The rest of the chapter is organized as follows: The next section explores literature review of existing intelligent city concepts, definitions and ideas. The subsequent section contains the case study of four European smart cities: Vienna, London, Amsterdam and Stockholm. The section titled 'Factors that Could Directly Impact Intelligent Cities of the Future' discusses factors such as climate change, data privacy and security, data transparency and public acceptance that could potentially affect the next generation of intelligent cities. The next section analyses the effect of various emerging technologies on the next generation of intelligent cities and the final section concludes the chapter. In this contribution, we will use the terms 'smart city' and 'intelligent city' interchangeably.

LITERATURE REVIEW

Although extensive research and experimentation have been carried out on the concept of intelligent cities with numerous pilot projects around the globe, there is no standard definition of a smart city. Many cities around the world have adopted intelligent projects and smart way of solving specific problems. Hence generalizing the term 'smart city' or 'intelligent city' to apply to all cities has been difficult. This has paved way for companies, governments and authors to adopt their own definitions of a smart city, depending on the context. A Smart City according to Giffinger et al. (2007) is "A City performing in a forward-looking way in economy, people, governance, mobility, environment, and living, built on the smart combination of endowments and activities of self-decisive independent and aware citizens". According to Caragliu and Nijkamp (2009), "A city can be defined as 'smart' when investments in human and social capital and traditional (transport) and modern communication infrastructure fuel sustainable economic development and a high quality of life, with a wise management of natural resources, through participatory action and engagement".

Historically, technology has been one of the prime driving force behind smart city initiatives. With advancements in information and communication technologies (ICT), the technology driven growth continues to flourish (Pablo, et al. 2018). However, high levels of globalization, free trade and changing global scenarios can directly impact the next generation of intelligent cities. Moreover, the integration of different sectors within the city and their interconnection are expected to increase substantially leading to the next generation of intelligent cities being highly integrated ensemble of these sectors. They will be characterized by seamless participation of citizens, organizations and government bodies in different levels.

Rudolf Giffinger, et al (2007) identifies six characteristics across different sectors that can be used to generalize and elaborate different types of smart cities. Although other classifications may exist, this is a robust attempt that accounts for a variety of domains and sub-domains.

Figure 1. Characteristics of intelligent cities across different sectors

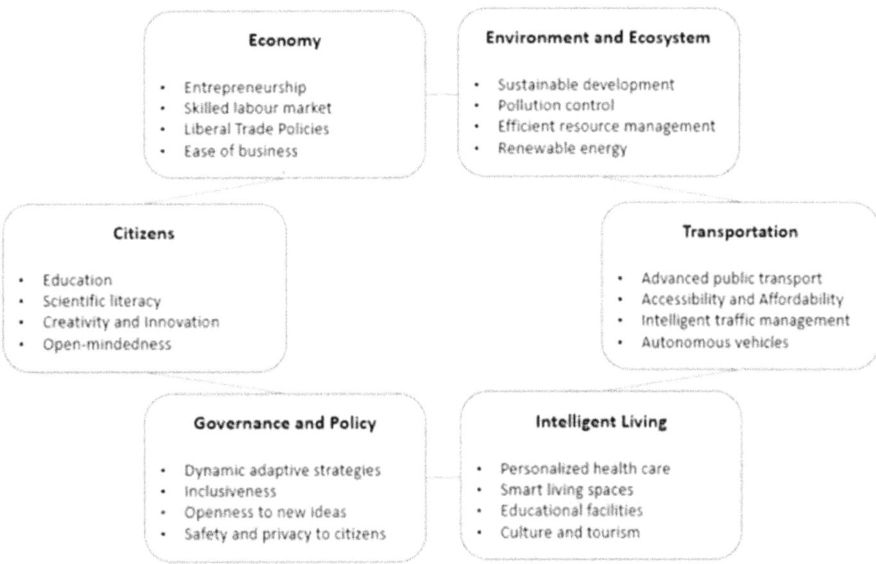

The key characteristics, given in Figure 1, include the economy, people, law and governance, transportation and mobility, environment and intelligent living ecosystem. Although the goals and desires of individual cities may vary, intelligent cities generally make advancements on all six of these characteristics with varying priorities among different sectors. The economic aspects include globalization, ease of trade, entrepreneurship, production, flexibility of the labour market and competitiveness in the global platform (Editors, 2011). The people's aspect revolves around education, scientific literacy, social interactions, community building and openness towards new ideas and initiatives. Smart governance encompasses the political landscape, active participation from citizens on public policy making and innovative governing techniques

Leydesdorff and Deakin (2011), introduce us to the Triple-Helix model of smart cities which emphasizes on the society's support for the transformation of the city as a key enabler for intelligent city initiatives. The Triple-Helix model considers cities as dense integrated networks among three sectors namely universities and academic institutions, industrial sector, and interactive democratic government. This comprehensive model demonstrates how technology and society co-evolve as a part of the 3 sectors of helix.

Inclusiveness in the form of citizen co-participation in policy drafting and decision making is a key aspect of the next generation of intelligent cities. They are characterized by transparency, collaboration, inclusion and engagement between and within various sectors of the city. This is formally called Smart City 3.0 in the 3rd Generation Smart City evolutionary phase model, as depicted in Figure 2. The first generation of smart cities were led by technology-driven initiatives and projects. Their goals and development plans revolved around the implementation or inclusion of specific technologies in the urban landscape (Thomas Mueller, 2017). This created an appealing environment for technology innovators, providing them with an ideal platform to showcase the possible pathways through which their technologies transform cities into highly efficient and digitalized ecosystems (Herman van den Bosch 2017). However, Boyd Cohen (2015) argues that cities known to aggressively implement technology driven solutions were ill-equipped to properly understand or handle the implications and outcomes. They seem to miss out on the key aspect of how intelligent cities interact with their citizens and influence their lives. Most intelligent city projects around the world fall into this category of smart city 1.0 including the city of Songdo in South Korea.

Smart City 2.0 represents a generation of smart city projects and initiatives that are led by local government bodies and city municipalities. These projects are advocated by municipal leaders and city administrators who were intended to define a smart future for their city. They often use technology as enablers to achieve their goals. Examples include the introduction of free public Wi-Fi, highly integrated public transport systems, smart meters and intelligent traffic management systems. It is argued in Smart City 3.0 (2018) that even this model might not be the ideal approach to fulfil the dynamic and specific demands of modern cities and their citizens. Most leading smart cities around the world can be categorized as smart city 2.0, including London, Stockholm, Singapore, etc.

Figure 2. The 3 generations of Smart cities

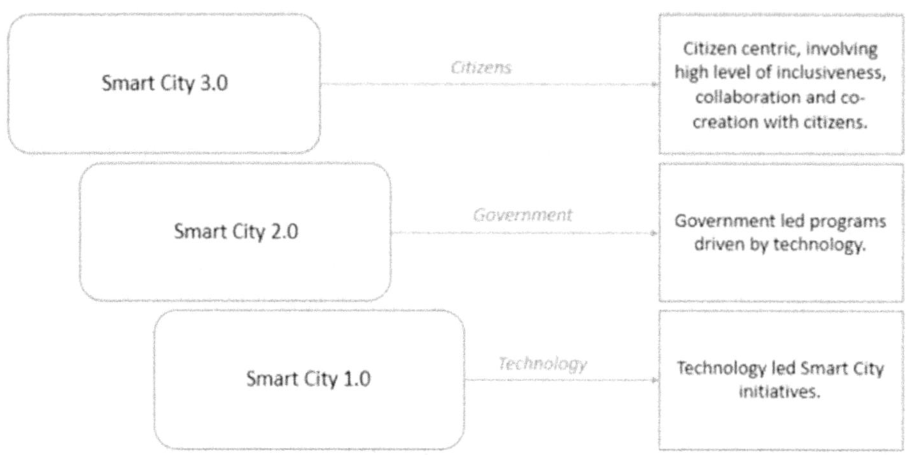

Smart city 3.0 consists of a new generation of cities that embrace a citizen-centric approach to the development of the smart city (Smart Cities Team, 2018). This involves citizen co-creation, active participation of citizens in policy making, transparency and collaboration. Specific needs of the city's inhabitants and different communities are taken into account and micro solutions are provided to cater the individuals. This is a highly democratized and complicated model that offers optimum user satisfaction and quality. Smart cities such as Amsterdam, Barcelona and Vienna are rapidly progressing towards Smart City 3.0. Amsterdam encourages active involvement of its citizens in deciding and implementing intelligent city initiatives. It prefers a large ensemble of small scale smart projects that cater to individual needs rather than large scale city wide implementations. Citizen participation is motivated by local apps, community meetings, frequent intelligent city workshops and competitions.

Vienna for example, includes citizens as investors in its local solar power plant projects to meet its 2050 renewable energy goals. Barcelona has opened its doors to ideas and innovations from its citizens by promoting the Barcelona Open Challenge. The push for inclusion and social equality in smart city 3.0 initiatives has paved way for local sharing activities among the inhabitants. This is in the form of local repair shops, renting out personal parking spaces to the public, lending out tools and carpooling services.

Although strict guidelines are not evident, the general philosophy of Smart City 3.0 may be defined as follows:

1. Identify stakeholder communities and gain insights from them on specific needs.
2. Combine these collected user needs into groups that are complimented by a consensus of citizens.
3. Quantify the consensus-based user needs into measurable functional requirements and devise an appropriate project plan.
4. Review the drafted plan to ensure all consensus-based needs are met, and that no extraneous requirements have been included.
5. Carefully consider the needs of the user over the entire project work lifecycle, from project conception to project delivery, expiration and disposal.
6. Maintain compliance with the three pillars of Sustainability,
    a. protecting the natural environment
    b. enhancing human quality of life
    c. maintaining fiscal prudence

Furthermore, it is argued that as time progresses, intelligent cities may move between different generations. Intelligent cities may move from one phase to another based on prevailing conditions and development plans. Some cities such as Kansas has jumped directly from a first generation city to a third generation city while cities like Singapore may always remain in the smart city 2.0 category. The ideal intelligent city generation is highly subjective and debatable. Smart city 2.0 offers structured administrative guidelines and plans for development with well-defined future goals for the city which may end up compromising on inclusiveness. Smart city 3.0 on the other hand, consists of highly specific and micro level initiatives that are driven by the needs of the citizens. However, they may quickly become overcomplicated and chaotic. Boyd Cohen (2015) suggests that an ideal model could be the one that is a combination of smart cities 2.0 and 3.0. It is imperative that city officials take initiatives about the future of their city while encouraging active citizen participation in decision-making and welcoming innovation and entrepreneurship.

CASE STUDIES OF LEADING SMART CITIES IN EUROPE

According to Roland Berger (2019), of 153 smart cities surveyed and studied, only 15 have a well-planned comprehensive smart city approach and framework. European cities present unique challenges and opportunities to the development of smart cities. The characteristics of the intelligent cities surveyed are depicted in Figure 3.

Figure 3. Characteristics and Dimensions of European Smart Cities

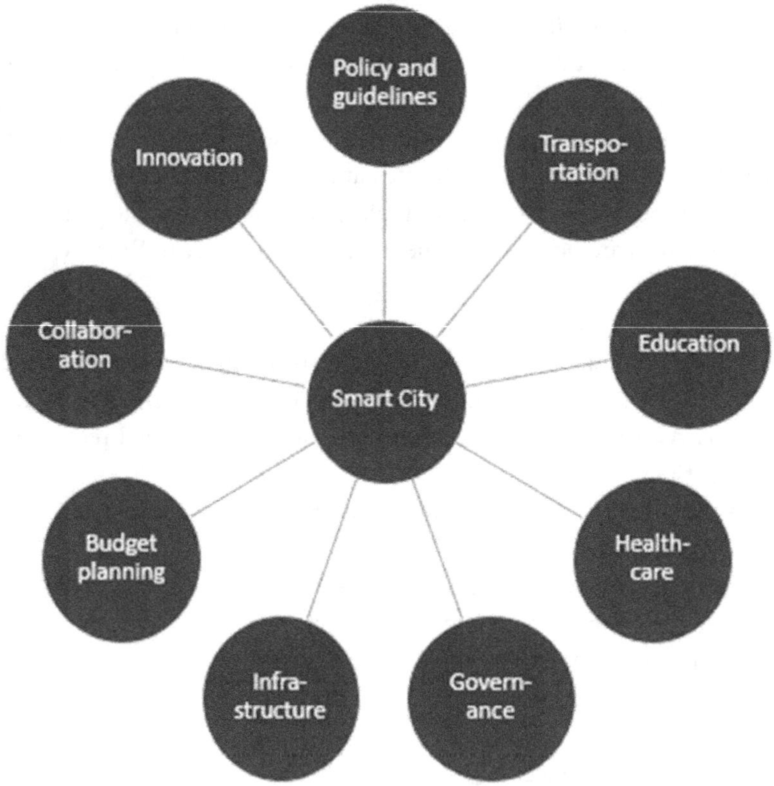

This is mainly due to the dense nature of the population in cities and the geographical proximity of cities with each other. This is radically different from that of, for example the USA. Most cities in USA are thoroughly planned and engineered with the future expansion and development in mind. They are relatively less densely populated and are often situated far from each other. Europe considers sustainable development, an imperative part of its future plans and has laid out both short term and long term goals for its members. With population growth, sustainable development and adoption of technology being some of the key factors driving the next generation of smart cities, the smart cities of Europe serve as excellent case studies.

We considered four leading smart cities as our choices of study: London, Amsterdam, Vienna and Stockholm. Each of these cities are unique in their needs, goals, strategies, existing supportive infra-structures and approaches. The subsequent sections look at these cities in detail.

London

London is the tech capital of Europe – by size, level of investment and the presence of more than a third of Europe's billion-dollar 'unicorn' companies. It is home to 46,000 tech companies, supporting 240,000 jobs in an ecosystem that is worth an estimated $44 billion. Smart City London is built on 7 strategies that include transport, environment, health inequalities, housing, culture, economic development and the

London plan. 'Smarter London Together' is a government led collaboratory effort to have better digital services, innovation, connectivity, digital inclusion, cyber-security and open data.

The smart city initiative in London is spearheaded by 5 missions declared by the Mayor of London in Smarter London Together (2018), who seeks to transform the city into a next generation smart city. The missions are as below:

1. More user-designed services
   Explore new approaches to digital inclusion to engage Londoners
   Engage citizens and communities using new civic platforms
   Ensure more diversity in the tech field to promote equality
   Ensure regulations in design with common standards to make it user-centric
   Promote Innovation challenges to invoke innovations from the tech sector
2. Better deal for city data
   Increase secure data sharing and collaboration within Londoners using the London Office for Data Analytics (LODA)
   Promote an open-ecosystem to promote data sharing and transparency
   Enforce data rights and accountability to regulate how public data is handled
   Counter cyber threats by developing city-wide cyber security solutions
3. World-class connectivity and smarter streets
   Promote common standards in tech to improve connectivity and scalability
   Develop intelligent infrastructure throughout the city
   Coordinate connectivity and 5G projects through the 'Connected London' initiative
   Enhance the connectivity, reliability and speed of public Wi-Fi throughout the city
4. Enhanced digital leadership skills
   Make public services open to innovation through the digital corridor
   Develop workforce upskilling and enable digital empowerment using the Mayor's Skills for Londoners Strategy
   Promote the use of technology in solving everyday problems
5. Improved city-wide collaboration
   Promote social healthcare and MedTech innovation to cater the needs of the citizens
   Explore new partnerships and tech models with different businesses
   Collaborate with other cities within UK and around the world
   Establish a London Office of Technology (LOTI) to define common standards for technology initiatives

This roadmap laid out by the authorities of London is being fulfilled by a variety of projects that have been initiated across different sectors. With the citizens in primary focus, these projects seek to make urban life easier and help London become a more intelligent city. The authorities have encouraged active participation of its citizens in policy making. With special emphasis on human capital for development. To empower its citizens and equip them with the latest skills, London launched the exclusive Mayor's Digital Talent Programme. This is expected to increase the number of Londoners employed in technological sector to around 200,000 by 2020.

The government of London has always stressed on the importance of openly available data and its analysis. They believe that reliable and accessible data could present exciting opportunities. They have also taken a variety of intelligent city initiatives that have largely been successful. Accessible high-speed internet connectivity has been successfully attained almost throughout the city. With 90% of Londoners using the internet, Wi-Fi connectivity in over 250 tube stations and over 78% of Londoners owning smartphones, the impact of technology and internet on the city's inhabitants is profound.

The London Datastore is an open source data base with over 700+ datasets. This program was initiated in public interest, to help address urban challenges and improve public services. This data can be analyzed and combined to gain insights about how the city functions. This integration of the technological sector with everyday activities in the city helps build intelligent solutions. This leads to better housing and development planning, the use of demographic data to understand the city's inhabitants and their needs and the use of air pollution sensors and monitors throughout the city to control air quality.

London has taken extensive efforts to monitor air quality throughout the city. London has one of the largest networks of air quality monitors in the world, with powerful modelling and forecasting. This is enabled by the 100s of air quality stations situated throughout the city. This helps the mayor of the city to identify hotspots of air pollution within the city and address them accordingly (The Future of Smart, 2016), Safety and security is another key aspect of any large metropolitan city. London has taken steps to implement technology and digitalize its police force by deploying over 22,000 body worn cameras. This is the largest such initiative in the world and has considerably increased accountability and evidence gathering. Furthermore, London's notorious knife crimes are being tackled by analyzing data related to the time and location of these occurrences, uncovering patterns in the data and planning appropriate patrols and surveillances to prevent such attacks.

The participation of the general public to instigate micro projects was also explored by London using the Open Works research project. It was aimed at accounting for individual ideas and opinions from the general public to shape the city. This successful project has eventually led to over 20 different projects being kick started throughout the city. The success of the Open Works initiative is partly associated with the innovative platform approach adopted by the research team. Considering the disadvantages of a top to bottom approach that is often followed in the initial intelligent cities, Open Works emphasized on the participation of all stakeholders and interaction with each other. They build a large network of citizens, organizational representatives and authorities collaborating and working together to work on initiatives together.

Amsterdam

Amsterdam is one of the pioneers of the smart city program in Europe and having successfully implemented multiple smart city projects across different sectors, it is one of the leading smart cities in Europe. It followed a holistic strategy, based on a quadruple helix concept that consists of the government, businesses, citizens and universities. Among large cities, it was one of the first to adopt a bottoms-up approach, giving citizens a vital role in shaping the future of the city. The Chief Technical Officer of Amsterdam's Smart City Initiative oversees the various smart city project initiatives and drives innovation. The government of Amsterdam has focused on inclusiveness and active participation of citizens and small scale startups over big corporations in driving smart city projects.

The bottom-up approach and citizen centered decision making adopted by Amsterdam has helped sustain a high quality of life and an inclusive environment. The cooperation between public and private sector has been testimonial in its success as a modern sustainable smart city. The city has taken extensive steps to encourage innovation, new ideas and debates (Lisa Smith, 2017).

Figure 4. Themes in Amsterdam smart city

Digital City	Data Infrastructure, Internet of Things, Big Data, drones and robots, algorithms, blockchain, AI, privacy and data openness
Energy	smart grid, smart lighting, bio fuel/ bio gas, wind energy, solar energy, low carbon footprint, smart buildings
Mobility	Accessibility, Logistics, public transport, autonomous vehicles, electric/ hybrid vehicles, bicycles, infrastructure
Circular City	Waste management, e-waste, circular economy and business models, water management, resource management
Governance & Education	Policy, living labs, participation and cooperation, entrepreneurship, digital democracy, healthy urban living
Citizens & Living	Public engagement and participation, sharing economy, inclusiveness, pollution, sustainable development

A collaborative ecosystem is set up for different parties to interact and take decisions. One key aspect of this is the Amsterdam smart city online platform that acts as a central forum to plan and discuss about smart city projects. The online platform has been extremely successful in taking smart city projects to the people and making the data transparent and useful. The platform categorizes ideas into 6 broad themes as shown in Figure 4.

Amsterdam has hundreds of projects underway across these themes with active citizen collaboration and contribution. This is further supported by frequent publications, conferences, hackathons and challenges that engage all stakeholders. Some of Amsterdam's popular and successful projects include the following:

- Klimaatstraat (Climate street):

This is an extended effort by over 40 entrepreneurs to transform one of Amsterdam's shopping street, the Utrechtsestraat, into a sustainable shopping street where innovative ideas and experiments are tested. This project is one of several such initiatives launched by the municipality of Amsterdam to ensure that the city is one of the leaders in sustainable development by 2040 (Stephanie Vahlkamp, 2016). This project consists of an ensemble of energy-saving and efficient technological solutions such as smart meters called toons, intelligent lighting and energy displays. Furthermore, Amsterdam was also a part of a Netherlands wide large-scale project to install photovoltaic roof panels in newly developed areas.

- Vehicle2Grid:

This pilot project is designed to tackle the various technical, economic, social and environmental impediments of electric vehicles with the help of cooperating smart project partners. With households rapidly installing solar panels to generate electricity, it is imperative that efficient means of storing and using excess generated electricity is established. Possible solutions include the use of electricity to charge the batteries of electric cars or to feed the energy back into Amsterdam's electricity grid. Vehicle2Grid seeks to instigate the seamless integration of electric cars and locally generated electricity (Somayya M, 2016).

- Smart Lighting:

With the help of the IoT paradigm, Amsterdam has been successful in implementing smart lights in its streets that are designed to save energy and improve security. These lights can be adjusted according to the weather, local needs or be used to regulate the movement of pedestrians and vehicles. Movement sensors are being used to vary the brightness of the lights. If the sensors detect the presence of a pedestrian or a bicyclist, the lights switch on along the path. This helps conserve energy and reduce light pollution.

- Sustainable heating:

Sustainable district heating projects in the city of Amsterdam includes the optimization of the waste and energy power station and the installation of cutting-edge daylight collectors as a new source of heat. Waste heat is recovered through a variety of innovative means, including from sewer waste to power waste heat recovery systems. Heating buildings and households consumes massive amounts of electricity during winter. Hence, such sustainable heating techniques will reduce emissions and lead to efficient use of energy.

Amsterdam is also unique in its effort to enable citizens to use smart apps catering to specific small-scale projects. This include applications such as renting out one's bicycle, renting out their personal parking space and community apps to solve local problems and challenges. These align well with Amsterdam intelligent city's emphasis on large number of small projects rather than large-scale projects. This offers an unparalleled level of customizability, collaboration, support and inclusiveness that will be key to advance Amsterdam into the future as one of the world's leading intelligent cities.

Vienna

Vienna, the capital of Austria is considered one of the most livable cities on the planet. This is derived based on evidence from a variety of factors including infrastructure, healthcare, safety, gender equality, liberty and education.

A key factor that sets Vienna apart from other smart cities is its unique well laid out long term objectives as a part of its smart city plan. The city has drafted specific goals across different sectors from energy to healthcare. Such long term initiatives provide much needed clarity and direction of development for the city planners. All new projects, from infrastructure to public transport are checked for adherence with the long term plans. This improves operational efficiency of the city, quality of life and optimizes various workflows with emphasis on sustainable development. Vienna is one of the few European intelligent cities with a well-defined intelligent city framework, strategy and well defined long term and short term goals (Pozdniakova Anna, 2018). This paves the way for high levels of integration among different sectors within the city. A general overview of selected intelligent city objectives and goals according to CeesJan Mol et al. (2015) is as follows:

- Reduction of per capita CO2 emissions from 3.1 tons to 1 ton (- 80% CO2 from 1990 to 2050).
- Plan of ensuring that 50% of Vienna's gross energy consumption will originate from renewable sources by 2050.
- Decrease of Motorized Individual Traffic (MIT) from 28% to 15% by 2030. By 2050 all vehicles within the municipal boundaries run without conventional propulsion technologies and will run on clean and renewable sources of energy.
- Develop the innovation triangle Vienna–Brno–Bratislava to be one of the most future-oriented cross-border innovation Regions in Europe by 2030.
- Ensure that the share of technology-intensive products in the export volume increase from 60% to 80% by 2050.
- Ensure that all the inhabitants of Vienna have a good neighborhood and safe life conditions irrespective of their background, physical and psychological condition, sexual orientation and gender identity.
- Provide affordable and accessible quality medical care to all.

Vienna is considered one of the prime candidates to pioneer the next generation of intelligent cities. This is due to its combination of well-structured long-term goals and the active inclusion of its citizens in intelligent city projects to work these targets. The general smart city approach is based on the need to optimize the usage of resources, reduce pollution, maintain a high quality of life and uphold active social participation by its citizens. A few such collaborative smart city initiatives include the Citybike initiative that is a mobility service that helps the citizens of Vienna. It consists of a large network of public bicycles that are used to cover short distances in the center of the city. The project has been extremely successful with over 11 million rides recorded already (Citybike Wein, 2019). Vienna is well renowned for its eHealth initiative which has digitalized health care by using electronic health records, remote health monitoring equipment and taking extensive measures to ensure data protection and security. Booking appointments, providing intuitive explanations about diseases, public health alerts and connecting with doctors are made easier using mobile healthcare systems.

The city of Vienna is continuously taking steps to adopt new ideas and democratize smart city project initiatives. An example is the Digital Agenda Vienna which actively involves citizens opinions and contribution in its plan.

Stockholm

Stockholm, the capital city of Sweden has a long history of environment conscious development, dating back to the 1990s. This has been a collaborative effort by the people, government and NGOs to strive for sustainable development. It has paid off with the per capita CO2 emissions falling by 25% since 1990 (Antonio Garrido-Marijuan, et al., 2017). Sweden has set a long-term goal of being fossil fuel free and using 100% renewable energy by 2040 (Sweden, 2018). It has also earned the title of the first green capital of Europe in 2010. This is impressive especially provided the fact that the city of Stockholm itself has been developing and growing rapidly. With a population of around 900,000 it is the most populous city in Scandinavia and is often termed the capital of Scandinavia. It is also considered one of the top 5 fastest growing cities in Europe. The growing population presents a unique set of challenges to smart city Stockholm.

Historically, Stockholm is world renowned for its extensive fiber optic cable network, ensuring reliable, fast and high speed connectivity throughout the city. The goal of this initiative was to invoke economic growth, innovation, facilitate research, collaboration and meet communication needs. The connectivity offered by this extensive fiber network has been a unique enabler for smart city initiatives and projects and will also be one of the most favorable factors for transforming Stockholm into a smart city of the future. Such a robust communication infrastructure directly supports other intelligent city initiatives such as Green IT and e-services. Moreover, Stockholm's Kista Science City is an ICT hub that has acted as a platform for companies, academia, authorities and the general public to collaborate on technologies.

The growing population of Stockholm puts additional stress on its transportation network as mobility is a key enabler in intelligent cities. The regional government has tried innovative approaches to deal with the ever growing demand on its transportation infrastructure (road vehicles, rail, tram) and traffic management systems with environment consciousness in mind. This includes the concept of congestion charge that was introduced to try and limit the number of vehicles entering the core city. This is done so by enforcing a charge on vehicles entering the city center during specific times of the day. This initiative has generated largely favorable outcomes with the city noticing a 30% reduction in traffic during peak hours and considerable decrease in carbon dioxide and nitrogen oxide emissions. Other outcomes include a reduction in the number of private vehicles with an increase in the use of bicycles and public transports (Trendsetter, 2006).

According to Giannakopoulou Lazarina (2015), the Stockholm Public Transport (SL) runs commuter trains, buses, subways and trams in the city. The transportation board along with the city council have taken several steps along the years to make public transport more environment friendly and clean. Stockholm also established 'environmental zones' in the city that prohibited heavy vehicles older than 8 years from entering the inner city. This resulted in companies changing to modern and cleaner vehicles. Presently, the environmental zones in the city are governed and controlled by the police force. This requires extensive patrolling and manual workforce deployment to monitor adherence to the law. However, with the advent of IoT based sensors and object detection models built into road surveillance cameras, smart solutions can be derived to automatically identify non-compliant drivers. This can also be used to drivers who park their cars illegally in the environment zone and take necessary action. Such implementations could decrease manual labour, increase compliance, reduce pollution and safeguard the environment.

Moreover, the prime objective of Stockholm in its clean transport system initiative is to limit and discourage the use of private vehicles, especially ones that run on fossil fuels and to promote the use of clean energy vehicles that run on biofuel, biogas and electricity and the use of public transport. The use of vehicles that run on clean fuel is encouraged by exempting them from restrictions in environmental zones, build better refueling infrastructure and have dedicated website for drivers of clean vehicles. The government has also experimented with the use of biogas-powered garbage collecting vehicles with positive feedback from the drivers and have also encouraged office transports to adopt clean vehicles.

Stockholm demonstrates high economic growth along with a high employment rate. Financial component of Stockholm Smart city strategy includes an attractive environment to make investments or establish a business. It boasts one of the best start-up locations in the world especially in digitalization. It offers a highly inclusive and skilled labour market. These have been the major factors, driving Stockholm towards advanced smart city initiatives and development.

FACTORS THAT COULD IMPACT INTELLIGENT CITIES OF THE FUTURE

Intelligent city projects often encounter dynamic challenges that arise due to the environment, growing needs of the inhabitants, technological needs and due to existing socio-economic situations. A variety of factors such as climate change, data privacy and security, public appeal play an important role in shaping the growth of intelligent cities. These factors must be considered when developing intelligent city plans and frameworks to administer intelligent development. This section explores a few important factors that could directly influence the next generation of intelligent cities.

Climate Change

Anthropic climate change (man-made climate change) refers to the adverse impact on planet Earth's climate by mankind. Climate change is characterized by global warming, as a consequence of which polar ice caps melt, sea levels rise, extreme weather events occur more often, global average temperatures increase and the habitats of flora and fauna are destroyed. Climate change is not a linear phenomenon. Once a certain threshold is crossed, the rate of climate change could accelerate and tip over the point of no return, putting the future of life on Earth at stake (Jennifer Chu, 2019). With cities representing majority of the human population and acting as the hub of human activities, it is imperative that they develop in a sustainable and environment friendly manner.

This has caused aggressive reforms from governing bodies and organizations such as the European Union (EU)'s 20-20-20 targets. The European Union set the 20-20-20 targets which obligates the EU members to instigate a 20% reduction in greenhouse gas emissions by 2020 to the levels as of 1990, to energy efficiency by 20% and to ensure the share of energy produced from renewable sources is at least 20%. Reforms are needed across a variety of different sectors by countries to meet the ambitious plans put forth by EU. This push towards sustainable development and environment consciousness majorly influenced and affected the direction of development of Intelligent Cities in this region.

European cities emphasize heavily on green energy initiatives, energy efficient designs and processes, technology driven sustainable development, heavy inclination towards renewable sources of energy and quality of life. Coastal cities like Amsterdam are under imminent threat from rising sea levels due to melting polar ice caps. It lies at an average elevation of 2 meters below sea level. Smart city planners should take adequate measures to control climate change and reduce its negative effects on people and environment.

Data Privacy

Smart cities are technologically advanced environments that gather, process and analyze huge volumes of data across different sectors. As more and more data is gathered and analyzed, city planners can provide better services to their citizens. This includes the need to collect personal data of the inhabitants as well. This personal data refers to unique identifiers that may be in the form of fingerprints, retina scans, face scans, social security numbers, personal government IDs and documents, etc. Furthermore, additional information such as location of the person, their name, age, contact details, bank account details and other demographic details are often collected through mediums such as smart devices, computers, IoT sensors, cameras, etc. Such confidential data often tends to be transmitted over public networks e.g. the

Internet and processed by third parties. Mishandling this data could lead to dire consequences including but limited to identity thefts, scams, exposure of personal information, breach of security, etc.

Furthermore, with intelligent cities of the future being highly sophisticated environments that contain millions of connected devices that are constantly generating and transmitting data, privacy and security are important issues. The use of IoT, advanced analytics and image analytics should not breach the freedom and privacy of individuals. Hence strict guidelines and protocols are necessary to govern and regulate the use of data. The policies may be drafted by the authorities, organizations and the citizens together to ensure transparency in the process, enable trust, promote inclusiveness and to ensure agreement on mutual interests. Efforts can be taken to prevent the misuse or theft of sensitive personal data by enforcing high security standards on hardware and software to avoid third party abuses. However, this could deteriorate the quality of personalized content that is provided to the users such as targeted marketing, personalized suggestions and offers, customer specific recommendations, etc. This calls for a logical compromise between privacy and quality of personalized services offered.

Cyber Security

Smart cities infrastructure generally consists of highly connected and integrated ecosystems, comprising millions of sensors, smart diverse devices and heterogeneous networks. Digital technology is an integral part of life in intelligent city controlling a variety of important functions, from controlling street lights to controlling traffic management systems. This constitutes a massive network of inter-connected devices that cover the entire city. Such a dense large scale is very complex and vulnerable to cyber attacks. The disadvantages faced by IoT devices and networks are also faced by intelligent city technologies as a breach in any one node (device or sensor) could cause chaos through the entire network, affecting millions of lives. Such a disruption could bring city life to a standstill, jeopardizing the economy. Furthermore, such security breaches are a threat to human life due to the highly integrated nature of technology with human activities.

The next generation of future intelligent cities will almost entirely be dependent on computers, smart processing devices and advanced AI and machine leaning based algorithms to monitor and control them. Authorities and urban planners must take adequate security measures to deal with looming security threats. Cyber-attacks such as ransomware have wide-spread implications and are extremely difficult to recover from, once infected. Hence it is imperative that appropriate tests are performed to check the integrity of hardware and software, vulnerability assessments are regularly performed, citizens and planners are made aware of imminent security threats and their consequences.

Data Transparency

With IoT based sensors, smart objects, smart data storing, and processing techniques, smart cities collect, process and store massive amounts of data. This data may be sourced from surveillance cameras, digital kiosks, sound detectors, social media analytics, IoT sensors, etc. Such data collecting and storing practices often happen without the consent or knowledge of the individual. This rapid growth in data availability has exceeded our ability to control our identity and privacy. This calls for the need to define new standards and protocols to regulate how personal data is stored, accessed and processed.

It is important that the next generation of intelligent cities take adequate steps to safeguard data and ensure that it is transparent. Many of the leading smart cities such as Amsterdam, Vienna and London have adopted strict guidelines and policies on the collection, storage and the use of data. Smart city Barcelona, for example, has set up a dedicated Data Transparency Lab (DTL) whose aim is to provide technical, financial and regulatory support to tools that can provide transparency about how user data is processed and stored. It seeks to create an international community of dedicated researchers, citizens and policy makers who can regulate and reform online user-data transparency.

Public Acceptance

Intelligent city projects often bring massive changes to the city's landscape in a short period of time. Care should be taken to continuously monitor and assess the impact of the new changes. It is important to get feedback from the city's inhabitants and ensure their satisfaction and acceptance. Public acceptance of such smart city initiatives is key and could lead to immediate backlash if overlooked. The Sidewalk Labs project in Toronto is an example of why communication and continuous engagement with the public is important. Starting out as a regular intelligent city project, it emphasized on the deployment of IoT sensors, waste management, delivery and logistics, reforms in housing and traffic policies, high speed public Wi-Fi, environment-friendly and user-friendly public transportation systems, etc. Called a global hub for urban innovation, the project was expected to generate over 3900 direct jobs with an initial infrastructure investment of around $1.6 billion (Wharton, 2019).

However, resistance quickly arose for the project when it was reported that Sidewalk Labs had plans to extend the project to over 350 acres. This backlash is indicative of the rising trend of people being skeptical about privatization. Although the company had taken adequate efforts to ensure data transparency and privacy, public satisfaction and communication gaps ultimately led to backlash. Therefore, it is important that intelligent city projects build trust with their stakeholders, especially the general public.

EMERGING TECHNOLOGIES FOR NEXT GENERATION INTELLIGENT CITIES

Unlike the first generation of smart cities, modern cities are not entirely based on the implementation of new technology. They are primarily driven by collaborative efforts from the citizens, government and third party organizations to facilitate development and solve problems. However, the role of technology in driving intelligent city projects will always be indispensable. This is especially the case due to advancements in computing and the rapid realization and implementation of emerging technologies. The Internet of Things (IoT) has played an influential role in propelling intelligent city initiatives. The notion of connected devices and sensors has revolutionized various sectors, from transportation to energy. IoT offers unparalleled optimization, control and customizability to cyber-physical systems. The use of AI has phenomenally influenced the everyday lives of residents in intelligent cities. Machine Learning and Deep Learning, coupled with computer vision have transformed object detection, traffic management, surveillance and security and self-driving vehicles. Advancements in technology also continue to make energy generation, cleaner and more efficient. More efficient solar PV cells, improved blade angles in Wind Turbines, stronger and lighter alloy materials, CANDU (Canadian Deuterium Uranium) and thorium nuclear reactors are some examples. This section explores a few prominent emerging technologies and their impact on next generation of intelligent cities.

Autonomous Vehicles

The number of autonomous vehicles (AV) on roads is expected to rapidly rise over the next decade. These vehicles are expected to have a large influence on the urban landscape and their subsequent development. Although intelligent cities have highly sophisticated traffic management systems, infrastructure and intelligent transportation, they are not generally designed by accounting for self-driving vehicles. Nikolaos Gavanas (2019) suggests that autonomous vehicles could transform the flow of traffic through dense urban areas, making it more stable, uniform and less accident prone. The realization of autonomous vehicles and related technologies is also expected to affect the locations of residences and companies, urban areas and roadways. This could lead to reformation of land usage and future infrastructure development projects and could directly shape the development of next generation of intelligent cities.

It is expected that large-scale induction of autonomous vehicles into current cities could transform a variety of sectors including mobility, emissions, energy efficiency, business models, location of residential zones, transportation infrastructure, etc. However, the lack of available data on the effects has been a major challenge for urban planners. Impact assessment and conclusions are difficult to arrive upon. Hence extensive studies must be carried upon small scale implementations of autonomous vehicle projects. Experimental implementations have already been performed in controlled environments where transportation involves repetitive and pre-planned tasks such as the operation of trucks in quarries and open pit mines. Volvo Trucks are already operating driverless trucks in Kristineberg mine in Sweden. Peter Campbell (2018) expects automation to have a major impact on the profession of truck driving both in case of long haulage and short haulage.

Europe is generally considered to be well-equipped to support autonomous vehicles compared to the rest of the world. The Netherlands is especially highlighted by Consultancy.eu (2019) to have adequate measures, policies, regulations, supporting infrastructure and awareness to sustain self-driving vehicles. It is continuing to explore future avenues to expand the use of autonomous vehicles including truck platooning and in public transportation systems. Truck platooning has already been experimented upon, with promising results. This is partly due to advancements in technologies such as cloud computing and the Internet of Things, enabling vehicles to communicate with each other in real time and coordinate throughout the journey. This helps in improving fuel economy, increase aerodynamic efficiency, coordination, trucker's safety and fleet management.

The impact of autonomous vehicles on European cities is especially interesting due to their high population density, historic unplanned free development and highly integrated public transportation. The next generation of intelligent cities will have to account for autonomous vehicles and their needs. Appropriate policies, safety rules and regulations must be drafted to ensure safety. Traffic laws may have to be revised to allow smooth integration of such vehicles into existing systems and Infrastructures have to be re-designed or tweaked for seamless integration. Although most autonomous vehicles at present mandate the presence of a human driver, completely autonomous vehicles without any human intervention are already undergoing road trials in the USA. Hence, urban planners and the civil authorities have to plan and design the city's infrastructure accordingly. The future of main-stream road transport is expected to become largely automated soon. The next generation of intelligent cities should be planned with emphasis on the inclusion of connected autonomous vehicles following the Internet of Vehicles (IoV) vision.

Immersive Technologies (AR and VR)

Immersive technologies are unique due to their virtual and augmented nature. Virtual Reality (VR) and Augmented Reality (AR) are slowly being accepted for commercial use. Although such immersive technologies were more often used for leisure, videos gaming and the immersive experience, they find excellent applications in the context of intelligent cities. The innate ability to simulate a real-world environment in a virtual world finds tons of applications. One of the main objectives of the next generation of intelligent cities is to enrich the lives of people by improving the quality of life, improve urban experience, make life easier and offer personalized experiences. This is possible with the help of AR and VR where the user is presented with a unique opportunity to explore and experience environments virtually. Although these two terms may be used interchangeably, AR super imposes images and virtual objects onto the real world while VR presents the user with a virtual environment. Some applications of AR and VR in the next generation of intelligent cities are explored below:

- Urban planning – AR and VR can be used to generate accurate and detailed models of buildings and the urban landscape. This could aid the architect and engineers in construction and could be potentially used to attract investors and buyers. Public sector infrastructure projects can also be modelled using AR/VR and feedback from citizens can be received. Immersive virtual models of the intelligent city paradigm may be created and displayed for public references and planning.
- Education – Immersive technologies can provide interactive and highly engaging experience to students using virtual environments and objects. With traditional classes and lectures seeing a digital overhaul lately, AR and VR can aid teachers and professors convey their ideas in a more desirable manner. It could also encourage students to interact with each other and with the virtual environment and invoke curiosity within them. It could also help students grasp unintuitive concepts and environments such as outer space, deep sea environments and the quantum world.
- Navigation – The use of an AR layer to assist drivers navigate their way to reach their destination. This could help to overcome the dangerous and inefficient practice of looking at maps in intelligent phones while driving. Such situations could make the driver lose concentration of their surrounds and could put them in a vulnerable situation. An augmented display layer with navigation can drastically improve navigation experience and overall safety (Dmitry Khramtsov, 2018).
- Tourism – AR and VR technology can present an amazing outlook to tourism and culture. They can help introduce the city's historic significance to the tourists with detailed models of monuments and places of interest. Moreover, the visitors can be given a tour of the intelligent city and an outlook into the upcoming intelligent city projects. It could also aid in their navigation and could give intuitive illustrations and guides to help them learn about the city's system such as integrated transportation, visitor management, tour guides, etc.

Immersive technologies are in the forefront of technology user interactions and are considered the front-end. They may very well be an integral part of daily life in the next generation of intelligent cities: both in terms of leisure and for making urban life easier.

Generative Adversarial Networks (GANs)

Object detection has been an area of extensive research in computer vision and deep learning. Generative Adversarial Networks (GAN) have emerged as one of the most powerful and fascinating concepts in deep learning, especially image analytics. They have been used for a wide range of applications including object detection, image generation, super-resolution, image noise removal, etc.

In intelligent cities, path finding and route optimization are areas of active research. In complex urban environments, efficient navigation is important. Mohammadi, et al. (2018) describe the use of GANs for path planning in support of indoor and outdoor navigation and mobility. Their proposal includes an architecture based on GANs to recommend accurate paths as a part of smart mobility. The system provides user with generated paths that helps them navigate through complex environments. This finds excellent use in urban navigation in intelligent cities and can be used for route optimization as a part of connected vehicles and intelligent traffic management initiatives. It can also be used for indoor applications such as emergency evacuation, floor planning and designing, etc.

Identifying pedestrians on the roads is a major challenge faced by a variety of applications in intelligent cities such as driverless autonomous cars, traffic management, surveillance, penalizing jay walking, and so on. Convolutional Neural Networks (CNN) have found plenty of success in this sector due to their ability to handle image data with ease. CNNs are excellent in detecting objects in videos and images, especially people and images. However, real-life applications are challenging due to their low level of tolerance for the occurrence of false negatives. An object detecting algorithm used in a self-driving car failing to identify a pedestrian could be catastrophic. This demands a failsafe system with high precision and accuracy to effectively identify objects in the field of view. Ranjith, et al. (2018) proposes an architecture based on DCGAN (Deep Convolutional Generative Adversarial Network) to tackle the problem of pedestrian detection in intelligent cities. DCGANs are effective in handling images with blurry objects due to the generative component in the architecture. The network essentially generates a

Figure 5. Architecture of the DCGAN for Pedestrian Detection

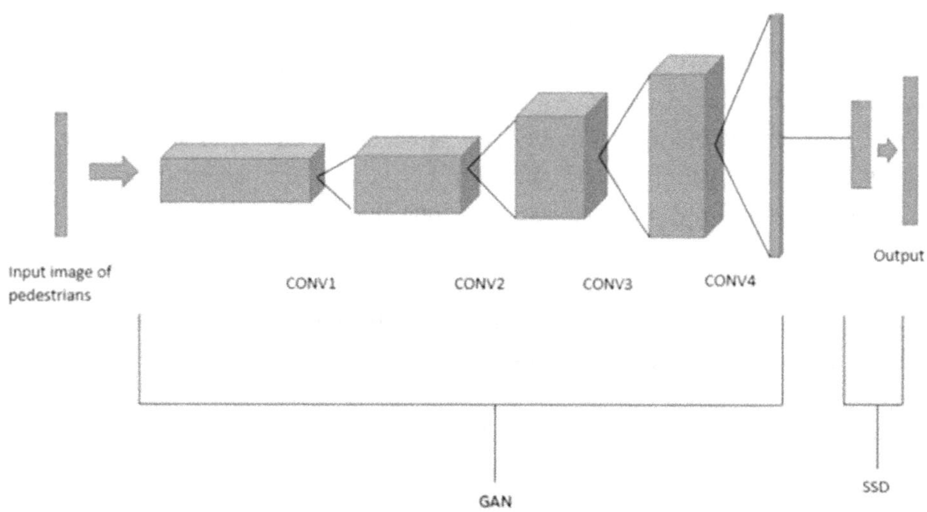

high resolution image of the blurred object for further analysis. This is especially useful in identifying pedestrians at a distance as they often tend to be blurry with minimum visual features.

The proposed architecture is based on cascading a DCGAN with a Single Shot Detector (SSD) as shown in Figure 5. The SSD used in the network is pre-trained on ImageNet data and the reconstructed images from the DCGAN are fed into it. The network was tested with the CIFAR-10 dataset and compared with a conventional SSD model. The results suggest an object detection rate of over 80% compared to the 35.5% of the standard SSD. This could potentially lead to substantial improvements in pedestrian safety in the next generation of intelligent cities, especially in the era of autonomous driverless vehicles. This can also be applied to penalize jaywalking, perform pedestrian behaviour analysis and analyze pedestrian movement. Pedestrian tracking could be important due to the emergence of driverless vehicles where jaywalking could be dangerous and the growing population density that might contribute to street crimes, decongest pedestrian movement in crowded areas, etc.

IoT and Machine Learning

Internet of Things (IoT) has become an integral part of intelligent cities with these smart urban environments consisting of millions of connected devices, sensors and IoT appliances. IoT has penetrated almost all sectors, from transportation to healthcare. It has paved way for Intelligent homes, ambient intelligence, intelligent and personalized healthcare, smart factories. Variations of the IoT have also found success e.g. Industrial IoT (IIoT) and Internet of Vehicles (IoV).

One important application of IoT is in addressing noise pollution. Smart cities strive towards sustainable development, with special importance on reducing pollution and making urban environments more livable. Regional governments and municipal bodies lay out strict guidelines regarding the control of pollution in intelligent cities. Although extensive efforts have been taken by intelligent city planners to curb air pollution and water pollution, the intrusion of technology to curb noise pollution is limited. The EU mandates its member countries to assess and analyze noise levels every 5 years. This discrete measurement of noise levels may not be an accurate representation of the dynamic nature of noise levels emanating from varying sources. Moreover, occurrence of short burst of loud noise can cause adverse health effects and such events are overlooked in the periodic monitoring methodology. This warrants the need for a system that continuously and accurately monitor noise levels.

Conventionally, efforts have been made to minimize and control the decibel levels of noise overall. However, sound at the same decibel level can be perceived as annoying noise or pleasant sound. Hence it is imperative that the next generation of intelligent cities adopt techniques that analyze more than just the decibel level. Yasser, et al. (2018) present a machine learning based noise classification IoT solution for intelligent cities. The authors extract Mel-frequency cepstral coefficients (MFCC) as audio features and apply Support Vector Machines (SVM) and K-Nearest Neighbors (KNN) models on the data. The approach adopted by the authors is shown in Figure 6.

The authors performed the experiment on data consisting of sounds from 8 different sources including sirens, car horns and gun shots. Segregation of different sounds enables more efficient control of urban noises by isolating different sources of sound. This allows monitoring, control and regulation of sources of noise in the urban environment. For example, drivers could be penalized for the noise emanating from their vehicles or for excessive use of horns. IoT sensors can be programmed to detect gun shots or sounds of people in distress to enable swift response from the authorities. Moreover, developing large

Figure 6. Noise Classification using Machine Learning

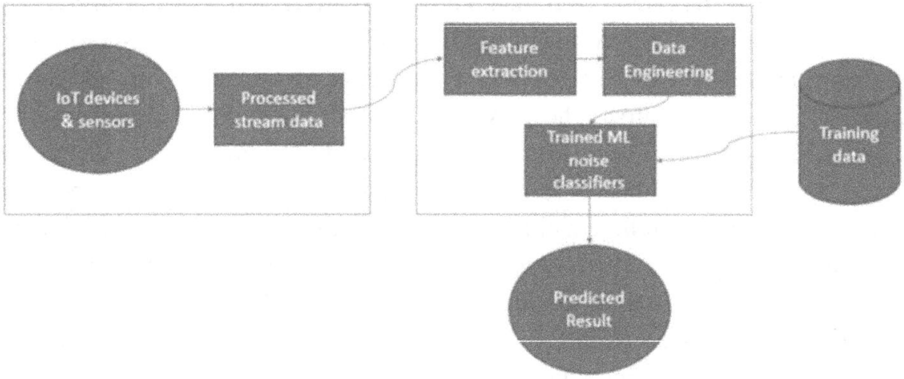

scale heat maps of the composition of noise in different parts of the city, along with the decibel level itself will help intelligent city planners and administrators.

Big Data

Big Data (BD) has played an important role in the evolution of intelligent city initiatives. Utilization of Big Data has led to the development of data-driven solutions. The evolution of IoT generates many terabytes (TB) of data every day. This massive volume of data comes from a variety of heterogenous sources and is often unstructured e.g. opinions in text from general public via social media. Although this data is collected and stored, efficiently accessing, processing and analyzing big data has always posed a challenge. The data may originate from sources such as sensors in traffic signals, weather monitoring systems, surveillance cameras, health care data such as patient diagnostics and prevailing health conditions and risks in the city, demographic data, vehicle generated data, public data, etc. A variety of architectures and framework models have been explored (Hashem, et al. 2016).

Conventional machine learning algorithms and frameworks are generally not designed to handle the volume and variety of data present in big data. This has paved for innovation in the area of big data analytics leading to the development of efficient algorithms and techniques. Furthermore, Database searching operations have been revolutionized with the advent of quantum algorithms such as the Grover's algorithm. This algorithm offers quadratic time speed up over the best classical algorithm (linear search) while searching unordered databases and lists. In highly integrated environments such as intelligent cities, the efficient storage, analysis and use of data is extremely important, as well as its security and privacy.

Quantum Annealing

Although quantum computing has been an area of active research for decades, it has been gaining rapid popularity lately. In addition to quantum computers with a few qubits already built and used to conduct experiments, quantum algorithms are being explored as potential solutions for real-life problems. Some of these algorithms are capable of functioning on hybrid computers that consist of both classical and quantum computing components. This has paved way for a wide range of possibilities with quantum computing.

Quantum annealing is a technique that is used to solve complex optimization problems by finding global minima of an objective function. Some applications of quantum annealing include de-congestion of air spaces and management of aircraft trajectories to optimize Air traffic Control (ATC) (Stollenwerk, et al., 2019), designing highly optimized and efficient logistical networks (Yongcheng, et al., 2019), solving routing problem of capacity constrained vehicles (Feld, 2019) and traffic optimization (Neukart, 2017). D-Wave Systems have designed Quantum Processing Units (QPUs) that are designed to facilitate quantum annealing. These systems are efficient in solving Quadratic Unconstrained Binary Optimization (QUBO) problems which are often encountered while dealing with traffic flows.

Growing traffic problems pose one of the biggest challenges of modern Smart cities. Almost every city has taken initiatives to try and manage urban traffic flow. Traffic congestions could lead to a variety of problems including air pollution, noise pollution, fuel expenses, economic losses and frustration for the commuters. Prominent traffic problems could cripple everyday life in smart cities and can severely hamper development. City administrators and government bodies have taken initiatives to address the issue of traffic congestion in cities. From the construction of flyovers, highways and additional infrastructure to cater to the growing number of vehicles on road, the efficient management of traffic flow has also been a key area of innovation and development. Although the use of IoT sensors and advanced machine learning solutions have been extensively researched, classical computers struggle to cope with optimizing solutions on the scale of an entire city.

Neukart, et al. (2017) suggest a solution to traffic flow regulation using the D-Wave system. The authors propose a hybrid classical-quantum system that can solve the traffic flow regulation problem by minimizing the time taken for a given set of vehicles to travel between their source and destination. Their experiment and case study on the city of Beijing could help to address similar problems in European cities. The city plan and layout are similar, due to their historic significance as cultural and administrative centers. With cities such as London, Amsterdam and Vienna having roads and streets that were not designed to handle the volume of vehicles seen in road today, QUBO based QPUs can optimize the traffic flow and suggest alternate routes for vehicles to avoid traffic congestions in busy roads and intersections.

Nuclear Energy

Sustainable development has been one of the prime focuses of smart cities, especially the ones in Europe. Cities consume a disproportionately large amount of electricity today, compared to the past. This Figure is set to increase as cities get smarter and more devices and sensors are connected to the grid. Given such a scenario, it is imperative that intelligent cities acquire their power from clean sources of energy to strive towards sustainable development and to adhere to the strict climate change protocols such as the one adopted by the European Union. Although solar energy and wind energy have gained popularity in the form of small scale implementations such as solar street lights, wind farms, solar power plants, solar panels in smart homes, they do not scale to the dynamic energy demands of an entire city. They often tend to be complicated and unreliable due to their low capacity factor. Capacity factor of Solar PV plants are around 10% to 20% depending on the region while that of wind farms in the United Kingdom are around 35% (Digest of UK Energy Statistics, 2019).

Nuclear power plants have some of the highest capacity factors among commercially viable power sources, at around 90%, depending on plant maintenance and re-fueling schedules (Nuclear Energy by the Numbers, 2019). Moreover, the land area required for power generation is one of the least, among all the power sources. This is especially important in the cities of the future where space will be a luxury.

Furthermore, due to the small area needed and safety by design, Thorium based reactors could be situated well in the heart of cities. This can considerably reduce the extensive infrastructure that is often needed to distribute power throughout the grid from remote power stations. The loss due to transmission is considerably reduced, thereby making the entire power cycle more efficient. Fast breeder Thorium reactors are promising solutions to address the energy needs of intelligent cities due to relative abundance of thorium compared to plutonium and uranium for fuel, lesser amount of radioactive waste generated, half-life of the generated waste is lower and the by-products of Thorium reactors cannot be used to produce nuclear weapons. Moreover, mining thorium is easier and safer as it is often mined from open-pit mines unlike uranium which is extracted from underground mines.

CONCLUSION

Over the past decade, Intelligent Cities, also known as Smart Cities, have undergone rapid development. The next generation of intelligent cities will be characterized, not by their technological breakthroughs and infrastructure projects but their ability to address the dynamic and specific needs of its citizens. This calls for innovation, diversity and inclusion, collaboration among different sectors and the active involvement of citizens in policy making. Rather than following specific frameworks and models, the next generation of intelligent cities could be an ensemble of multiple such concepts.

European cities such as Vienna, London, Amsterdam and Stockholm, being some of the prime candidates for the next generation of intelligent cities are examined in detail in this chapter. Moreover, although each city has its own specific demographic, geographic and cultural needs and challenges, factors such as climate change, sustainable development, growing urban population, emerging technologies, digital privacy, cyber security and AI Automation could have a direct impact on the growth of intelligent cities. Although the next generation of cities will involve more citizen participation and decision making in the process, technological advancements will continue to be one of the prime drivers in the realization and implementation of intelligent city projects. Technologies such as IoT, IoV, Computer Vision and Big Data are being widely used in intelligent cities while autonomous vehicles are being swiftly introduced into public roads. AR, VR, quantum computing, and fast breeder nuclear reactors are other emerging technologies that could play an important role in the next generation of intelligent cities.

REFERENCES

Alsouda, Pllana, & Kurti. (2018). *A Machine Learning Driven IoT Solution for Noise Classification in Smart Cities*. Academic Press.

Campbell, P. (2018). *Trucks Headed for a Driverless Future.* https://www.ft.com/content/7686ea3e-e0dd-11e7-a0d4-0944c5f49e46

Chamoso, P., González-Briones, A., Rodríguez, S., & Corchado, J. M. (2018). Tendencies of Technologies and Platforms in Smart Cities: A State-of-the-Art Review. Wireless Communications and Mobile Computing.

Chu, J. (2019). Breaching a "carbon threshold" could lead to mass extinction. *MIT News*. http://news.mit.edu/2019/carbon-threshold-mass-extinction-0708

Cohen, B. (2015). *The 3 Generations of Smart Cities*. https://www.fastcompany.com/3047795/the-3-generations-of-smart-cities

Consultancy.eu. (2019). *Europe best prepared for autonomous driving, Netherlands leads*. https://www.consultancy.eu/news/2609/europe-best-prepared-for-autonomous-driving-netherlands-leads

Digest of UK Energy Statistics (DUKES): Electricity. (2019). *DUKES chapter 5: statistics on electricity from generation through to sales*. Department for Business, Energy and Industrial Strategy.

Dinakaran, R. (2018). *Deep Learning Based Pedestrian Detection at Distance in Smart Cities*. doi:10.1007/978-3-030-29513-4_43

Ding, Y., Chen, X., Lamata, L., Solano, E., & Sanz, M. (2019). *Logistic Network Design with a D-Wave Quantum Annealer*. arXiv:1906.10074

Feld. (2019). A Hybrid Solution Method for the Capacitated Vehicle Routing Problem Using a Quantum Annealer. *Frontiers in ICT, 6*.

Garrido-Marijuan, A., Pargova, Y., & Wilson, C. (2017). The Making of a Smart City: Best Practices across Europe, Belgium 2017. Academic Press.

Giffinger, R. (2007). Smart Cities: Ranking of European medium-sized cities. Centre of Regional Science, Vienna UT.

Ibrahim, Victor, & Nor, Adewole, Ibrar, Abdullah, … Haruna. (2016). The Role of Big Data in Smart City. *International Journal of Information Management, 36*. doi:10.1016/j.ijinfomgt.2016.05.002

Khramtsov, D. (2018). *Augmented reality applications in smart cities*. https://medium.com/coinmonks/augmented-reality-applications-in-smart-cities-8f511fe5895

Lazarina. (2015). *Describing a Smart City, Case Study: City of Stockholm*. Politechnico Di Milano, School of Architecture and Society.

Loet, L., & Mark, D. (2011). The Triple-Helix Model of Smart Cities: A Neo-Evolutionary Perspective. *Journal of Urban Technology*, *18*(2), 53–63. doi:10.1080/10630732.2011.601111

Mehdi, M. (2018). Path Planning in Support of Smart Mobility Applications using Generative Adversarial Networks. *2018 IEEE International Conference on Internet of Things (iThings) and IEEE Green Computing and Communications (GreenCom) and IEEE Cyber, Physical and Social Computing (CPSCom) and IEEE Smart Data (SmartData).*

Mol, Khan, Aalders, & Schouten. (2015). *Report: A spotlight on Smart City Eindhoven*. Venturespring B.V.

Mueller, T. (2017). *Redefining the Smart City Concept: A New Smart City Definition*. https://hub.beesmart.city/strategy/en/towards-a-new-smart-city-definition

Neukart, Compostella, G., Seidel, C., von Dollen, D., Yarkoni, S., & Parney, B. (2017). Traffic Flow Optimizer using a Quantum Annealer. *Frontiers in ICT*, *4*, 29. doi:10.3389/fict.2017.00029

Nuclear Energy by the Numbers. (2019). Nuclear Energy Institute.

Pozdniakova, A. (2018). Smart city strategies "London-Stockholm-Vienna-Kyiv": in search of common ground and best practices. *Acta Innovations*. doi:10.32933/ActaInnovations.27.4

Smart Cities Team. (2018). *Smart City 3.0: Building Compelling Smart City Business Cases*. https://www.arcweb.com/blog/smart-city-30-building-compelling-smart-city-business-cases

Smith, L. (2017). *Amsterdam Smart City: A World Leader in Smart City Development*. https://hub.beesmart.city/city-portraits/smart-city-portrait-amsterdam

Somayya, M., & Ramaswamy, R. (2016), Amsterdam Smart City (ASC): fishing village to sustainable city. *Proceedings of the 11 International Conference on Urban Regeneration and Sustainability*, 831-*842*. 10.2495/SC160681

Stollenwerk, T. (2019). Quantum Annealing Applied to De-Conflicting Optimal Trajectories for Air Traffic Management. [*IEEE Transactions on Intelligent Transportation Systems*, 1–13.

Street-Savvy - Meeting the biggest challenges starts with the city. (2011). *Scientific American*, *305*(3), 27–29.

Sweden. (2018). *Energy Use in Sweden*. https://sweden.se/society/energy-use-in-sweden/

The Future of Smart. (2016). Harnessing digital innovation to make London the best city in the world. Greater London Authority.

Together, S. L. (2018). The Mayor's roadmap to transform London into the smartest city in the world. Greater London Authority.

Trendsetter. (2006). *Evaluation Report- Trendsetter*. Trendsetter Report No 2005:16, Trendsetter Deliverable No 4.7, Jun 2006.

Vahlkamp, S. (2016). *de-klimaatstraat-duurzaam-ondernemen-en-wonen*. http://buurtwinkels.amsterdammuseum.nl/783/nl/de-klimaatstraat-duurzaam-ondernemen-en-wonen

van den Bosch, H. (2017). *Smart Beyond Technology Push*. https://smartcityhub.com/technology-innnovation/smart-beyond-technology-push/

Wein, C. (2019). *10 Millionen Fahrten bei Citybike Wien!* https://www.citybikewien.at/en/about-citybike/press/596-neuer-meilenstein-bei-citybike-wien-erreicht-2

Wharton. (2019). *What's Fueling the Smart City Backlash?* https://knowledge.wharton.upenn.edu/article/whats-behind-backlash-smart-cities/

ADDITIONAL READING

Berger, R. (2019). *Report: The Smart City Breakaway: How a small group of leading digital cities is outpacing the rest*. Roland Berger.

Bhalavat, K. (2019), Smart City projects and technologies to keep an eye on, https://pluto-men.com/insights/smart-city-projects-and-technologies-to-keep-an-eye-on/,Pluto-Men

KEY TERMS AND DEFINITIONS

Anthropic Climate Change: It refers to the impact of human activities on the Earth's climatic conditions. Also termed as Global warming, it alludes to the increase in global surface temperature and melting glaciers due to emission of greenhouse gases. The frequency and severity of extreme weather events such as super cyclones, floods, heat waves and droughts are on rise around the world, as a direct consequence of man-made climate change.

Autonomous Vehicles (AV): Autonomous vehicles or self-driving vehicles or driverless vehicles are vehicles that operate without the need of a human driver. These vehicles are driven by advanced inter-connected systems of cameras, sensors, instruments and integrated circuits. They are powered by advanced technologies such as artificial intelligence, machine learning and deep neural networks.

Breeder Nuclear Reactor: These are a class of reactors that generate more radioactive fissile material than what it needs to use. It works by irradiating a fissile material such as Thorium-232 that is loaded along with the fissile fuel and thereby creating more neutrons.

Capacity Factor: This is defined as the ratio of the electrical energy output over a given period of time to the maximum possible electrical energy output over the same period.

Convolutional Neural Networks (CNN): They are a class of deep neural networks that are generally used to analyze image data. They use convolution instead of simple matrix multiplication in a few layers of the network. They have shared weights architecture and have translation invariant characteristics.

Generative Adversarial Network (GAN): It refers to a type of neural network that consists of a generative and a discriminative network that contest with each other especially in a game scenario. They are used to generate new data that are statistically similar to the training data.

Industrial Internet of Things (IIoT): It refers to an interconnected network of sensors, IoT devices, IoT gateways and instruments in the manufacturing and industrial sector. This connectivity enables the digitalization of the industry by facilitating data collection, data storage and analysis leading to improvement in efficiency and productivity.

PV Cell: This a Photo Voltaic cell, also called a solar cell, that generates electricity from the energy of light using the photovoltaic effect. Series of PV cells are combined together to form solar panels.

Quadratic Unconstrained Binary Optimization (QUBO): This is a pattern matching technique that is used in machine learning, especially with quantum annealing. It can be formulated as minimizing a quadratic polynomial over binary variables.

Quantum Annealing: Quantum annealing is used to optimize discrete binary optimization problems by finding the global minima of a given function. It is useful in solving problems where there are large numbers of local minima present. D-Wave Systems introduced the first quantum annealer in 2011, named D-Wave One.

Single Shot MultiBox Detector (SSD): It is a method used to detect objects in images using a deep neural network. It is especially powerful when used to detect multiple objects in the same image, providing excellent speed and accuracy.

Chapter 5
A Smart City for the Blind: Marburg as a Case Study

Dago Schelin
Philipps University of Marburg, Germany

Péricles Varella Gomes
UniBrasil University Center, Brazil

Verônica Isabela Quandt
https://orcid.org/0000-0003-0006-9852
Universidade Positivo, Brazil

ABSTRACT

In this chapter, the authors present the German city of Marburg as an alternative model for the conception of a smart city. Marburg's historical relation with its visually impaired citizens has shaped not only its infrastructure but also its human framework. Generally, smart cities are equated with world-class major metropolitan areas, with international airports, use of high band internet, internet of things, and other IT infrastructures. However, Marburg might be considered a smart city according to other criteria. This case study articulates the uniqueness of what Marburg has been able to achieve using a diverse approach of cultural acceptance of the blind, becoming a world reference for other small cities. The authors suggest that Marburg can serve as a model for other cities. This hypothesis was reached through critical investigations into concepts of smartness and disability, intersected with the insights obtained in a focus group interview.

INTRODUCTION: MARBURG - CITY OF THE BLIND

There are several reasons for someone to visit the German city of Marburg. Being a university town, many students and academics come here in order to attend conferences at the Philips University of Marburg[1]. Many visitors discover the city's touristic appeal as Marburg bears several noteworthy historical sites[2]. But, although the early gothic architecture from the Elisabeth Church is hard to ignore, perhaps

DOI: 10.4018/978-1-7998-5062-5.ch005

it is even more unlikely that one will overlook the number of blind people walking on the sidewalks, crossing a street, or waiting for a bus. Some use a white cane, some follow a guide dog, most are busy with their smartphones. For many years, Marburg has been known for being the capital city of the blind[3]. Newspapers often like to refer to Marburg as the *Mecca for the blind* (Coordes, 2016b; Fittkau, 2017). Currently, one-third of all university-level legally blind students in Germany attend Philipps University of Marburg (Coordes, 2016a, p. 5). Nowhere else do blind and visually impaired students have so many subject-areas to choose from in an environment which is also accommodating to their specific needs. But, how did this come to be?

During World War I, Germany had to provide support for soldiers who were injured in the battles. Specific cases were assigned to specialized hospital units. After the war, about three thousand soldiers had undergone medical care at the Eye Clinic of Philipps University of Marburg. Most had lost or damaged their sight as a result of mines, hand grenades, splintering projectiles or poison gas. At the time, Professor Alfred Bielschowsky (1871-1940), director of the Eye Clinic, realized that medicine alone would not be enough to readjust these young men into society. Among the returnees were also many high school and university students and academics for whom Bielschowsky organized braille courses, took care of their accommodation, and provided general every day resources. He appointed a student, Carl Strehl (1886-1971), who himself had lost much of his sight, to start teaching the war-blind. To enable further education, the Association of Blind Academics[4] and the German Institute for the Blind[5] (also known as Blista) were founded in 1916. Bielschowsky became honorary director of Blista and Strehl took over its management. From 1927, Strehl worked as Blista's director, a post he carried out for almost 40 years.

Carl-Strehl-Schule, Germany's first Gymnasium[6] for the blind and visually impaired, is still the core of Blista. There, around 280 blind and visually impaired young people attend classes and prepare for professional and academic life. As part of the institution, the youth live in around 40 assisted shared residences distributed throughout the city. Part of the essentials of the program also include activities such as taking the bus, visiting friends, eating without spilling, cooking, washing clothes, vacuuming, and cleaning.

In the class, many lessons are learned through (literally) hands-on explorations. Students who have no sight perception at all learn through a combination of the senses, especially those of touch and hearing. There are models to understand earthquakes, human-formed puppets that can be taken apart, and molecule structures that can be assembled and disassembled. Electricity and colors are expressed through tones, while nitric oxide is recognized through sniffing, etc. (Coordes, 2016b).

Some Numbers and Statistics

In 2015, the population of Marburg was 72,453[7]. According to the Participation Report on the Life Situation of People with Disabilities in the City of Marburg (*Teilhabebericht zu den Lebenslagen von Menschen mit Behinderungen in der Universitätsstadt Marburg*, 2015, pp. 55, 212), there were 12,940 people with disabilities living in the city. Of those, 397 were deemed blind. By adding the visually impaired (legally blind), the number goes up to 637.

Philipps University of Marburg is considered the university with by far the largest number of severely disabled students in Germany, with about 150 blind or severely visually impaired students and at least 40 students who are wheelchair bound. An additional unique feature of Philipps University of Marburg is that it allows disabled students to study almost any subject they wish (for example, as of 2015, 55 courses were attended by blind and low-vision students). At other German universities, special courses for the blind are usually limited to only a few courses (*Teilhabebericht zu den Lebenslagen von Menschen mit Behinderungen in der Universitätsstadt Marburg*, 2015, p. 67).

Currently around 45,000 blind and 500,000 visually impaired people live in Germany. Even if one does not suspect it in Marburg because of the high number of younger people with visual impairment, most of the significantly visually impaired people in Germany (visual acuity 0.3 = 30%) is older than 60 years of age. About 20,000 people become visually impaired every year. Of the 45,000 blind people, the youth (0 to 30 years of age) make up 13%, the middle-aged (31 to 60 years old) 17%, and senior citizens (60 years of age or older) about 70% (*Teilhabebericht zu den Lebenslagen von Menschen mit Behinderungen in der Universitätsstadt Marburg*, 2015, p. 212). These numbers illustrate that the probability of losing all or part of one's eyesight increases exponentially as one gets older. Such findings will be important to make a case for disability being a potential defining term for most human beings.

These numbers become even more relevant in a world whose older population is growing dramatically (He et al., 2016). As the authors intend to discuss, it might seem less obvious to the layperson that disability is a condition that eventually catches up to most human beings sooner or later. In aligning this fact to the concept of Smart City to actual non-abstract citizens, the issue of special needs shall prove to be directly and indirectly relevant to everyone.

SMART CITY: A BROADER CONCEPT

To find a satisfactory definition of a Smart City, also referred to as Intelligent City, is a challenging endeavor given the ever renewing terms and rules in the context of urbanism and society. Common sense might understand Smart City as a potentially positive idea to be implemented in a world in which more than half of humanity lives in urban areas (United Nations, 2019, p. 56). The mere adjective has an evident positive ring to it. Who would want to counter-argue the benefits of smartness? From a marketing perspective, "smart" seems more user-friendly than the elitist "intelligent" (Nam & Pardo, 2011).

City seems to have a mixture of both positive and negative connotations. Some might deem it the only way forward, equating it with progress. Others may see all kind of misfortune brought on by the phenomenon of overcrowding populations. According to United Nations (2019, p. 10) statistics, while currently 55% of the world population lives in urban areas, the expectation is that by 2050, this number will reach around 68%.

In order to get more specific Smart City definitions, Albino et al. (2015) present a useful summary drawn from studies of the last twenty years. By narrowing down key dimensions valued by each researcher, a distilled importance is given to concepts such as information technology education, IT infrastructure, IT economy, and quality of life (Mahizhnan, 1999); economy, mobility, environment, and people governance (Giffinger et al., 2007); technology, economic development, job growth, and increased quality of life (Eger, 2009). Terms begin to overlap when Thuzar (2011) focuses a concept of smartness of a city on the quality of life, sustainable economic development, management of natural resources through participatory policies, convergence of economic, social, and environmental goals. Kourtit & Nijkamp's (2012) key concepts are measured in human capital, infrastructural capital, social capital, and entrepreneurial capital.

According to Kresin (2016), current smart city concepts urgently need an upgrade. The focus ought to be on citizens rather than only on the chain of technological innovation. While citizens find themselves impacted with ever more complex systems integrated in their life in smart cities, they have less and less understanding of the actual output. Kresin (2016) thus defends a participative of smartness by giving an example of how, in 2014, Amsterdam conducted an experiment with its citizens. They distributed hundreds of Smart Citizen Kits consisting of small sensors that yielded data on exposure to ozone particles. The result was wide participation and enthusiasm (Kresin, 2016, p. 183). Nevertheless, data-collection technology is still a fresh and delicate matter, not the focus of this chapter.

If one were to count the words mostly used in the myriad definitions presented by Albino et al. (2015), one would find that there is much emphasis given firstly to the term *information*, then to *infrastructure*, and thirdly to both *technology* and *sustainability*[8]. Each of these terms is loaded with ideological meaning which, in turn, carries various possibilities of political ideas (and agendas), economic interests, doctrines, myths, and symbols. The way in which the terminology is actually lived out as a Smart City will depend on the ideological regime. Most smart cities analyzed by the above-mentioned researchers stem from capitalist contexts in which a competitive market with hopes for profit is a matter of principle (Fuchs, 2019; Shapiro, 2006). To the system itself, it might seem counterintuitive to value sustainability unless there is a surplus of capital guaranteed at the end. And if even human beings are seen as capital, as expressed by Kourtit & Nijkamp (2012), it is crucial to raise the issue of disability. A look into smart cities in the context of disability might point to blind spots intrinsic to the prevailing worldview. A critical eye on the technological determinism of the Smart City is, therefore, of crucial importance.

MARBURG AND SMART DISABILITY

Aside from characteristic Intelligent City implementations such as an interconnected solar powered grid in commercial buildings and private homes, a robust bike sharing system along with good accessibility common to a standard German city[9] (Geoportal, 2019), an integrated intelligent camera network aimed at citizen safety[10] (*LiSA" geht im Jägertunnel in den Testbetrieb*, 2018), and free Wi-Fi citywide, Marburg has a myriad of accessibility systems for the blind which have gradually developed throughout the last 100 years.

In Marburg (and indeed almost anywhere), the blind and visually impaired would likely be the first to declare their appreciation for living in a largely connected digital world. Smartphones actually live up to their adjective, especially what concerns their audio and haptic interface. While walking through town, it is not uncommon to hear lose phrases coming out of someone's phone (usually several times an audiobook's regular speed rate). Tapping and listening are a major part of a blind person's engagement with his/her smart phone apps. It is an acquired skill in which a user can choose at what speed s/he would like to have the text read back to them[11]. Reading an email becomes listening to it and can be done in a matter of seconds thanks to the revolutionary voice-over technology. One can choose out of a number of voice types. Also, now smartphones receive voice commands à la Alexa for functions such as deleting emails. There are more than a hundred apps that make everyday life easier for people with visual impairments. For most young people with no or low vision, the smartphone is, allegorically, a digital Swiss army knife (Sticht, 2017).

While nowadays many blind and visually impaired can use smart devices in order to accomplish daily tasks, the city of Marburg offers several other conveniences, some of which are not unique only to Marburg. There are accessibility points throughout the city's traffic structure. A paving with nubs crosswise to the walking direction is located on the sidewalk at the end of which the traffic light mast can be found. At many crossroads, there is a pilot tone consisting of a faint clattering sound to locate the traffic light mast, which becomes louder with increasing traffic noise. The apparatus also emits acoustic signals during the green phase of pedestrian traffic lights in order for the user to know when the lights are green. A vibrating metal knob with a tactile arrow on the traffic light mast also facilitates maneuvering in traffic. During the green phase of the pedestrian light, the upper part of the metal knob vibrates. The tactile arrow is used to ascertain which direction the street is to be crossed. In many public places throughout the city there are guide stripes on sidewalks and walking areas on which white canes can easily find directional orientation. There are also at least 13 buildings with tactile floor plans for the blind/visually impaired to perceive through touch what departments are located in which part of the floors.

In addition to these structural systems, Marburg offers services such as tandem bicycle rentals, general train assistance, gastronomy with Braille menus, and "talking" ATMs. *Marburg from another point of view*[12] is the motto of the city's specialized city tours for the blind/visually impaired. Moreover, more informal arrangements and facilities-services have emerged in the city: cinema with audio description, nature walks, special sports, and dark dinners[13], among others.

Beyond the technology-centered standard – although far from being anti-technology – Marburg appears to specialize in an alternative mode of smartness, not necessarily unplugged (Calzada & Cobo, 2015) but certainly human-oriented. The idea of a city thought for the differently-abled involves a natural progress of a city's history and social context along with creative civic and institutional integration, all conditions that Marburg fulfills.

For other-needs policies to be implemented and practiced, such as those seen in case of this Hessian city, a paradigm shift should be envisioned. The valued characteristics of the typical Intelligent City – IT-based supra-mechanistic structures intended to yield economic turnover – ought to function contingently to a differentiated model, one that focuses on disability. This shift might entail a counterintuitive adjustment of preconceptions. In one sense, the shift is simply a natural continuation of the city's history. But at the same time, it can also be an opportunity to challenge an old ontological problem, namely that of ableism.

Disability: A Challenging Conception

The term disability encompasses several degrees of signification. There is a relatively older, though still very much prevalent, conception of disability based on so called ableist frameworks (Golledge, 1993). Ableism stems from the morpheme present in words such as "ability" and "able" combined with the Greek suffix "ism" that designates devotion or adherence. It refers to a prejudiced view of individual and societal values based on physical and mental capabilities. Also referred to as handicapism, ableism has been criticized for generating forms of social oppression and political discrimination (Imrie, 1996, p. 398). In its essence, an ableist society considers persons with disabilities as inferior to the non-disabled. Beyond a mere medical definition, disability has thus come to designate a minority grouping (Hahn, 1994).

In historical terms, ableist practices can be perceived in apparently good-intentioned folklore all the way to extremist acts. The former is manifest in presuppositions of cause and consequence explained by myths and beliefs[14], the latter in para-sociomedical justifications of normalcy[15]. Looking at the situation today, it seems that even though there has been a shift in practices[16] with changes in policies, there is still an ableist regime grounded in what Campbell (2001) calls disability ontology, in this case a clearly negative ontology with sub-texts always running in the background. According to Campbell, *the ongoing stability of ableism, a diffuse network of thought, depends upon the capacity of that network to 'shut away', to exteriorize and unthink disability and its resemblance to the essential (ableist) human self*

(Campbell, 2001, p. 44). Even if such an understanding brings pragmatic results in cases of laws favoring the disabled, it does not solve the ontological problem.

The field of ontology[17] might seem foreign to mainstream discussions on smart cities, since it is more commonly used in the confines of philosophy. Perhaps distanced from most techno-quantitative research areas, ontology draws from a branch in philosophy that deals with more abstract notions such as metaphysics. It is concerned with the nature of things, of existence, of being itself and its relation to the world. An example of an ontological question might start with: in the face of the ability-disability binary, is it possible for one to be? Be the answer yes, no, or somewhere in-between, it is a starting point for any debate about one's place in the world. This kind of question follows Plato's rationale (Fogelin, 1971, p. 378) of what constitutes reality all the way to more recent developments, especially with Martin Heidegger's seminal *Being and Time* (Heidegger, 1962) in which he poses the problem from the onset:

Do we in our time have an answer to the question of what we really mean by the word "being"? Not at all. So it is fitting that we should raise anew the question of the meaning of Being. But are we nowadays even perplexed at our inability to understand the expression "Being"? Not at all. So first of all we must reawaken an understanding for the meaning of this question (Heidegger, 1962, p. 19).

There is something intrinsic to disability that sets it apart from other terms such as impairment or illness. This something is its ontology. It implies a constant state of being. Paraphrasing Carol Thomas, disability has less to do with what one can or cannot do than with what one can or cannot be (Thomas, 1999, p. 60). The dualistic Cartesian manner of sorting abled and disabled should be questioned at the core of being. It is far from the scope of this contribution to critique – indeed to claim to know – all of the intricacies of the ableist position. But it seems reasonable to discuss Smart City conceptions while still acknowledging the pervasiveness of ableism.

The authors' point in bringing up the ontological problem of disability is to show that difficulties run much deeper than (just) discrimination towards those with physical/mental impairment. The difficulties start well before, at the level of being. However, Hughes (1999, p. 155) reminds us that this manner of seeing disability is *constructed – not discovered – in the non-disabled gaze*. Hopefully the reader will notice that there are emancipatory examples for paradigm shifts within the way of being of a city in which disability can be de-constructed and re-constructed.

The discussion around disabilities (such as the case of blindness in the German context) has come a long way in the past hundred years. Policies, political correctness, technology, all have definitely brought new possibilities for those deemed persons with disabilities. Blind citizens of the city of Marburg, for example, cross most any street with ease since they can rely on reliable technology. But are the confines of being and doing even addressed (let alone resolved) through technology? A human framework might tell the difference between working toward bridging the gap of accessibility using technology (Illich, 1985) and blind faith in computerization, automation, electronics, mechanization, programing, cybernetics, and robotics.

Countering ableist arguments, Torn (2001, p. 27) claims that it is frailty that is the common basis for humanity since sooner or later every human being suffers limitations due to circumstances, accident, or old age. Although, he partly criticizes this interpretation of disability for being superfluous, Hughes (2007, p. 677) paraphrases Torn's argument stating that *if it is frailty that confers humanity, then non-disabled people who fail to recognize their vulnerability fail to recognize their own humanity.*

Newer social media practices seem to prefer the filtering out of imperfection, especially that of physical representation. The younger smartphone generation (Gafni & Geri, 2013) chuckles at the incapacity of more aged users who don't know that taking a *selfie* from a higher angle hides the enlarged jowl. Though online self-image is not the focus of this study, it is worth mentioning that certain visual expectations also run in the ontological background. Examples of the economies of such ontologies are recent films and series that portray blind heroes. Television series such as *In the Dark* (Kingsbury, 2019) and *Daredevil* (Goddard, 2015-2018), both starring blind protagonists (whose actors are not actually blind in real life), invest heavily on physical beauty of their main characters – in the case of *Daredevil*, in which the hero is a man, he is surrounded by attractive women. In the case of *In the Dark*, the heroine is a very attractive young woman (Perry Mattfeld). From a feminist theoretical point of view, the voyeuristic male gaze is present even in spite of disability (Mulvey, 1975). One might even argue that the element of bodily beauty (ability) is necessary to compensate for the lack of sight (disability). Such compensation is present in many a narrative, be it in films or novels featuring disabled characters. Either they embody the role of the supercrip or of the pity-worthy. The inclusion of imperfection into representation should be at the interest of the average human being, after all, mortality, disease, loss, and hardship are presupposed (Aries, 1974). The challenge might be to transpose such a notion beyond oneself, onto the reaches of a city.

To recognize common vulnerability seems to be a positive first step in challenging ableist positions. This recognition has taken form and been applied to several subareas of city dynamics such as pro-environmental and healthful commuting alternatives in which, for instance, the "weaker" party in traffic receives extra protection from the law in countries such as Germany and the Netherlands (Pucher & Buehler, 2008, p. 512).

In the context of blindness and visual impairment, it should also be added that ableism is acted out through the eyes of the non-disabled. It follows several schools of thought but stems from the same relation of power. A feminist approach would call it an objectifying gaze (Mulvey, 2013, p. 9) evident in the much debated male gaze. Foucault (1973) might call it a medical or surveying gaze with the eyes being instruments of power. Ivan Illich's critical-historical point of view might draw upon notions of an institutionalizing gaze (Illich, 1995a, 1995b). Even film studies can offer several manners of exerting an ableist look of pity (Schelin, 2019, p. 69) which sometimes translates into morbid curiosity. The subtraction of the one who looks by the one who is looked at is the far-encompassing ableist gaze.

As will be argued, disability bears varied layers and issues relevant to urbanism and geo-social planning. Any subsequent analysis of disability and city life should beware of ableist dominance. Its politics and practices are all-pervasive: in religion, medicine, society, all the way to the idiosyncratic view towards disability that every individual develops.

RESEARCH METHODOLOGY: GROUP INTERVIEWS

In preparation for a survey, with the over 200 blind and visually impaired students of Carl-Strehl-Schule[18], the authors met with a selected group of blind citizens of Marburg in a group interview setting[19]. Since it might have been too broad an attempt to make an investigation on behalf of those with disabilities, let alone the whole population of Marburg, the main point of the talks was to find ways to assess the human framework of the city in relation to its blind population. The chosen method, the group interview, also referred to as focus group interview (Frey & Fontana, 1991), served as a testing ground for several assumptions by local newspaper articles about Marburg and its blind/visually impaired residents (Coordes, 2016b; Fittkau, 2017; Kautzleben, 2010; Sticht, 2017).

The group interviews served as a first filter to a pre-formulated questionnaire. The questionnaire had already undergone several phases of preparation and pre-testing with specialists in the area of accessibility for the blind[20]. The group, therefore, helped to clarify, adapt, and even debunk predetermined assumptions about the daily life of people with visual impairment. Mainly because of the lack of scientific information on the subject, it was found necessary to invest time in these initial exploratory questions in order to arrive at a better appreciation of the social context, and to determine subtleties in the research hypotheses.

Several advantages were perceived in opting for a group interview setting. First of all, following Gina Wisker's propositions, the group context provided a manner for acquiring information based on *emotions, feelings, experiences*, as well as for obtaining *insider experience* and *privileged insights* (Wisker, 2010, p. 192). The blind/visually impaired are, in this case, the source of sociological data that point to phenomenological issues not easily found through statistics.

In his work *Symbolic Interactionism*, Herbert Blumer points out that having a group of insightful participants who are acquainted with the social circumstances being investigated *is more valuable many times over than any representative sample* (Blumer, 2009, p. 41). Such is the case with the interviewees in question[21].

Being blind[22] and having lived more than 10 years in Marburg, the four interlocutors had many insights to share, both conceptual as well as experiential. Issues concerning being blind in Marburg would be raised by the interviewer, discussed by the group and resumed again by the interviewer. The discussion usually followed a dialectical dynamic in which an issue would be argued by one interviewee, then counter-argued by another, several times back and forth, until the matter was settled. Such was the case when the group was asked if Marburg was a good city for the blind and visually impaired specifically for being relatively small. Ms. Carvalho tended to agree mentioning that when she lived in Porto Alegre, Brazil, everything was far away. Mr. Flach agreed partly since, in his opinion, it really depends where you live in Marburg. Mr. Ivanic was skeptical whether the size of a city is an important criterion for being good for the blind. For him, if size and distances are an issue, then transportation should be evaluated. He went on to explain how in Marburg the bus drivers are attentive to blind people waiting at the bus stop, usually stopping with the opening door right in front of the waiting commuter and letting him/her know which bus number just stopped.

At major bus stops and stations in Marburg, there is also an audio knob that when pressed informs users about the upcoming busses (number and time of arrival), though this technological aspect of the city's bus system was not mentioned by any of the interviewees. Mr. Flach went on to mention that bus drivers also inform blind passengers, upon entering, which seat near them is vacant.

It would be too self-congratulatory to label Marburg residents as more altruistic than those of any other city. If it is true that people in Marburg are more helpful and more attentive to their fellow blind and visually impaired citizens, it may simply be because of a series of circumstances. This was one of the major issues discussed in the group interview. The rationale follows that, in comparison to other cities, on average there are more blind and visually impaired young people who come to Marburg. They come in order to attend the school (Carl-Strehl-Schule) or the university (Philipps University of Marburg). In a city of about 70 thousand inhabitants in which about 1% are legally blind, it is only a matter of time for one to chance upon the issue. It is also a matter of time duration, namely over 100 years of social presence. The debate distilled the issue into a range of terms. The comparative form of the adjective *behindertenfreundlicher* (more disabled-friendly) was debunked early on. The most recurring term was *Gewohnheit* which in English usually translates to *habit* or *custom*. After further deliberation, a consensus was reached: Marburgers are *used to* the blind community. The meaning of the term aligns with Henri Bergson's notion of habituation being a *somatic* acquisition (Simon, 1986, pp. 46–47). The apparently-passive process of habituation has transformed the physical disposition of the city. Among the facets of *Gewohnheit*, the visually impaired have their own very practical challenge of getting used to infrastructures. According to Nancy Worth, by learning how to move to, from, and within social and geographical spaces through touch, sound, etc, the visually impaired develop an intuitive *bodily knowledge* which gives them a sense of agency (Worth, 2013, p. 576).

Gewohnheit, in the context of the group interview, seems to imply a neutral feeling bordering passive indifference. This negative connotation, however, sheds positive light on the aforementioned problem of ableism. It subverts the norm and disrupts stereotypes, even if modestly. In the following selection of the interview, Mr. Batmaz and Mr. Flach exemplify this issue as follows:

Batmaz: if you are blind you are not a bad person, but you're not a better person either. That is, if you walk five times faster and step on people's feet, then I think it's quite reasonable for someone to say, "Hey, watch out!" Even if you are blind.

Flach: That is actually an important part of inclusion. It is a sign of being taken seriously. (Schelin et al., November 15, 2019)

Implied in the statements above is the idea that Marburg is structured in such a way that the blind can come and go with such ease that they may abuse the advantage. In other words, the more a city is familiar with its minority group, the greater the chances for balance. The participants of the group interview considered such a scenario as a fair compromise. Marburg seems to be an example of such awareness. In their project titled *Smart City: Ranking of European medium-sized Cities* in which Giffinger et al. (2007) delve into extensive data, structure, and statistical analysis to choose, evaluate, and rank European cities in terms of their smart status, several important characteristics are highlighted as being primary for a city to acquire the reputation.

Especially awareness seems important for a smart city as certain potentials can only be mobilised if inhabitants, companies or the administration are aware of the cities' position – knowing the city from the inside but also being aware of the surroundings and the system of cities the city is located in (Giffinger et al., 2007, p. 10).

In sum, the discussion has come back to a dialectical framework. On the one hand, significance is given to being aware of people's needs. On the other hand, there is *Gewohnheit*, being *used to* the blind citizens (being unaware of stigma).

CONCLUSION

In this chapter, the authors have attempted to make a case for Marburg as an alternative model for the conception of a smart city. The city's historical relationship with the blind and visually impaired since World War I has shaped not only its infrastructure but also its human framework. This hypothesis was reached through critical investigations into concepts of smartness of a city and disability intersected with the insights of a focus group interview.

If the ontological problem of disability is difficult to grasp – let alone to solve – the manner in which a difference may be noticed is though urban pragmatisms such as the ones mentioned in this article. Some shifting has happened and has been noticed in the last one hundred years. If the German Nazi regime between World Wars I and II embraced ableism in its uttermost malevolence, time has shown other possibilities in practical and visible ways. The ideal of perfection, especially that of physical and genetic, translated in the Nazi rejection of Modern art of the time for being averse to traditional representations of bodies, has itself been overcome and become tainted. In writing on the aesthetics of disability, Siebers (2005) points out that in contrast to the Nazi ideal:

Disability aesthetics do not embrace genius, bodily integrity, and health. Nor does it support the aversion to disability required by traditional conceptions operate both as a critical framework for questioning aesthetic presuppositions in the history of art and as a value in its own right important to future conceptions of what art is (Siebers, 2005, p. 546).

The concept of disability seems to be miles away from recent terms such as "defective" and "deviant." But the ontology is invisible, detectible only indirectly. An underpinning disinterest toward these ontologies seems to be a force used to *justify discrimination against people whose bodies, minds, desires, and practices differ from the unmarked norm* (Kafer, 2013, p. 17). The concept (and actuality) of a smart city for the blind might serve as testimony to reassure those plighting for or part of populations who have historically been under oppression. What better way to reshape the invisible aspects of an urbane being through a smart kind of city for the blind?

Far from being ideal – then again, closer to reality – Marburg, the capital city of the blind, might serve as a reference for prospective smart cities. At the very least, the issue of disability might come to be one of the criteria for a city's smartness status. Marburg seems to have, at least to some extent, overcome the negative ontology of blindness. The advantages of such a position benefit not only the blind community, but also other minorities who normally transit the structured prejudice of the un-neutral city. Normative expectations are lifted (again, to some extent) in a place where being blind – as well as being among the blind and even having blind friends – is the new normal.

ACKNOWLEDGMENT

The authors would like to thank Andrea Katemann, Kerstin Hühnlein, Jan-Bernd Röllmann, Bahaddin Batmaz, Ulrike Schönhagen, Markus Biber, Mirien Carvalho, Jens Flach, Savo Ivanic, and Siegrun Nagel.

REFERENCES

Albino, V., Berardi, U., & Dangelico, R. (2015). Smart Cities: Definitions, Dimensions, Performance, and Initiatives. *Journal of Urban Technology*, *22*(1), 2015. doi:10.1080/10630732.2014.942092

Aries, P. (1974). *Western attitudes toward death: From the Middle Ages to the present*. Johns Hopkins University Press.

Bauman, Z. (1991). *Modernity and the Holocaust*. Polity Press.

Bevölkerung und Bevölkerungsvorgänge der hessischen Gemeinden am 30. Juni 2019. (2019). [Statistical]. Hessisches Statistisches Landesamt.

Blumer, H. (2009). *Symbolic Interactionism: Perspective and Method*. University of California Press.

Bürgerservice: Blinde und Sehbehinderte. (2019). [Citiy Administration]. Leben in Marburg. https://www.marburg.de/portal/seiten/blinde-und-sehbehinderte-900000287-23001.html

Calzada, I., & Cobo, C. (2015). Unplugging: Deconstructing the Smart City. *Journal of Urban Technology*, *22*(1), 23–43. doi:10.1080/10630732.2014.971535

Campbell, F. A. K. (2001). Inciting Legal Fictions: Disability's Date with Ontology and the Ableist Body of the Law. *Griffith Law Review*, *10*, 42–62.

Coordes, G. (Ed.). (2016a). *Blindenstadt Marburg*. Magistrat der Universitätsstadt Marburg. https://www.marburg.de/downloads/datei/

Coordes, G. (2016b, July 5). Marburg, das "Mekka der Blinden." *Ärzte Zeitung*. https://www.aerztezeitung.de/panorama/article/914262/marburg-mekka-blinden.html

Denscombe, M. (2007). *The good research guide for small-scale social research projects* (3rd ed.). McGraw-Hill.

Eger, J. M. (2009). Smart Growth, Smart Cities, and the Crisis at the Pump: A Worldwide Phenomenon. *I-WAYS-The Journal of E-Government Policy and Regulation*, *32*(1), 47–53. doi:10.3233/IWA-2009-0164

Fittkau, L. (2017, October 19). Hauptstadt der Blinden. In *Länderreport*. Deutschlandfunk Kultur. https://www.deutschlandfunkkultur.de/marburg-hauptstadt-der-blinden.1001.de.html?dram:article_id=398614

Fogelin, R. J. (1971). Three Platonic Analogies. *The Philosophical Review*, *80*(3), 371. doi:10.2307/2184102

Foucault, M. (1973). *The Birth of the Clinic: An Archaeology of Medical Perception*. Pantheon Books.

Frey, J. H., & Fontana, A. (1991). Article. *The Group Interview in Social Research.*, *28*(2), 13.

Fuchs, C. (2019). Beyond Big Data Capitalism, Towards Dialectical Digital Modernity: In C. Fuchs & D. Chandler (Eds.), Digital Objects, Digital Subjects (pp. 43–52). University of Westminster Press.

Gafni, R., & Geri, N. (2013). Generation Y versus Generation X: Differences in Smartphone Adaptation. *Proceedings of the Chais Conference on Instructional Technologies Research*, 6.

Geoportal. (2019). https://www.marburg-biedenkopf.de/dienste_und_leistungen/auto_und_mobilitaet/_Auto-Mobilitaet.php

Giffinger, R., Pichler-Milanović, N., Fertner, C., Kramar, H., Kalasek, R., & Meijers, E. (2007). *Smart Cities: Ranking of European Medium-Sized Cities*. Centre of Regional Science, Vienna University of Technology. www.smart-cities.eu

Goddard, D. (2015). *Daredevil (TV Series)*. Netflix.

Golledge, R. G. (1993). Geography and the disabled: A survey with special reference to vision impaired and blind populations. *Transactions of the Institute of British Geographers*, *18*(1), 63–85. doi:10.2307/623069

Hahn, H. (1994). The minority group model of disability: Implications for medical sociology. *Research in the Sociology of Health Care*, *11*, 3–24. doi:10.4135/9781412939645.n69

Hansen, R., & King, D. (2013). Sterilized by the State: Eugenics, Race, and the Population Scare in Twentieth-Century North America. Cambridge University Press. doi:10.1017/CBO9781139507554

He, W., Goodkind, D., & Kowal, P. (2016). *International Population Reports*. P95/16-1, An aging world: 2015. US Census Bureau. https://www.census.gov/content/dam/Census/library/publications/2016/demo/p95-16-1.pdf

Heidegger, M. (1962). *Being and Time*. Harper & Row.

Hughes, B. (1999). The Constitution of Impairment: Modernity and the aesthetic of oppression. *Disability & Society*, *14*(2), 155–172. doi:10.1080/09687599926244

Hughes, B. (2007). Being disabled: Towards a critical social ontology for disability studies. *Disability & Society*, *22*(7), 673–684. doi:10.1080/09687590701659527

Illich, I. (1985). *Tools for conviviality*. Marion Boyars.

Illich, I. (1995a). Guarding the Eye in the Age of Show. *Anthropology and Aesthetics*, *28*(Autumn), 47–61. doi:10.1086/RESv28n1ms20166929

Illich, I. (1995b). *The Scopic Past and the Ethics of the Gaze: A Plea for the Historical Study of Ocular Perception. Working Papers No. 6*. University Park, PA: Science, Technology, and Society Studies. http://www.davidtinapple.com/illich/1998_scopic_past.PDF

Imrie, R. (1996). Ableist Geographies, Disablist Spaces: Towards a Reconstruction of Golledge's "Geography and the Disabled.". *Transactions of the Institute of British Geographers*, *21*(2), 397. doi:10.2307/622489

Kafer, A. (2013). *Feminist, Queer, Crip*. Indiana University Press.

Kautzleben, V. (2010, October 15). Eine Stadt mit vielen weißen Stöcken. *Oberhessische Presse*. https://www.op-marburg.de/UNIversum/Das-pralle-Leben/Eine-Stadt-mit-vielen-weissen-Stoecken

Kingsbury, C. (2019). *In the Dark*. CBS Television Distribution.

Kourtit, K., & Nijkamp, P. (2012). Smart Cities in the Innovation Age. *Innovation (Abingdon)*, *25*(2), 93–95. doi:10.1080/13511610.2012.660331

Kresin, F. (2016). 22. Smart cities value their smart citizens. In G. Marini (Ed.), *Urban Europe* (pp. 181–186). Amsterdam University Press. doi:10.1515/9789048535811-023

LiSA geht im Jägertunnel in den Testbetrieb. (2018). https://www.marburg.de/portal/meldungen/-lisa-geht-im-jaegertunnel-in-den-testbetrieb-900004262-23001.html

Mahizhnan, A. (1999). Smart cities: The Singapore case. *Cities (London, England)*, *16*(1), 13–18. doi:10.1016/S0264-2751(98)00050-X

Mulvey, L. (1975). Visual Pleasure and Narrative Cinema. *Screen*, *16*(3), 6–18. doi:10.1093creen/16.3.6

Mulvey, L. (2013). *Fetishism and Curiosity: Cinema and the Mind's Eye*. Palgrave Macmillan.

Nam, T., & Pardo, T. A. (2011). Conceptualizing Smart City with Dimensions of Technology, People, and Institutions. *Proceedings of the 12th Annual International Digital Government Research Conference: Digital Government Innovation in Challenging Times*, 282–291. 10.1145/2037556.2037602

Proctor, R. N. (2002). *Racial hygiene: Medicine under the Nazis*. Harvard University Press.

Pucher, J., & Buehler, R. (2008). Making Cycling Irresistible: Lessons from The Netherlands, Denmark and Germany. *Transport Reviews*, *28*(4), 495–528. doi:10.1080/01441640701806612

Schelin, D. (2019). *Vision and Blindness in Film*. Büchner Verlag.

Schelin, D., Batmaz, B., Carvalho, M., Ivanic, S., & Flach, J. (2019, November 15). *Group Interview: The Relation Between the City of Marburg and its Blind/Visually Impaired Citizens* (D. Schelin, Interviewer) [Audio recording]. Academic Press.

Shapiro, J. M. (2006). Smart Cities: Quality of Life, Productivity, and the Growth Effects of Human Capital. *The Review of Economics and Statistics*, *88*(2), 324–335. doi:10.1162/rest.88.2.324

Siebers, T. (2005). Disability Aesthetics. *PMLA*, *120*(2), 542–546.

Simon, Y. R. (1986). *The Definition of Moral Virtue* (V. Kuic, Ed.). Fordham University Press.

Sticht, C. (2017, June 6). Smartphone als Segen für Sehbehinderte. *Frankfurter Rundschau.*

Thomas, C. (1999). *Female Forms: Experiencing and understanding disability*. Open University Press.

Thuzar, M. (2011). Urbanization in Southeast Asia: Developing smart cities for the future? *Regional Outlook*, 96.

Torn, S. (2001). The social model of disability: An outdated ideology? In W. Nicholas, S. N. Barnartt, & B. M. Altman (Eds.), *Exploring Theories and Expanding Methodologies: Where we are and where we need to go* (Vol. 2, pp. 9–28). Emerald Group Publishing Limited. doi:10.1016/S1479-3547(01)80018-X

United Nations. (2019). *World Urbanization Prospects: The 2018 Revision (ST/ESA/SER.A/420)*. Department of Economic and Social Affairs, Population Division. https://population.un.org/wup/Publications/Files/WUP2018-Report.pdf

Wisker, G. (2010). *The postgraduate research handbook: Succeed with your MA, MPhil, EdD and PhD*. Palgrave Macmillan.

Worth, N. (2013). Visual Impairment in the City: Young People's Social Strategies for Independent Mobility. *Urban Studies (Edinburgh, Scotland)*, *50*(3), 574–586. doi:10.1177/0042098012468898

ADDITIONAL READING

Allwinkle, S., & Cruickshank, P. (2011). Creating Smart-er Cities: An Overview. *Journal of Urban Technology*, *18*(2), 1–16. doi:10.1080/10630732.2011.601103

Barnes, C. (2012). *The problem body: Projecting disability on film* (Vol. 27). The Ohio State University Press; doi:10.1080/09687599.2012.695521

Barrionuevo, J. M., Berrone, P., & Ricart, J. E. (2012). Smart Cities, Sustainable Progress. *IESE Insight*, *14*(14), 50–57. doi:10.15581/002.ART-2152

Benin, D., & Cartwright, L. (2006). Shame, Empathy and Looking Practices: Lessons from a Disability Studies Classroom. *Journal of Visual Culture*, *5*(2), 155–171. doi:10.1177/1470412906068871

Blanton, S., & Reiss, B. (2019). The Shifting, Elusive, and Sometimes Contradictory Language of Disability—Insights from Keywords for Disability Studies. *AJOB Neuroscience*, *10*(2), 88–90. doi:10.1080/21507740.2019.1618415

Blindengeld. (2018, July). *Deutscher Blinden- und Sehbehindertenverband e. V.* https://www.dbsv.org/blindengeld.html

Bommersheim, E., & Chandra, O. (2015). Innovation through inclusion: A pilot project. *Local Economy*, *30*(5), 584–590. doi:10.1177/0269094215589562

Brennan, T., & Jay, M. (Eds.). (1996). *Vision in Context: Historical and Contemporary Perspectives on Sight*. Routledge.

Cameron, A. (1995). Dissimulations: Illusions of Interactivity. *Millenium Film Journal*, *28*, 33–47.

Campbell, F. A. K. (2008). *Refusing Able(ness): A Preliminary Conversation about Ableism*. 4.

Clear, M. (1999). The "Normal" and the Monstrous in Disability Research. *Disability & Society*, *14*(4), 435–448. doi:10.1080/09687599926055

Deakin, M., & Al Waer, H. (2011). From Intelligent to Smart Cities. *Intelligent Buildings International*, *3*(3), 140–152. doi:10.1080/17508975.2011.586671

Debord, G. (1983). *Society of the Spectacle* (Vol. 4). Rebel Press; doi:10.1080/10509200500536041

Denery, D. G. I. I. (2005). *Seeing and Being, Seen in the Later Medieval World, Optics, Theology and Religious Life*. Cambridge University Press. doi:10.1017/CBO9780511496462

Gehl, J. (2013). *Cities for People*. Island Press.

Hayhoe, S. (2016). The epistemological model of disability, and its role in understanding passive exclusion in eighteenth and nineteenth century Protestant educational asylums in the USA and Britain. *International Journal of Christianity & Education*, *20*(1), 49–66. doi:10.1177/2056997115620621

Heron, J. (1996). Co-operative Inquiry: Research into the Human Condition. *Sage (Atlanta, Ga.).*

Hine, J., Swan, D., Scott, J., Binnie, D., & Sharp, J. (2000). Using Technology to Overcome the Tyranny of Space: Information Provision and Wayfinding. *Urban Studies (Edinburgh, Scotland)*, *37*(10), 1757–1770. doi:10.1080/00420980020080381

Hollands, R. G. (2008). Will the real smart city please stand up?: Intelligent, progressive or entrepreneurial? *City*, *12*(3), 303–320. doi:10.1080/13604810802479126

Kleege, G. (2005). Blindness and Visual Culture: An Eyewitness Account. *Journal of Visual Culture*, *4*(2), 179–190. doi:10.1177/1470412905054672

Kudlick, C. (2005). The Blind Man's Harley: White Canes and Gender Identity in America. *Signs (Chicago, Ill.)*, *30*(2), 1589–1606. doi:10.1086/423351

Kudlick, C. J. (2003). Disability History: Why We Need Another "Other.". *The American Historical Review*, *108*(3), 763–793. doi:10.1086/529597 PMID:12964564

Lazar, J., & Stein, M. A. (Eds.). (2017). *Disability, human rights, and information technology*. University of Pennsylvania Press. doi:10.9783/9780812294095

Lefebvre, H. (1991). The Production of Space (D. Nicholson-Smith, Ed.; Vol. 9). Blackwell. doi:10.1027/1618-3169/a000129

Levin, D. M. (1993). *Modernity and the Hegemony of Vision*. University of California Press.

Linders, D. (2012). From e-government to we-government: Defining a typology for citizen coproduction in the age of social media. *Government Information Quarterly*, *29*(4), 446–454. doi:10.1016/j.giq.2012.06.003

Miller, P., Parker, S., Gillinson, S., & Demos (Organization :London, E. (2004). Disablism: How to tackle the last prejudice. *Demos (Mexico City, Mexico).*

Mainka, A. (2018). *Smart World Cities in the 21st Century*. De Gruyter; doi:10.1515/9783110577662

Meier, M. (Ed.). (2015). *Teilhabebericht zu den Lebenslagen von Menschen mit Behinderungen in der Universitätsstadt Marburg* (Sozialplanung Der Universitätsstadt Marburg). Magistrat der Universitätsstadt Marburg. https://www.marburg.de/portal/seiten/teilhabe-900001185-23001.html

Meijer, A., & Bolívar, M. P. R. (2016). Governing the smart city: A review of the literature on smart urban governance. *International Review of Administrative Sciences*, *82*(2), 392–408. doi:10.1177/0020852314564308

Paterson, M. (2011). More-than visual approaches to architecture. Vision, touch, technique. *Social & Cultural Geography*, *12*(3), 263–281. doi:10.1080/14649365.2011.564733

Retief, M., & Letšosa, R. (2018). Models of disability: A brief overview. *Hervormde Teologiese Studies*, *74*(1). doi:10.4102/hts.v74i1.4738

Roy, S. (2016). The Smart City Paradigm in India: Issues and Challenges of Sustainability and Inclusiveness. *Social Scientist*, *44*(5/6), 29–48.

Schelin, D. (Ed.). (2016). *Cinema Invites Other Gazes: Analyzing a film from different scientific angles*. Editora Prismas.

Snyder, S. L., & Mitchell, D. T. (2007). *Cultural Locations of Disability* (Vol. 9). University of Chicago Press.

KEY TERMS AND DEFINITIONS

Ableism: The term refers to a worldview that considers people with disabilities less valuable than those without any disabilities.

Blind: This adjective defines the inability to see; lack of sight. However, there are several degrees and levels of blindness which degrees severity, viz: legal blindness, visual impairment, low-vision, significantly visually impaired, etc.

Disability: This is a term usually used to describe a condition, physical and/or mental, which a person might have. The authors propose that this term, which contains the negative prefix "dis" connected to "ability" carries a political, social, and ontological weight.

Group Interview: It refers to a methodology of quantitative research, usually used to conduct surveys, especially in the social sciences.

Intelligent or Smart City: Whereas most definitions of an Intelligent or Smart City focus on technological interconnectedness of smart objects and electronic services, the authors have chosen to focus on other aspects that also define a smart city. Smart City is a definition given to urban areas in which citizens' lives are enhanced by the systems intertwining the geographical and historical areas of the community.

Ontology: This term refers to a branch of philosophy that studies the meaning of being. In computer science and information science, an ontology is a data model that represents a set of concepts within a domain and the relationships between those concepts. The focus of this study is on the former meaning.

Visual Impairment: In contrast to blindness, visual impairment usually means that the person has some residual sight, also referred to as low vision.

ENDNOTES

1 Philipps University of Marburg was established in 1527. It is considered the oldest protestant university in the world. Currently the university has about 25 thousand students and over 4,500 employees including professors, academic and administrative staff. Many notable names have passed through the University of Maburg. Perhaps the most well known students were Jacob and Wilhelm Grimm (1802-1805) who, later in life, came back to collect their famed fairy tales. In sum, so far 11 scholars who worked or studied in the University of Marburg have received a Nobel Prize.

2 The city's Old Town (Oberstadt) with many historical half-timbered houses, the Elizabeth Church (1235), the authentic Lutheran Church (Lutherische Pfarrkirche, also known as St.-Marien-Kirche

or Stadtpfarrkirche), the medieval Castle (also known as Landgrave Castle) where notables such as Martin Luther made a mark.

3 In German, "Hauptstadt der Blinden."

4 In German, "Verein blinder Akademiker."

5 In German, "Deutsche Blindenstudienanstalt."

6 In the German education system, there are three kind of schools as of fifth grade: "Realschule", "Hauptschule", and "Gymnasium". "Gymnasium" is the most advanced of the three, highlighting academic learning. It is comparable to the British grammar school system or to prep schools in the United States.

7 According to current reports of the Hessian Statistical Office (*Bevölkerung und Bevölkerungsvorgänge der hessischen Gemeinden*, 2019, p. 534), the population in Marburg is currently 76,590. For the sake of this study, the authors decided to use the official numbers of 2015 since they include more details about citizens with disabilities.

8 This is an admittedly uncritical comparison based on Albino et al. (2015) summary, the result of which was found through an online word frequency counter, which counts the frequency usage of each word in a given text.

9 The city offers services such as a geoportal for commuters in general, but especially for cyclers.

10 A new intelligent video system known as LiSA was recently installed in a trial run. The security system is intended for commuters who feel unsafe in passing through an 80-meter-long tunnel where a few years prior a crime took place.

11 The average conversational voice speed rate in English is 120 to 150 words per minute (wpm). Audiobooks range from 150 to 160 wpm. Referentially, when it comes to smart devices, the English voices Microsoft David, Microsoft Zira, and Microsoft Mark offer a rate of approximately 800 words per minute.

12 "Marburg aus einer anderen Sicht" (*Bürgerservice: Blinde und Sehbehinderte*, 2019).

13 In Marburg known as Dinner in the Dark, dark dinners offer sighted people the chance of experiencing meals in pitch darkness. In this restaurant a staff of blind cooks and waiters prepare and serve dinners.

14 Take one of several samples of biblical texts such as the gospel of John chapter 9 in which Jesus heals a blind man. It starts with a presumption of a moral fault on behalf of the blind man or his family as having sinned: *As he went along, he saw a man blind from birth. His disciples asked him, 'Rabbi, who sinned, this man or his parents, that he was born blind?'* (New International Version, John 9:1,2)

15 Aktion T4 is an example of an ableist worldview in its severest form. It refers to the official procedure of involuntary euthanasia of around 300 thousand men, women, and children with disabilities in Nazi Germany. In 1933 so-called *hereditary defects* started having to undergo compulsory sterilization (Hansen & King, 2013, p. 121). In 1939, the Reich Committee for the Scientific Registering of Hereditary and Congenital Illnesses was founded in order to categorize what it deemed defective infants, also referred to as *lebensunwertes Leben*, or life unworthy of life (Proctor, 2002, p. 10). For a more in-depth look into modernity's extremism of bodily ideal and the holocaust see Zygmut Bauman's *Modernity and the Holocaust* (Bauman, 1991).

16 In Germany, for instance, a blind person receives a so-called compensation for disadvantages (Nachteilsausgleich), also known as blind money (Blindengeld). It is a benefit that ranges from 300.00 to 636.00 Euros per month, depending on what region you live in ("Blindengeld," 2018).

In addition to the direct financial compensation, a blind person has several rights that facilitate mobility, participation is social life and public services.

17 Not to be confused with ontology in the field of computer science.

18 As a follow-up to the current investigation, a quantitative study is in its exploratory phase. Chosen as target population, blind/visually impaired students from Carl-Strehl-Schule are to answer a survey on their relation with the city of Marburg.

19 Martyn Denscombe offers a succinct explanation of the group interview. Concentrating on social research he defines focus group as *a small group of people who are brought together by a trained moderator (the researcher) to explore attitudes and perceptions, feelings and ideas about a topic* (Denscombe, 2007, p. 115).

20 Special thanks for this part of the project to Ulrike Schönhagen and Monika Häusler of the Deutsche Blindenstudienanstalt (Blista).

21 Bahaddin Batmaz (aged 45), Mirien Carvalho (50), Jens Flach (39), and Savo Ivanic (46) met with Dago Schelin for a Group Interview that took place at the Deutsche Blindenstudienanstalt, Am Schlag 8, 35037 Marburg, Germany, on November 15th, 2019, from 2 PM to 4 PM. The interview was 94 minutes long. It was recorded in audio, then transcribed (German) and translated into English. To access the full text of the interview, please contact the authors.

22 Actually, already the term *blind* could be qualified and subdivided into categories of its own. Ms. Carvalho was born blind without any sensitivity to light of any kind. Mr. Flach was born with glaucoma, and so has always been visually impaired, though his visual perception can roughly tell light from dark. Mr. Ivanic has a high degree of visual impairment, border-blindness, but can still read when wearing strong lenses and/or devices. Mr. Batmaz is 100% blind but still had some sight until about 15 years ago.

Chapter 6
Transforming Urban Slums:
Pathway to Functionally Intelligent Cities in Developing Countries

Darrold Laurence Cordes
Curtin University, Australia

Pornpit Wongthongtham
Curtin University, Australia

Greg Morrison
Curtin University, Australia

ABSTRACT

Cities in developing countries are increasingly under stress through urbanization, which leads to the expansion of slum areas or informal settlements due to demand for low-cost housing. This chapter presents the social, environmental, and economic realities facing slum dwellers and discusses their redevelopment into intelligent cities. The concept of 'function accompanying intelligent' is introduced for the transformation of slums to functional intelligent cities. In this context, a city is intelligent if it serves both the functional and social needs of its entire population. The chapter overviews an approach to integrated data collection, data analytics, and user access to information. Geospatial analysis of demographic, economic, social, and environmental data is introduced to help delineate slums, and to monitor the outcomes of urban planning initiatives and the progress of social wellbeing. The city of Accra in Ghana is discussed as a potential slum city to functional intelligent city transformation.

INTRODUCTION

The World Cities Report (2016) provides comprehensive insights into the challenges facing urban slum dwellers, policy makers, and urban planners. Slums exist everywhere, but they are more evident in developing countries exacerbated by rural-urban migration, poor urban planning, and weak governance. Slums are highly correlated with poverty. Mahabir et al. (2016) have provided an historical account of

DOI: 10.4018/978-1-7998-5062-5.ch006

the tried and failed policy initiatives to eradicate slums over the last 50 years. They refer to the *sites and services* approach during the 1970s that led to the relocation of slum dwellers to new sites and the former slums were demolished. Displaced people were required to pay for the new infrastructure developments and housing. Without adequate resources, the slums continued to flourish. For example, the Brazilian government effort in the 1970s, to forcibly displace favela (slum) dwellers into public housing without adequate planning and support, resulted in the creation of new favelas (Brown University, 2019). Mahabir et al. (2016) then describe the policy initiative of the 1980s to redevelop the slums in-situ. This approach also failed primarily because of the lack of financial commitment and other reasons relating to the failure to recognize the importance of addressing the economic conditions of the slum dwellers, governance issues, and land tenure issues (Mahabir et al., 2016). The next policy initiative was to bestow land ownership on the slum dwellers based on the assumption that the structure within which they lived was owned by the occupier. This assumption proved to be incorrect because occupiers could still be evicted by landlords. Mahabir et al. (2016) then discuss the more recent Nelson Mandela inspired Cities Without Slums initiative which acknowledges the link between poverty and slums. It appears that this initiative has given way to the United Nations Sustainable Development Goal (SDG) 11 which was revised extensively in 2018 to *make cities and human settlements inclusive, safe, resilient, and sustainable* (United Nations, 2019b).

In the following sections, we discuss the case of city of Accra in Ghana in western Sub-Saharan Africa.

Sub-Saharan Africa

The United Nations (2019a) and World Bank (2019b) aim to eliminate poverty by 2030, but efforts to reduce poverty are being hampered by persistent problems in Sub-Saharan Africa. A study by Kedir, Elhiraika, et al. (2017) reveals that Africa requires an economic growth rate of 16.6 per cent per year between 2015 and 2030 to meet this goal. This is a significant challenge for West African countries who are facing a poor investment climate and insufficient technical capacity to take advantage of further foreign direct investment (FDI) (Eregha, 2017). Sub-Saharan Africa suffers profound and abject poverty in many areas (Beegle, Christiaensen, et al. 2016). Patel (2018) estimates that in 2015 there were 413 million extremely poor people in Sub-Saharan Africa living off less than $1.90 per day. The World Bank (2019b) estimates that by 2030, nine out of ten extremely poor people in the world will live in that region, and that the number of people living in extreme poverty in Sub-Saharan Africa is increasing. The United Nations (2019c) estimate that by 2050, the two central African nations of Nigeria and the Democratic Republic of Congo will be among the ten most populous countries in the world.

The problems in Sub-Saharan Africa have triggered an elevated sense of urgency among the African Union (African Union, 2019). The Institute for Security Studies (2019) states that, while the African Union is pursuing a more realistic target to reduce poverty to below 3% by 2063 under the African Union Agenda 2063 initiatives for peace, prosperity, and the eradication of poverty, the diversity of nations on the African continent stands in the way of a uniform approach to poverty eradication. The United Nations Development Programme (UNDP) estimated the growth elasticity of poverty reduction in Sub-Saharan Africa to be -0.7 (Bhorat and Naidoo, 2014). Hence, for every 1% increase in gross domestic product (GDP), this results in a 0.7% reduction in poverty. The task of poverty reduction is, therefore, daunting. Bhorat and Naidoo (2014) also note that inequality will act to reduce the growth effects on poverty. The UNDP observations also highlight the challenges in dealing with slum rehabilitation.

Disease, terrorism, high fertility rates, regime instability, and the impact of climate change on food supply are setting the stage for an unimaginable human disaster unless urgent action is taken immediately. This action is going to be a challenge in Sub-Saharan Africa under the constraints of weak governance, corruption, inadequate legal structures, and problems with enforcement of the rule of law. In many instances this will preclude viable public-private partnerships in any redevelopment strategies including transport, water supply, electricity, and housing development. This may also limit the extent of involvement of socially responsible multi-national entities. Pro-poor tourism has been advanced as an opportunity for poverty reduction in Sub-Saharan Africa, but the presence of autocratic and kleptocratic regimes may preclude real transformational benefits for the poor (Saito, Ruhanen, et al., 2018; Chayes, 2015).

West Africa

Twerefou (2012), in a report on West Africa, commissioned by the United Nations Economic Commission for Africa, noted that economic growth was the result of donor inflows and not of domestic savings. Twerefou (2012) also reported that the region was not experiencing a period of sustainable development and hence the lack of improvement in the wellbeing of the broader population. Foreign direct investment (FDI) had failed to bring transformational change to West Africa and job creation was not widespread across the region (Twerefou, 2012). Resource extraction is capital intensive and, therefore, industries such as telecommunications and finance have offered limited employment opportunities.

Ghana

Ghana is a country of about 29.6 million people (as estimated in 2018), located in West Africa; its main export being oil, cocoa and gold. The World Bank (2019a) predicts economic growth of 7.6% in 2019 while acknowledging problems with fiscal consolidation, currency depreciation, and the energy sector. Ghana ranks high in Africa for freedom of speech and press freedom, but ranks 78/180 in the world for corruption - where ranking of 1/180 denotes the least corrupt (Transparency International, 2019). Ghana has been a model among the Economic Community of West African States (ECOWAS) for its social programs to reduce poverty and for economic growth. However, the World Bank (2007) Ghana Country Environmental Analysis report states that economic growth has come at a significant cost to the environment and to the health and wellbeing of its population. The report also comments on the state of Ghana's environment. Of the original forest area, 50% has been cleared and is now used for agricultural purposes, but land productivity is reduced because of soil erosion, crop yields have declined, river fish stocks have decreased, other forest and non-forest stocks are also in decline. Water shortage is being experienced in coastal towns during the dry season. Wildlife population and biodiversity are diminishing rapidly. The human population is experiencing serious health concerns because of pollution, poor water quality, and inadequate sanitation.

The World Bank (2007) also stated that Ghana's gross domestic product (GDP) growth should reflect the cost of the decline in its natural resources, and the cost of human and social capital. They claim that as of 2007, 9.6% of GDP is lost annually due to the reduction in natural resources, and increased health costs related to poor water and air quality, and sanitation. In a report for UNICEF on poverty in Ghana, (Cooke, Hague, et al. 2016) summarized their findings as follows:

- Poverty declined from a high of 56.5% in 1992 to 24.2% in 2013, enabling Ghana to reach its United Nations Millennial Development Goal to cut poverty by 50% by 2015.
- Urban poverty (10.6%) is substantially lower than rural poverty (37.9%), and there are regions, especially in the north-west of Ghana, where poverty reduction has not followed the rest of the country.
- Inequality, as measured by the Gini index, exhibits a widening of the rich/poor gap (Gini =37 in 1992 compared with 42 in 2013), with the highest levels of inequality in the Greater Accra Region of Ghana.
- There are approximately 6.4 million people, including 3.65 million children, living below the UNICEF poverty line of $312/year/adult.

City of Accra

Accra is the capital city and main commercial centre of Ghana. The Ghana Statistical Service (2019a) states that the Accra metropolis had a population of 2,052,342 in 2019. Engstrom et al. (2015) state that the Accra Metropolitan Assembly (AMA) had a population, in 2015, of 1.6 million of which 38.4% (or approximately 630,000 people) lived in slums. On the assumption that this percentage is still applicable in 2019, the population of slum dwellers in Accra in 2019 is approximately 788,000 people.

Objectives of This Chapter

The purpose of this chapter is to delineate a pathway for the transformation of slum cities in developing countries to become functional intelligent cities. Much of the literature on intelligent cities focuses on cities in the developed world and, while there have been many studies on slums, little attention has been given to the transformation of slums as the pivotal initiative leading to intelligent cities in the developing world. The World Economic Forum (2018) outlines the concept of a slum area being designated as a microcity or special development zone to be developed as the *framework for urban reform*. The pathway for the transformation of slum cities in developing countries presented in this chapter share the core concept of the microcity and expand on it as the genesis for intelligent cities of the future. At the centre of this approach is the recognition that sustainable economic development is the primary motivator for slum dwellers to engage in the transformation process.

The following section discusses the identification of slums in Accra using a combination of inputs to produce maps of likely slum neighbourhoods. This highlights the complexity of the slum environments and profiles the physical challenges to redevelopment. Mapping of slum areas helps to delineate land areas in a country where land ownership can be contested because no accurate records exist. An integrated environmentally sustainable socio-economic model that combines commercial activities and associated community infrastructure is then introduced. This is followed by discussions on imperatives within this model for manufacturing, energy production, food production, housing construction, water, sanitation and health, transport, and education. Finally, an insight is provided into the types of data that could be collected through a variety of techniques, the processing of the data, and the reporting of information to households and businesses as the first increment towards a functional intelligent city.

LOCATING SLUMS OF ACCRA

Every city and its slums are different. The UN-Habitat (2018) definition of slums suggests extensive household surveys and census data to correctly classify localities as slums. This is an expensive process and the infrequency of these surveys and censuses does not support continuous monitoring of any slum transformational efforts. Under the intelligent city concepts, continuous self-reporting or crowd-sourcing of data can contribute additional inputs to remote sensing or satellite imaging of cities. Data analytics plays a big role in the monitoring and evaluation of the slum city transformation. Data analytics may be used to develop comprehensive understandings of the slum environments, social structures, and the wellbeing of the slum population. These understandings are crucial to redevelopment plans and in the creation of opportunities for community involvement in such plans. Geospatial data (geographic and demographic data) and economic data may be combined with data sourced from the slum populations on issues ranging from food, transport, health, education, and access to government services to create visualizations to aid social, economic and infrastructure programs. Data analytics will continue to play an on-going role in monitoring the progress of the redevelopment, social engagement, economic growth, and contributions to environmental rehabilitation.

Defining a Slum City

There does not appear to be any formal definition of a slum city. One simple approach is to use the percentage of the total metropolitan area that is declared as slums and then establish a threshold above which the city may be viewed as a slum city. Another way is to take the percentage of the total population within the metropolitan area that are slum dwellers and then establish a threshold above which the city may be deemed a slum city. Or, there could be a combination of both. At a nominal threshold of 50%, slum dwellers are occupying 50% of the total metropolitan land area or 50% of the population live in slums. Considering the formula stated by Engstrom et al. (2015) that 61.7% of urban populations live in slums in Sub-Saharan Africa, we can conclude that all cities in Sub-Saharan Africa would exceed the 50% population threshold. It is also stated that 38.4% of the Accra Metropolitan Assembly (AMA) population live in slums occupying 15.7% of the land area. While not exceeding the nominal 50% threshold, the extent and degradation of conditions in slums in Accra, both in land area and population, would deem it to be a slum city.

Identifying Slums

Mahabir et al. (2016), when referring to UN Habitat 2018, state that approximately 1 billion people globally are slum dwellers and they project that this will grow to 2 billion people by 2030 and 3 billion by 2050. Engstrom, Pavelesku, et al. (2019) state that Sub-Saharan African slums are growing at a rate of 4.5% per annum. Using the 2019 estimate of 788,000 people living in slums in Accra, this would mean that the slum population would swell to 2,951,310 by 2050 unless mitigating interventions were adopted immediately. The growth in slum populations comprises the normal population increase and the rural-urban migration rate.

The UN-Habitat (2018) defines a slum household as a group of individuals living under the same roof in an urban area who lack one or more of the following:

- Durable housing of a permanent nature that protects against extreme climate conditions.
- Sufficient living space which means not more than three people sharing the same room.
- Easy access to safe water in sufficient amounts at an affordable price.
- Access to sanitation as a private or public toilet shared by a reasonable number of people.
- Security of tenure that prevents forced evictions.

Mahabir et al. (2016) state that work undertaken in slums has been mainly concentrated in the following areas:

- Socio-economic and policy.
- Methods for determining the physical characteristics of slums using remote sensing.
- Modelling of slums using artificial intelligence tools (machine learning) in attempts create a more complete understanding of the extent of the slums in urban areas.

Engstrom et al. (2015 and 2019) and Mahabir et al. (2016) acknowledge that the two-dimensional remote sensing view needed other information from household surveys, census data, other aerial maps, and boundary maps to obtain a more detailed view of slums. Large slum areas are not easy to hide in the urban landscape, but smaller pockets of slums with different spatial characteristics make them harder to discern.

Figure 1, by Marlenenapoli from commons.wikimedia.org/w/index.php?curid=14680987, shows a small part of the well-known Sodom and Gomorrah slum in Accra which, according to Owusu-Ansah, Tagbor and Togbe (2016), had a population of between 25,000 and 40,000 people over 146 hectares. To illustrate the difficulty of determining the population and area of slums, the Rockefeller Foundation (2019) estimate the population of Sodom and Gomorrah to be 80,000 occupying only 35 hectares. Other areas of Accra that are deemed slums do not have the same appearance and crowding of homes.

The following is an historical account of some initiatives to map the extent of the Accra slums using various techniques.

Mapping of Accra Metropolitan Assembly (AMA)

Mapping the AMA was undertaken by Verutes, Fiocco, et al. (2012) who sought to create street maps to map health related survey data to the environmental conditions of the survey respondees. They found that even rudimentary street maps did not cover their areas of interest. Street addresses in Accra were virtually non-existent and locational techniques (such as when applying for a bank account) were small hand drawn maps that attempted to show address locations relative to landmarks, named and unnamed streets, and photographs of the actual dwelling. This technique does not work well in slums because of the density of buildings and the lack of distinguishing features. Verutes et al. (2012) created digital maps of census areas that are used by the Ghana Statistical Service for census data collection. They then combined the census maps with geospatial data from satellite images to create relatively accurate street maps together with population densities. They were then able to locate earlier survey respondees and to place them more precisely within their environment.

In 2017, Ghana Post launched a service to assign digital addresses using latitude and longitude derived from the global positioning system (GPS) (GhanaPost, 2019). A mobile telephone application is used to exactly locate a physical address (latitude and longitude) and this is translated into an alphanumeric

Figure 1. Image of Sodom and Gomorrah (Agbogbloshie, Accra, Ghana)

code like a postcode. GhanaPost and other service organizations (ambulance, fire, police) could then use this for accurate delivery of postal items and emergency services. The GhanaPost address system will contribute to the demarcation of slum locations and this system is likely to be used in future censuses.

Several other methods for determining the extent of the slum areas in the AMA have also been developed. Engstrom et al. (2015) used a combination of remote sensing data, census data, and field survey data to create maps that were then compared with field derived slum maps produced by the United Nations and the AMA. The resulting map showed a close relationship with the United Nations AMA designated slum areas shown in Figure 2.

In a more recent study by Engstrom et al. (2019), utilizing household survey data, high resolution geospatial imagery, census data, and multiple methodologies, they were able to produce a machine learning slum index and a consumption poverty map. They observed that household size, elevation, and population density were the most significant variables in the creation of a slum index. Refer to Figure 3, which is the machine learning slum index where the higher index values indicate areas of higher slum incidence.

Figure 4 is a consumption-based map from which Engstrom et al. (2019) conclude that low variability of building orientation is more likely to be correlated with higher consumption. Slum areas are often characterized by random building orientations. These researchers observed that occupancy in slum dwellings is for relatively long periods from which it may be concluded that other factors such as linkages to business opportunities and jobs, social networks, and cultural/religious associations may contribute more to longevity of stay than monetary considerations. This is an important observation when considering any redevelopment plans for slums.

Figure 2. Remote Sensing of Slum Areas in Accra, Ghana

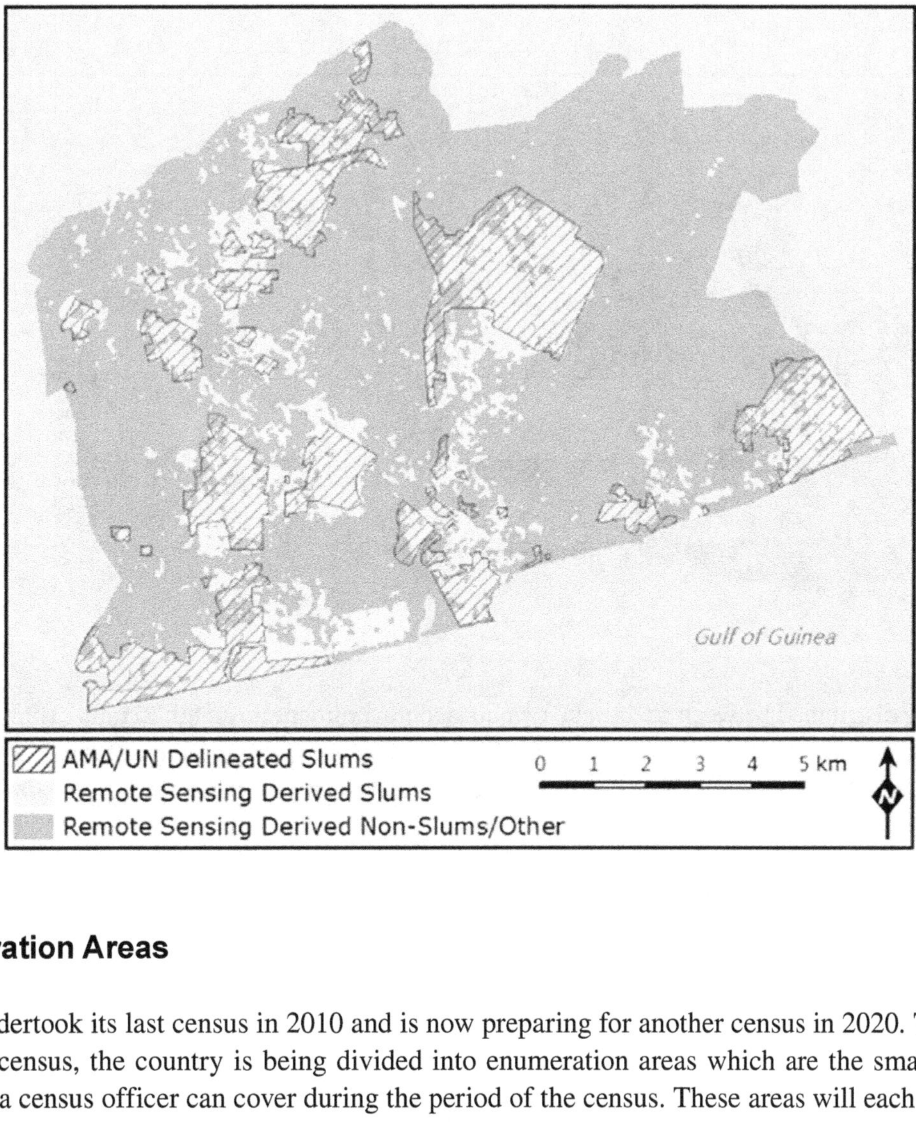

Enumeration Areas

Ghana undertook its last census in 2010 and is now preparing for another census in 2020. To facilitate the 2020 census, the country is being divided into enumeration areas which are the smallest spatial units that a census officer can cover during the period of the census. These areas will each contain approximately 600 people in 150 households. The number of enumeration areas is expected to be around 50,000. Extensive demographic and other data will be collected during the census.

The enumeration areas are an ideal spatial unit for testing slum and other low-income sustainable development initiatives and the alignment with census data in the future will help develop a comprehensive understanding of community wellbeing. Figure 5 by Ghana Statistical Service (2019b) is a map of the enumeration areas in the Ashanti Region showing population density. The dark area near the centre of the map is the city of Kumasi which has a population like Accra of 2,096,053 people.

Tenure of Occupancy

Regarding the occupancy, a relevant question is: who owns the land and structures on the land? Possible answers are that it is traditionally owner owned land or it is government owned land. Land rights

Figure 3. Machine Learning Slum Index of Accra, Ghana

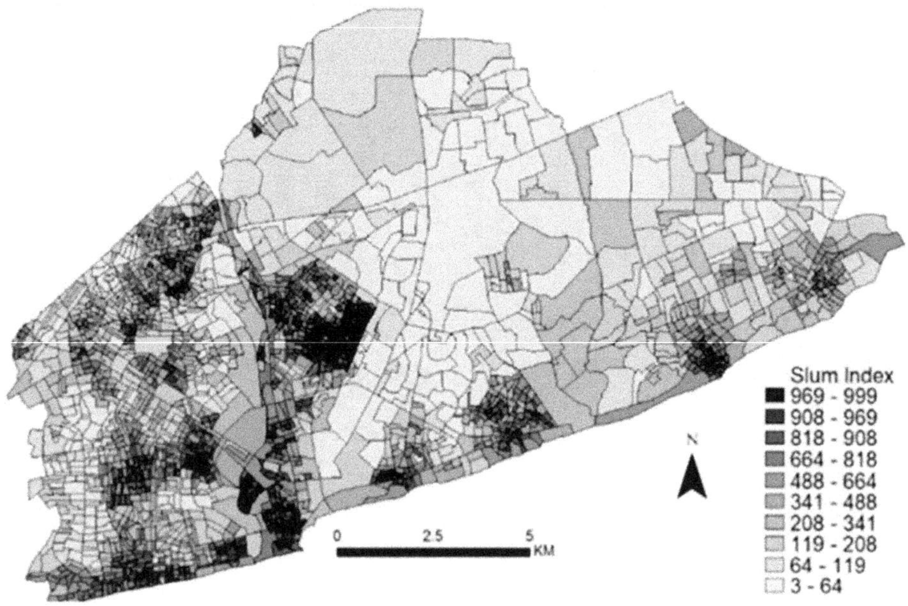

in Ghana are distributed between traditional owners and government on an 80:20 ratio. Traditional land is referred to as skin or stool land and is often subject to disputes over who has authority or delegation over its use or disposal (Stacey, 2018). It is also stated that the processes for the administration of land

Figure 4. Poverty Map by Consumption for Accra, Ghana

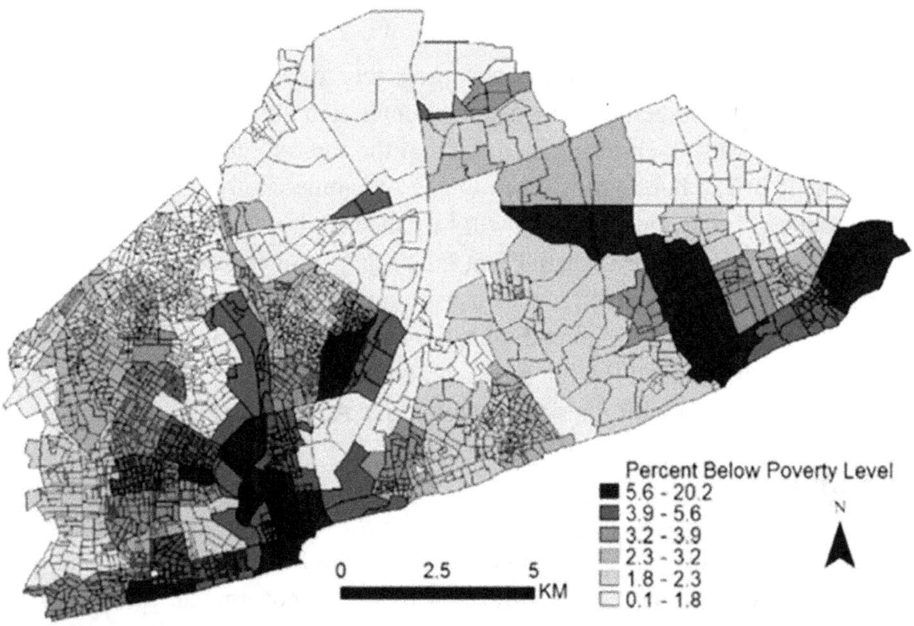

Figure 5. Ashanti Region - Enumeration Areas - Population Density

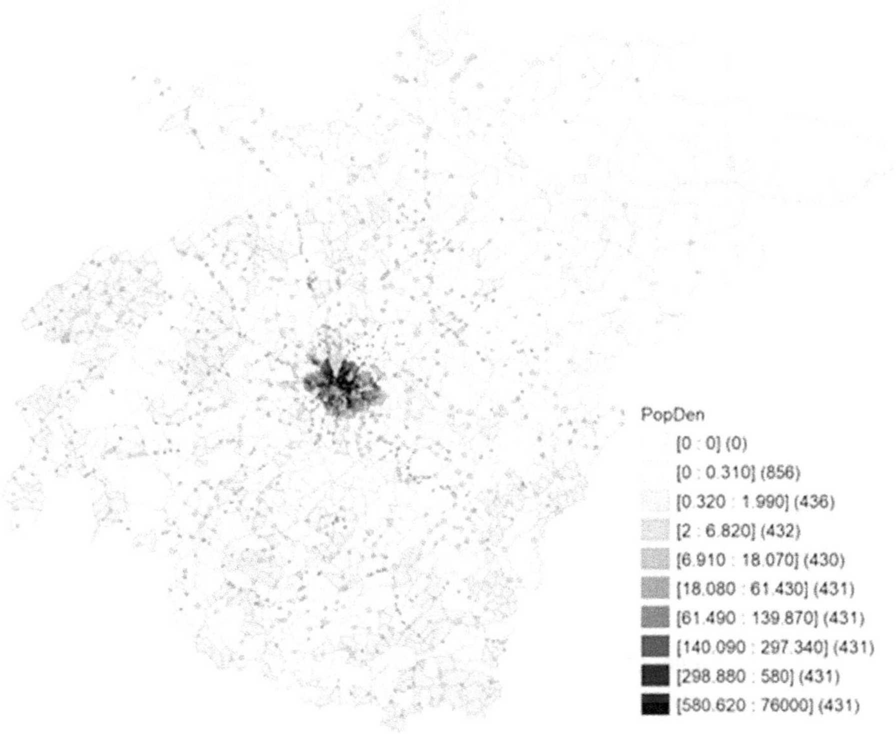

entitlements in Ghana are not working well and this is further exacerbated in dealing with slum land because of the presence of squatters. Traditional land is not often registered with the Land Title Registry and disputes over the demarcation of boundaries often occur too many times.

Issues of land ownership and occupancy entitlements must be resolved prior to redevelopment of the slums. This will instil confidence in all stakeholders that slum city transformation can proceed with the establishment of a built environment that provides affordable housing, small business facilities, and community infrastructure without the need for relocation.

SLUM CITY TRANSFORMATION (SCT)

The pathway for slum cities to become functional intelligent cities is by first addressing the urban slums and the enduring issues that are at the core of the people who inhabit them. It is not a simple matter of just rebuilding with better quality houses (World Economic Forum, 2019). Housing is an important issue but any attempt to improve the quality of life of slum dwellers must recognize the broader social, economic, and environmental conditions of the people. This means that basic infrastructure such as transport, roads, water and electricity supply, appropriate housing, and other services enjoyed by the more affluent sectors of society must also be made available to the slums. Economic opportunities must be integrated into the slum redevelopment fabric to encourage entrepreneurial initiatives and to create job opportunities while preserving and enhancing the social fabric.

In developing countries in Sub-Saharan Africa, the infrastructure of the non-slum portions of cities may also be inadequate but, in terms of intelligent city objectives, this does not preclude the integrated upgrading of all urban services and the incremental implementation of intelligent city strategies. Accra is a city that has experienced considerable growth, but it lacks an efficient integrated transport structure with its seaport Tema and the regional centres. The government of Ghana has proposed a new international airport north east of the city, and a railway link to the Akosombo region of Ghana is under construction. Ghana has ambitious plans for its regional rail network. Road access into Tema is congested and the movement of goods in and out of the port is slow and inefficient. The AMA reports that every day more than 2 million people enter Accra for work, leisure or other activities. This creates significant traffic congestion on the arterial links to the surrounding region and pollution levels are high.

The concept for slum transformation embraces economic, social and environmental factors. Recognizing that slum dwellers have formed social networks and supporting situational frameworks (there are strong links between the rural areas of Ghana and sections of Accra slums), any redevelopment should take place in-situ. Environmental factors should address the need to improve physical and emotional health, including access to clean water, proper sanitation, health care and a quality education. Air quality and resilience to climate change are important markers of a more secure environment. Economic stimulus is essential.

The reason why rural migrants often gravitate to urban slums, is that often they cannot afford to love anywhere else. The reason that slum dwellers resist resettlement, is that they often cannot afford the cost of resettlement. These concepts and limitations can apply to all new urban plans. Other factors can be the needs of people inhabiting new developments, that need to be adequately met without the need to commute long distances.

Figure 6 is a conceptual modular design that can be implemented incrementally across a much larger slum community. There is no need to immediately demolish the whole slum, instead the incremental approach offers the opportunity to gather inputs and support from local people to contribute to more robust transforming strategies. Figure 6 addresses the following factors:

- **Economic infrastructure** – which includes an urban farm (aquaponics) that is capable of intensive fish and vegetable production in a significantly compressed footprint using much less water than other farming techniques. Aquaponics can be scaled to serve the needs of families in small home-based systems or larger community systems under community management arrangements. Manufacturing can address the general needs of the community. One such need is for paper products (printer/copy paper, toilet paper, paper towels, tissues, and sanitary products) which are consumed daily. Both initiatives provide employment and street vendor opportunities for residents.
- **Environment** – which includes the provision for local processing and recycling of waste products through an environmentally sustainable down-draft gasification process that converts organic matter into usable products. These usable products may include a synthetic gas used for electricity generation and other gases (such as hydrogen for industrial use). Small community-based systems would mean rapid clean-up of the solid waste discarded by households as shown in Figure 7. The bags of waste have economic value and can be traded for cash at processing points. Municipal solid waste generated in other areas of the city may also be processed, leading to eventual elimination of large landfill sites while bringing income to the processing facility. Distributed solar installations can be installed throughout the communities to further aid energy resilience and autonomy while creating economic opportunities.

- **Social** – the provision of affordable functional housing is essential for the wellbeing of its inhabitants. The United Nations Sustainable Development Goal 11 acknowledges the urgent need for affordable housing to address the challenges of rapid urbanization (United Nations, 2019a). 3D printing can quickly construct a variety of dwellings to suit single and multiple families, and for other purposes. Claims that a single-family home can be constructed within 24 hours for less than $4,000 need to be tested and the cost verified (CNBC, 2018). These homes can be printed to include toilets, aquaponics facilities, and solar panels. They can also be equipped with sensors to report air quality, fire detection, water consumption, electricity consumption and generation, emergency service alerts, and security alerts. Each home can be equipped to provide access to the whole range of intelligent city services. High speed access to the Internet provided by 5G telecommunications services can provide access to interactive education, entertainment, and information services.

Figure 6. Conceptual Modular Structure of Slum Transformation

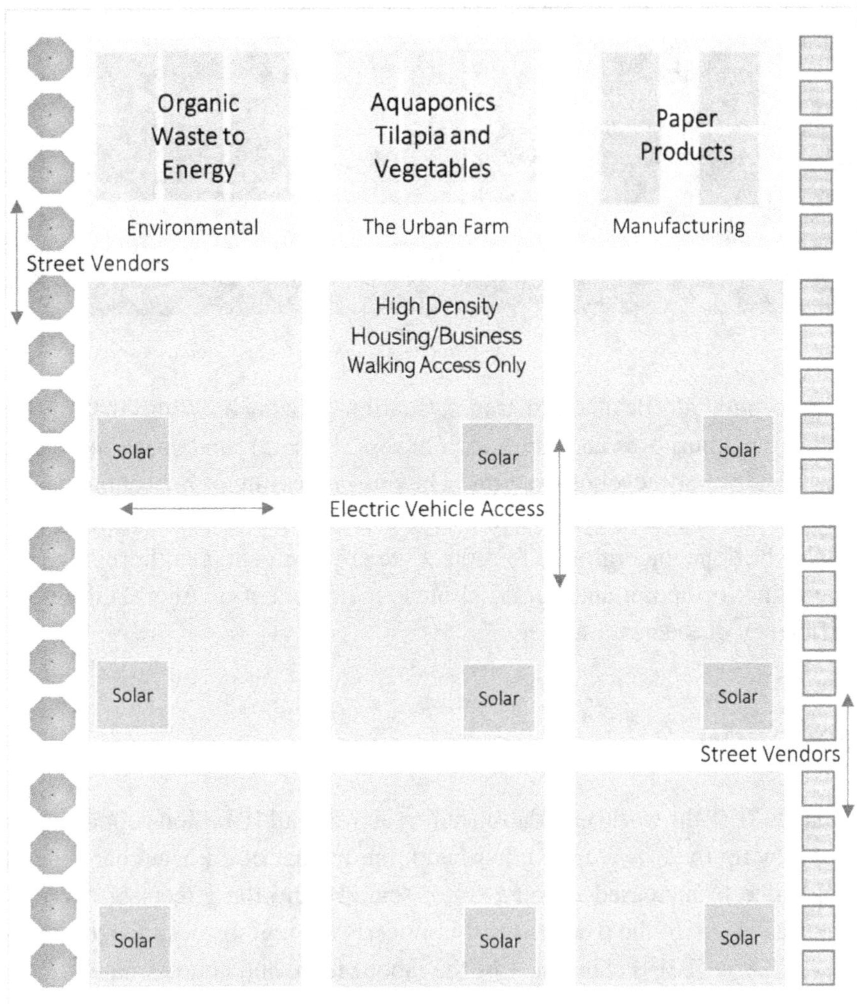

Figure 7, from atcmask.com, shows how household trash is disposed of in the slum district of Agbogbloshie in Accra. In many slums, there are no waste collection services. Organic matter is left to decompose, and plastics get blown away by the wind or washed away in floods. This has an impact on the surrounding areas and contributes to overall environmental degradation and serious health issues for slum dwellers. The household waste can be feedstock for the environmentally sustainable down-draft gasification energy conversion process.

Figure 7. An Image of Agbogbloshie, Accra, Ghana

Figure 8 is a conceptual profile of a slum transformation showing buildings that are multi-level to free up space for recreational purposes and for ease of access. Underground space may be used for storage tanks for rainwater and water recycling systems. The pre-processing of household waste where organic matter is separated from recyclables (metal cans) and solid inorganic matter (bricks, and other building materials) may also be done underground to protect the environment. Geothermal heat transfer may be used for space cooling in the hot and humid climate environment of Accra. Grounder water cooling improves the efficiency of the heat transfer.

FOOD PRODUCTION

It is estimated that by 2050, the world population may reach around 10 billion people requiring an increase in food production by up to 98%. On the other hand, the impact of climate change could significantly reduce crop yields due to increased salinity, water scarcity and the effects of ozone. USAID (2019) note that West Africa is one of the world's most impoverished regions despite a richness in natural and human resources. USAID (2019) claim 60% of the labour force and approximately 35% of the region's GDP are attributed to the agriculture sector in West Africa. However, agricultural productivity suffers from the inability of farmers to connect with markets, the cost and availability of high-quality seeds and

Figure 8. Conceptual Profile of Slum Transformation

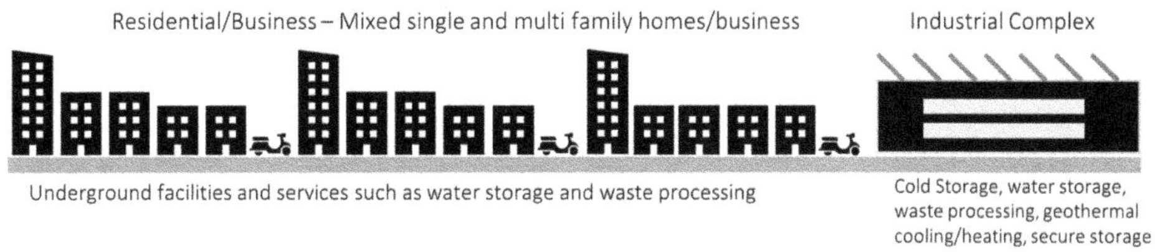

fertilizer, and insufficient information on new technologies and farming best practices. Furthermore, high transportation costs and lack of support for regionally agreed policies on cross border trade further exacerbate sector growth.

According to USAID (2019), the region experiences low per hectare crop yields and an estimated 13.5% of the regional population is undernourished. This makes the agricultural sector in West Africa extremely vulnerable to the effects of climate change. Sultan, Defrance, and Iizumi (2019) found that there is already evidence of crop losses from global warming and conclude that further losses will be experienced unless crop adaptation measures are implemented.

In a study by Asante and Amuakwa-Mensah (2015), it was observed that temperatures across Ghana are rising and that rainfall is becoming more erratic and variable. Naab et al. (2019), in a field study, through focus group discussions and key informant interviews, concluded that climate change was manifesting in shorter rainy seasons, longer dry seasons and loss of soil fertility. Asante and Amuakwa-Mensah (2015) forecast dire consequences of climate change in Ghana with periods of extremely high temperature and low rainfall, increasing desertification, and increased sea-surface temperatures affecting Ghana marine fisheries. They also predict a narrowing of the range of suitable weather for cocoa production and lower yields from rice, cassava, and other rooted crops.

The Ministry of Finance (2019) report to the Ghanaian parliament have advised a sharp decline in the agriculture sector growth rate for the first quarter of 2019 from 4.7% in 2018 to 2.2% in 2019. The Ministry report also announced several initiatives to stimulate the agriculture sector but made no mention of the impact of climate change.

The Urban Farm – Aquaponics

Aquaponics combines hydroponics and aquaculture in a single integrated system. It is introduced as a mitigating strategy for the effects of climate change in the area of fish and vegetable production. The aquaponics industry globally is in its infancy, but it offers scalability from domestic to large scale commercial systems that are suitable for urban and rural environments.

In this scenario, the fish waste adds nutrients to the water which then becomes food for the plants. The plants purify the water which is then recycled back to the fish tanks. The main inputs to the system are the fish fingerlings, fish feed which can be organic feed (Nelson & Pade, 2019a), water (to top up the recycled water to a constant level), and electricity to drive pumps, conveyors and lighting. The recirculating system depicted in Figure 9 is energy efficient and can be adapted to different fish species and a wide range of plant types to suit market needs (Jena, Biswas & Saha, 2017). When combined with a renewable energy source, such as solar or biomass waste-to-energy, environmental impact can be minimized.

Figure 9. Aquaponics Process

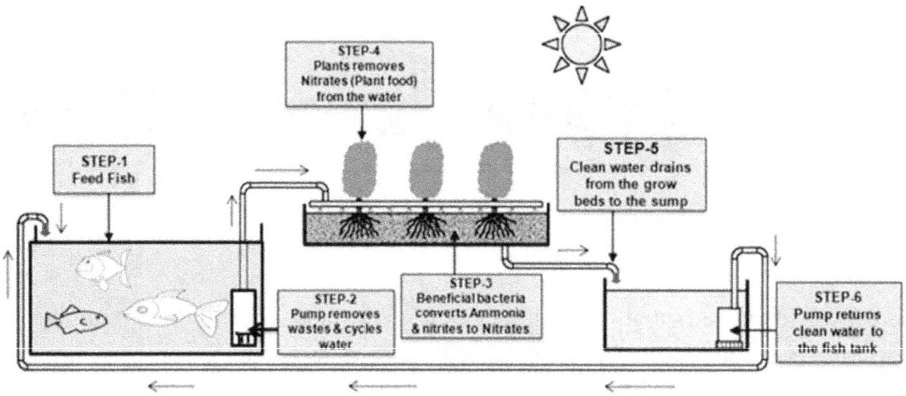

According to Nelson and Pade (2019b), some advantages of aquaponics include:

- Yield per hectare is 8 times greater than traditional agriculture with 1/6th of the water usage. This is a significant benefit in dry climates and in areas facing degradation due to climate change. It also means that aquaponics systems can be introduced as viable urban farms producing fresh fish and vegetable products. Storage and transportation requirements are minimized.
- No requirement for herbicides, pesticides, antibiotics and growth hormones.
- System is non-seasonal, food can be grown year-round in a stable environment.
- No manufactured or mined fertilizers are required.

The United States Foreign Agricultural Service (USDA, 2019) report on Ghana Fish and Seafood acknowledges the importance of seafood to the national diet and to the agriculture sector. The report estimates annual consumption of about 775,000MT or about 25.8Kg per capita. There is a shortfall of an estimated 370,000MT over domestic production. USDA also comment on the importance of the aquaculture sector in bridging this gap. Aquaponics can contribute to the reduction of seafood imports with the establishment of domestic, community and commercial aquaponics systems in rural and urban areas of Ghana.

Domestic Aquaponics

Domestic aquaponics are systems that can be installed or integrated into home designs with production capacity for a typical household. These systems would produce up to 150Kg of fish per year together with 1,500 heads of lettuce or other vegetables. Current stand-alone costs for these systems render them too costly for low-income households because of the capital cost of the equipment and the floor area of approximately 30sqm. This can change significantly if the aquaponics systems can be mass 3D digitally printed.

There are currently no industry initiatives pursuing the 3D printing of aquaponics systems. However, under the scenarios for urban redevelopment presented in this chapter, there is motivation to explore the design and development of a range of low-cost aquaponics systems. Each system requires one or more household members with knowledge of aquaponics to ensure continuous productive operation. Excess products may be available for sale to other households. In close-proximity communal style living, community managed aquaponics may be more appropriate.

Community Aquaponics

Community aquaponics can be more reliably operated and managed to ensure consistent supply of quality fish and vegetable products. They can also be configured to operate as a profit centre through sale of excess product into the larger community.

Commercial Aquaponics

The use of intensive aquaponics production can result in the return of previously farmed land to the natural habitat or the reduction in the rate of diminution of the natural environment. Commercial aquaponics can be undertaken in the redevelopment of urban slum areas and in rural areas to create employment. Commercial aquaponics can be focal points for food processing, packaging, and marketing of local traditionally grown produce as well as the aquaponics output. The creation of a hub also improves logistics, product quality, branding and reputation. Aquaponics is suitable for all regions of Ghana.

The aquaponics industry has evolved out of small hobby systems for domestic needs and there are few examples of larger commercial installations. These systems require a substantial initial capital investment compared with traditional pond aquaculture or agriculture. In an urban environment, high volume fish and vegetable production with yields per hectare of more than 50MT of fish and more than 1,000,000 lettuce plants per year are achievable. Commercial aquaponics creates zero waste, produces fresh food, and is environmentally sustainable. The absence of pesticides, herbicides, and antibiotics and mined fertilizers makes it perfect for any situation, especially an urban environment. Production is a year-long process and can provide employment for the community in production management, harvesting and packaging, marketing, administration, systems maintenance, and facilities maintenance.

Solar energy combined with biomass downdraft gasification provide the energy for the production system, water filtering and purification, and general use. In the modular system shown in figure 9, all aquaponics production will be electronically monitored and supervised from a central facility. Supervision will include training and regular inspections. Aquaponics sales will be e-commerce enabled and will support business-to-customer (B2C) and business-to-business (B2B) sales without intermediaries.

RENEWABLE ENERGY

The demand for energy grows rapidly as urban regions expand due to population growth. This impacts on the centralized generation capability and on the capabilities of the distribution networks. Ghana has a relatively high access rate for electricity with potentially 100% of the population having access by 2022 (Kumi, 2017). Ghana's energy generation profile is approximately 57% thermal with the balance provided by hydro-electric power from the Akosombo Dam facility on the Volta River and other locations. However, the national distribution grid contributes to a high level of loss and the poor tariff structure provides financial uncertainty for electricity utilities. The installed capacity is approximately 4,650MW but availability is limited to about 2,400MW because of aging and failing infrastructure, seasonal rainfall variations affecting hydro-electric power generation, and unreliable fuel supplies for thermo-electric power generation (Power Africa, 2019a).

The energy sector in Ghana is in crisis because, under the current contract structure, payment must be made for the full capacity of 4,650MW even though peak demand in 2019 was only 2,700MW. This results in a sector shortfall and payment of $1,268 million for unused capacity (Ministry of Energy, 2019). Furthermore, the energy sector has an accumulated debt backlog of approximately $2,748 million. The Ministry of Finance (2019) reports that this is an unsustainable burden on the emerging Ghanaian economy and have implemented the Energy Sector Reform Program to overhaul the energy sector, to renegotiate current contracts, and to increase emphasis on renewable sources (Ministry of Energy, 2019). Initiatives were also announced to increase demand, but the delivery infrastructure needed a significant overhaul to increase efficiency. One such initiative is the revival of the aluminium industry under the Ghana Integrated Aluminium Development Corporation (Ghanaweb, 2019a). This was the industry that was used to justify the building of the Akosombo Dam hydro-electric power facility in the early 1960s. Almost 60 years later, the Ghanaian government now see this as an industry that can contribute to economic growth (Ghanaweb, 2019a). Aluminium production requires significant energy, and this will contribute to the increase in demand for energy, and the reduction of the supply/demand deficit.

The consequences of the crisis in the energy sector do not bode well for the redevelopment of urban slums and low-income communities across Ghana. Access to low-cost renewable energy is vital for the redevelopment of these communities. In this section, microgrids delivering energy from solar and biomass offering greater developmental autonomy, lower costs, economic opportunities under energy trading schemes, and lower carbon footprints for a more sustainable environment are discussed.

Microgrids

Microgrids were mainly used in remote rural settings that had no access to the electricity grid. They have now emerged as an integral component of urban planning strategies. Microgrids servicing groups of households and businesses within the slum environment, facilitate autonomous or near autonomous operation of shared renewable energy sources. Microgrids can be monitored for the supply and demand for electricity, the environmental impact of energy use, and usage patterns. These can be built into new urban developments to complement grid supply during peak times, and they can be operated in island mode (off grid) during extended outages.

Solar Energy

Advances in solar photovoltaic panel technology have resulted in a dramatic drop in the capital cost of domestic installations. While solar photovoltaic panels are popular for small installations, solar thermal systems are viable for a wide range of commercial applications and utility scale electricity generation. An excellent example of the use of solar thermal for hydroponics is the Sundrop Farms installation in South Australia (Sundrop Farms, 2019). Solar power provides the energy to maintain a uniform greenhouse environment to produce vegetables and fruit, and to generate freshwater from seawater. There are many other applications for solar energy supporting SCT initiatives e.g.:

- Powering essential 24/7 services in conjunction with base load power generation.
- Powering all aquaponics and other production systems.

- Street vendor stands and small business operations.
- Recharge points for all electric vehicles.
- Street lighting and other community lighting.
- Shared energy sources for high density community housing in urban areas.

These systems range in power output of a few watts to many kilowatts but become a part of an integrated energy delivery system in the SCT. Solar, even with battery storage, has low entry-level costs compared with other forms of renewable energy. These initiatives complement the Power Africa goals to increase power generation by 30,000MW with 60 million new connections by 2020 to reach 300 million Africans (Power Africa, 2019b).

BioMass Materials

This refers to organic matter used as a fuel, especially in a power station, for the generation of electricity. Such materials are readily available and accessible sources of energy. Examples of such organic material include:

- Wood – including offcuts, municipal tree trimmings, construction wood waste, and packaging materials.
- Biomass waste from food manufacturing and processing.
- Sludge from municipal sewerage treatment, domestic septic, and industrial wastewater systems.
- Agricultural crop and animal residues.
- Municipal Solid Waste (MSW) or household waste.

The disposal of MSW to landfill or to various forms of incineration or other processing is problematic in developing countries. There are significant environmental concerns over air pollution due to the methane generated from organic waste decomposition, and surface and ground water integrity. The accumulation of MSW due to inadequate recovery and disposal systems poses serious health hazards. MSW typically includes food waste, plastics, glass, paper, metal cans, and garden trimmings. Recycling can recover only approximately 25% of household waste with the remainder going to landfill (EPA, 2019).

MSW also has economic value in respect of recycling and the production of other derived products including chemicals, fertilizers, transportation fuels, and synthetic gas. The organic component of MSW has high energy content and represents a feed stock for biomass Waste-to-Energy (W2E) generation (Seo, Alam, & Yang, 2018). It is estimated that one ton of MSW can produce up to 1,000 KWh of electricity using gasification technology (Global Syngas, 2019). Downdraft gasification converts organic material into a synthetic gas (syngas) through the application of extreme temperature in an oxygen starved environment. This is an environmentally superior solution to landfill and incineration. The Global Syngas (2019) can then be used for electricity generation with minimal environmental impact. The oxygen deficient downdraft gasification process prevents the formation of contaminants such as highly toxic dioxins and furans. Figure 10 illustrates the downdraft gasification process.

Biomass Waste-to-Energy (W2E) systems are appealing for low-income urban areas for the following reasons:

Figure 10. Downdraft Gasification Process

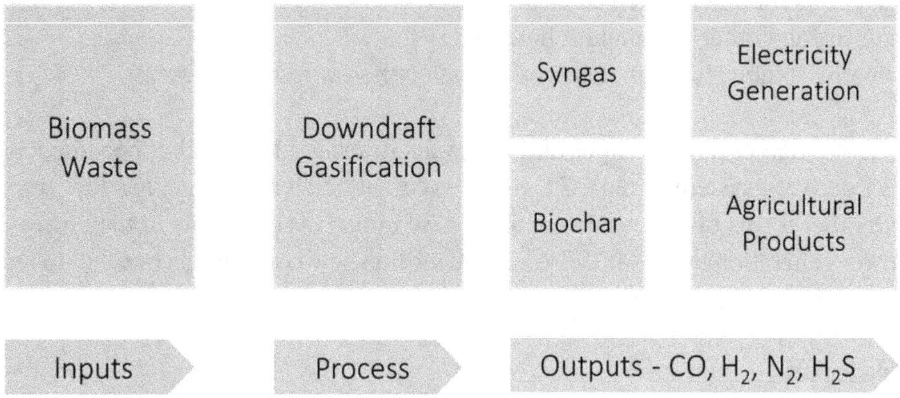

- **Urban clean-up** – urban waste including recyclable materials have economic value. Collection can be achieved at low cost while providing a source of income for residents. Transportation costs are minimized resulting in further reduction of greenhouse gas emissions.
- **Environmental impact** – unprocessed organic material releases methane gas and can cause pollution of surface and ground water. Urban clean-up results in a healthier environment.
- **Renewable energy** – a clean syngas can be used in combined heat and power systems to reduce reliance on fossil fuel electricity generation and to contribute to a carbon neutral environment.
- **Recyclable materials** – can be recovered, sorted, and packaged for sale, hence adding another revenue stream.

Downdraft gasification systems can be constructed as modular systems which allows for incremental expansion under expenditure budgets or to meet the growth needs of the community.

Peer-to-Peer Energy Trading

Peer-to-Peer (P2P) energy trading enables small producers/consumers (or prosumers) to trade their excess energy to local and external consumers. P2P trading platforms are suitable for domestic consumers and for businesses.

These and other trading concepts are being trialled around the world. For example, the government of Western Australia is the first government to allow P2P energy trading across the state-wide grid through the RENeW Nexus project. The project is funded by the Australian Government through the Smart Cities and Suburbs Program (Sundararajan, 2017). The aim of the project is to examine the value and efficiency of P2P energy trading in the City of Fremantle. Intelligent electric meters were installed at 50 houses across the City of Fremantle. The energy data was collected at frequent time intervals (down to 5 seconds) and uploaded via the Internet to cloud storage for processing. Data was collected on energy imported from the grid, energy exported to the grid, energy generated from rooftop photovoltaic systems, and energy consumption in the households. The results from the RENeW Nexus project are summarised as follows:

- The longest P2P trading period was 13.5 hours while the shortest was 9 hours.
- People tend to consume more energy on hot days for cooling and on cold days for heating. The highest P2P energy trading was therefore observed during winter and summer months, with the lowest during the spring.
- P2P trading was more active between 8AM and 10AM, and again at 3.30PM.
- It is cheaper to buy electricity through a P2P network.

The trial used blockchain to record all energy transactions. Blockchain is a type of distributed ledger technology that can be used to securely store digital transactions. Blockchain has demonstrated great uptake potential in P2P energy trading with a growing number of start-up companies, pilots, trials, and research projects adopting the technology within their business model. The revolution of blockchain encourages innovation and enables a transition to low-carbon sustainable energy systems (Mattila et al., 2016). According to reports by Deloitte (Grewal-Carr & Marshall, 2016) and PWC (Hasse et al. 2016), blockchain has the potential to disrupt the energy sector using energy commodities as digital assets to be traded.

Blockchain facilitates direct payment of transactions between consumers and prosumers without the need for a central billing authority. Payments can be made more frequently such as every five seconds, in contrast to the more common monthly or bi-monthly billing periods. Energy micro-payments and pay-as-you-go solutions are not only beneficial to consumers but also to utility companies (Andoni et al., 2019). The creation of a time-stamped trading transaction is verifiable and transparent. Energy trading transactions are recorded in multi-location data structures, thereby providing resilience and robustness to the blockchain system. Security of energy supply can be achieved by increasing the volumes of renewable energy and implementing efficient and flexible electricity trading markets (Mylrea & Gourisetti, 2017).

In summary, the potential impacts and benefits of blockchain for P2P energy trading include the following:

- Efficient, flexible, and robust renewable electricity trading market on a virtual P2P basis.
- Energy security, both cyber security and security of energy supply.
- Energy cost reduction through energy optimization.
- Sustainability, by facilitating renewable energy generation and low carbon solutions.

These benefits offer considerable potential for slum redevelopment and the establishment of economic centres for energy trading within the SCT.

HOUSING

Ghanaweb, an online media entity, has reported that the Ghana government estimated a shortfall of 2 million homes in Ghana by 2020 (Ghanaweb, 2019b). On the conservative assumption that each household comprises 5 people, this equates to 10 million people, or about 1/3 of the population of Ghana, without adequate housing. Ghanaweb also reported a government initiative to raise $3.1 billion to finance 200,000 new homes at about $15,500 each using conventional building techniques. On the same assumption for occupancy, this investment would provide housing for about 1 million people at a total cost of ownership that is well in excess of $15,500 per home. These homes would not be affordable by low-income slum dwellers and, therefore, an alternative approach to redevelopment of slum housing is required.

It was earlier estimated that approximately 788,000 people live in slums in Accra city. This would require 157,600 new home constructions in Accra at a substantially reduced total cost of ownership. Cooke, Hague, and McKay (2016), in their report for UNICEF on poverty and inequality in Ghana, have claimed that in 2016, there were 6.4 million people (including 3.65 million children) living on less than $312/adult/year. This sector of the population would remain without adequate housing unless there was a significant improvement in their economic status. Thus, a whole new approach to housing for low-income households is required.

Slums and poverty are related, and slums are often the only place to find affordable housing by people migrating from rural areas. Subsidized housing would not be enough to draw people out of the slums because the operational costs of living in an enhanced environment would still be a challenge for these home seekers.

One potential solution to lowering housing costs is the 3D printing of homes. New Story, a USA non-profit organisation, has focused on ending global homelessness, and claimed that they can produce low cost housing that is affordable for people living on less than $200 per month. This is a significant improvement over conventional construction techniques and opens opportunities for lower cost subsidized housing. New Story are planning to print a village of homes in Latin America in less than 24 hours. The 3D printing of large structures such as homes and commercial buildings has been proven viable and will become more affordable under economies of scale (All3DP, 2019). The cost of construction will also benefit from the ability to use locally sourced construction materials.

The 3D printing industry has developed rapidly. Small complex objects using different materials such as bionic eyes, skin, antibacterial teeth, bionic ears, bones, and even ovaries, have been printed (Murphy, 2018). 3D printing of larger structures is attracting much interest and there are examples of the successful printing of houses and industrial buildings. All3DP (2019), a magazine dedicated to reporting 3D printing developments and innovations, claims that 3D printed houses offer many advantages, such as the following:

- Environmentally friendly - suitable for a wide variety of building materials including mud, organic waste, rice husks, and plastic.
- Inexpensive – residential houses to industrial buildings can be constructed at a relatively low price.
- Shorter construction periods - the printing robot can be programmed for the house design and then set in operation.
- Uncommon shapes - 3D printing can output complex shapes. One interesting possibility is the printing of domestic aquaponics systems, solar energy systems, and toilets.

Figure 11, from iconbuild.com, illustrates the type of a low-cost printed house that New Story is planning to construct in Latin America.

A window to the vision of 3D printing for buildings can be seen through the achievements of Apis Cor, a USA company specializing in the design and development of specialized 3D printing equipment for the construction industry (Apis Cor, 2019a and 2019b). They have successfully printed commercial, residential, and affordable housing structures. The company is aiming to be one of the first companies to construct buildings on the Moon and Mars, after being a winner of the National Aeronautics and Space

Figure 11. Example of Digitally Printed House

Administration 3D-Printed Habitat Competition (Apis Cor, 2019b). Figure 12 illustrates an image of a commercial building printed by Apis-Cor.com in Dubai.

WATER SUPPLY AND MANAGEMENT

Supply of clean water in slum districts is a major issue. In Accra, many areas are not serviced with piped water. This means that water must be transported by truck to homes and businesses where it is pumped to polytanks. In slums, this is not possible because of access difficulties and the cost of polytanks. People must carry water from delivery points to maintain daily fresh supplies in their homes.

Urban water provision is failing to meet the demands of the population in Sub-Saharan Africa. Well-water is often problematic because of potential pollution effects from inadequate water protection practices (Grönwall & Oduro-Kwarteng, 2017). In high humidity regions, atmospheric water can be harvested to further reduce the environmental impact. Atmospheric Water Generation (AWG) requires only air and electricity to produce potable water after filtration (Aboueleta, 2019). AWG systems are highly portable and can be located anywhere but are currently limited to small quantities of potable water. Cost is the current limiting factor for widespread use of AWG. AWG systems may be digitally printed into building structures and utilize existing solar or other renewable energy to lower costs.

The average annual rainfall in Accra is 809mm (WorldData, 2019). Rainwater capture can be facilitated for domestic use when homes are digitally printed to incorporate roof drainage and tank storage. For industrial buildings that have significant roof area, storage tanks can supply water for building operations and with appropriate water recycling strategies they can reach self-sustainability. For example, aquaponics is an intensive ecosystem that employs an almost closed loop system to grow fish and vegetables with

Figure 12. Example of Digitally Printed Commercial Building

only 1/6th the water consumption of open land farming (Nelson & Pade, 2019b). Commercial buildings that are air-conditioned use systems that dehumidify the input air stream and generate water. This water can also be collected and stored and used for toilets, and for general use after treatment.

Future systems will need to consider ground water remediation or desalination using developing technologies such as carbon nanotubes (Ihsanullah, 2018). The extension of the city water supplies received from the Akosombo dam to slum regions will incur substantial capital cost and water levels in the Volta are declining because of climate change (Ndehedehe, Awange, et al. 2017).

SANITATION AND HEALTH

The Bill and Melinda Gates Foundation are sponsoring global efforts to address the chronic situation of sanitary diseases arising from inadequate toilet systems (Bill and Melinda Gates Foundation, 2013). They claim that 2.5 billion people do not have access to safe and affordable sanitation and are funding and promoting the *Water, Sanitation & Hygiene: Reinvent the Toilet Challenge* which aims to develop a new toilet system for use in developing and developed countries globally. The challenging criteria require that the toilet must:

- Remove germs from human waste and recover valuable resources such as energy, clean water, and nutrients.
- Operate off the grid without connections to water, sewer, or electrical lines.
- Cost less than US$.05 cents per user per day.

- Promote sustainable and financially profitable sanitation services and businesses that operate in poor urban settings.
- Is a truly aspirational next-generation product that everyone will want to use in developed as well as developing nations.

Selected responses to this challenge, listed in Table 1, are being currently field tested in Durban in South Africa.

Table 1. Toilet System Trials

Name of System	Developer	Description
Nano-membrane Toilet	Cranfield University	Household toilet uses no external energy or water.
Eco San	Caltech	Solar powered public toilet system.
NEWGenerator	University of South Florida	Suitable for use in homes and shared toilet facilities.
Neighbourhood Treatment System	Duke University	Public toilet system using supercritical water oxidation.
Firelight Home Toilet	Sedron Technologies	Household toilet converts faecal matter into electricity.

These developments have significant ramifications for the SCT (Slum City Transformation). Owusu-Ansah, Tagbor, and Togbe (2016) researched access to health care in slums in the Sodom and Gomorrah district and observed that malaria and diarrhoeal diseases linked to poor sanitation conditions in the district were most prevalent. In the slum areas of Accra, pay-per-use public toilets are the main facility but open defecation is common, especially among children (Emory, 2016). Owusu-Ansah et al. (2016) observed that slum dwellers who were enrolled in the National Health Insurance Scheme (NHIS) in Ghana were more likely to seek medical attention. Others sought advice and treatment from pharmacists and other sources. They also noted that the cost of the initial joining fees was the main deterrent to NHIS membership registration. There is a public health issue in the slums that requires the development of greater awareness of available health care services. The SCT can provide information on health risks, preventative measures, and community health service providers. New homes in the SCT can be printed with in-house toilet facilities, and the public toilets under trial in Durban, South Africa, hold great promise for wider community access to safe environmentally sustainable toilets.

TRANSPORTATION AND ACCESS

Slums are typically highly dense configurations of buildings where people live. The ability to move around is limited by the space between buildings which is often almost impassable because of accumulated trash in walkways. The design of the SCT is limited by the amount of land available and the density of the population. Some separation can be achieved with multi-level buildings but the wide suburban streets visible in modern urban planning are not possible.

This suggests an opportunity to encourage the use of other forms of transport. Improved walking paths will make it easier to move in and out of more delineated housing layouts, and bicycle paths encourage pedal-powered transport. All motorized vehicles are prohibited and only electric vehicles for people movement, deliveries of bulky items, and service vehicles are permitted. The objective is to provide a fast and less strenuous means to arrive at transport hubs for longer journeys or to escort children to and from school. Solar equipped houses and power poles from the micro-grids serving the communities will be accessible for vehicle recharging. All community vehicle assets can be managed and safeguarded using GPS to monitor movements.

EDUCATION

School absenteeism is much higher for children living in slums in Ghana than for children not living in slums (Glenn, 2014). The main reason is usually economic. Children often miss school to assist parents in the market place. They are given a choice of going to school and not having food or going to the market where they can help sell items for food. Schooling in Ghana has been free for all students since the 2017 introduction of the free secondary high school program. However, the costs of uniforms, books and other materials, and various fees must be paid by parents. This places economic stress on families and even if donor funds are available these funds may not necessarily be used for the intended purpose (Maslin Nir, 2019). It is noted that there are exceptions with evidence that some children from slums are enrolled in private preschools (Felsman, 2014). There is insufficient evidence to conclude that this results from parents' ability to pay or some external donor source. There is a high correlation between poverty and slums. Improvement of slum conditions must not only lead to improved economic circumstances but also lead to integrative local partnerships and social programs to ensure that no child is left behind.

IOT AND DATA ANALYTICS

Much is written about the Internet of Things (IoT), big data, and data analytics, and how this is going to change the nature of our cities. Mahabir et al. (2016) and Bettencourt (2014) present the idea that cities are a collection self-organizing social structures and that the city is shaped by these structures and the individuals who live and work there. Prior to the age of ICT, the Internet, and intelligent mobility devices, print media disseminated local, regional and global news. Interpersonal communications were facilitated by written exchanges delivered by the postal service, or verbally via the telephone (Bettencourt, 2014). These large interconnected social networks benefited from information sharing which gave voice to community ideas and guidance to the shaping of the environment by urban planners. New technologies for information collection and sharing are supportive and enabling these *social dynamics* (Bettencourt, 2014). Cities assume a character that is reflective of its culture and shared beliefs, which also includes a voice from slum dwellers. IoT, big data and data analytics are terms that refer to the data collection, storage and processing, and visualization of information. Technology has made it possible to collect and process vast amounts of data quickly. The following sections present an overview of the information framework, the basic data classifications for the SCT, the data collection methods, and the cloud processing platforms.

Information Framework

The transition from a slum city to a functional intelligent city is an incremental one. Trash-can sensors to indicate the need for emptying, heads-up displays in vehicles for traffic routing, autonomous vehicles for public transport, intelligent street lighting, gaming and virtual reality, or even video surveillance of traffic intersections and high crime locations should not be contemplated at the outset. Instead, careful consideration should be given to the establishment of the data sharing structures so that government agencies, industry and individuals can submit data for processing, and then receive information in a variety of visualizations. In this sense, transforming of slum cities to functional intelligent cities can have significant advantages because the intelligent systems are being built bottom-up against a top-down design. The fundamental structure of identifying the appropriate data types, the collection of data, storing and processing, disseminating information to the community, and interacting with the community should be the basic goal of intelligent city strategies, especially in developing countries.

While the data collection, storage, processing, and visualization is relatively straightforward for some data streams, problems arise when integrating the dataflows from different sources. Common cross-platform semantics will enable consistent interpretation of the data and ease of processing. Data exchange models are being tested in different cities globally. Alternatively, as depicted in Figure 13, information can be accessed through well-defined Application Programming Interfaces (APIs) to various entity platforms, such as government agencies. This then enables the compilation of dashboard presentations for mobility devices and computers to suit the intelligent city end user.

Figure 13. API between Service Provider Systems and Smart City Applications

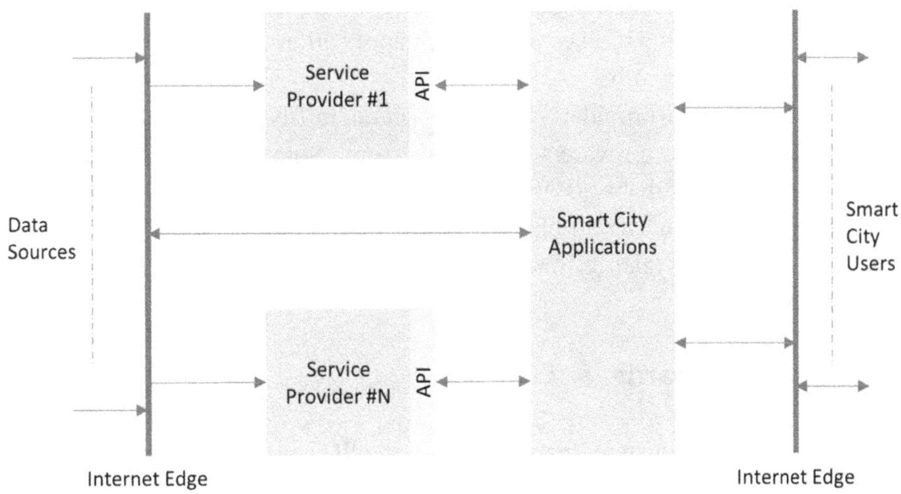

Data Classification

There are six broad classifications of data that relate to the intelligent city model:

- **Economic Data** – this is generated from the commercial and household income data.
- **Social Data** – it relates to the general wellbeing of the community and includes both quantitative (such as school attendances, and clinic visits) and qualitative (social structures, interest groups, and cultural events) data.
- **Environmental Data** – it is a wide range of critical data including weather, air quality, water quality, sanitation, and land usage.
- **Structural Data** – this refers to energy sources and distribution, road types and quality, transport systems, food sources and supply, schools, and clinics.
- **Institutional Data** – this relates to the presence and engagement of institutions such as those agencies involved in welfare, health, education, aid agencies, non-government organizations (NGOs), and research programs.
- **Remote Sensing Data** – this is data from satellite and other aerial platforms of the city metropolitan area, including the slum under transformation.

Each entity within the SCT is capable of collection of a diverse range of data. Table 2 provides a broad view of the types of data to be collected under each classification and the entity primarily responsible for the collection. Table 3 shows the sources of such data and how these sources are accessed.

Data Processing

The Internet Cloud has made it possible to access vast storage capabilities and serverless computing. This allows all SCT data to be stored in the cloud and processed on demand for minimal cost. Furthermore, the cloud-based environment eliminates the cost of in-house servers, their upgrades, and operational management. Data backup and data integrity are managed within the cloud for minimal cost. Several providers charge only for the time it takes to run each application. High speed Internet is the minimal requirement for SCT to perform well.

Ghana has access to several providers who have adequate fibre capacity. There are several well-established cloud-based service providers to choose from. Some of the best-known providers are Microsoft Azure, BigQuery, Alibaba, IBM Cloud, and Amazon Web Services. Figure 14 illustrates the general structure of the cloud-based platforms. These cloud-based service providers offer similar services including fast database queries, machine learning capabilities, data analytics, and pay-as-you-use application processing.

Smart City User Dashboards

Smart city users comprise all registered participants of the SCT including individuals, registered businesses, street vendors, and all other service provider nominees. Each dashboard is tailorable to the requirements of each participant. Individuals can monitor their home environment, including electricity consumption, water usage, security status, and financial transactions from their smartphones, tablets or computers. Businesses report total sales by product and price, and employee or agent payments. It is possible to determine the extent of cash generated and distributed to slum dwellers.

Table 2. Data to be Collected and Collection Entities

Data Classification	Description	Reporting Entity(s)
Economic	Total revenues	SCT Commercial Enterprise
	Number, age and gender of employees	SCT Commercial Enterprise
	Household incomes	Households
	Household members, gender, and age	Households
	Mobile phone credit top-ups[1]	Households/Individuals
	Mobile money transfers/receipts[1]	Households/Individuals
Social	School attendance	Schools
	Gender of attendees	Schools
	Age of attendees	Schools
	Number of health clinic visits	Clinics
	Gender of patients	Clinics
	Type of treatment sought	Clinics
	Population	Ghana Statistical Service
Environmental	Water source and usage	SCT
	Electricity source and usage	SCT
	Weather and air quality	SCT Weather Stations
	Field observations (flora and fauna)	Schools
Structural	Energy supply and distribution	SCT
	Road types and quality	SCT
	Transport systems	SCT
	Food sources and supply	SCT
	Schools and facilities	SCT
	Clinics	SCT
Institutional	Number of institutions	SCT
	Location of each institution	SCT
	Number of people in each institution	SCT
	Primary activity of each institution	SCT
Remote Sensing	Satellite image data	Maxar (https://www.maxar.com)
Note 1:	*The automatic collection of mobile phone credits and mobile money transfers/receipts requires the permission of the account owner to automatically report this data using a downloaded purpose-built App.*	

CONCLUSIONS

Slums are a persistent and increasingly serious problem for cities in the developing world. Linked with poverty, they have resisted attempts to eliminate them through resettlement, redevelopment, self-development through land ownership, forced eviction, and other tried and failed approaches. In Sub-Saharan Africa, the population of urban slums is often more than 50% of the total urban population.

Table 3. Methods of Data Collection

Data Classification	Description	Source	Accessed
Economic	Revenues and costs	API	On-demand
	Number, age and gender of employees	API	On-demand
	Household incomes	App/Kiosk	On-demand
	Household members, gender, and age	App/Kiosk	On-demand
	Mobile phone credit top-ups	App	Automatic/event
	Mobile money transfers/receipts	App	Automatic/event
Social	School attendance	App	As available
	Gender of attendees	App	As available
	Age of attendees	App	As available
	Number of health clinic visits	App	As available
	Gender of patients	App	As available
	Type of treatment sought	App	As available
	Population	API	When available
Environmental	Water source and usage, electricity source and usage	App	As available
	Weather and air quality	SCT Weather Station	Streaming data
	Field observations (flora and fauna). School programs.	App	As available
Structural	Energy supply and distribution	App	Scheduled
	Road types and quality	App	Scheduled
	Transport systems	App	Scheduled
	Food sources and supply	App	Scheduled
	Schools and facilities	App	Scheduled
	Clinics	App	Scheduled
Institutional	Number of institutions	App	Scheduled
	Location of each institution	App	Scheduled
	Number of people in each institution	App	Scheduled
	Primary activity of each institution	App	Scheduled
Remote Sensing	High resolution imagery	Download	Biannually
	Mosaic imagery	Download	Biannually
	Infrared imagery	Download	Biannually
Legend			
API	*Application Programming Interface.*		
App	*A purpose-built application for mobility devices (smartphones, tablets, and notebook computers) that collects the designated data and sends it automatically, scheduled, or on-demand to the SCT for processing.*		
Kiosks	*At SCT facilities that enable input of data as a convenience/alternative to mobility Apps.*		
SCT Weather station	*A comprehensive weather station collecting temperature, humidity, rainfall, wind speed and direction, solar radiation, and air pollutant information.*		
On-demand	*Data collection is initiated as required for processing.*		
Automatic/event	*Details of each event are transmitted as they happen.*		
As available	*Data is sent by the person collecting the information when it is available.*		
Streaming data	*Data, such as from weather stations and video sources are streamed in real-time.*		
When available	*Data is collected from external sources when a notice of availability has been received.*		

Figure 14. General Structure of Cloud-Based Platforms

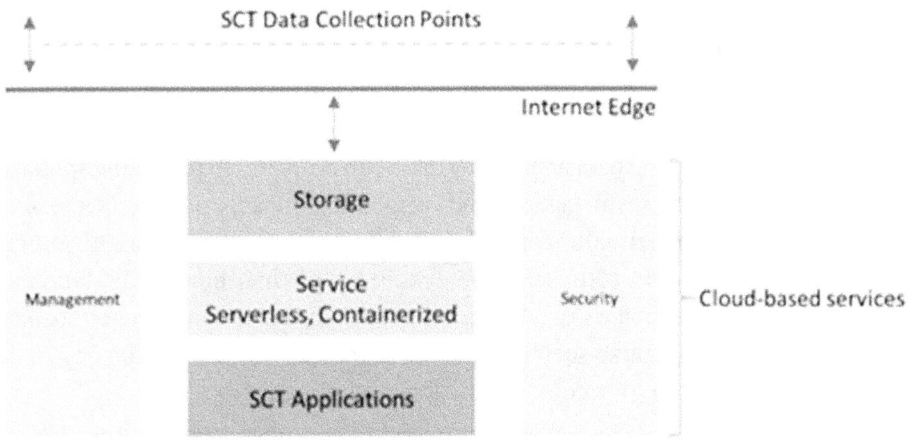

There is no clear definition for a slum city, but the magnitude of the problem requires major policy initiatives to provide alternatives to the deprivation of living in slums. However, policy initiatives may not be the most efficient approach in countries where there is weak governance, corruption, autocracy, or regime instability.

Any rehabilitation activity must address the economic prospects for slum dwellers. The approach taken in this chapter is an incremental process of replacing segments of the slum with a new livelihood paradigm with employment and small business opportunities arising from products and services produced from within the new communities. Each increment is highly interactive with the slum community it replaces, while being conscious of the social fabric and broader community environment. There is necessarily a cost to all stakeholders. However, the benefits are improved wellbeing through better living conditions, improved incomes, and a cleaner and safer environment. This, in turn, has a positive impact on the GDP of the city.

Numerous income-generating ideas have been introduced. Aquaponics, as an urban farm, provides a connection with the previous rural livelihoods of many of the slum dwellers. The collection of organic waste, which is a visible marker of the slum environment, and its conversion to electricity or other products has both environmental and economic benefits. Many daily in-use commodity products are imported. Local manufacture not only serves as import replacement but offers employment and other economic opportunities. The problem of providing housing for its residents is not just an issue for cities in developing countries but also a challenge for developed countries. On some estimates, there is an accumulating shortfall of housing of about 4,000 homes per hour globally. The cost of home ownership is being forced beyond the reach of many young people.

This chapter has introduced the idea that functional homes can be digitally printed. While this process is in its early development phase, it offers great promise for customizing homes that meet the needs of people emerging from poverty. They can be configured to provide improved sanitation and to include income earning systems. As rehabilitation progresses, the home designs can be adjusted based on lessons learned from the initial constructions. Ownership of the land being redeveloped is also an issue in many Sub-Saharan countries. This chapter has given some insights into the challenges of land ownership in

Ghana. It will take a major policy effort to redefine urban land entitlements to give confidence to home owners and business developers that their investments are secure and are protected by the law.

The slum city transformation is a partnership between government, private investors and the slum dwellers. Each stakeholder has an equally important role to play to bring success to the project. Data collection and processing provides an on-going measure of performance. The IoT is introduced to highlight the ease of attaching a wide range of sensors and other Internet edge devices. Remote sensing imagery, combined with other near terrestrial data and survey data, can be useful in providing spatial and temporal visualizations of the slum city transformation, and to monitor impacts outside of the slum areas. Big data and the IoT applied are the stimulus for the implementation of broader intelligent city strategies. All stakeholders are provided with insights to the conditions of their household, community and city through easy to access data dashboards that are tailored to their needs. This open transparent access to information helps define and organize social structures in the city and allows the city to develop under its own unique culture and vision to become a functional intelligent city.

REFERENCES

Aboueleta, M. (2019). *Atmospheric water harvesting: Our one environmental chance against a water crisis*. Retrieved from https://www.quantumrun.com/article/atmospheric-water-harvesting-our-one-environmental-chance-against-water-crisis

Africa, P. (2019a). *Ghana. Energy sector overview*. Retrieved from https://www.usaid.gov/sites/default/files/documents/1860/Ghana_-_November_2018_Country_Fact_Sheet.pdf

Africa, P. (2019b). *Two out of three people in Sub-Saharan Africa lack access to electricity*. Retrieved from https://www.usaid.gov/powerafrica

African Union. (2019). *Our aspirations for the Africa we want*. Retrieved from https://au.int/en/agenda2063/aspirations

All3Dp. (2019). *About Us*. Retrieved from https://all3dp.com/about/

Andoni, M., Robu, V., Flynn, D., Abram, S., Geach, G., Jenkins, D., ... Peacock, A. (2019). Blockchain technology in the energy sector: A systematic review of challenges and opportunities. *Renewable & Sustainable Energy Reviews*, *100*, 143–174. doi:10.1016/j.rser.2018.10.014

Asante, F. A., & Amuakwa-Mensah, F. (2015). Climate Change and Variability in Ghana: Stocktaking. *Climate (Basel)*, *3*(1), 78–99. doi:10.3390/cli3010078

Beegle, K., Christiaensen, L., Dabalen, A., & Gaddis, I. (2016). *Poverty in a rising Africa*. Retrieved from https://www.un.org/africarenewal/sites/www.un.org.africarenewal/files/Poverty%20in%20a%20Rising%20Africa%20Overview.pdf

Bettencourt, L. (2014). The Uses of big data in cities. *Big Data*, *2*(1), 12–22. doi:10.1089/big.2013.0042 PMID:27447307

Bhorat, H., & Naidoo, K. (2014). *Drivers of inequality in the context of the growth-poverty-inequality nexus in Africa: An overview of key issues*. Retrieved from https://www.undp.org/content/dam/rba/docs/Reports/undp-rba_Income%20Inequality%20in%20SSA_Chapter%203.pdf

Bill and Melinda Gates Foundation. (2013). *Water, sanitation & hygiene: reinvent the toilet challenge fact sheet*. Retrieved from https://docs.gatesfoundation.org/Documents/Fact_Sheet_Reinvent_the_Toilet_Challenge.pdf

Brown University. (2019). *Favelas in Rio de Janeiro, past and present*. Retrieved from https://library.brown.edu/create/fivecenturiesofchange/chapters/chapter-9/favelas-in-rio-de-janeiro-past-and-present/

Chayes, S. (2015). *Thieves of State*. New York, NY: W. W. Norton & Company.

CNBC. (2018). *New Story is a nonprofit that hopes to use 3-D printers to build homes in poor communities*. Retrieved from https://www.cnbc.com/2018/03/15/new-story-nonprofit-will-3-d-print-homes-in-poor-communities.html

Cooke, E., Hague, S., & McKay, A. (2016). *The Ghana Poverty and Inequality Report*. Retrieved from https://www.unicef.org/ghana/media/531/file/The%20Ghana%20Poverty%20and%20Inequality%20Report.pdf

CorA. (2019a). *Home*. Retrieved from https://www.apis-cor.com

Cor, A. (2019b). *Top five teams win a share of $100,000 in virtual modeling stage of NASA's 3D-Printed Habitat Competition*. Retrieved from https://www.nasa.gov/directorates/spacetech/centennial_challenges/3DPHab/five-teams-win-a-share-of-100000-in-virtual-modeling-stage

Emory. (2016). *Should public toilets be part of urban sanitation solutions for poor families living in slums?* School of Public Health. Retrieved from http://sanipath.org/wp-content/uploads/2016/04/SaniPath-Public-Latrin-Policy-Brief.pdf

Engstrom, R., Pavelesku, D., Tanaka, T., & Wambile, A. (2019). Mapping poverty and slums using multiple methodologies in Accra, Ghana. *2015 Joint Urban Remote Sensing Event. JURSE, 2019*. doi:10.1109/JURSE.2019.8809052

Engstrom, R., Sandborn, A., Yu, Q., Burgdorfer, J., Stow, D., Weeks, J., & Graesser, J. (2015). Mapping slums using spatial features in Accra, Ghana. *2015 Joint Urban Remote Sensing Event. JURSE, 2015*. doi:10.1109/JURSE.2015.7120494

EPA. (2019). *Facts and Figures about Materials, Waste and Recycling*. Retrieved from https://www.epa.gov/facts-and-figures-about-materials-waste-and-recycling/national-overview-facts-and-figures-materials

Eregha, P. B. (2015). Foreign direct investment inflow, volatility, and domestic investment in West Africa. *Journal of Developing Areas*, *49*(2), 273–294. doi:10.1353/jda.2015.0018

Farms, S. (2019). *Innovation. Sundrop is the future*. Retrieved from https://www.sundropfarms.com/innovation/

Felsman, C. (2014). *Schools in Slums: A surprising number of the poorest kids are enrolled in private preschools*. Retrieved from https://educationinnovations.org/blog/schools-slums-surprising-number-poorest-kids-are-enrolled-private-preschools

Ghana Statistical Service. (2019a). *Population by regions - Greater Accra*. Retrieved from http://www.statsghana.gov.gh/regionalpopulation.php?population=MTM0NTk2MjQzOS4yMDE1&&Greater%20Accra&regid=3

Ghana Statistical Service. (2019b). *Population by regions - Greater Accra*. Retrieved from https://www.statsghana.gov.gh/regionalpopulation.php?population=MTI5MzE3OTU5OC40NDg1&&Ashanti&regid=1

GhanaPost. (2019). *Ghana's official digital addressing system*. Retrieved from https://ghanapostgps.com/mapview.html

Ghanaweb. (2019a). *Akufo-Addo wants vibrant Ghana aluminium industry*. Retrieved from https://www.ghanaweb.com/GhanaHomePage/business/Akufo-Addo-wants-vibrant-Ghana-aluminium-industry-727527

Ghanaweb. (2019b). *Ghana's housing deficit hits two million – Housing Minister*. Retrieved from https://www.ghanaweb.com/GhanaHomePage/business/Ghana-s-housing-deficit-hits-two-million-Housing-Minister-739235

Glenn, G. (2014). *In Ghana, school absenteeism and poverty go hand-in-hand*. Retrieved from http://news.wypr.org/post/ghana-school-absenteeism-and-poverty-go-hand-hand#stream/0

Grewal-Carr, V., & Marshall, S. (2016). *Blockchain: Enigma. Paradox. Opportunity*. Deloitte. Retrieved from https://www2.deloitte.com/content/dam/Deloitte/uk/Documents/Innovation/deloitte-uk-blockchain-full-report.pdf

Grönwall, J., & Oduro-Kwarteng, S. (2017). Groundwater as a strategic resource for improved resilience: A case study from peri-urban Accra. *Environmental Earth Sciences*, *77*. doi:10.100712665-017-7181-9

Hasse, F., von Perfall, A., Hillebrand, T., Smole, E., Lay, L., & Charlet, M. (2016). *PwC global power & utilities, Blockchain - an opportunity for energy producers and consumers?* Retrieved from https://www.pwc.com/gx/en/industries/assets/pwc-blockchain-opportunity-for-energy-producers-and-consumers.pdf

Ihsanullah, I. (2018). Carbon nanotube membranes for water purification: Developments, challenges, and prospects for the future. *Separation and Purification Technology*, *209*, 307–337. doi:10.1016/j.seppur.2018.07.043

Institute for Security Studies. (2019). *Reducing poverty in Africa: huge potential, but more realistic goals needed*. Retrieved from https://issafrica.org/about-us/press-releases/reducing-poverty-in-africa-huge-potential-but-more-realistic-goals-needed

Jena, A. K., Biswas, P., & Saha, H. (2017). Aquaponics- a step towards urban agriculture. Innovative Farming. *International Journal of Agriculture.*, *1*, 163–167. Retrieved from https://www.researchgate.net/publication/316191741_Advanced_Farming_Systems_in_Aquaculture_Strategies_to_Enhance_the_Production

Kedir, A., Elhiraika, A., Chinzara, Z., & Sandjong, D. (2017). Growth and development finance required for achieving sustainable development goals (SDGs) in Africa. *African Development Review, 29*(S1), 15–26. doi:10.1111/1467-8268.12230

Kumi, E. N. (2017). *The electricity situation in Ghana: Challenges and opportunities*. CGD Policy Paper. Washington, DC: Center for Global Development. Retrieved from https://www.cgdev.org/sites/default/files/electricity-situation-ghana-challenges-and-opportunities.pdf

Mahabir, R., Crooks, A., Croitoru, A., & Agouris, P. (2016). The study of slums as social and physical constructs: Challenges and emerging research opportunities. *Regional Studies. Journal of Regional Science, 3*(1), 399–419. doi:10.1080/21681376.2016.1229130

Maslin Nir, S. (2019). *In Ghana, free high school brings opportunity and grumbling*. Retrieved from https://www.nytimes.com/2019/06/15/world/africa/ghana-free-senior-high-school-brings-chaos.html

Mattila, J., Seppälä, T., Naucler, C., Stahl, R., Tikkanen, M., Bådenlid, A., & Seppälä, J. (2016). *Industrial Blockchain Platforms: An Exercise in Use Case Development in the Energy Industry*. ETLA Working Papers 43, The Research Institute of the Finnish Economy. Retrieved from https://ideas.repec.org/p/rif/wpaper/43.html

Ministry of Energy. (2019). *Energy sector recovery program*. Retrieved from http://energycom.gov.gh/files/2019%201111%20ESRP%20ESTF_Clean_v3.0redacted%20final.pdf

Ministry of Finance. (2019). *Mid-year fiscal policy review of the 2019 budget statement and economic policy & supplementary estimates of the government of Ghana for the 2019 financial year.* Retrieved from https://drive.google.com/file/d/10bSdWq1woHs-IjK2HAz-UYKs7SlDjTym/view

Murphy, J. (2018). *7 amazing body parts that can now be 3D printed.* MDLinx. Retrieved from https://www.mdlinx.com/internal-medicine/article/2668

Mylrea, M. E., & Gourisetti, S. N. G. (2017). Blockchain for smart grid resilience: Exchanging distributed energy at speed, scale and security. *Resilience Week (RWS)*, 18-23. doi:10.1109/RWEEK.2017.8088642

Naab, F. Z., Abubakari, Z., & Ahmed, A. (2019). The role of climate services in agricultural productivity in Ghana: The perspectives of farmers and institutions. *Climate Services, 13*, 24–32. doi:10.1016/j.cliser.2019.01.007

Ndehedehe, C. E., Awange, J. L., Kuhn, M., Agutu, N. O., & Fukuda, Y. (2017). Analysis of hydrological variability over the Volta river basin using in-situ data and satellite observations. *Journal of Hydrology. Regional Studies, 12*, 88–110. doi:10.1016/j.ejrh.2017.04.005

Nelson and Pade. (2019a). *A non-gmo fish food for aquaponics*. Retrieved from https://aquaponics.com/product-category/aquaponic-growing-supplies/fish-food/

Nelson and Pade. (2019b). *Aquaponics*. Retrieved from https://aquaponics.com/

Owusu-Ansah, F. E., Tagbor, H., & Togbe, M. A. (2016). Access to health in city slum dwellers: The case of Sodom and Gomorrah in Accra, Ghana. *African Journal of Primary Health Care & Family Medicine, 8*(1), e1–e7. doi:10.4102/phcfm.v8i1.822 PMID:27247151

Patel, N. (2018). *Figure of the week: Understanding poverty in Africa.* Retrieved from https://www.brookings.edu/blog/africa-in-focus/2018/11/21/figure-of-the-week-understanding-poverty-in-africa/

Report, W. C. (2016). *Urbanization and development: Emerging Futures.* Retrieved from http://wcr.unhabitat.org/

Saito, N., Ruhanen, L., Noakes, S., & Axelsen, M. (2018). Community engagement in pro-poor tourism initiatives: Fact or fallacy? Insights from the inside. *Tourism Recreation Research, 43*(2), 175–185. doi:10.1080/02508281.2017.1406566

Seo, Y., Alam, M. T., & Tang, W. (2018). *Gasification of Municipal Solid Waste.* doi:10.5772/intechopen.73685

Stacey, P. (2018). Urban development and emerging relations of informal property and land-based authority in Accra. *Africa, 88*(1), 63–80. doi:10.1017/S0001972017000572

Sultan, B., Defrance, D., & Iizumi, T. (2019). Evidence of crop production losses in West Africa due to historical global warming in two crop models. *Scientific Reports, 9*(1), 12834. doi:10.103841598-019-49167-0 PMID:31492929

Sundararajan, S. (2017). *Australian Government grants $8 million for blockchain energy pilot.* Retrieved from https://www.coindesk.com/australian-government-grants-8-million-for-blockchain-energy-pilot

Syngas, G. (2019). *Waste to energy gasification.* Retrieved from https://www.globalsyngas.org/syngas-production/waste-to-energy-gasification/

The Rockefeller Foundation. (2019). *Accra.* Retrieved from https://nextcity.org/informalcity/city/accra

Transparency International. (2019). *Ghana.* Retrieved from https://www.transparency.org/country/GHA

Twerefou, D. K. (2012). *Report on Sustainable Development Goals for the West Africa Subregion.* United Nations Economic and Social Council Economic Commission for Africa. Retrieved from https://www.uneca.org/sites/default/files/uploaded-documents/SDG/2013/sdg2013_draft-sdgs-report-west-africa_en.pdf

UN-Habitat. (2018). *SDG Indicator 11.1.1 Training Module: Adequate housing and slum upgrading.* United Nations Human Settlement Programme (UN-Habitat), Nairobi. Retrieved from https://unhabitat.org/wp-content/uploads/2019/04/Indicator-11.1.1-Training-Module.pdf

United Nations. (2019a). *Sustainable development goals.* Retrieved from https://sustainabledevelopment.un.org/sdgs

United Nations. (2019b). *Sustainable Development Goal 11. Make cities and human settlements inclusive, safe, resilient and sustainable.* Retrieved from https://sustainabledevelopment.un.org/sdg11

United Nations. (2019c). *Growing at a slower pace, world population is expected to reach 9.7 billion in 2050 and could peak at nearly 11 billion around 2100.* Retrieved from https://www.un.org/development/desa/en/news/population/world-population-prospects-2019.html

USAID. (2019). *Agriculture and food security.* Retrieved from https://www.usaid.gov/west-africa-regional/agriculture-and-food-security

USDA. (2019). *Ghana Fish and Seafood Report*. Retrieved from https://apps.fas.usda.gov/newgainapi/api/report/downloadreportbyfilename?filename=Fish%20and%20Seafood%20Report_Accra_Ghana_3-8-2019.pdf

Verutes, G. M., Fiocco, M. B., Weeks, J. R., & Coulter, L. L. (2012). Health, poverty, and place in Accra, Ghana: Mapping Neighbourhoods. *Journal of Maps*, *8*(4), 369–373. doi:10.1080/17445647.2012.744365 PMID:23505395

World Bank. (2007). *Ghana country environmental analysis*. Retrieved from http://documents.worldbank.org/curated/en/719681468032166880/pdf/369850GH.pdf

World Bank. (2019a). *Overview*. Retrieved from www.worldbank.org/en/country/ghana/overview

World Bank. (2019b). *Poverty overview*. Retrieved from https://www.worldbank.org/en/topic/poverty/overview

World Economic Forum. (2018). *Slums could inspire the cities of the future. Here's how*. Retrieved from https://www.weforum.org/agenda/2018/01/this-startup-is-turning-slums-into-microcities/

World Economic Forum. (2019). *What slums can teach us about building the cities of the future*. Retrieved from https://www.weforum.org/agenda/2019/03/why-slums-could-provide-the-housing-inspiration-of-the-future/

WorldData. (2019) *Climate for Greater Accra (Ghana)*. Retrieved from https://www.worlddata.info/africa/ghana/climate-greater-accra.php

ADDITIONAL READING

Chayes. S. (2017). *Thieves of State: Why Corruption Threatens Global Security*. YouTube, 18:15. Retrieved from www.youtube.com/watch?v=jeO2p0JfOoo

Fleming. S. (2017). *Aquaponics and a New Way of Thinking*. TEDxCharlotte, 15:46. https://www.youtube.com/watch?v=df5OrJptsWU

Gates. W. (2018). *Bill Gates talks toilets*. YouTube, 2:17. Retrieved from https://www.youtube.com/watch?v=82pmo53FQIY

Insider, T. (2018). *3D-Printed Home Can Be Constructed for Under $4,000*. YouTube, 3:12. Retrieved from https://wwnw.youtube.com/watch?v=wCzS2FZoB-I

Keller. A. (2013). *Pop-up houses improve South African slums*. TEDxWWF, 17:09. Retrieved from https://www.youtube.com/watch?reload=9&v=p_YgOQp2uVM

Monney, I., & Antwi-Agyei, P. (2018). Beyond the MDG water target to universal water coverage in Ghana: The key transformative shifts required. *Journal of Water, Sanitation, and Hygiene for Development: a Journal of the International Water Association*, *8*(2), 127–141. doi:10.2166/washdev.2018.176

Murray. S. (2017). *A playful solution to the housing crisis*. TED@Westpac, 10:42. Retrieved from www.ted.com/talks/sarah_murray_a_playful_solution_to_the_housing_crisis/transcript?language=en

Waste, E. (2018). *Waste to Energy by Advanced Gasification*. YouTube, 4:26. Retrieved from https://www.youtube.com/watch?v=vVvCEkKxWs0

KEY TERMS AND DEFINITIONS

3D Printing: Three-dimensional printing of objects as directed by a set of software functions. Materials used for the printing are varied, thus allowing for intricate shapes and sizes to be printed from plastics, cement, carbon fibre, silver, steel, titanium, and other materials. Printers range in size from small desktop printers to large robotic machines that are used to print buildings by progressively layering materials under direction from the software. Also referred to as additive manufacturing.

Accra Metropolitan Assembly (AMA): One of 254 metropolitan, municipal and district assemblies in Ghana, and the political and administrative body for the city of Accra. It has a population of about two million people which is added to by another two million people who commute to the city daily for work, leisure and other purposes.

Agbogbloshie: A slum location in the city of Accra, commonly referred to as Sodom and Gomorrah. It is the location of a large e-waste dump where old computers and consumer electronics are processed to recover valuable materials such as copper, gold and other materials. The recycling process often involves burning which releases highly toxic gases into the atmosphere. With a population of about 45,000 mainly rural immigrants, Agbogbloshie has extremely poor living conditions and high crime rate.

Agriculture: Encompasses all activities related to the cultivation of the soil to grow crops and to raise livestock. It also includes the processing of plant and animal products into forms that are suitable for human consumption such as food, clothing, construction materials, and paper products.

Akosombo Dam: The location of a hydro-electric plant on the Volta River, Ghana. The construction of the Akosombo Dam flooded part of the Volta River Basin and created Lake Volta, the largest man-made lake in the world by surface area and third largest by water volume covering about 3.6% of Ghana's land area. The Akosombo Dam was originally intended to provide electricity for the aluminium industry and to the neighbouring countries of Togo and Benin.

Aquaculture: Can be likened to agriculture where fish is grown instead of plants and livestock. Fish is cultivated for human consumption. Fish farms may be constructed in ocean waters, rivers, ponds and tanks. When combined with hydroponics, fish and vegetables can be grown together in a fresh water aquaponics environment. According to some estimates, 50% of all seafood consumed today comes from aquaculture.

Aquaponics: A system for growing fish, vegetables, and fruits in a controlled environment that uses only water and fish food. Fish provide nutrients for the vegetables and fruits which in turn cleanse the water for the fish. Aquaponics can be scaled from small home-based systems to large commercial systems. Aquaponics is contained within controlled greenhouse environments for year-round farming and does not use herbicides or pesticides. It consumes much less water and occupies significantly less land area than conventional agriculture. Aquaponics enables sustainable food production in areas susceptible to climate change.

AWG: Atmospheric water generation. The extraction of water from the atmosphere using electricity.

Biomass: Any organic material that derives from plants and animals including wood and wood processing waste, agricultural crops and waste materials, household waste, and sewerage solid waste. Biomass contains energy that can be released under different techniques to produce other products including synthetic gases, fertilizers and biofuels. Municipal solid waste is a large source of accessible energy that can be used as feedstock for processing with minimal environmental impact.

Blockchain: A type of highly distributed database or system of ledgers that ia linked using cryptography. A blockchain network maintains permanent records of transactions which are managed through peer-to-peer protocols requiring a majority level of consensus for any change. Blockchain was originally developed as the highly distributed secure framework for cryptocurrency trading. It has more recently been applied to the secure tracking of food from producer to consumer, to the tracking of high value assets, and to other applications where high levels of documentation create elevated transactional risk.

Crowd-Sourcing: The sourcing of goods, services, and money from the public, usually via the Internet. In this contribution, it refers to the collection of data. Crowd-sourcing of specific data can be facilitated by purpose-built mobility applications for smartphones and tablets to provide greater survey capabilities over larger populations than traditional survey techniques. Crowd-sourcing of data in a more general sense is that captured by social media platforms such as Facebook and Twitter, and large search engines such as Google and Yahoo.

Dashboard: A computer or mobility device display that provides a summary of information of interest to the user and allows access to more detailed information. In the intelligent city context, the user may choose what government services (such as welfare, health, education) are most important, public transport schedules, community activities of interest, employment opportunities, locations of family members, local news feeds, and school schedules for dashboard display.

Data Analytics: A rapidly emerging field of information science arising from the explosion of data generated by many Internet based applications and services. Data analytics embodies a sequential process of descriptive, diagnostic, predictive and prescriptive analytics. Each type has a different purpose and requires different techniques to gain meaningful outcomes. The latter two often employ machine learning to gain valuable insights and directional guidance in decision making, such as in self-driving automobiles.

Downdraft Gasification: A chemical process for the breakdown of organic material into its constituent gas and solid components. Gas components can then be combined into synthetic gases or other molecular structures to suit requirements. Synthetic gas may be used in electricity generation. It is environmentally friendly, scalable, and adaptable to a wide range of organic feedstocks including municipal solid waste.

E-Commerce: An exchange of goods and services electronically. It must involve an exchange of value. E-commerce is available to individuals, businesses, and government through a wide range of Internet services. Typically, a product or service is offered on a website and promoted through social media and other media. Payment can be made using a variety of trusted intermediaries such as banks, card services companies and mobile wallets.

ECOWAS: The Economic Community of West African States comprising Nigeria, Benin, Togo, Core d'Ivoire, Senegal, Burkina Faso, Mali, Niger, Guinea, Ghana, Guinea-Bissau, Sierra Leone, Mauritania, Liberia, and Carbo Verde. The goal of ECOWAS is to create a single trading bloc through normalization of finance, trade, and immigration between member countries, and to establish a framework for collective security.

Enumeration Areas: Geographic areas that are the smallest spatial unit a single enumerator (census person) can cover during a census period. Ghana is currently preparing for the 2020 census with the designation of about 50,000 enumeration areas. With an estimated population of about 30 million people in 2020, this corresponds to about 600 people in about 150 households per enumeration area.

FDI: Foreign direct investment is a controlling investment made in a project, asset, or business in one country by investors in another country.

GDP: Gross domestic product is the market value of all goods and services produced in a year.

Geothermal: Relates to the thermal characteristics of the earth that can be used for heating and cooling purposes. The sub-surface temperature is more constant than surface temperatures and heat exchanges can be used for space cooling during hot months and space heating in cold months. Geothermal energy is efficient, environmentally clean, and cost-effective. Geothermal energy may be used to manage greenhouse environments for aquaponics.

Global Positioning System (GPS): Comprises a network of low-earth orbit satellites that provide location information (latitude and longitude) for terrestrial objects with high accuracy. The ability to track the movement of goods and to monitor the location of resources is vital to commerce. GPS is extensively used by individuals for navigation and tracking purposes.

Horticulture: A branch of agriculture that focuses on the growing, marketing and processing of high-value plants for food, and for ornamental and therapeutic purposes. Horticulture is intensive small-scale farming of such products. For example, cut flowers have high value in certain markets and can be transported easily.

Hydroponics: A production method where the plants are grown in a nutrient rich solution in greenhouses rather than in the ground. Special greenhouse designs are needed to handle high temperatures and humidity in tropical regions.

Informal Settlements: Settlements that are characterized by poor housing, lack of facilities, overcrowding and insecure tenancy or residential status. The living conditions are generally poor, giving rise to health issues and crime. Informal settlements are more commonly referred to as slums.

Intelligent or Smart Cities: Cities that have embraced the Internet of Things (IoT) servicing a wide range of data collection, processing and information disseminating services for residents. No agreed quantitative definition of an intelligent city exists. These cities vary considerably in terms of the level of services and the interaction with the population. In developing countries intelligent cities may only offer basic services.

Internet Cloud: A term used to describe the amorphous nature of the Internet and remotely connected computers and services. It is all there, but it cannot be observed because it is in the cloud.

Internet of Things (IoT): Includes all devices including sensors, computers, and other electronic devices that can be connected to the Internet so that data can be sent to a central location for processing. The Internet of Things has created an explosion in data and the development of new methods for analysis, interpretation, and visualization of the data.

KWh: A measure of energy. 1KW is one 1000 watts of power and when multiplied by one hour becomes 1KWh of energy. For example, a 2,000 watt device (2KW) would use 1KWh of energy in 30 minutes.

Machine Learning: A process of machine modelling of some behavior based on the accumulation of information that best describes the required outcome. In the detection of slums, machine learning derived from remote images of known slums can identify other slums.

Municipal Solid Waste (MSW): Waste generated by households and industry in a city. Household waste contains recyclable materials (metal cans, glass, and plastics) and organic materials (food scraps, paper, and plastics) from which the energy content can be recovered and used to generate electricity, heat, and other products. It can also include solid sewerage waste.

MW: A measure of power. 1MW is equivalent to one million watts.

Nanotubes: Tiny tubes of carbon that are as small as one nanometre, or one billionth of a metre. Nanotubes have potential widespread application for structural reinforcement and emerging applications in water purification, additives to paint and plastics, medical sensors, and tumour detection.

Peer-to-Peer (P2P): Or person-to-person, connects two parties engaged in a transaction of equitable exchange without the need for a trusted intermediary such as a bank.

Photovoltaic: Refers to the conversion of photons into electrical energy. This is the action of solar panels. The amount of electrical energy generated is a function of the area of the panel (the amount of sunlight captured), the strength of sunlight irradiating the panel, and the conversion efficiency of the panel.

Prosumers: Someone who produces something that they consume and then make available the surplus for sale to others. Grid connected domestic solar power generation is an example.

Remote Sensing: The sensing of some action, event, or natural phenomena using a range of remote sensors. For example, images of the earth taken from an orbiting satellite use remote sensing devices such as cameras to provide high resolution colour or infrared images of the earth.

Serverless Computing: A service provided by a cloud provider who manages the servers and allocates resources depending on client needs. The servers physically exist but the cloud obscures them.

Skin or Stool Land: Names given to land claimed by traditional owners in Ghana. The term skin or stool tends to mean the same thing except that their use connects with the cultural background of the land and ownership.

Slums: Usually urban regions of a city that are of sub-standard quality, lacking basic services, and quality of housing. Usually associated with poverty. Also, known as informal settlements.

Smart City Transformation: A term created for this chapter that describes the process of transformation of a city's slums as the genesis of creating a more functional intelligent city that best serves all of its residents.

Sodom and Gomorrah: The name given to the slum district of Agbogbloshie, Accra, Ghana, that describes the horror of human deprivation in that district.

Spatial Data: Data that has a geographical or spatial context, such as location on a map represented by latitude and longitude coordinates (point) or within an area on a map (polygon) or a line (streets or railways). Also referred to as geospatial data.

Sub-Saharan Africa: The southern region of African that is separated from northern Africa by the Saharan desert. The World Bank estimates that Sub-Saharan Africa has a combined GDP of $1.7 trillion and an estimated 2.6% economic growth for 2019. The World Bank also estimates that there are more than 400 million people living in extreme poverty in Sub-Saharan Africa and that number is rising.

UNICEF: The United Nations Children's Fund deals with issues relating to children.

United Nations: The intergovernmental organization responsible for maintaining international peace and security, developing friendly relations among nations, achieving international cooperation, and being a centre for harmonizing the actions of nations.

Urbanization: The growth of urban population as a result of immigration from rural areas and from natural population growth within the urban areas. In developing countries, unsustainable urban development places stress on services and housing with serious consequences for public health due to poor sanitation, lack of safe water, inadequate waste disposal, and a degraded environment.

Volta or Volta River: The main river system in Ghana and feeds the Akosombo hydro-electric power plant. The Volta Lake was created by the damming of the Volta River and is the largest man-made freshwater lake in the world by surface area.

World Bank: An international financial institution that provides loans and grants to the governments of poorer countries for the purpose of pursuing capital projects.

Chapter 7
Air Pollution Monitoring in Intelligent Cities Using Weighted Association Rule Mining

Goksu Tuysuzoglu
https://orcid.org/0000-0002-2926-4267
Graduate School of Natural and Applied Sciences, Dokuz Eylul University, Turkey

Derya Birant
https://orcid.org/0000-0003-3138-0432
Department of Computer Engineering, Dokuz Eylul University, Turkey

ABSTRACT

Through the use of internet of things-based sensors in air quality monitoring stations, concentration of different pollutants and meteorological parameters can be regularly measured. In case of unusual conditions (e.g., increased levels of dangerous pollutants), a smart assessment system can produce warning so that appropriate air quality management process can be initiated. In this context, the objective of this study is to discover relationships and patterns among air pollution features and characteristics. In this case, determination of frequently observed association rules can trigger an appropriate background smart environment system when a critical situation is detected. In the experimental studies in the current project, traditional association rule mining and weighted association rule mining methods have been employed using real-world datasets collected from 21 monitoring stations in Turkey. In consequence, useful and outstanding association rules exceeding the user-defined support and confidence levels were obtained that can form basis for further research.

DOI: 10.4018/978-1-7998-5062-5.ch007

INTRODUCTION

With the emergence of newer technologies in recent years, various types of electronic data are being collected via sensors and smart devices, and analyzed through the use of intelligent data mining systems, to ensure that assets, resources and processes are managed more efficiently. Environmental science is one of the relevant areas that use intelligent systems to produce solutions for environmental problems and to protect the environment against factors such as pollution, adverse climate change and environmental degradation. In this context, observing air pollution data and analyzing and processing them are among the useful studies in this field.

In the current study, it is aimed to design an intelligent environment system that provides air quality control in cities by processing air pollution data collected from IoT sensors attached to smart devices in monitoring stations. For this purpose, Association Rule Mining (ARM) process is applied, which is one of the popular data mining methods. According to the rules formed by different combinations of the specified features of the available dataset, normal conditions are determined in terms of air quality factors to establish an intelligent system to detect the critical and exceptional situations. One drawback of the traditional ARM algorithms is that each item or factor is assumed to have the same significance even though some of them have greater importance compared to others within a transaction or within the whole item space. Therefore, the setting of the traditional ARM models should be organized in a weighted manner according to importance levels. In this respect, Tao et. al. (2003) developed an improved version of ARM, which is called Weighted Association Rule Mining (WARM). In this study, because different air pollutants/meteorological factors have different effects on air pollution, the rules are generated using WARM by taking account of item weights.

In the literature, several WARM models with different properties have been proposed. Wang et. al. (2000) introduced an algorithm, called WAR (Weighted Association Rules), which consists of two main parts. In the first part, the frequent item sets are generated through traditional ARM algorithms by ignoring the weights. Then post-processing is applied on the frequent item sets during rule-generation by introducing the intervals of item weight. WUARM (Weighted Utility ARM) developed by Khan et. al. (2008) considered both the significance of items and their utility as frequency of occurrences in transactions. Some of the WARM models were implemented using frequent pattern growth algorithm such as WIP (Weighted Interesting Pattern) mining and WFIM (Weighted Frequent Itemset Mining) as reported by Yun and Leggett (2005) and Yun and Leggett (2006). While WFIM was applied by creating ascending weight ordered prefix tree, WIP used a new concept of a weighted hyperclique pattern which considered weight confidence to discover frequent patterns with similar levels of weights and the h-confidence to determine affinity patterns with strong support. The proposed model used in this chapter was implemented using Apriori approach because pattern growth algorithms take huge amount of memory while storing in tree structure, which may not fit in memory when data size is very large.

A number of studies have focused on fuzzy set theory combined with WARM to mine weighted boolean and quantitative data (Wang and Zhang, 2003; Yue et. al., 2000; Lu and Qian, 1999; Gyenesei, 2000; Muyeba et. al., 2008). While FWARM (Fuzzy Weighted Association Rule Mining) proposed by Muyeba et. al. (2008) resulted in a valid downward closure property, others could not guarantee the validation. However, fuzzy algorithms have several disadvantages such as requiring knowledge about the problem solved and difficulties in setting the rules and choosing membership functions. Our algorithm differs from many other reported in the literature in terms of easy implementation and application.

The novelty of this study is the implementation of WARM in the field of environmental sciences. This is something that has not been done or reported in a satisfactory manner. The research in this contribution is also original in that it assesses air pollution parameters in Turkey using this method. In the experimental studies, traditional ARM and WARM methods were compared on the real-world dataset that were collected from 21 monitoring stations in Turkey between the dates of 1 January 2018 and 1 May 2019.

The potential contributions of the current research can be listed as follows:

- Determination of important relationships among different pollutants (PM_{10}, SO_2, NO_2 and O_3), meteorological parameters (temperature, relative humidity and air pressure) and monitoring stations (urban, rural, and industrial) are identified.
- Understanding that, as in the subject of air pollution, the other environmental criteria such as water pollution parameters and soil contamination factors can also be evaluated.
- Demonstration of the applicability of WARM in an environmental problem situation and proven advantages compared to traditional ARM methods.
- Development of a system for intelligent cities, that helps to monitor and manage air quality control.

The chapter is organized in five main sections as follows: The following section presents a review of the methodologies related to association rule mining methods, specifically about weighted association rule mining (WARM). The application areas of WARM are also investigated. The next section gives a detailed explanation of the applied techniques (traditional ARM and WARM algorithms) used in this study. In the section after this, the dataset description is presented. The next section explores both the experimental setting and the obtained results in which significant association rules among air pollutant factors are determined. Finally, future directions and concluding remarks including a brief summary are provided.

RELATED WORKS

Association Rule Mining (ARM) methods have been used in many studies to find high quality relationships among the attributes of a problem domain. Weighted association rule mining (WARM) which is the sub-branch of ARM has also been used in many research areas. Some of these studies will be briefly discussed in this section.

Tao et. al. (2003) proposed the weighted scheme of association rule mining in discovering significant relationships among the items of transactions. In traditional association rule mining applications, all items were given the same weights even though level of their importance within the transactions or within the item spaces was different. In their algorithm, each item has its own weight according to its role in the given scenario; also, different item sets or transactions may have specific itemset/transaction weights. Besides, downward closure property, which is a rule of ARM that if an itemset is large, then all of its subsets are also large, is also validated by this method. Weighted downward closure property is taken into account by using weighted support in the process of finding significant itemsets.

In order to construct a personalization recommendation system based on user focus, Zhang et. al. (2004) developed an algorithm for text information retrieval which considers a user's entire query requests at a given period of time. The algorithm was implemented using weighted association rule mining with

FP-Growth (frequent pattern growth). Their proposed method increased the efficiency in the subject of time requirement.

Wang and Zhang (2003) combined fuzzy set theory with weighted association rule mining by assigning weights to the relative attributes having different significance. Kaya and Alhajj (2004) expanded the research by mixing fuzziness, weighted association rules and genetic algorithms. Fuzzy quantitative association rules were determined by employing a clustering method based on a genetic algorithm, which dynamically adjusts and optimizes membership functions.

Li (2005) implemented another work of weighted association rule mining using fuzzy set theory to obtain user-guided association rules to analyze data of application loan including 650 instances of applicants for appliance loans. Their method resulted in flexible, natural and understandable rules in the final stage.

Yun and Leggett (2005) used weighted association rule mining with FP-Growth algorithm in which a minimum weight restraint and a weight range were specified and items were supplied with different weights.

Sun and Bai (2008) presented a new method, different from other classical WARM models. This method did not require the pre-assigned item weights. A support metric was defined for the itemsets in the database which only contained binary attributes to find interesting patterns. The weight of each transaction was evaluated by HITS (Hyperlink-Induced Topic Search) algorithm. Based on the computed weights, the new support was defined to give the significance of itemsets. Their model outperformed the link-based methods.

Dhanalakshmi and Porkodi (2018) applied their proposed WARM model to analyze Tobacco smoke effect on the placenta dataset. After the calculation of weighted support, they separately implemented one existing (based on FP-Growth) and two novel ARM techniques to obtain association rules. In the final step, association rules were filtered depending on the weighted support and significant gene ontology terms. Their proposed WARM model provided more optimal results compared to existing ones.

Another application area was the analysis of the performance of a cricket team using a WARM method (Saraswat et. al., 2018). The resulting output was aimed at being used in the planning of game winning. The important relationships among attributes such as the effect of toss, venue condition, batting first or second, and the final game result (win/loss) were detected using weighted support and weighted confidence.

Qiu et. al. (2019) introduced their study to automatically identify the building operational strategies (on/off status, sequencing and coordinated control). The experimental results were validated by a power metering system data. They applied two data mining methods: decision tree (CART) to determine on/off and sequencing control strategies and WARM to specify coordinated control strategy. Their proposed framework worked accurately.

Table 1 presents some recent studies which use WARM models. The bold notation in "Methods" column indicates algorithms that outperformed the other methods. It is apparent that WARM methods provide good performance in several different areas of expertise, including bioinformatics (Lakshmi and Vadivu, 2019; Indhumathy et. al., 2018), recommendation systems (Cagliero et. al., 2018; Bhavithra and Saradha, 2018), and activity recognition (Hui et. al., 2019; Wen et. al., 2015).

The abbreviations used in Table 1 mean as follows:

- PCID: Predicting Disease Comorbidity based on Integration of Multi-Scale Data
- WARBCP: Weighted Association Rule Based Disease Comorbidity Prediction
- WTARM: Weighted Temporal Apriori Rule Mining

Table 1. Recent Studies Utilizing WARM Methods in Main Methodology

Ref. Sorce	Aim	Data	Methods
(Lakshmi and Vadivu, 2019)	Disease comorbidity prediction	Protein-protein interaction data, Pathway data and Gene Ontology annotation data	PCID and the proposed model: **WARBCP**
(Hui et. al., 2019)	Classification of video scenes by discovering the relationship between action semantic and scene	Daily human activities data	The proposed model: **WTARM**, SVM, ARM, and TARM
(Rahimi et. al., 2019)	Identification of sentiment shifters in reviews	A medical drug corpus and a corpus including reviews of products, restaurants and movies	SentiWordNet, Domain Specific Lexicon, ML-unigram, Proposed-DM-Dependency, Proposed-DM-SRL, Proposed-ML and **Proposed-ML + Shifter**
(Cagliero et. al., 2018)	Recommendation of additional reviewers for papers with critical assignments	Conference and journal datasets	WAR
(Bhavithra and Saradha, 2018)	Personalized web page recommendation	Web query log data	K-NN, CF, CBR and the proposed model: **CBR with WARM**
(Indhumathy et. al., 2018)	Prediction of potentially new HCV-Human protein interactions	HCV-human protein interactions data (HCVpro)	The proposed model: WARL
(He et. al., 2017)	Root cause identification to find the hidden cause-related relationships	Big data of product infant failure	ARM and **WARM**
(Wang et. al., 2016)	Discovering correlated alarm sequences	Quantitative alarm sequences data	Fuzzy weighted alarm association rules mining
(Wen et. al., 2015)	Finding the frequent patterns of human activities	The sensor readings data from smart environments	Naïve Bayesian, Decision Tree, HMM and **the proposed weighted frequent patterns mining model**
(Cao et. al., 2014)	Software reliability qualitative evaluation	Databases about unpredictable cases in the software development process and software reliability	Delphi Hierarchy Process Based on Cloud Model and WARM

- SVM: Support Vector Machine
- TARM: Temporal Apriori Rule Mining
- DM: Data Mining
- SRL: Semantic Role Labelling
- CF: Collaborative filtering
- WARL: Weighted Association Rule Mining
- K-NN: K Nearest Neighbor
- WAR: Weighted Association Rules
- CBR: Case-Based Reasoning
- HMM: Hidden Markov Model

In the area of environmental sciences, a number of studies exist that used association rule mining techniques. Among them, one of the most studied topics is air pollution or air pollutant factors. The major aim is either finding association rules between air pollutant factors and a specific disease (Toti et. al., 2016; Sadat et. al. 2015; Karimipour and Kanani-Sadat, 2016) or detecting the significant relationships between air pollutants and industrialization (Li et. al., 2019; Ran et. al., 2016). The most applied techniques are Apriori (Othman et. al., 2018; Zeng and Xiang, 2017; Toti et. al., 2016; Karimipour and Kanani-Sadat, 2016; Isikli et. al., 2015; Qin et. al., 2015) and FP-Growth (Li et. al., 2019; Wu et. al., 2017) to discover the relations.

Li et. al. (2019) presented a related study on the subject of relationship between industrialization level of 31 regions of China and a total of annual NO_2, SO_2 and PM_{10} concentrations by applying FP-Growth method. They detected ranges of six factors for industrialization related to low and high air quality.

Othman et. al. (2018) applied both association rule mining and clustering techniques in order to explore the main factors which influence O_3 formation. In the first step, K-Means clustering was implemented to discover the appropriate ranges for the selected variables, and then Apriori technique was used to find the hidden patterns among the chosen variables.

Zeng and Xiang (2017) explored the pattern rules of air quality from historical data by applying Apriori algorithm and then they selected the sampling parameter based on Q-learning in order to develop an adaptive sampling framework through participatory sensing for urban air quality. According to the experimental results, their system predicted air quality with acceptable accuracy.

Wu et. al. (2017) implemented both Voronoi-based clustering method and association rule mining technique based on FP-Growth to develop a data-driven urban air pollution partitioning system. They carried out experiments on the hourly concentrations of air pollutants such as $PM_{2.5}$, NO_2, SO_2, PM_{10}, O_3 and CO and meteorological conditions such as wind, temperature and relative humidity. Association rule mining was utilized to investigate the relationship between weather conditions and air quality index. They grouped 348 cities of China in terms of air pollution levels into 28 clusters three of which included the largest number of cities.

A fuzzy association rule mining method was applied for the detection of the specific correlations between SO_2 emission concentration patters and operation characteristics of coal-fired power plants (Ran et. al. 2016). An optimal strategy was developed by reducing the coal consumption rate and SO_2 emission concentration.

Karimipour and Kanani-Sadat (2016) searched for the relationship between air pollution factors such as PM_{10}, CO, NO_2, SO_2, $PM_{2.5}$, and O_3 and spatial distribution of allergy prevalence by deploying the fuzzy spatial association rule mining method based on Apriori. They also considered the patients' distance to the parks and roads. They conducted experiments for two months. At the end, they detected that SO_2 and O_3 had no effect on allergy prevalence while distance to roads and parks in addition to CO_2, NO_2, PM_{10} and $PM_{2.5}$ have a direct relation with it. On the other side, later in the study, CO, SO_2 and O_3 were found to be ineffective.

Isikli et. al. (2015) conducted a study for finding the impacts of environmental risk indicators on the city life cycle. They implemented association rules based on Apriori, classification trees, MANOVA and LDA to data including air quality, demographical factors and socioeconomic development taken from the cities of Turkey. The general purpose was to determine the direct relationship between the well-being of human-being and air quality. The state of city development, suicide rate and the number of people who migrate to and from the city were employed as city life cycle parameters while air quality factors were used as environmental risk indicators.

After an intense literature study, it was noticed that the Weighted Association Rule Mining (WARM) method plays an active role in drawing up important association rules in the applied areas – also that it has never been used in the extraction stage of the relations among air pollution factors. For this purpose, the aim of our study, as reported in this contribution, is to extract significant association rules from the dataset which is comprised of daily concentrations of different air pollutants and three meteorological factors measured at different types of monitoring stations by the application of WARM method.

METHODOLOGY

In this section, traditional Association Rule Mining, Weighted Association Rule Mining and the proposed methodology used throughout this study is introduced in detail as well as the performance measures such as support, confidence or lift metrics used in the evaluation of the models.

Association Rule Mining

Identification of interesting patterns in order to obtain useful knowledge is the basic purpose of data mining. By analyzing the data, hidden structures are inferred from the problem domain so that predictive models can be generated by classification or clustering tasks to take critical decisions for any future problems. In addition, another main task of data mining, called Association Rule Mining (ARM), allows to discover important relationships among data samples by using association rules.

ARM models consider a number of transactions, $T = \{T_1, T_2, ..., T_n\}$, each of which contains a set of items from the itemset, I, where $I = \{i_1, i_2, ..., i_m\}$, where n is the number of transactions and m is the number of items. In this scenario, the question is: "What does the value of one feature tell us about the value of another feature?". The basic purpose of this approach is the determination of the significant relationships among these items i.e. finding a set of items that frequently appear together by taking whole transactions into consideration. ARM methods are often utilized in many different application areas such as market basket analysis, discovering patterns in biological databases, fraud detection in credit card transactions etc. Most of the well-known methods introduced under ARM are either Apriori, FP-Growth or Eclat.

To illustrate the method, in the following example in Table 2, we present seven items in the itemset $I = \{A, B, C, D, E, F, G\}$ and five transactions T_i's which include items from I. The aim is to extract all association rules using these recorded transactions from the past and select the most important ones from among them. An association rule is the combination of two concepts - antecedent and consequent - both of which are list of items. As an example, if $\{X\} \rightarrow \{Y\}$ is an association rule where both of X and Y are itemsets then X is known as *antecedent* and Y as *consequent*. This rule implies that transactions which contain itemset X also contain itemset Y.

Various metrics are considered in the evaluation process to understand the strength of associations among each item. *Support* is one of these metrics which tells us the fraction of the total number of transactions where a specific itemset occurs. It is calculated as shown in (1):

$$\text{Support}(\{X\} \rightarrow \{Y\}) = \frac{\text{Transactions containing both X and Y}}{\text{Total number of transactions}} \quad (1)$$

In the given scenario in Table 2, $\text{Support}(\{A\} \rightarrow \{B\}) = 2/5 = 0.4$. In the same way, support values of all other association rules can be calculated. Association rules with higher support values than the given support value by the user are called *frequent* and they are worth considering for further analysis because they provide enough information on the relationships between their items.

Table 2. An Example Itemset and Related Transactions

Original Itemset (*I*)	**A, B, C, D, E**
Transaction T_1	A, C, E
Transaction T_2	A, B, D, E
Transaction T_3	B, E
Transaction T_4	A, B, D, E
Transaction T_5	E

Another important evaluation metric is known as *confidence*. It is defined as the likeness of occurrence of consequent on the cart given that the cart already has the items in the antecedents. By using the formula as shown in (2), $\text{Confidence}(\{A\} \rightarrow \{B\}) = 2/3 = 0.667$ according to the scenario in Table 2. Association rules which satisfy both minimum support and minimum confidence thresholds are called *strong*.

$$\text{Confidence}(\{X\} \rightarrow \{Y\}) = \frac{\text{Transactions containing both X and Y}}{\text{Transactions containing X}} \tag{2}$$

The general problem of association rule mining can be reduced into a two-step process: *finding all frequent itemsets* using minimum support threshold and *detection of strong association rules from these frequent itemsets* using minimum confidence threshold (Han et. al., 2011).

Other three essential measures are *lift, leverage* and *conviction* as defined in equations (3), (4) and (5), respectively. *Lift* shows the interestingness of a rule in other words how many times more often antecedent and consequent occur together than expected if they were statistically independent. When calculating the *conviction*, the probability of antecedent appearing without the consequent is calculated; if the antecedent and consequent are dependent, the probability that the antecedent is seen independently of the consequent is calculated. We can think of the *leverage* as the difference between the probability of the set which includes both antecedent and the consequent and the sum of their individual probabilities.

$$\text{Lift}(\{X\} \rightarrow \{Y\}) = \frac{\text{Support}(X \cup Y)}{\text{Support}(X)*\text{Support}(Y)} \tag{3}$$

$$\text{Leverage}(\{X\} \rightarrow \{Y\}) = \text{Support}(X \cup Y) - \text{Support}(X)*\text{Support}(Y) \tag{4}$$

$$\text{Conviction}(\{X\} \rightarrow \{Y\}) = \frac{1\text{-Support(Y)}}{1\text{-Confidence}(\{X\} \rightarrow \{Y\})} \tag{5}$$

Weighted Association Rule Mining

In Weighted Association Rule Mining (WARM) framework, all items are assigned predefined weights which represent the significance of items within a transaction. This approach is differently from traditional Association Rule Mining (ARM) methods. The general WARM framework is implemented as below:

1. Assign all item weights as $W = \{ w_1, w_2, \ldots, w_m \}$, where m is the number of items
2. Normalize all item weights as $W' = \{ w'_1, w'_2, \ldots, w'_m \}$
3. Calculate transaction weights as $W_t = \{ w_{T1}, w_{T2}, \ldots, w_{Tn} \}$ by using normalized item weights, W'
4. Normalize transaction weights as $Wt' = \{ w_{T1}', w_{T2}', \ldots, w_{Tn}' \}$, where n is the number of transactions in the dataset
5. Calculate weighted support values of each itemset
6. Eliminate redundant itemsets which are under the predetermined support value and put the remaining ones into frequent itemsets
7. Discover strong association rules which are above the predefined confidence threshold.

According to the given scenario in Table 2, the first step is to assign corresponding weights to the items of I. Then, all weights are normalized for the later processing. Table 3 (a) displays the initially given weights by the user and their normalized versions. After the calculation of the normalized item weights, the next step is to find the weights of each transaction T_i as w_{T_i} according to equation (6), where $i = \{1, 2, \ldots, n\}$ and n is the number of transactions. The relative importance of each transaction is determined in this way. As in the calculation of item weights, transaction weights are also normalized as shown in Table 3 (b).

$$\text{Weight}(T_i) = w_{Ti} = \frac{\sum_{j=1}^{|T_i|} w_j}{|T_i|} \tag{6}$$

The next, perhaps the most essential, step is to find all frequent itemsets which occur together very often. In the beginning, the user should introduce the weighted support threshold to control the selection of itemsets. Here, the weighted support is given as 0.3. All itemsets from Level 1 to Level 5 are evaluated that the ones which are equal or above 0.3 are marked as *frequent* and they are shown as bold and gray shaded in the Table 3 (c).

In ARM methods, if an itemset is found as frequent, then its subsets are also assumed as frequent. This assumption is called *downward closure property*. Here, an important difference from traditional ARM is attracted attention that downward closure property does not have to hold any more. For instance, if the itemset {A, B, C} is classified as frequent, then {A, B} may not be included in frequent itemsets

Table 3. (a, b, c). An Example Scenario Using WARM to Find Frequent Itemsets

Items Weight (W) Normalized Weight (W')
A 1 0.04
B 5 0.20
C 4 0.16
D 5 0.20
E 10 0.40
Total 25 1.00
Table 3b.
Transactions Transaction Weight (W_T) Normalized Transaction
Weight (WT')
T_1 A, C, E 0.200 0.172
T_2 A, B, D, E 0.210 0.181
T_3 B, E 0.300 0.258
T_4 A, B, C, D, E 0.200 0.172
T_5 B, C, E 0.253 0.217
Total 1.163 1.00
Table 3c.

Level 1	**A**	**B**	**C**	**D**	**E**					
Weighted Support	0.525	0.828	0.561	0.353	1.0					
Count	3	4	3	2	5					
Level 2	**A, B**	**A, C**	**A, D**	**A, E**	**B, C**	**B, D**	**B, E**	C, D	**C, E**	**D, E**
Weighted Support	0.353	0.344	0.353	0.525	0.389	0.353	0.828	0.172	0.561	0.353
Count	2	2	2	3	2	2	4	1	3	2
Level 3	A, B, C	**A, B, D**	**A, B, E**	A, C, D	**A, C, E**	**A, D, E**	B, C, D	**B, C, E**	**B, D, E**	C, D, E
Weighted Support	0.172	0.353	0.353	0.172	0.344	0.353	0.172	0.389	0.353	0.172
Count	1	2	2	1	2	2	1	2	2	1
Level 4	A, B, C, D	A, B, C, E	**A, B, D, E**	A, C, D, E	B, C, D, E					
Weighted Support	0.172	0.172	0.353	0.172	0.172					
Count	1	1	2	1	1					
Level 5	A, B, C, D, E									
Weighted Support	0.172									
Count	1									

because the probability of the selection of {A, B, C} as frequent itemset can be arisen from the high weight of item C.

After finding all frequent itemsets, strong association rules are determined using the initially given weighted confidence threshold. According to the scenario in Table 3 (c), we note that {A, E} is a frequent itemset. Two association rules can be produced from {A, E} as A → E and E → A. Weighted confidence values of these two rules do not have to be the same according to equation (4), because they express different sequence of itemsets. While Weighted Confidence (A → E) = Weighted Support ({A, E}) / Weighted Support ({A}) = 1, the Weighted Confidence (E → A) = Weighted Support ({A,

E}) / Weighted Support ({E}) = 0.525. In the scenario of Table 3 (a, b, c), the user defined weighted confidence threshold is 0.6. According to that, while A → E is a strong association rule, E → A is not.

$$\text{Weighted Confidence}(\{X\} \rightarrow \{Y\}) = \frac{\text{Weighted Support XE "Weighted Support"(X,Y)}}{\text{Weighted Support(X)}} \quad (7)$$

Using equation (7), all association rules are evaluated and the ones providing the required weighted confidence value are chosen as strong association rule in the final step.

PROPOSED APPROACH

In this study, an intelligent environmental system is developed by using the WARM methodology. Figure 1 displays the general framework of the proposed system. The concentrations of air pollutants and some meteorological parameters from three different types of monitoring stations (urban, rural and industrial) are recorded, on a daily basis, in data storage/servers and Cloud. Each log data is treated as a transaction in WARM scenario.

The data management team is in charge of analyzing and processing all recorded data in order to obtain useful information related to air quality factors. For this purpose, the framework is implemented using WARM methodology by the data management team. Unlike the example in Table 3 (a, b, c), there are more than one value of each item showing the degree of amount as $item_i = \{ item_{i\_1}, item_{i\_2}, \ldots, item_{i\_k} \}$. For example, the concentration of each pollutant has six different degrees such as *Good, Moderate, Sensitive, Unhealthy, Bad,* and *Dangerous*. Instead of only giving weights to items themselves, the proposed system assigns weights to item values as in Table 4 (a, b) in order to correctly apply the methodology. Remaining parts of the original WARM method are applied in the same way by calculating transaction

Table 4 (a, b). Weight Assignment to Item Values

Table 4 (a)							
PM_{10}		**SO_2**		**NO_2**		**O_3**	
Good	w_1	Good	w_7	Good	w_{13}	Good	w_{19}
Moderate	w_2	Moderate	w_8	Moderate	w_{14}	Moderate	w_{20}
Sensitive	w_3	Sensitive	w_9	Sensitive	w_{15}	Sensitive	w_{21}
Unhealthy	w_4	Unhealthy	w_{10}	Unhealthy	w_{16}	Unhealthy	w_{22}
Bad	w_5	Bad	w_{11}	Bad	w_{17}	Bad	w_{23}
Dangerous	w_6	Dangerous	w_{12}	Dangerous	w_{18}	Dangerous	w_{24}
Table 4 (b)							
TEMP		HUM		PRES		STATION	
Low	w_{25}	Low	w_{28}	Low	w_{31}	Urban	w_{34}
Moderate	w_{26}	Moderate	w_{29}	Moderate	w_{32}	Industrial	w_{35}
High	w_{27}	High	w_{30}	High	w_{33}	Rural	w_{36}

weights using the weights of item values. After that, frequent itemsets and strong association rules are found as in the traditional WARM method.

In the following example, according to the given scenario, three strong rules were identified. For example, in Rule 1, when the concentration of SO_2 is moderate and the concentration of NO_2 is high in addition there is very high temperature, then it is likely to observe the station type as urban with 40% of support and 75% of confidence. Other rules are also interpreted in the same way.

Rule 1:

SO2=moderate NO2=sensitive Temp=low → Station=urban (40% support, 75% confidence)

Rule 2:

PM10=good O3=moderate SO2=good NO2=good → Station=rural (25% support, 85% confidence)

Rule 3:

PM10=moderate SO2=sensitive Pres=high → Station=industrial (30% support, 80% confidence)

After the exploration of all significant association rules, the phase of reporting and presentation takes place in order to express the experimental results of the applied methodology in a user-friendly manner. The system allows accessing the results using both mobile devices and personal computers. According to the outcomes, the system puts the air quality management team on alert in case of the detection of an unexpected situation in order to take the required precautions.

The general advantages of the proposed system are as follows:

- Intelligent cities are designed to monitor changing air conditions and provide appropriate warning systems.
- Ecological development is sustained and reinforced which is good for the environment.
- Living things in the fresh air environment continue to live in a healthier and prosperous environment.
- A cleaner environment is obtained by inspecting the air quality in a regular and controlled manner.

DATASET DESCRIPTION

Air quality network of Turkey, which is composed of 313 monitoring sites, measures a variety of air pollutant factors such as PM_{10}, SO_2, NO_2, NO, O_3 and CO to take precautions in case an unexpected situation in terms of pollutant levels occurs. In this study, daily concentrations of air pollutants (PM_{10}, SO_2, NO_2 and O_3) and various meteorological factors (temperature, relative humidity and air pressure) taken from 21 monitoring stations were considered. The monitoring stations are listed in Table 5; the data were collected during the period: 1 January 2018 and 1 May 2019. These monitoring stations are

Figure 1. General Framework with Potential Impacts of the Proposed Intelligent Environmental System

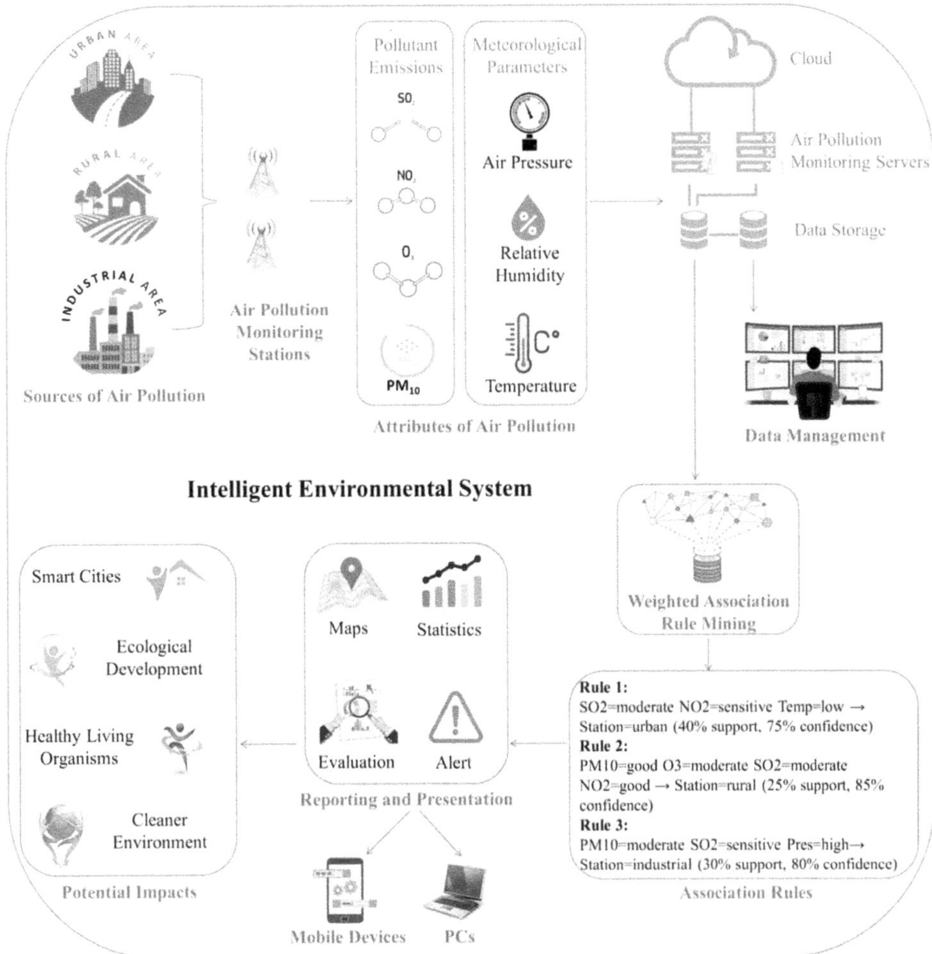

from three different regions of Turkey: Blacksea, Eastern Anatolia and Marmara. The network types of stations are urban, rural and industrial. In total, 9980 instances are taken into consideration. To facilitate public access to information on air quality of Turkey, the monitoring results were published online at: http://laboratuvar.cevre.gov.tr

All attributes are categorized such that daily concentrations of pollutants are labeled as *Good, Moderate, Sensitive, Unhealthy, Bad,* and *Dangerous* in terms of the level of air pollution and meteorological factors are also labeled in the same manner. Table 6 shows the sample dataset used in this study.

The proposed model was applied in such a way that a number of transactions with the itemsets, which are full of air pollution data, are conducted to infer association rules and observe useful relations. The itemset of the mentioned scenario is given in Figure 2.

Table 5. Information about the Selected Air Quality Monitoring Stations

ID	AQMS NAME	CITY	COUNTY	LONG.	LAT.	TYPE	THE MEASURED POLLUTANTS
1	Artvin	Artvin	Merkez	41.8182	41.1752	Urban	PM_{10}, SO_2, NO_2, NO, O_3, NO_X
2	Bayburt	Bayburt	Merkez	40.2255	40.2558	Urban	PM_{10}, SO_2, NO_2, NO, O_3, NO_X
3	Canakkale - Biga Icdas	Canakkale	Biga	27.1072	40.4173	Industrial	PM_{10}, SO_2, NO_2, NO, O_3, NO_X, CO
4	Canakkale - Can - MTHM	Canakkale	Can	27.0498	40.0293	Urban	PM_{10}, SO_2, NO_2, NO, O_3, NO_X
5	Edirne - Kesan - MTHM	Edirne	Kesan	26.6352	40.8511	Urban	PM_{10}, SO_2, NO_2, NO, O_3, NO_X, $PM_{2.5}$
6	Erzincan	Erzincan	Merkez	39.4950	39.7430	Urban	PM_{10}, SO_2, NO_2, NO, O_3, NO_X
7	Erzurum - Pasinler	Erzurum	Pasinler	41.5721	40.0335	Rural	PM_{10}, SO_2, NO_2, NO, O_3, NO_X
8	Igdir	Igdir	Merkez	44.0536	39.9261	Urban	PM_{10}, SO_2, NO_2, NO, O_3, NO_X
9	Igdir - Aralik	Igdir	Aralik	44.6209	39.7868	Rural	PM_{10}, SO_2, NO_2, NO, O_3, NO_X, $PM_{2.5}$
10	Istanbul - Basaksehir -MTHM	Istanbul	Basaksehir	28.7898	41.0954	Industrial	PM_{10}, SO_2, NO_2, NO, O_3, NO_X, CO
11	Istanbul - Esenyurt - MTHM	Istanbul	Esenyurt	28.6688	41.0192	Urban	PM_{10}, SO_2, NO_2, NO, O_3, NO_X
12	Kocaeli - Gebze - MTHM	Kocaeli	Gebze	29.4365	40.8108	Urban	PM_{10}, SO_2, NO_2, NO, O_3, NO_X
13	Kocaeli - Korfez - MTHM	Kocaeli	Korfez	29.7888	40.7461	Industrial	PM_{10}, SO_2, NO_2, NO, O_3, NO_X, $PM_{2.5}$, $PM_{2.5}$ Flow Rate
14	Kocaeli - Yenikoy - MTHM	Kocaeli	Basiskele	29.8844	40.7042	Urban	PM_{10}, SO_2, NO_2, NO, O_3, NO_X
15	Rize	Rize	Merkez	40.5328	41.0217	Urban	PM_{10}, SO_2, NO_2, NO, O_3, NO_X
16	Rize - Ardesen	Rize	Ardesen	41.0475	41.1273	Rural	PM_{10}, SO_2, NO_2, NO, O_3, NO_X, $PM_{2.5}$
17	Trabzon - Akcaabat	Trabzon	Akcaabat	39.5923	41.0143	Urban	PM_{10}, SO_2, NO_2, NO, O_3, NO_X, CO
18	Yalova - Armutlu- MTHM	Yalova	Armutlu	28.7845	40.5292	Rural	PM_{10}, SO_2, NO_2, NO, O_3, NO_X, $PM_{2.5}$
19	Zonguldak - Eren Enerji Lise	Zonguldak	Catalagzi	31.8801	41.4964	Industrial	PM_{10}, SO_2, NO_2, NO, O_3, NO_X, CO, $PM_{2.5}$
20	Zonguldak - Eren Enerji Santral	Zonguldak	Catalagzi	31.8964	41.5105	Industrial	PM_{10}, SO_2, NO_2, NO, O_3, NO_X, CO, $PM_{2.5}$
21	Zonguldak - Eren Enerji Tepekoy	Zonguldak	Catalagzi	31.9374	41.5269	Industrial	PM_{10}, SO_2, NO_2, NO, O_3, NO_X, CO, $PM_{2.5}$

EXPERIMENTAL RESULTS

Table 6. Sample Dataset

Season	PM_{10}	SO_2	NO_2	O_3	TEMP	HUM	PRES	CLASS
Autumn	Moderate	Good	Good	Good	Moderate	Moderate	Medium	Urban
Summer	Good	Good	Good	Good	High	Moderate	High	Urban
Summer	Moderate	Good	Good	Good	Low	Low	High	Urban
Winter	Good	Sensitive	Unhealthy	Good	High	High	High	Industrial
Spring	Good	Good	Good	Good	High	Moderate	High	Rural

Two main methodologies were implemented in the experimental studies: traditional ARM model based on Apriori and WARM. The predefined confidence threshold was 0.6 and the support threshold ranged from 0.2 to 0.9 to show the change of the number of k-itemsets for both the applied models. The WARM model was implemented in Visual Studio 2019 and the results of Apriori model was obtained using WEKA application (Holmes et. al., 1994).

The first step before applying ARM models was to discretize all numeric values into their respective intervals. Therefore, the daily concentration values of all air pollutant factors were categorized into the

Figure 2. Itemset used in the transactions of air pollution dataset

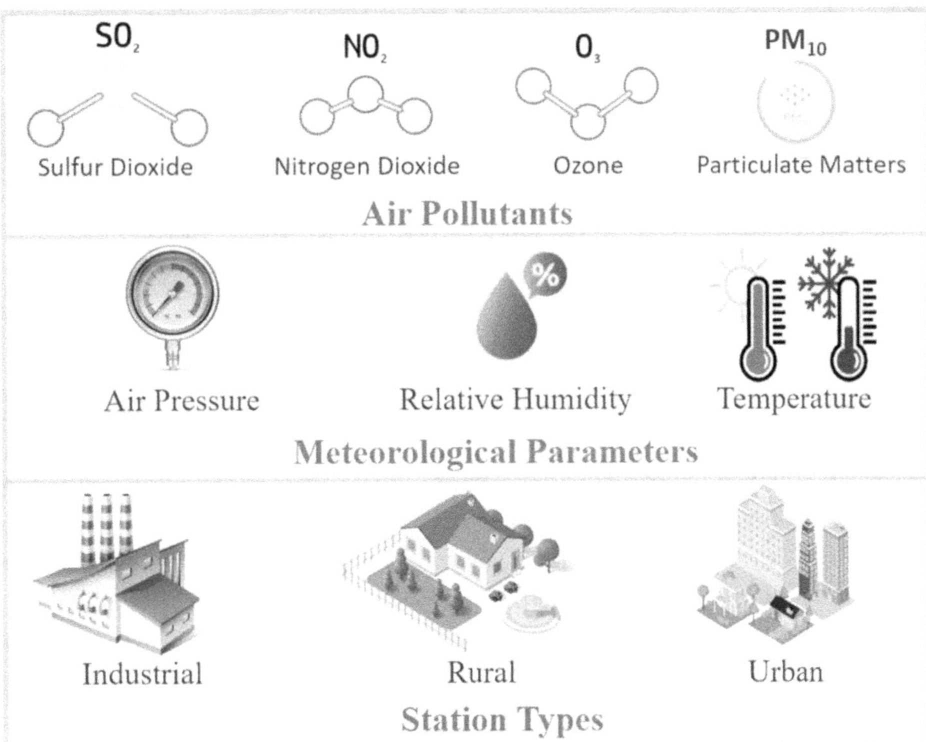

corresponding index categories "Good, Moderate, Sensitive, Unhealthy, Bad, Dangerous" using National Air Quality Index Breakpoints defined at the website of http://laboratuvar.cevre.gov.tr as shown in Table 7.

Table 7. National Air Quality Index Breakpoints for the Air Pollutants (SO_2, NO_2, O_3 and PM_{10})

Index	SO_2 [μg/m³]	NO_2 [μg/m³]	O_3 [μg/m³]	PM_{10} [μg/m³]
Good	0-100	0-100	0-120	0-50
Moderate	101-250	101-200	121-160	51-100
Sensitive	251-500	201-500	161-180	101-260
Unhealthy	501-850	501-1000	181-240	261-400
Bad	851-1100	1001-2000	241-700	401-520
Dangerous	>1101	>2001	>701	>521

Meteorological parameters were also categorized as shown in Table 8 (b) according to the equal-sized intervals between their minimum and maximum values displayed in Table 8 (a) where Temp, Hum and Pres are the abbreviations of Temperature, Relative Humidity and Air Pressure, respectively.

Table 8. (a and b). Breakpoints Determined for the Meteorological Parameters

Table 8 (a)					
minTemp [°C]	**maxTemp [°C]**	**minHum [%]**	**maxHum [%]**	**minPres [mbar]**	**maxPres [mbar]**
-31.44	34.23	0	100.95	713.6	1051.38
Table 8 (b)					
Index	**Temperature Breakpoints**		**Relative Humidity Breakpoints**	**Air Pressure Breakpoints**	
Low	< -10		< 35	< 827	
Moderate	-10 - 10		35 - 70	827 - 940	
High	> 10		> 70	> 940	

The season/time as to when the measurements were performed were also added separately according to the month of the date. The values of some air pollutants rise in some seasons. For example, the concentration of air pollutants increases as more fuel is burned due to the need for heating in winter. With this study, the relationship between season and pollutant concentrations can also be observed.

Several examples of strong association rules was found using WARM with support threshold = 0.2, by displaying their confidence, lift, leverage and conviction values as below:

Examples of rules with 2 itemsets:

Rule 1:

SEASON=Spring → SO2=Good (Confidence: 0.907, Lift: 0.9704, Leverage: -0.0087, Conviction: 0.703)

Rule 2:

PM10=Good → TEMP=High (Confidence: 0.6252, Lift: 1.0514, Leverage: 0.0205, Conviction: 1.0815)

Rule 3:

STATION=Urban → SO2=Good (Confidence: 0.9135, Lift: 0.9774, Leverage: -0.0111, Conviction: 0.7563)

Rule 4:

NO2=Good → PM10=*Good* (Confidence: 0.6716, Lift: 1.0009, Leverage: 0.0006, Conviction: 1.0018)

Examples of rules with 3 itemsets:

Rule 1:

O3=Good, HUM=Moderate → STATION=Urban (Confidence: 0.6183, Lift: 1.1732, Leverage: 0.03, Conviction: 1.2392)

Rule 2:

SEASON=Winter, TEMP=Moderate → O3=Good (Confidence: 0.9739, Lift: 1.0003, Leverage: 0.0001, Conviction: 1.0125)

Rule3:

PM10=Good, HUM=Moderate → SO2=Good (Confidence: 0.9737, Lift: 1.0418, Leverage: 0.0086, Conviction: 2.4839)

Rule4:

TEMP=High, STATION=*Urban* → SO2=*Good* (Confidence: 0.9573, Lift: 1.0243, Leverage: 0.007, Conviction: 1.5312)

Examples of rules with 4 itemsets:

Rule1:

SO2=Good, PRES=High, STATION=Industrial → O3=Good (Confidence: 0.9746, Lift: 1.001, Leverage: 0.0003, Conviction: 1.0397)

Rule2:

O3=Good, TEMP=High, HUM=Moderate → SO2=Good (Confidence: 0.9754, Lift: 1.0437, Leverage: 0.0095, Conviction: 2.6587)

Rule3:

SEASON=Spring, SO2=Good, PRES=High → O3=Good (Confidence: 0.9879, Lift: 1.0147, Leverage: 0.003, Conviction: 2.1851)

Rule4:

STATION=Urban → O3=*Good*, PRES=*High*, NO2=*Good* (Confidence: 0.6376, Lift: 0.9274, Leverage: -0.0263, Conviction: 0.8623)

Examples of rules with 5 itemsets:

Rule1:

NO2=Good, O3=Good, PRES=High, STATION=Industrial → SO2=Good (Confidence: 0.9818, Lift: 1.0505, Leverage: 0.0126, Conviction: 3.5937)

Rule2:

SEASON=Winter, SO2=Good, NO2=Good, TEMP=Moderate → O3=Good (Confidence: 0.9831, Lift: 1.0098, Leverage: 0.002, Conviction: 1.5677)

Rule3:

SO2=Good, O3=Good, TEMP=High, HUM=Moderate → NO2=Good (Confidence: 0.9699, Lift: 1.0338, Leverage: 0.0072, Conviction: 2.0534)

Rule4:

SO2=Good, NO2=*Good*, O3=*Good*, STATION=*Urban* → PM10=*Good* (Confidence: 0.6346, Lift: 0.9457, Leverage: -0.0164, Conviction: 0.9003)

Examples of rules with 6 itemsets:

Rule1:

PM10=Good, SO2=Good, NO2=Good, HUM=High, PRES=High → TEMP=High (Confidence: 0.6514, Lift: 1.0955, Leverage: 0.0202, Conviction: 1.1629)

Rule2:

SO2=Good, NO2=Good, HUM=High, PRES=High, STATION=Urban → O3=Good (Confidence: 0.988, Lift: 1.0149, Leverage: 0.0031, Conviction: 2.2127)

Rule3:

PM10=Good, SO2=Good, TEMP=High, HUM=High, PRES=High → NO2=Good (Confidence: 0.956, Lift: 1.0191, Leverage: 0.0043, Conviction: 1.4067)

Rule4:

PM10=*Good*, SO2=*Good*, NO2=*Good*, O3=*Good*, TEMP=*High* → HUM=*High* (Confidence: 0.6085, Lift: 1.0185, Leverage: 0.0043, Conviction: 1.0282)

Examples of rules with 7 itemsets:

Rule1:

PM10=Good, SO2=Good, O3=Good, TEMP=High, HUM=High, PRES=High → NO2=Good (Confidence: 0.9558, Lift: 1.0188, Leverage: 0.0042, Conviction: 1.3997)

Rule2:

PM10=Good, NO2=Good, HUM=High, PRES=High → TEMP=High, SO2=Good, O3=Good (Confidence: 0.6242, Lift: 1.0989, Leverage: 0.0206, Conviction: 1.1496)

Rule3:

PM10=*Good*, SO2=*Good*, NO2=*Good*, O3=*Good*, TEMP=*High*, HUM=*High* → PRES=*High* (Confidence: 0.9662, Lift: 1.2984, Leverage: 0.0526, Conviction: 7.5779)

As shown in Table 9, the number of k-itemsets was analyzed with varying minimum support threshold values from 0.2 to 0.9 in increments of 0.1. According to results, the maximum and the minimum

Table 9. The variations of k-itemsets according to different support thresholds

Support	Size of Itemset						
	1	2	3	4	5	6	7
0.2	14	61	108	96	45	10	1
0.3	12	34	50	38	16	3	0
0.4	8	23	30	16	3	0	0
0.5	8	18	16	5	0	0	0
0.6	5	9	7	2	0	0	0
0.7	4	5	1	0	0	0	0
0.8	3	3	1	0	0	0	0
0.9	3	2	0	0	0	0	0

number of itemsets were found as 108 and 1 for $k=3$ and $k=7$, respectively, for support threshold value of 0.2. There were no 8-itemsets generated from the attributes given in Table 6.

Figure 3 shows the variations of the number of k-itemsets, for the support value of 0.2. It can be observed that the number of the frequent 2-itemsets and 5-itemsets are close to each other. The WARM algorithm generally constructs a form quite similar to the bell curve. However, the skewness and kurtosis of the curve can change according to the minimum support threshold.

As shown in Figures 4, 5 and 6, when the minimum support value decreases, the number of frequent rules increases almost exponentially. When the minimum support value is too high, the ARM and WARM algorithms cannot discover enough information from data. On the other hand, a large amount of frequent rules are generated when the algorithms are executed with small support values. For this reason, a reasonable threshold can be set to mine valuable and sufficient frequent rules.

Figure 3. The variations in the number of k-itemsets as found by the WARM model

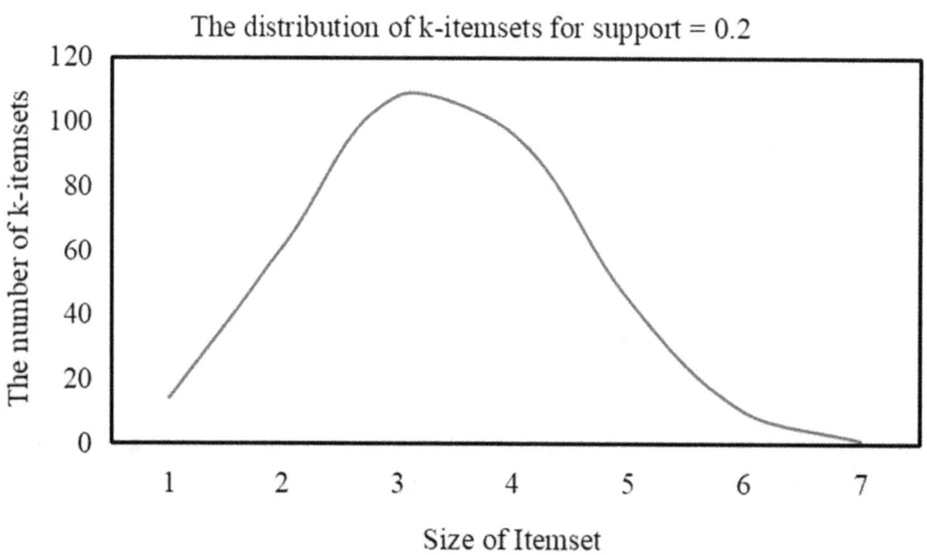

Figure 4 plots the comparison results of the applied method Apriori and the proposed WARM model at the confidence threshold of 0.6. For different support thresholds, the number of strong association rules was determined. In this configuration, the difference between the value of item weights is one. As an example, assume that the weights of the attribute values of PM_{10} being "Good", "Moderate", "Sensitive", "Unhealthy", "Bad", and "Dangerous" are determined as 1, 2, 3, 4, 5, and 6, respectively, with only one difference among different item values. According to the results, WARM is able to detect more association rules than Apriori. It means that WARM explored other important rules that the Apriori method could not detect. For example, when the minimum support is 0.45, WARM discovers 324 frequent rules, while standard ARM finds 280 rules for the same dataset. From this point of view, WARM and ARM algorithms have comparable effects on the number of rules. The effect of adding item weights into the generation of frequent itemsets contributes much to find the undiscovered association rules. If the weights related to air pollutants and meteorological factors are considered, then some missing, probably important, rules can be detected.

Figure 4. Comparisons of WARM and Apriori according to the number of rules in terms of different support thresholds at 0.6 confidence

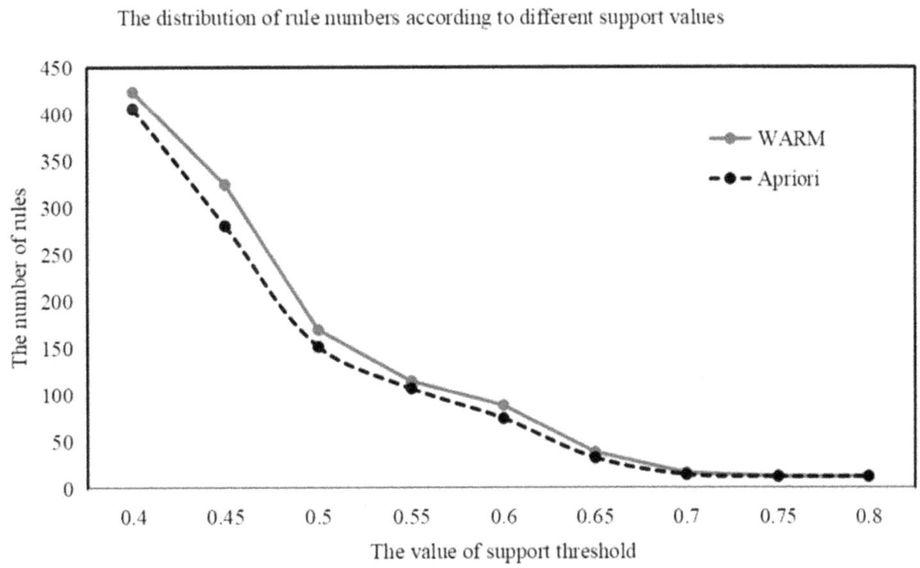

In order to interpret the impact of different confidence levels in the generation of association rules, experiments were also done at the confidence threshold of 0.9. Figure 5 displays the comparison results which were obtained with the same item weights as in Figure 4. As expected, increasing the confidence value from 0.6 to 0.9 decreased the total number of found rules. For instance, 180 rules were now detected by WARM at the support value of 0.4 while the model generated 423 rules when the confidence value was 0.6. The same condition was valid for Apriori. Additionally, WARM performed even better than Apriori at 0.9 in finding more association rules when compared with the results of 0.6; because the proportional difference between them increased. To illustrate, the ratios of the difference of the number of found association rules among two models at the support value of 0.55 are 7.02% and 13.85% in

Figure 5. Comparisons of WARM and Apriori according to the number of rules in terms of different support thresholds at 0.9 confidence

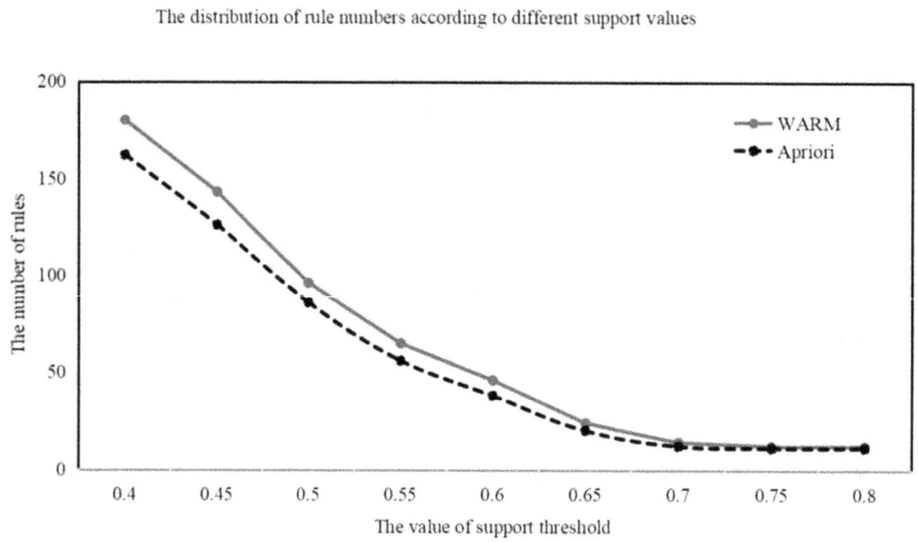

confidence thresholds of 0.6 and 0.9, respectively. It is apparent that the detection of the important rules was performed by the improvement with approximately 7%.

The other important issue is the effect of different weight assignments. To show the difference, experiments were repeated with double the item weights. For example, assume the weights of the attribute values of PM_{10} being "Good", "Moderate", "Sensitive", "Unhealthy", "Bad", and "Dangerous" are now determined as {1, 2, 4, 8, 16, 32} according to this scenario. Now, Figure 6 demonstrates the results

Figure 6. Comparisons of WARM and Apriori according to the number of rules in terms of different support thresholds at 0.9 confidence with doubled item weights

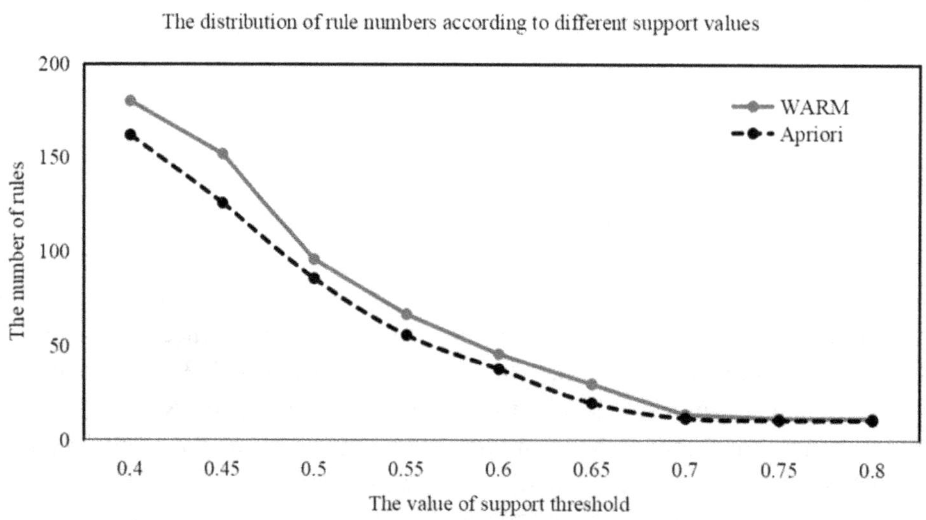

according to this case at the confidence value of 0.9. There is an increase in the number of association rules at the support values of 0.45, 0.55 and 0.65. The change of item weights positively affected the output. Therefore, the determination of weights properly is very important for the WARM method.

CONCLUSION

Presently, intelligent systems are being used to ensure that city operations are made more efficient, fuel and energy savings are achieved, waiting time in traffic is much reduced, city information is available easily, living things are less affected by natural disasters and especially a cleaner environmental life is enjoyed. For this purpose, an intelligent environmental system is designed and introduced in this study, for the purposes of monitoring pollution in the air. Air quality monitoring data and meteorological parameters of 21 stations in Turkey, collected during the period of 1 January 2018 and 1 May 2019, were analyzed and processed using association rule mining and the proposed WARM method. The aim was to develop a smart environmental system to monitor the status of air pollution factors, for triggering an alarm mechanism when a certain critical condition is detected, and to carry out appropriate policy by the air quality management team. In this study, the WARM model was used for the first time to address an environmental problem.

In this study, three experiments were performed to evaluate the proposed approach. Firstly, the change of the number of frequent k-itemsets according to different support thresholds was monitored using WARM model and a pattern similar to bell curve was observed from 1-itemset to 7-itemset. In the second case study, experiments were performed at different confidence thresholds of 0.6 and 0.9. While the number of association rules decreased as the confidence value increased, the ratio of difference between WARM and Apriori also rose in favor of WARM. In the third experiment, the effects of using different item weights were discovered. When the item weight was doubled, WARM discovered even more rules. Since, determining appropriate item weights is crucial to the performance of the WARM model, as a result of the experiments, the proposed model showed a remarkably better performance. According to the experimental results, the most essential association rules were detected and the WARM method discovered the rules which could not be explored by Apriori.

As for the future work, this study can be expanded to analyze other environmental factors such as water quality or soil contamination. Besides, WARM model in this study can be updated using fuzzy sets theory in the process of assigning attribute values.

REFERENCES

Bhavithra, J., & Saradha, A. (2018). Personalized web page recommendation using case-based clustering and weighted association rule mining. *Cluster Computing*, 1–12.

Cagliero, L., Garza, P., Pasini, A., & Baralis, E. M. (2018). Additional reviewer assignment by means of weighted association rules. *IEEE Transactions on Emerging Topics in Computing*.

Cao, B. Y., Ma, S. Q., & Cao, H. H. (Eds.). (2014). *Ecosystem assessment and fuzzy systems management*. Springer International Publishing.

Gyenesei, A. (2000, September). Mining weighted association rules for fuzzy quantitative items. In *European Conference on Principles of Data Mining and Knowledge Discovery* (pp. 416-423). Springer. 10.1007/3-540-45372-5_45

Han, J., Pei, J., & Kamber, M. (2011). *Data mining: concepts and techniques*. Elsevier.

He, Y., Zhu, C., He, Z., Gu, C., & Cui, J. (2017). Big data oriented root cause identification approach based on Axiomatic domain mapping and weighted association rule mining for product infant failure. *Computers & Industrial Engineering*, *109*, 253–265. doi:10.1016/j.cie.2017.05.012

Holmes, G., Donkin, A., & Witten, I. H. (1994). *Weka: A machine learning workbench*. Academic Press.

Hui, K., Chen, C., & Zhan, Y. (2019, May). A Novel Approach to Classify Video Scenes via Mining Weighted Association Rules of Actions in Temporal Sequence. In *Proceedings of the 2019 4th International Conference on Multimedia Systems and Signal Processing* (pp. 145-152). ACM.

Indhumathy, M., Nabhan, A. R., & Arumugam, S. (2018). A weighted association rule mining method for predicting HCV-human protein interactions. *Current Bioinformatics*, *13*(1), 73–84. doi:10.2174/1574893611666161123142425

Isikli, E., Ustundag, A., & Cevikcan, E. (2015). The effects of environmental risk factors on city life cycle: A link analysis. *Human and Ecological Risk Assessment*, *21*(5), 1379–1394. doi:10.1080/10807039.2014.957953

Khan, M. S., Muyeba, M., & Coenen, F. (2008, September). A weighted utility framework for mining association rules. In *2008 Second UKSIM European Symposium on Computer Modeling and Simulation* (pp. 87-92). IEEE.

Lakshmi, K. S., & Vadivu, G. (2019). A novel approach for disease comorbidity prediction using weighted association rule mining. *Journal of Ambient Intelligence and Humanized Computing*, 1–8. doi:10.100712652-019-01217-1

Li, T., Li, Y., An, D., Han, Y., Xu, S., Lu, Z., & Crittenden, J. (2019). Mining of the association rules between industrialization level and air quality to inform high-quality development in China. *Journal of Environmental Management*, *246*, 564–574. doi:10.1016/j.jenvman.2019.06.022 PMID:31202021

Li, Y. (2005, September). A user-guided association rules mining method and its application. In *The Fifth International Conference on Computer and Information Technology (CIT'05)* (pp. 150-156). IEEE.

Lu, J., & Qian, Z. (1999). Mining Boolean and General Fuzzy Weighted Association Rules database. *Mathematica Applicata*, *12*, 38–43.

Muyeba, M., Khan, M. S., & Coenen, F. (2008, May). Fuzzy weighted association rule mining with weighted support and confidence framework. In *Pacific-Asia Conference on Knowledge Discovery and Data Mining* (pp. 49-61). Springer.

Othman, Z. A., Ismail, N., Bakar, A. A., Latif, M. T., & Abdullah, S. M. S. (2018). Rules discovery of high ozone in Klang areas using data mining approach. *International Journal on Advanced Science. Engineering and Information Technology*, *8*(6), 2683–2689.

Qin, S., Liu, F., Wang, C., Song, Y., & Qu, J. (2015). Spatial-temporal analysis and projection of extreme particulate matter (PM10 and PM2. 5) levels using association rules: A case study of the Jing-Jin-Ji region, China. *Atmospheric Environment, 120*, 339–350. doi:10.1016/j.atmosenv.2015.09.006

Qiu, S., Feng, F., Li, Z., Yang, G., Xu, P., & Li, Z. (2019, April). Data mining based framework to identify rule based operation strategies for buildings with power metering system. [). Tsinghua University Press.]. *Building Simulation, 12*(2), 195–205. doi:10.100712273-018-0472-6

Rahimi, Z., Noferesti, S., & Shamsfard, M. (2019). Applying data mining and machine learning techniques for sentiment shifter identification. *Language Resources and Evaluation, 53*(2), 279–302. doi:10.100710579-018-9432-0

Ran, P., Wang, Y., Liu, P., & Li, Z. (2016). A Fuzzy Association Rule Mining Based Approach to Identifying SO 2 Emission Concentration Patterns of Coal-fired Boilers. *Chemical Engineering Transactions, 52*, 631–636.

Saraswat, D., Dev, V., & Singh, P. (2018, September). Analyzing the performance of the Indian Cricket Team using Weighted Association Rule Mining. In *2018 International Conference on Computing, Power and Communication Technologies (GUCON)* (pp. 161-164). IEEE. 10.1109/GUCON.2018.8675115

Sun, K., & Bai, F. (2008). Mining weighted association rules without preassigned weights. *IEEE Transactions on Knowledge and Data Engineering, 20*(4), 489–495. doi:10.1109/TKDE.2007.190723

Tao, F., Murtagh, F., & Farid, M. (2003, August). Weighted association rule mining using weighted support and significance framework. In *Proceedings of the ninth ACM SIGKDD international conference on Knowledge discovery and data mining* (pp. 661-666). ACM. 10.1145/956750.956836

Toti, G., Vilalta, R., Lindner, P., Lefer, B., Macias, C., & Price, D. (2016). Analysis of correlation between pediatric asthma exacerbation and exposure to pollutant mixtures with association rule mining. *Artificial Intelligence in Medicine, 74*, 44–52. doi:10.1016/j.artmed.2016.11.003 PMID:27964802

Wang, B. Y., & Zhang, S. M. (2003, November). A mining algorithm for fuzzy weighted association rules. In *Proceedings of the 2003 International Conference on Machine Learning and Cybernetics (IEEE Cat. No. 03EX693)* (Vol. 4, pp. 2495-2499). IEEE. 10.1109/ICMLC.2003.1259932

Wang, J., Li, H., Huang, J., & Su, C. (2016). Association rules mining based analysis of consequential alarm sequences in chemical processes. *Journal of Loss Prevention in the Process Industries, 41*, 178–185. doi:10.1016/j.jlp.2016.03.022

Wang, W., Yang, J., & Philip, S. Y. (2000). *Efficient mining of weighted association rules (WAR). IBM Thomas J.* Watson Research Division.

Wen, J., Zhong, M., & Wang, Z. (2015). Activity recognition with weighted frequent patterns mining in smart environments. *Expert Systems with Applications, 42*(17-18), 6423–6432. doi:10.1016/j.eswa.2015.04.020

Wu, X., Wang, Q., Liu, Y., & Li, Y. (2017, December). Data-driven regional analysis of urban atmosphere pollution based on density clustering. In *2017 IEEE 2nd Information Technology, Networking, Electronic and Automation Control Conference (ITNEC)* (pp. 412-417). IEEE. 10.1109/ITNEC.2017.8284764

Yue, J. S., Tsang, E., Yeung, D., & Shi, D. (2000, October). Mining fuzzy association rules with weighted items. In *Smc 2000 conference proceedings. 2000 ieee international conference on systems, man and cybernetics.'cybernetics evolving to systems, humans, organizations, and their complex interactions'(cat. no. 0)* (Vol. 3, pp. 1906-1911). IEEE.

Yun, U., & Leggett, J. J. (2005, April). WFIM: weighted frequent itemset mining with a weight range and a minimum weight. In *Proceedings of the 2005 SIAM international conference on data mining* (pp. 636-640). Society for Industrial and Applied Mathematics. 10.1137/1.9781611972757.76

Yun, U., & Leggett, J. J. (2006, April). WIP: mining Weighted Interesting Patterns with a strong weight and/or support affinity. In *Proceedings of the 2006 SIAM International Conference on Data Mining* (pp. 624-628). Society for Industrial and Applied Mathematics. 10.1137/1.9781611972764.73

Zeng, Y., & Xiang, K. (2017). Adaptive Sampling for Urban Air Quality through Participatory Sensing. *Sensors (Basel)*, *17*(11), 2531. doi:10.339017112531 PMID:29099766

Zhang, Z., Chen, E., Wang, J., & Wang, X. (2004, April). Enabling personalization recommendation with WeightedFP for text information retrieval based on user-focus. In *International Conference on Information Technology: Coding and Computing, 2004. Proceedings. ITCC 2004.* (Vol. 1, pp. 272-276). IEEE. 10.1109/ITCC.2004.1286465

ADDITIONAL READING

Bellinger, C., Jabbar, M. S. M., Zaïane, O., & Osornio-Vargas, A. (2017). A systematic review of data mining and machine learning for air pollution epidemiology. *BMC Public Health*, *17*(1), 907. doi:10.118612889-017-4914-3 PMID:29179711

Siwek, K., & Osowski, S. (2016). Data mining methods for prediction of air pollution. *International Journal of Applied Mathematics and Computer Science*, *26*(2), 467–478. doi:10.1515/amcs-2016-0033

Tuysuzoglu, G., Birant, D., & Pala, A. (2018, February). Ensemble Methods in Environmental Data Mining. In *Data Mining* (p. 1). BoD–Books on Demand. doi:10.5772/intechopen.74393

KEY TERMS AND DEFINITIONS

Air Pollution Monitoring: It is the process of measurement of air pollutant factors by sensors embedded in different monitoring sites to check whether or not the measurement values comply with the air quality standards. Aim is to monitor and manage the sir pollution situations.

Apriori: This is one of the most popular association rule mining techniques that was developed by Agrawal and Srikant in 1994 to find frequent itemsets and to obtain strong association rules which are inherent in the dataset.

Association Rule Mining (ARM): It is one of the major tasks of data mining methods in which important relationships among the attributes of a dataset are explored and frequently observed patterns are found.

Data Mining: This is the process of discovering useful information and trends in a given data set by extracting hidden patterns through applying the mechanisms of classification, clustering and association rules analysis.

Intelligent (or Smart) City: This is the name given to a city that implements environmentally friendly projects involving information and communication technologies using data collected from a diverse range of sensors and smart devices to improve the quality and performance of services by reducing the consumption of energy, cutting down on waste and total costs.

Weighted Association Rule Mining (WARM): This refers to a modified version of the traditional association rule mining method where each item has its own weights that express the significance of it within transaction.

Weighted Confidence: This is also one of the performance metrics of weighted association rule mining. It expresses the likeness of occurrence of consequent on the cart, given that the cart already has the items in the antecedents using their weighted support values.

Weighted Support: This is one of the performance metrics of weighted association rule mining method. It refers to the sum of the transaction weights for a specific itemset.

Chapter 8
Connected Vehicles:
A Vital Component of Smart Transportation in an Intelligent City

Zaigham Mahmood
https://orcid.org/0000-0001-7411-7496
Northampton University, UK

ABSTRACT

An intelligent city, also known as smart city or cognitive city, is a new vision for urban development to integrate the various sectors and functions of the society through the use of internet of things and sensor-based technologies. The main objective is to seamlessly integrate and efficiently manage a city's resources such as transportation, healthcare, commerce, manufacturing, education, utilities, law enforcement, as well as city's various e-government information systems. The aim is to improve the quality of life of citizens and efficiency of city's services and processes, through the use of latest distributed technologies. In this chapter, the authors discuss the connected vehicles vision, as a core element of the internet of vehicles paradigm, as part of the intelligent transportation system (ITS). In doing so, they discuss the underlying concepts, technologies, types of vehicular connectivity, and mobile and vehicular ad hoc networks (i.e., MANET and VANET).

INTRODUCTION

An Intelligent City, also known as Smart City or Cognitive City, is a new vision for urban development to integrate the various sectors and functions of the society through the use of Internet of Things (IoT) and sensor-based technologies. The objective is to integrate and manage a city's resources and relevant processes such as transportation, healthcare, commerce, education, utilities, law enforcement, as well as city's various e-government information systems. The ultimate aim is to improve the quality of life of citizens and efficiency of city's services and processes, ensuring that citizens' day to day needs are met and their quality of life and working environments are improved.

DOI: 10.4018/978-1-7998-5062-5.ch008

Relevant computing and IoT-based technologies, social media offerings, distributed computing paradigms, mobile technology, communication protocols, and device connectivity frameworks are already well established and understood. These technologies and smart systems are readily available to support the following, and more:

- To empower the citizens to provision and access the e-services provided by the social and commercial sectors of the intelligent cities
- To allow city managers to interact directly with the citizens and to effectively monitor the digital and smart environments of the modern cities
- To develop smart environments where citizens can enjoy improved quality of life working and living in digital intelligent cities.

In this chapter, the focus is on Intelligent Transportation Systems (ITS) which is an essential component of the Internet of Vehicles (IoV) vision. The main objectives of ITS are as follows:

- to provide digitally based services and products relating to different modes of transport (e.g. cars, buses, etc.) and traffic management (e.g. planning, control, management, and movement of vehicles)
- to enable road users (e.g. drivers and pedestrians) to be better informed (e.g. regarding road conditions, road incidents ahead, etc.) and make their journeys safer and more coordinated, to allow them to experience 'smarter' use of transport networks (e.g. infrastructures that allow safe movement of connected and autonomous vehicles).
- to improve the quality of city transportation in a number of ways e.g. increased efficiency of road transport, better traffic management, more efficient mobility, less road incidents, safer driving experience, etc.

This chapter is organised as follows. In the following three sections, we introduce the concepts and technologies relating to the Intelligent Cities (IC) vision. In the next two sections, we discuss Intelligent Transportation System (ITS) as an essential component of IC, and explore the Connected Vehicles paradigm as an important aspect of the ITS. Following this, we discuss in detail the technologies that enable the Connected Cars vision, as well as, the inherent issues and limitations of the relevant technological systems.

INTELLIGENT CITY: CHARACTERISTICS AND DIMENSIONS

An Intelligent City (IC), sometimes also called Digital City or Smart City, refers to an intelligent physical environment that embeds information and communication technologies including sensor based inter connected systems to develop interactive spaces and smart objects therein, to bring computation speed and accuracy into the physical world (Steventon and Wright, 2006). It is an attractive and highly useful application of the Internet of Things (IoT) paradigm.

The idea behind an Intelligent City is to provide an improved quality of life to the people living and working in the city. Since the deployment of technology, and the priority of e-services to be made available to citizens differs from city to city and country to country, the conceptualisation of an Intelligent

City depends on the level of development by way of amount of automation, digital innovation, willingness to implement technological changes by the city management, relevant resources put in place by the government, and aspirations of the city dwellers (Smart Cities Mission, 2019). Frost & Sullivan in Sing (2014) identify certain key aspects that define a Smart city. These include smart governance, smart energy, intelligent buildings, smart mobility, intelligent transportation, digitally based infrastructure, smart technology, intelligent healthcare, technologically based industry, and smart citizen. Deakin and Al Wear (2011) list four factors that contribute to the definition of a Smart city. These are:

- Application of a diverse range of digital and sensor based technologies to communities and the industrial sectors in the city.
- Use of ICT and distributed computing technologies to transform life and working environments within the city.
- Deployment of technologies (as mentioned above) in governmental information systems and provision of e-services.
- Territorialisation of practices that bring technologies and people together to enhance the innovation and knowledge that they offer.

We can also add human resource, artificial intelligence and machine learning to the above list. In fact, the intelligence of cities resides in the effective combination of digital telecommunication networks, seamless connectivity of digital devices, ubiquitously embedded intelligence in smart objects, sensor-based networks, and cognitive competence built in the information systems.

The development, integration, and adoption of an Intelligent City capabilities require a framework that can be divided into four dimensions (Figure 1) viz:

- Technology framework: this is the core element that consists of a number of infrastructures including digital, intelligent, ubiquitous, wired, wireless, hybrid, and citywide
- Human framework: this is essential as it is the residents who are the recipient of any city-wide provision. It includes human elements such as creativity, learning, humanity, and knowledge
- Institutional framework: this is required for the development, management, updating and deployment of the above and the related inherent technologies
- Data management framework: this is necessarily required to support the collection, storage and analysis of the Big Data that is generated by the sensors, other smart objects, and by the humans.

The Intelligent Community Forum (2006) suggests a framework for the development of an Intelligent City Intelligent City, that bring together the following:

- Deployment of broadband communication for industry, commercial sector, government facilities and citizens
- Necessary education, training and expertise to perform knowledge-based tasks for city management as well as the user community i.e. general public
- Government policies and services that promote digital democracy to bridge the digital divide to ensure that all sectors of the society and citizens benefit accordingly
- Innovation in the public and private sectors, and efforts on part of the city management, to fund and manage the development of new ventures

Figure 1. Dimensions of an Intelligent City

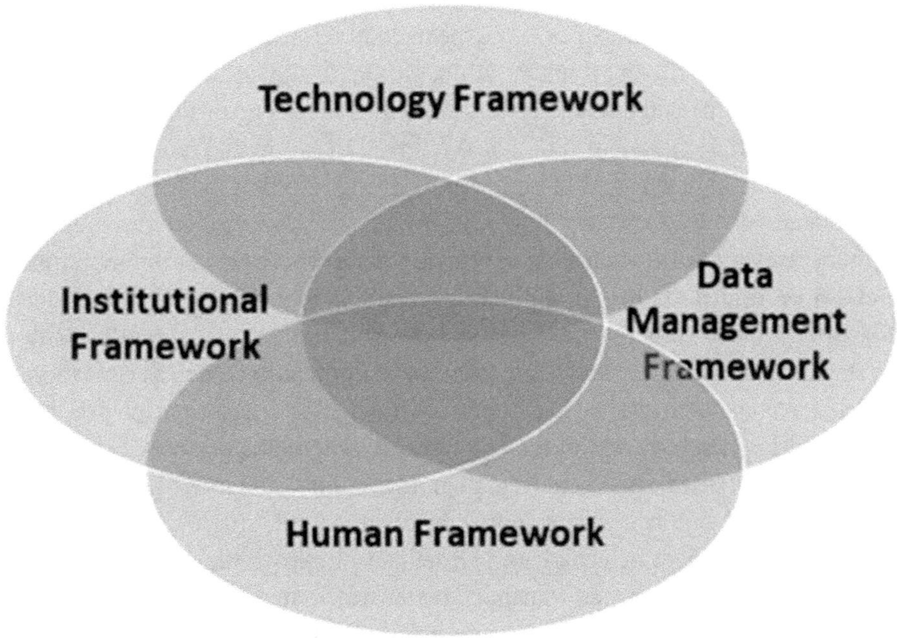

- Economic development and marketing to attract talented employment and investments; and policies to retain and further development of the available expertise.

In an Intelligent City, the way people work and live, and conduct economy, education, health, entertainment, and transportation, etc. is more technologically based. As Honeywill (2019) suggests, an Intelligent City can be visualised as having the following five dimensions:

- Being a cultural capital: this refers to having attractive economical and cultural elements that put a city on the global map to attract tourism and make economic gains
- Providing smart environments for work, living and entertainments: this relates to people in the IC and refers to developing and enhancing their imagination, inventiveness and creativity
- Ensuring diversity of social and communal solutions: this refers to diversity of cultures within the IC and celebration of ethnicity within the physical and metaphysical boundaries
- Utilising social intelligence at the highest level: this suggests that it is the socially intelligent minority of citizens in the Intelligent City that delivers the Peak Value
- Developing digital innovative landscape: this refers to embedding into the IC environment, communication infrastructure, digital interfaces and problem-solving tools for the public.

INTELLIGENT TRANSPORTATION

As mentioned above, there are a number of key aspects that define an Intelligent City e.g. smart governance, intelligent mobility, intelligent transportation, etc. In this chapter, we focus mainly on the Intelligent Transportation System, as the other aspects are out of the scope of this chapter.

An intelligent transportation system (ITS), also referred to as Smart Transportation within an Intelligent City scenario, is an application of the Internet of Things (IoT) in the form of seamless connectivity of transport related vehicles (e.g. cars, buses, trains, etc.), transport related roadside smart systems (e.g. traffic lights), relevant infrastructure, and relevant enterprise information systems. ITS is having a significant effect on transportation in applications such as electronic toll collection, ramp meters, traffic monitoring and speed cameras, traffic-signal coordination, transit signal priority and traveller-information systems. Its adoption is expected to increase in applications such as fleet monitoring, tolling management, ticket management, arrival and departure timetables, transportation pricing, telematics, as well as traffic monitoring. Primary aim of the ITS is to provide innovative services relating to different modes of transport and traffic management (e.g. planning, control and movement of vehicles) to enable users to be better informed and make safer, more coordinated, and 'smarter' use of transport networks (structures that allows movement of vehicles). ITS helps to improve the quality of transportation in a number of ways e.g. more efficient road transport, better traffic management, more efficient mobility, less road incidents, real time warning systems, safe driving experience, etc.

In the following sections, we elaborate on different varieties of vehicle connectivity (e.g. V2V, V2I, V2C, V2P, etc.), types of ad hoc networks (e.g. VANET and MANET), and the IoV.

CONNECTED VEHICLES

A Connected Vehicle (CV) is one that is linked to other vehicles via the Internet and capable of sharing data with smart devices inside other connected cars as well as with smart objects in the environment e.g. traffic lights, and sensors embedded in the roads, etc. Here, a vehicle refers to all manner of transport systems on the roads including cars, buses, tractors, long vehicles, etc.

A connected vehicle is an indispensable node in the Internet of Vehicles (IoV) vision that is an essential component of the Intelligent Transportation System (ITS). IoV and ITS are both applications of the IoT in an Intelligent City scenario. In 2020, some of the leading manufacturers of cars, active in further development of the CV vision, include: General Motors, Google, BMW, Audi, Mercedes Benz, Tesla, Jaguar, Porche, and Nissan (Technavio, 2019). By 2020, CMSWire (2019) expect to see around 380 million cars on the roads. Business Insider (2019) expect 94 million connected cars to be shipped in 2021 and 82% of these will be wirelessly connected.

Much work has been done with respect to connectivity of vehicles. In the following sub-sections, we discuss different types of vehicle connectivity in the IoV vision.

Types of Vehicular Connectivity

There are several approaches to vehicle communication including: Vehicle-to-Vehicle (V2V), Vehicle-to-Infrastructure (V2I), Vehicle-to-Cloud (V2C), Vehicle-to-Pedestrian (V2P) and Vehicle-to-Everything

(V2E). These are further explored in the following sub-sections. It should be noted that, in what follows, we use the term 'vehicle' to denote all types of road vehicles including cars.

Vehicle-to-Vehicle (V2V) Connectivity

This refers to a vehicle on the road, whether stationary or moving, being connected to other vehicles in the environment. V2V connectivity aims to determine location, position, speed, direction of travel, and, in general status of vehicles in the area. The relevant V2V technology allows vehicles to transmit and receive omni-directional messages creating a 360-degree awareness of other vehicles in proximity. Messages are generally exchanged every 5-10 second, to assess mainly the potential crash threats ahead and traffic flow and congestion situations, if any and as they develop. The underlying technology uses dedicated short-range communication (DSRC), with a possible frequency of 5.9 GHz, that is normally used by a WiFi network. In this context, the technology can employ visual, audible, and tactile alerts, or a combination of these, to warn vehicle drivers to take suitable actions to avoid crashes, e.g. to slowdown or change lane, etc. The general aim is to provide safer experience for drivers and comfortable travel by passengers; and also to reduce possibilities of accidents and fatalities.

Vehicle-to-Infrastructure (V2I) Connectivity

V2I refers to a vehicle in communication with highway infrastructure including smart devices and traffic related systems such as traffic monitors, traffic signals, lane sensors, speed cameras, communication satellites, GPS, parking meters, bus stops, and other relevant mobile devices in the vicinity. The underlying technologies include sensors such as RFID, radars, navigation systems, and speed cameras. Similar to the V2V connectivity, communication takes place using dedicated DSRC frequencies. The objective is to capture data from inter connected devices (possibly, including other vehicles on the network) as part of the road infrastructure systems to gain real-time advisory information for vehicles (and drivers) to change speed, change lane, apply brakes, follow diverted roots or to avoid dangerous situations ahead. Like V2V, the aim is to have better and safer experience for drivers and passengers; and to reduce possibilities of accidents and injuries. Difference between Vehicle-to-Vehicle and Vehicle-to-Infrastructure is that, whereas V2V requires other vehicles in the vicinity to be also connected, V2I needs connectivity mainly between the infrastructure and the vehicle itself.

Vehicle-to-Cloud (V2C) Connectivity

This refers to a vehicle being linked up to a cloud distributed computing environment. Since, vehicles can produce upwards of 560 GB data per vehicle per day, this Big Data needs to be stored, managed and analysed. In this respect, V2C communication provides support for the following: 1) to store and retrieve data in real time; 2) to share and analyse information for decision making as and when needed; and 3) to provision, consume and process other cloud-based services of all varieties. V2C is an attractive idea as cloud technologies can help and support in a number of other ways as well e.g.: 1) to reduce the size of central consoles and storage space within vehicles; 2) to reduce the amount of hardware under the dash board of the vehicle; and 3) to perform data-intensive tasks as activities outsourced to remote cloud-based servers (Ashwini, 2017). This, in turn, means reduction in expensive data storage elements within the vehicles and thus possibly reduced prices of increasingly computerised cars. Cloud related capabilities can also enhance automotive safety as well as augmented reality in vehicles.

Vehicle-to-Pedestrian (V2P) Connectivity

This type of connectivity refers to direct communications between vehicles and pedestrians, within close proximity of the vehicles. Pedestrians comprise a broad set of road users including people walking nearby or attempting to cross the roads, children walking or being pushed in buggies, people using wheelchairs or mobility devices, passengers waiting on bus stops, people getting on or off the buses, and those riding bicycles. The objective is to increase safety of pedestrian and to avoid collisions involving road users, as well as to help improve safety of vehicle drivers and passengers, and reduce fatalities. An important facet of V2P is the pedestrian detection system (ITS, 2019) that can be implemented in a number of ways, viz: 1) having the technology embedded within the vehicles for blind spot warning, incident ahead warning, etc.; 2) having smart devices embedded within the roadside infrastructure e.g. for lane closure warnings; and 3) having technology based smart objects carried by pedestrians e.g. smart devices for warning purposes (e.g. smart phones and smart watches, etc.). Car manufacturers are already well-advanced regarding field testing of V2P related technology. One example is Telstra Cars in collaboration with Cohda Wireless, who have successfully conducted V2P technology experiments in South Australia, using common every day scenarios such as a car and a cyclist approaching a blind corner, over a mobile network (Auto Connected Car News, 2017).

Vehicle-to-Anything (V2X) Connectivity

V2X, or Vehicle-to-Everything (V2E), is an umbrella term for vehicular communication system that uses WLAN technology. This, relatively emerging market, refers to the technology and processes that allow vehicles to communicate with other vehicles, roadside traffic system around them, as well as other sensor operated smart devices in the vicinity. In this respect, V2X is a context aware system that includes V2V, V2I and V2P communication (Investopedia, 2017). In addition to improvement in the safety for all, V2X technology offers a range of other convenient benefits e.g. payments at toll stations and parking meters, etc. (Mahmood, 2020). However, in order for the Vehicle-to-Anything connectivity to become more ubiquitous, other vehicles and objects in the traffic system and in the immediate environment must also have V2X technology embedded in them. It is necessary that pedestrians and cyclists must also carry V2X-enabled smart devices to receive relevant information and alerts. It is for this reason that V2X needs to advance much more and reasonably quickly. Many V2X suppliers such as Delphi Automotive, Daimler, Audi, Qualcomm, TomTom, and Continental are working hard towards V2X integration plans.

Mobile Ad Hoc Network (MANET)

MANET is a type of ad hoc network that allows change of location for its nodes and can configure itself on the fly. It is used for communication between vehicles and roadside smart infrastructure and other intelligent systems. In this network, also known as Wireless Ad Hoc Network (WANET) or Ad hoc Wireless Network, nodes (i.e. vehicles on the roads) are autonomous and generally mobile in nature and free to move independently in any direction at varying speeds. Some MANETs are restricted to a local area of wireless devices, while others may be connected to the Internet. Note that WANET is different from VANET which we discuss in the next sub-section.

MANETs are self-configuring infrastructure-less network, with the following features: dynamic topology, variable capacity links, energy constrained operation, and limited physical security (Mahmood, 2020). In MANET, every node is a potential router. Here, the routing protocols often used include the following technologies: DSR (Dynamic Source Routing), ADOV (Ad Hoc On demand Distance Vector), TORA (Temporally Ordered Routing Algorithm) and OLSR (Optimized Link State Routing).

MANETs are decentralised, peer-to-peer, self-forming, self-healing and, therefore, more robust than centralised networks. Since the MANET architecture evolves with time as the nodes (i.e. vehicles) change locations, it has the built-in flexibility to resolve issues such as node disconnection, node isolation or slow transmission of Internet based data. Also, since the MANET topology changes with changing location of its nodes, these networks are scalable and more flexible, which, in turn, entails lower administration effort and costs. However, the negative is that with the time evolving nature of the network and changing mobility patterns of vehicles and devices, there can be variations in the network performance and the quality of service. There are often, also, issues relating to signal protection, reliability of nodes, limitation of processing power, and even the adequacy of power supply. Still, as mentioned before, flexibility of MANETs is a huge positive attribute.

Vehicular Ad Hoc Network (VANET)

VANET may be considered as a subclass of MANET communication, applicable specifically to vehicles e.g. cars. It consists of a network of communicating vehicles, whether stationary or moving at high speeds on the roads. Until recently, VANETs were used mainly to provide comfort and safety to drivers in vehicular environments. Now, however, these networks are seen as infrastructure and traffic management mechanism for an Intelligent Transportation System (ITS) for an increasing number of autonomous vehicles on the roads. One of the distinguishing characteristics of VANET is the content-centric distribution - the content being more important than the source. This is in marked contrast to the Internet where agents demand information from specific sources.

VANET communication protocols are similar to those used in wired networks, where each host has a unique IP address. However, assigning such addresses to moving vehicles is much more complex and often requires a Dynamic Host Configuration Protocol (DHCP) server, a heresy for ad hoc networks that operate without any infrastructure, using self-organization protocols. One difference between VANET and MANET is that the routing protocols of MANET are not exactly suitable for VANET architecture which generally consist of three types of categories:

- Cellular and WLAN network where fixed gateways and WiMaX and WiFi Apps are used for connection to the Internet for routing and getting traffic information
- Pure ad hoc variety i.e. between vehicles and fixed gateways
- Hybrid type i.e. a combination of both infrastructure and ad hoc networks.

Some of the popular architectures of VANET include the following:

- WAVE (Wireless Access in Vehicular Environment) by IEEE
- C2CNet (Car-to-Car Network) by C2C Consortium

- CALIM (Continuous Air Interface for Long to Medium range) by ISO.

VANETs can be seen as systems consisting of entities that can be divided into three domains as mentioned below:

- Mobile domain: that consists of two sub-parts: vehicle domain (i.e. cars, buses, etc) and mobile device domain (i.e. handheld devices, smart phones, navigation system, etc)
- Infrastructure domain: that also comprises two sub-parts: roadside infrastructure domain (i.e. roadside units and traffic lights) and central infrastructure domain (i.e. traffic and vehicle management centres)
- generic domain: that, in turn, consists of Internet infrastructure domain and the private infrastructure domain.

Internet of Vehicles (IoV)

Internet of Things (IoT) paradigm is paving the way for the evolution of Internet of Vehicles (IoV) vision, which is a vision that integrates VANET, IoT and Mobile Cloud Computing technologies. IoT and VANET have already been elaborated in some detail above; the Cloud Computing model is the on-demand availability and provision of computer system resources (software and virtualised hardware, software development platforms and environments, etc.), especially data storage and computing power, on a pay-as-you-go approach. In Vehicle-to-Cloud (V2C) communication, any vehicle in VANET can directly communicate with the cloud environment, and access cloud-based services including the storage facilities.

The IoV may be regarded as ITS integrated IoT. It refers to data exchanges between networked vehicles and between vehicles and roadside units (traffic monitors, traffic signals, lane sensors, speed cameras, etc.), in real time. It integrates networked intelligent sensors, wireless and mobile communication technology, vehicle navigation systems, and information platforms in the environment. It also fully supports distributed data, information processing, and Internet technology for efficient exchange of information and sharing it between humans and vehicles, for V2V, and between vehicle and roadside infrastructure. In this respect, IoV can be regarded as an extension of Vehicle-to-Vehicle (V2V) connectivity that enhances driving aids for fully autonomous driving by furthering the vehicles' artificial intelligence awareness of the surrounding environment and machine learning capabilities. The essential components of IoV include the following (Indu, 2019):

- Vehicles (as nodes of the ad hoc network, e.g. cars),
- Roadside Units (RSU) and ITS infrastructure (e.g. road and traffic related sensors, signals and smart traffic related objects)
- Personal devices (e.g. smart phones and PDAs)
- Human operators (e.g. drivers and pedestrians).

Therefore, we can envisage the following interaction (Mahmood, 2020)s:

- Vehicle-to-Vehicle interaction
- Vehicle and Infrastructure interaction

- Vehicle and sensors interaction
- Vehicle and RSU interaction
- Vehicle and human interaction
- Personal devices and vehicle interaction
- Cloud and vehicle interaction
- Device-to-device (D2D) interaction.

To organise the various interactions within the IoV, several architectures have been proposed by various researchers e.g. Nanjie (2011), Bonomi (2013), Wan & Zhang (2014), Raj (2016), and Gandotra (2017). These are all layered architectures that incorporate a variety of interactions, as suggested above. In a recent article, Sherazi & Khan, et. al. (2019) proposed an architecture that incorporates multiple wireless interfaces (e.g. WAVE), long-range WiFi, and 4G/LTE installed on-board units to establish a context-aware network connectivity. This is mainly to attempt to meet the requirements of pervasive connectivity for VANETs to make them scalable and more adaptable to the IoV vision. A well organised architecture promises benefits such as better connectivity, scalability, adoptability, fast response for decision making in real time, and increased efficiency of communication between interconnected smart devices.

IoV is especially useful for autonomous self-driving vehicles. However, it is a highly complex integrated network system mainly because of the following characteristics:

- Number of nodes in the network (i.e. vehicles) can increase and decrease in real time
- Nodes are mobile in nature that have varying speeds and changing location in real time
- Distances between nodes change in real-time.

Notwithstanding the limitations as mentioned above, the IoV and connected vehicles paradigm promises many highly attractive benefits including the following: improved road safety, accident prevention, better driver experience, real time information about road conditions, efficiency of driver related processes, economical saving, reduced fuel emission and waste, etc. As in any network, there are also issues relating to trust, privacy and confidentiality of data, as well as cyber threats to comprise the network in various ways. However, industry are aware of these and mechanisms are being put in place to reduce the possibilities of cyber threats; and protect privacy and confidentiality of data being transmitted across the networks.

ENABLING TECHNOLOGIES

Technologies that enable the connectivity of vehicles in the IoV paradigm already exist. Some of these are briefly introduced in this section. For more detail, refer to Mahmood (2020).

- IoT: this is the main underlying paradigm where 'things' are inter-connected smart objects; these being vehicle of all varieties
- Cloud Computing: this refers to a computing environment that provides on-demand availability of virtualised hardware and software; as well as storage and other technology related services
- AI and Machine Learning: these provide embedded intelligence in smart devices (e.g. mobile phones and sensors) and relate to smart calculating devices to mimic cognitive functionality

- 5G and DSRC: 5G is the latest cellular network technology for mobile Internet connectivity; and DSRC supports interoperability to serve as basis for V2V and V2I communication
- VOIP: this refers to Voice over Internet Protocol that provide man-machine interaction via the Internet. It treats the voice signals same as other digital packets
- Telematics: this is an interdisciplinary subject encompassing telecommunication, vehicular technologies, and computer science. It involves sending/receiving information using telecommunication Devices.

CONCLUSION

Technologies based on the Internet of Things (IoT) paradigm are helping to build intelligent cites to further improve resources utilisation, enhance safety on the roads, provide better healthcare, and support smart living, etc. Intelligent Transportation Systems (ITS), as part of the IoV vision, is also an essential aspect of a well-developed and successful smart city, which further entails development of connected vehicles technologies in terms of V2V, V2I, V2P and V2X infrastructure and communication.

Much work and experimentation has already been undertaken and the connected vehicles vision has moved from hype to reality. Even the smart driverless vehicles also exist, though they are not that common on the roads at the moment. The Google self-driving car project is probably the most advanced. They have successfully conducted over 5 million miles of road testing. Gartner estimate, as reported in Ismail (2017), that the number of such cars will increase to around 250 million by the end of 2020. In Pittsburgh, in the US, 50 road intersections are already equipped with smart traffic signals with a view to reducing traffic congestion. In the UK, in February 2020, a Nissan Leaf autonomous vehicle successfully completed a 230-mile self-navigated journey (HumanDrive, 2020). Austin transportation department in the US, have tested smart parking meters as part of its smart mobility project.

Although, there are a number of technological issues with respect to the connectivity of vehicles in the IoV vision, these are being investigated and will, hopefully, soon be resolved. The future of IoV is, indeed, highly attractive.

REFERENCES

Ashwini, A. (2017). *What is the technology stack used in IOT product (Internet of Things)?* Available at: https://www.quora.com/What-is-the-technology-stack-used-in-IOT-product-Internet-of-Things

Auto Connected Car News. (2017). *Vehicle to Pedestrian (V2P) and Bicycle Demoed.* Available at: https://www.autoconnectedcar.com/2017/07/vehicle-to-pedestrian-v2p-and-bicycle-demoed/

Bonomi, F. (2013). The Smart and Connected Vehicle and the Internet of Things, Synchronization. *Telecommunication Systems*, *2013*, 1–53.

CMSWire. (2019). *Connected Car Experiences in 2019.* Available at: https://www.cmswire.com/digital-experience/connected-car-experiences-in-2019-exploring-the-possibilities/

Deakin, M., & Al Waer, H. (2011). From Intelligent to Smart Cities. *Journal of Intelligent Buildings International: From Intelligent Cities to Smart Cities.*, *3*(3), 140–152. doi:10.1080/17508975.2011.586671

Gandotra, P., Kumar Jha, R., & Jain, S. (2017). A survey on device-to-device (D2D) communication: Architecture and security issues. *Journal of Network and Computer Applications*, *78*, 9–29. doi:10.1016/j.jnca.2016.11.002

Honeywill, R. (2019). *What is an Intelligent City?* Available at: https://rosshoneywill.com/articles/what-is-an-intelligent-city/

HumanDrive. (2020). *Nissan Leaf Completes the UK's Longest and most Complex Autonomous Journey.* Available at: https://humandrive.co.uk/nissan-leaf-completes-the-uks-longest-and-most-complex-autonomous-car-journey/

Indu, S. K. (2019). Internet of Vehicles (IoV): Evolution, Architecture, Security Issues and Trust Aspects. *Int J Recent Technology and Engineering*, *7*(6), 2019.

Insider, B. (2019). *Automotive Industry Trends: IoT Connected Smart Cars and Vehicles.* Available at: https://www.businessinsider.com/internet-of-things-connected-smart-cars-2016-10?r=US&IR=T

Intelligent Community Forum. (2006). *What is an Intelligent Community.* Available at: https://www.intelligentcommunity.org/what_is_an_intelligent_community

Investopedia. (2017). *V2X (Vehicle-to-Vehicle or Vehicle-to-Infrastructure).* Available at: https://www.investopedia.com/terms/v/v2x-vehicletovehicle-or-vehicletoinfrastructure.asp

Ismail, N. (2017). *The Present and Future pf Connected Car Data.* Available at: https://www.information-age.com/present-future-connected-car-data-123466309/

ITS. (2019). *Connected Vehicles: Vehicles to pedestrian communication.* Available at: https://www.its.dot.gov/factsheets/pdf/CV_V2Pcomms.pdf

Mahmood, Z. (2020). Connected Vehicles in the IoV: Concepts, Technologies, and Architectures. In Z. Mahmood (Ed.), *Connected Vehicles in the Internet of Things. Springer.* doi:10.1007/978-3-030-36167-9_1

Nanjie, L. (2011). *Internet of Vehicles: Your next connection.* Huawei WinWin.

Raj, D. (2016). Fog. *IEEE 8th Int. Conf. Cloud Comput. Technol. Sci.*, 90–93.

Sherazi, H. H. R., & Khan, Z. A. (2019). A Heterogeneous IoV Architecture for Data Forwarding in Vehicle to Infrastructure Communication, Mobile. *Information Systems*, *2019*, 3101276.

Smart Cities Mission. (2019). *What is Smart City.* Available at: http://smartcities.gov.in/content/innerpage/what-is-smart-city.php

Steventon, A., & Wright, S. (Eds.). (2006). *Intelligent spaces: The application of pervasive ICT.* London: Springer. doi:10.1007/978-1-84628-429-8

Technavio. (2019). *Top 10 Connected Car Companies Leading the Global Connected Car Market in 2019.* Available at: https://blog.technavio.com/blog/top-10-connected-car-companies-leading-global-connected-car-market

Techopedia. (2012). *Cloud Computing for Vehicles: Tomorrow's High-Tech Car.* Available at: https://www.techopedia.com/2/28137/trends/cloud-computing/cloud-computing-for-vehicles-tomorrows-high-tech-car

Wan, J., Zhang, D., Zhao, S., Yang, L., & Lloret, J. (2014). Context-aware vehicular cyber-physical systems with cloud support: Architecture, challenges, and solutions. *IEEE Communications Magazine, 52*(8), 106–113. doi:10.1109/MCOM.2014.6871677

ADDITIONAL READING

Abu Talib, M., Abbas, S., Nasir, Q., & Mowakeh, M. F. (2018). Systematic literature review on Internet-of-Vehicles communication security. *International Journal of Distributed Sensor Networks, 14*(12). doi:10.1177/1550147718815054

Adly, A. S. (2020). Seamless V2I Communication in HetNet: State-of-the-Art and Future Research Directions. In Z. Mahmood (Ed.), *Connected Vehicles in the Internet of Things, Springer, 2020.*

Al-ani, R., Zhou, B., Shi, Q., & Sagheer, A. (2018), A Survey on Secure Safety Applications in VANET, 2018 IEEE 20th International Conference on High Performance Computing and Communications; IEEE 16th International Conference on Smart City; IEEE 4th International Conference on Data Science and Systems (HPCC/SmartCity/DSS), Exeter, United Kingdom, 2018, pp. 1485-1490

Al-Sultan, S., Al-Doori, M., Al-Bayatti, A., & Zedan, H. (2014). A comprehensive survey on vehicular Ad Hoc network, Elsevier. *Journal of Network and Computer Applications, 37*, 380–392. doi:10.1016/j.jnca.2013.02.036

Anusha, A., & Ahmed, S. M. (2017, July). Vehicle tracking and monitoring system to enhance the safety And security driving using IOT. In *2017 International Conference on Recent Trends in Electrical, Electronics and Computing Technologies (ICRTEECT)* (pp. 49-53). IEEE 10.1109/ICRTEECT.2017.35

Bali, P., Parjapati, G., Aswal, R., Raghav, A., & Gupta, N. (2015). *Intelligent Traffic Monitoring System (ITMS)*. Progress in Science and Engineering Research Journal.

Bojan, T. M., Kumar, U. R., & Bojan, V. M. (2014, December). Designing vehicle tracking system-an open source approach. In *2014 IEEE International Conference on Vehicular Electronics and Safety* (pp. 135-140). IEEE. 10.1109/ICVES.2014.7063737

Bojan, T. M., Kumar, U. R., & Bojan, V. M. (2014, December). An internet of things based intelligent transportation system. In *2014 IEEE International Conference on Vehicular Electronics and Safety* (pp. 174-179). IEEE. 10.1109/ICVES.2014.7063743

Bwalya, K. J. (2020). Spatial Intelligence and Vehicle-to-Vehicle Communication: Topologies and Architectures. In Z. Mahmood (Ed.), *Connected Vehicles in the Internet of Things, Springer, 2020.* doi:10.1007/978-3-030-36167-9_2

Bwalya, K. J. (2020). Protocols and Design Structures for Vehicular Networks. In Z. Mahmood (Ed.), *Connected Vehicles in the Internet of Things, Springer, 2020.* doi:10.1007/978-3-030-36167-9_5

Carter, J. (2015), Challenges Aplenty: What are the Roadblocks facing the Connected Car? Available at: https://www.techradar.com/news/world-of-tech/challenges-aplenty-what-are-the-roadblocks-facing-the-connected-car-1287028

Chang, J., Yao, W., & Li, X. (2017). The design of a context-aware service system in intelligent transportation system. *International Journal of Distributed Sensor Networks*, *13*(10). doi:10.1177/1550147717738165

Chang, T. H., Lee, D. H., Lin, C. Y., Hao, S., & Lin, S. C. (2018, February). A practical dynamic positioning and tracking on the vehicular ad-hoc network. In 2018 IEEE 4th World Forum on Internet of Things (WF-IoT) (pp. 338-341). IEEE 10.1109/WF-IoT.2018.8355138

Chaurasia BK and Verma S, (2011), Infrastructure based authentication in VANETs, J Multimedia Ubiquitous Eng 2011; 6(2).

China Electric Vehicle Charging Infrastructure Promotion Alliance. (2017). *Zhongguo diandong qiche chongdian jichu sheshi fazhan niandu baogao 2016-2017 ban* [China Electric Vehicle Charging Infrastructure Development Annual Report 2016-2017 Edition]. Beijing, China: National Energy Administration.

Coraggio, G. (2014), Connected cars – legal issues and hurdles, Available at: https://www.gamingtechlaw.com/2014/10/connected-cars-legal-issues.html

Desai, M., & Phadke, A. (2017, February). Internet of Things based vehicle monitoring system. In *2017 Fourteenth International Conference on Wireless and Optical Communications Networks (WOCN)* (pp. 1-3). IEEE 10.1109/WOCN.2017.8065840

Elsadig, M. A., & Fadlalla, Y. A. (2016). VANETs security issues and challenges: A survey. *Indian Journal of Science and Technology*, *9*(28), 2016.

Epam, (2019), Open Source, Cloud Computing and the Connected Car, Available at: https://www.epam.com/about/newsroom/in-the-news/2019/open-source-cloud-computing-and-the-connected-car

FierceWireless. (2018), 5G Americas delves into cellular V2X/5G versus DSRC, Available at: https://www.fiercewireless.com/wireless/5g-americas-delves-into-cellular-v2x-and-5g-versus-dsrc

Guerrero-Ibez, J. A., Flores-Corts, C., & Zeadally, S. (2013). Vehicular adhoc networks (VANETs): Architecture, protocols and applications. In *Next-Generation Wireless Technologies* (p. 4970). London, U.K.: Springer. doi:10.1007/978-1-4471-5164-7_5

Ignite (2018), 5 Approaches to Connected Car App Development, Available at: https://igniteoutsourcing.com/automotive/connected-car-app-development/

Intellias, (2019), Technologies Empowering Connected Cars, Available at: https://www.intellias.com/7-technologies-empowering-connected-cars/

Islam, M., Chowdhury, M., Li, H., & Hu, h. (2018). Cybersecurity Attacks in Vehicle-to-Infrastructure Applications and Their Prevention. Transportation Research Record, 2672(19), 66-78. Retrieved from https://journals.sagepub.com/doi/abs/10.1177/0361198118799012

Kaura, N., & Kad, S. (2016), A Review on Security Related Aspects in Vehicular Ad hoc Networks, International Conference on Security and Privacy (ICSP 2015), 78 (2016) pp 387-394

Kim, S. H., & Lee, I. Y. (2014). A secure and efficient vehicle-to-vehicle communication scheme using bloom filter in VANETs. *International Journal of Security and Applications*, *8*(2), 9–24. doi:10.14257/ijsia.2014.8.2.02

Kim, T. (2015), Assessment of Vehicle-to-Vehicle Communication based Applications in an Urban Network. Unpublished PhD Thesis. Virginia Polytechnic Institute and State University, USA

Kim, Y., & Kim, I. (2013), Security issues in vehicular networks, The International Conference on Information Networking 2013 (ICOIN), Bangkok, 2013, pp. 468-472

Kristiana, A., Schmitt, C., & Stiller, B. (2015), Investigating a Reliable Inter-vehicle Network in a Three-Dimensional Environment. Proceeding of the 3rd GI/ITG KuVS Fachgespräch Inter-Vehicle Communication (FG-IVC 2015). At Ulm, Germany. 13(3), 16-20

Kumar, D. A., & Bansal, M. (2017), A review on VANET security attacks and their countermeasure, 2017 4th International Conference on Signal Processing, Computing and Control (ISPCC), Solan, 2017, pp. 580-585

Lee, S., Tewolde, G., & Kwon, J. (2014, March). Design and implementation of vehicle tracking system using GPS/GSM/GPRS technology and smartphone application. In 2014 IEEE world forum on internet of things (WF-IoT) (pp. 353-358). IEEE.

Mai, A., & Schlesingerv, D. (2011), A Business Case for Connecting Vehicles. https://www.cisco.com/c/dam/en_us/about/ac79/docs/mfg/Connected-Vehicles_Exec_Summary.pdf

McShane C. (1999), The Origins and Globalization of Traffic Control Signal, Journal of Urban History. 25 (3): 379-404 at p. 382

Mishra, K. S., & Chand, N. (2012). Applications of VANETs: Present & Future. *Communications and Network*, *5*, 12–15. doi:10.4236/cn.2013.51B004

Mokhtar B and Azab M, (2015), Survey on Security Issues in Vehicular Ad Hoc Networks, Alexandria engineering journal, 2015, Elsevier, 54, 1115–1126

Mondal, S., & Mitra, S. (2020). Security Issues in Vehicular Ad Hoc Networks for Evolution Towards Internet of Vehicles. In Z. Mahmood (Ed.), *Connected Vehicles in the Internet of Things, Springer, 2020.* doi:10.1007/978-3-030-36167-9_10

Ozkul, M., Capuni, I., & Domnori, E. (2018). Context-Aware Intelligent Traffic Light Control through Secure Messaging. *Journal of Advanced Transportation*, *2018*, 1–10. doi:10.1155/2018/4251701

Paruchuri, V. (2011). *Inter-vehicular communications: Security and reliability issues. ICTC 2011* (pp. 737–741). Seoul, South Korea: IEEE; Retrieved from https://ieeexplore.ieee.org/abstract/document/6082689

Pendor, R. B., & Tasgaonkar, P. P. (2016, April). An IoT framework for intelligent vehicle monitoring system. In *2016 International Conference on Communication and Signal Processing (ICCSP)* (pp. 1694-1696). IEEE 10.1109/ICCSP.2016.7754454

Peng, S., Lee, G., Klette, R., & Hsu, C. (2017). Internet of Vehicles. Technologies and Services for Smart Cities 4th International Conference, IOV 2017, Kanazawa, Japan, November 22-25, 2017, Proceedings (1st ed. 2017.). 10.1007/978-3-319-72329-7

PostScapes. (2019), Internet of Things (IoT) History, Available at: https://www.postscapes.com/internet-of-things-history/

Qu, F., Wu, Z., Wang, F., & Cho, W. (2015, December). A Security and Privacy Review of VANETs. *IEEE Transactions on Intelligent Transportation Systems, 16*(6), 2985–2996. doi:10.1109/TITS.2015.2439292

Quora, (2019), What are the main differences between artificial intelligence and machine learning? Is machine learning a part of artificial intelligence? Available at: https://www.quora.com/What-are-the-main-differences-between-artificial-intelligence-and-machine-learning-Is-machine-learning-a-part-of-artificial-intelligence

Rawat, D. B., Bista, B., & Yan, G. (2016). *Securing vehicular ad-hoc networks from data falsification attacks, IEEE Region 10 Conference* (pp. 99–102). Singapore: TENCON.

Sachin, S., Muhammad, A., & Mohan, S. (2018), Smart vehicular hybrid network systems and applications of same, U.S. Patent Application 15/705,542, filed March 29, 2018.

Salagar P G and Tangade S S, (2015), A Survey on Security in VANETs, (2015), International Journal For Technological Research In Engineering Volume 2, Issue 7, March-2015

Sharma, S., & Mohan, S. (2020). Cloud Based Secured VANET with Advanced Resource Management and IoV Applications. In Z. Mahmood (Ed.), *Connected Vehicles in the Internet of Things, Springer, 2020*. doi:10.1007/978-3-030-36167-9_11

Singh, P. K. (2020), Singh R, Nandi SK, wt al Seamless V2I Communication in HetNet: State-of-the-Art and Future Research Directions, in in Mahmood Z (Ed) Connected Vehicles in the Internet of Things, Springer, 2020

Singh, S. (2014). "Smart Cities – A$1.5 Trillion Market Opportunity". Forbes, Available at: https://www.forbes.com/sites/sarwantsingh

Smartcity. (2017). Intelligent transportation systems for smart cities. Retrieved from https://www.smartcity.press/intelligent-transportation-system-for-smart-cities

Su, K., Li, J., & Fu, H. (2011), Smart city and the applications, in Proc. IEEE Int. Conf. Electron., Commun. Control (ICECC), Zhejiang, China, Sep. 2011, pp. 1028-1031.

Sumra, A., Hasbullah, H., & Manan, J. A. (2011), VANET security research and development ecosystem, *National Postgraduate Conference*, Kuala Lumpur, 2011, pp. 1-4

Tangade, S. S., & Manvi, S. S. (2013), A survey on attacks, security and trust management solutions in VANETs, *2013 Fourth International Conference on Computing, Communications and Networking Technologies (ICCCNT)*, Tiruchengode, 2013, pp. 1-6 10.1109/ICCCNT.2013.6726668

TeckInAsia. (2015), Challenges Faced by Connected Cars, Available at: https://www.techinasia.com/talk/top-8-challenges-roadblocks-faced-connected-cars

Upeksha, W. K. A. Fernando, et al. (2020), Smart Transportation Tracking Systems Based on the Internet of Things Vision, in in Mahmood Z (Ed) Connected Vehicles in the Internet of Things, Springer, 2020

Wei, Y. (2014). An Anonymous Routing Protocol with Authenticated Key Establishment in Wireless Ad Hoc Networks. *International Journal of Distributed Sensor Networks, 2014*, 212350.

KEY TERMS AND DEFINITIONS

5G Network: This is next-generation of cellular mobile communications standard, scheduled to follow 4G networks, capable of vastly higher data rates, lower latency, higher system capacity, massive device connectivity, and with energy saving features.

Intelligent City: This refers to a new vision for urban development, using IoT based technologies and tools, to integrate and effectively manage a city's resources and processes (e.g., transportation, healthcare, commerce, utilities, city's e-government functions), to improve the quality of life of citizens.

Intelligent Transportation System (ITS): This is where technology-based services, including sensing, analysis and control, relating to management of vehicles on the roads (e.g. buses, cars, lorries, etc.) to enable safe and smarter use of transport networks for more comfortable driving and travel experience.

Internet of Vehicles (IoV): This is a highly useful application of the IoT paradigm, where the moving network of IoT enabled vehicles (cars, buses, etc.) share information collected via the smart devices and sensors within and outside of the vehicles to help in decision making.

Mobile Ad-Hoc Network (MANET): This is a type of ad hoc network that can change locations and configure itself on the fly. It is used for communication between vehicles and roadside smart devices and intelligent systems. This network is also known as Wireless Ad Hoc Network (WANET)

Smart Homes: These are smart living spaces where status of heating, lighting and security, etc. can be remotely monitored and controlled (e.g., switching on/off lights/heating, and monitoring of security status, etc.) using IoT and sensor based smart objects (e.g., mobile phones).

Smart Manufacturing: An important component of Industrial IoT, this refers to the use of computer-integrated manufacturing processes and tools resulting in better adoptability, rapid design changes and intelligent automation of industrial level manufacturing systems.

Smart Roads: These are roads and streets where sensory systems embedded in roadside units (e.g., traffic lights, lane markings, road sign boards, etc.) inform and warn vehicles of road conditions (e.g., incidents ahead), and provide information regarding traffic management and flow.

V2C Communication: This refers to communication between vehicles on the roads and the cloud computing environment for storage and distributed processing of traffic related information for vehicles to take decisions in real time regarding road conditions and situations.

V2I Communication: This refers to IoV communication between vehicles on the roads and the ITS systems (traffic lights, speed cameras, lane markings, parking meters, etc.) in the environment, to provide advanced information on road situations and available facilities, to the vehicles and occupants.

V2P Communication: This refers to IoV communication between vehicles on the roads and pedestrians in the vicinity, to increase pedestrian and road users' safety (using pedestrian detection system) and to avoid accidents involving road users (e.g., cyclists near a blind corner).

V2V Communication: This refers to communication between vehicles on the roads (cars, buses, lorries, etc.), connected via the IoV and IoT based technologies, in relation to their location, speed, direction, and status etc. to develop 360-degree awareness of each other's presence for safer travel.

V2X Communication: This refers to vehicle-to-anything communication, where 'everything' refers to vehicles, roadside units, cloud environment, as well as smart objects (e.g., mobile phones, handheld devices) in the environment. It is an umbrella term for a variety of IOV connectivity platforms such as V2V, V2I, V2P, etc.

Vehicle Telematics: It is a method of monitoring the status of a vehicle, in terms of speed and location etc. On-board diagnostics systems together with GPS systems are used to monitoring status related parameters, including how the car is behaving internally.

Vehicular Ad-Hoc Network (VANET): It is a special form of MANET formed specifically by vehicles equipped with IoT-based smart devices, so that vehicles can share information with each other for their incident free travel. These networks act as infrastructure for intelligent transportation systems (ITS).

Chapter 9
Intelligent Cities:
A Compendious and Multidisciplinary Approach – Issues and Opportunities

Kwesi Atta Sakyi
ZCAS University, Zambia

ABSTRACT

This chapter focuses on a systematic and integrated approach to managing cities using multidisciplinary and technology-based approach and drawing on integrated knowledge from different fields. The chapter aims to discuss and analyse issues in an organic and holistic approach. It explores theory and combines praxis, applications, and futuristic conjectures. Emphasis is placed on issues of ethics, human rights, environmental friendliness, sustainability, and compliance with the 17 UN Sustainable Development Goals (SDGs). While the chapter has considerable input from existing scholarship, it adopts a narrative that is innovative and creative and engages with novel ideas and propositions for future progress. It also charts the breakthrough in transportation, commerce, medicine, education, the world of work, and contemporary health, with particular emphasis on issues relating to safety, confidentiality, privacy, facial recognition, city policing, crime detection, monitoring and evaluation, forecasting, and future of AI.

INTRODUCTION

This chapter discusses the broad issues and opportunities offered by intelligent cities, also known as smart cities. It takes a multidisciplinary and technology-based approach to issues of concern affecting smart city environments and living and working therein. The chapter is organised into sub-headings to give form and structure to the content. The author employs ideas and concepts in Economics, Geography, Social Sciences, Management, Public Administration, Psychology, and Business Communication, among others. Therefore, the discussion in this chapter is drawn from many knowledge fields.

The concept of an intelligent city has become a buzz term in the era of other contemporary terminologies such as globalisation, climate change, smart technologies e.g. ICT, social media, converged platforms, Internet, cloud computing, Internet of Things (IoT), machine learning and artificial intelligence, among

DOI: 10.4018/978-1-7998-5062-5.ch009

many others. The word convergence means coming together and its opposite is divergence. Convergence in this discussion means having multiple media on one platform such as the use of a smart-phone which enables the owner to watch TV, read newspapers, communicate via Twitter, get on Facebook to comment on people's walls or use messenger to engage in interactive messaging or Skype chat, view picture postings on Instagram, and to upload content on Youtube, among a variety of diverse online applications. This helps not only in cost-saving, but it is also extremely convenient for producers and consumers or prosumers (a combination of the terms producers and consumers).

There is, therefore, a lot of integration and convergence in online activities which are based on the concept of interoperability. Motta et al. (2019) define interoperability as the ability of things to interact and act as one for a specific purpose or objective. Modern manufacturers of gadgets and industrial products such as Microsoft, Huawei, Tesla, Space-X, Apple, and Samsung, among others ensure that their offerings are not just stand-alone gadgets but that these gadgets can talk and interact with other such devices embedded in other systems and environments; all done for the convenience of the end-user. However, a few companies and competitors are not too keen on interoperability as they want to lock their customers in to their niche markets for their customer retention and maximum benefits. However, this behaviour of some firms which do not want their gadgets to be interoperable with other gadgets cannot hold much longer in the future; for in the long run, consumers determine what is produced and owned by them.

The concept of intelligent cities has come about because of the realisation that, by the year 2050, 75 per cent of human beings will be living in cities (UN, 2008). It is also because there is a need to optimize efficient use of resources in the face of growing population and dwindling natural resources. While natural resources are decreasing in size, human population keeps increasing rapidly, especially in the developing countries of the world (Bavel, 2013). According to Chakravarty et al. (2009) cited in Bavel (2013), 50% of the world's carbon emission is caused by less than 18% of the world's population. There is therefore an urgent need to avoid wastage of global scarce resources in order to arrive at equilibrium. There is also greater need now to ensure that urban dwellers enjoy excellent living conditions; that urban workers enjoy leisure whilst working, and that smog in mega-cities such as Beijing, New Delhi, Mumbai and many other places is minimised by improving air quality for city dwellers (Reuters, 2018).

Other hazards encountered in cities include: emission of greenhouse gases (GHG), increase in garbage, traffic congestion, poor drainage and sanitation in slums, increase in urban crime and anti-social vices, outbreak of communicable diseases, increased pressure on social amenities such as potable water supply and electricity, and increases in injuries from domestic and industrial accidents.

Highly integrated cities can offer early warning signals and can also help with monitoring of dangerous situations in order to prevent accidents in the city. Integrated city systems enable city managers and policy makers to make efficient and effective policies for the greater good for a greater number of city dwellers. For example, transport by rail, cable car, city coaches, tube, Uber and Ola taxicabs, can be highly integrated and coordinated to offer commuters precise and seamless transport services. Emergency services in health care, fire fighting, and crime busting can be integrated with city systems and individual cell phone networks to provide speedy and tailor-made services for city dwellers.

City traders can use e-commerce to push their wares through pull-through demand systems without having hassles in the supply chain. Waste products from both production and consumption (prosumer waste) can be turned around quickly through integrated supply chain management in order to recycle them into new products. Many city waste management systems have already exceeded their capacity to deal with urban waste as most waste products are non-degradable plastics and metals (smart cities dive, 2017).

An intelligent (or smart) city is one where ICT facilities and latest technologies are deployed on a large scale to solve human problems and make their lives more comfortable (Williams, 2018). These may include, for example, decentralised school systems whereby students' study from their homes, and smart working environments many workers become telecommuters working from home. This provides much flexibility for workers and there is no concept of unemployment as everyone is connected on the grid and is monitored as well as processed and serviced in a centralised computer system.

An intelligent city is a smart and cognitive city which is able to make intelligent decisions and think for itself for the welfare and well-being of city-dwellers by making accurate predictions and forecasts using Big Data collected by interconnected smart objects, and Data Analytics. There is a paradigm shift in the way business is done in the city as these intelligent cities exhibit a lot of disintermediation by using ICT applications in attaining interoperability and inter-connectivity of city activities. Big Data analytics and Crowd-sourcing become the order of the day by using Cloud Computing capacity and capability to process, store, communicate, and share data to large number of people in split seconds, and thus saving costs (Armhein & Quint, 2009).

Computing uses common protocols and automation by supercomputers in remote locations to facilitate access to data for all city dwellers and for city transactions. Money stops changing hands as block-chain technology allows safe online monetary transactions using bit-coins and electronic money allocations based on predictive social and economic worth. Petrol and gas filling stations are replaced by e-charge ports or e-charge stations and there are drone ports or drone pads. In the midst of all this, questions are raised about trust, ethics, privacy and human rights of city-dwellers. These themes are further discussed in subsequent sections of this chapter.

LITERATURE REVIEW

According to Albino et al. (2013) an intelligent city (also referred to as a green, smart, digital or cognitive city) is one where there is extensive use of ICT facilities as well as the use of emergoing advanced technology in a caring, human-centred, and ecologically-friendly community. It is stated that there are currently about 50 per cent of the world's population of nearly seven billion living in cities (Albino et al., 2013). With the increase in urbanisation as a result of rural-urban migration and better opportunities, the cities have become the centre of attention. Many of the UN 17 Sustainable Development Goals (SDGs) aim at some of the many challenges of our time which are caused by the fast-pace of life in cities, such as issues of irresponsible consumption and production, and unsustainable lifestyles. Refer to Goals 11 and 12 in UN SDGs (2015) and United Nations (2014). According to Albino et al. (2013), the cities have become the focal point of global policy makers because majority of the pollution in the world today now lives in cities such as Cairo, Johannesburg, Jakarta, Manila, Mumbai, Mexico City, New York, New Delhi, Lagos, and Rio de Janeiro, among others. In view of the menace posed by urban sprawl, urban pollution, urban litter, and urban implosion, the Paris Agreement was signed on 4 November 2016 by 55 countries which contribute 55% of the greenhouse gas emissions (UN Climate Change, 2016).

The skyline of a modern city shows classic buildings, skyscrapers, and a well-lit city at night. Such cities are centres of culture, commerce, industry, finance, education, and entertainment. Dubai and Shanghai are typical examples of such modern cities which act as magnets to attract tourists from all over the world.

Green movements and environmental activists have brought pressure to bear on global citizens for them to review their lifestyles e.g. by eating say less beef and other livestock in order to save dwindling resources and reduce carbon emissions (Guardian, 2018). Many global citizens in such cities are adapting to global climate change imperatives by changing their lifestyles and becoming vegetarians or vegans as their own personal contribution and sacrifice towards achieving the greater good for the greater number. Massive tree planting campaigns have also been undertaken in countries such as Philippines, China, India, Madagascar, and Ethiopia, among many others in order to mitigate the negative effects of global warming (Tree Nation, 2018; Nguyen, 2018).

The Nobel Peace Prize of 2004 was awarded to the late Kenyan professor, Wangari Maathai for her efforts in tree planting (Green Belt Movement, 2019). Similar awards have been given to others from different parts of the world. There is a trend that as life becomes unbearable in towns and villages due to abject poverty, most people from these poor areas flee and troop to the city to seek the bright life and good life of the city (Green Belt Movement, 2019).

It is believed that city environment is generally such that prosperous life is only found in the cities. After all, the city performs many higher order functions. Thus, cities have been known for years to receive Guest workers who are then inducted into the city life in a process of cultural diffusion and acculturation that integrates the new arrivals. After some time, the new arrivals invite their relatives from back home to join them in a phenomenon called chain migration (Adepoju, 2004). In future, many of the city functions will be performed by robots in tasks such as conducting city tours. Intelligent robots can serve as tour guides and narrators in the form of robotours and roboraps. Old people and the infirm could be assisted by robotic helpers as they arrive in homes as drones when requested to perform household chores and they (i.e. city management) get paid automatically through city super-ledger accounting system which is automated.

In future, intelligent machine helpers (Protothrops or robothrops) could be summoned in homes with single adults to help give companionship to depressed people. Intelligent cities can connect with other cities across the globe to learn best practices and to exchange commodities without human intervention. The concept of functional cities changes the way cities interact as ICT facilitates, and increases the volume of trade among them. However, village settlements still appeal to those who prefer rustic rural life. Towns and villages can be enriched by their connectivity to the super-network of different settlement types.

Intelligent cities function as change agents and centres of excellence, as they help transform other cities based on the ideas of primate cities, growth poles, and Myrdal's cumulative concentration thesis (Saha, 2019). Using Von Thunen's model of 4-stage concentric circles or geo-economic rings World Atlas (2019), it can be seen that activities start at the centre at the core or central business district (CBD) and spreads outwards to incorporate residential areas, light industrial zones, heavy industries, and lastly farmland and green belts (World Atlas, 2019). Von Thunen, a German economist of the 19th century, came up with his formula for land use as:

$$L = Y\,(P\text{-}C) - Y\,(DF)$$

where L is locational rent, Y is yield of the land, P is market price of produce, C is cost of production, D is distance to the market, and F is transport cost.

Von Thunen's model (World Atlas, 2019) reflects residual profit after all costs have been factored in, showing price differentials based on distance and locational factors. The model assumes price differentials due to varying costs in different locations which could be rural or urban. In an intelligent city, such differential costs can be whittled away by information symmetry or synchronisation of information availability such that the Y (DF) of Thunen's equation becomes irrelevant, thereby having a homogenous market without price differentials. Von Thunen envisaged an ideal state based on farming practices reminiscent of the French Physiocrats (World Atlas, 2019).

The sector or radial model of settlement proposed by Hoyt (1939) shows a hub from which transport routes fan out in different directions. The radiating sector is superimposed on a Burgess concentric pattern to form a hybrid pattern of settlement. This model is propelled by economic, racial, industrial, social class and cultural factors whereby one could discern the coalescing of certain ethnic and racial groups, fanning out in a particular sector while the old patterns of concentric rings form the backcloth. New York has this pattern in zones such as Bronx, Queens, Manhattan, Staten Island, and Jersey, among others.

Another locational or settlement theory of note is that of a German, Walter Christaller in Planning Tank (2016) who came up with the Central Place Theory that was related to how size and spacing of settlements affect land use patterns as well as spatial relations among settlements (Planning Tank, 2016). The theory examined markets and settlement patterns where the market served as a magnetic pull directing other dependent activities. The Central Place Theory is similar to the Primate City model as well as the Growth pole model (Planning Tank, 2016). Emerging smart cities such as London and Bristol in UK; Chicago, San Francisco, Philadelphia, and Charlotte in the USA; Barcelona in Spain; New Delhi in India and Cologne in Germany (EU Commission, 2017; National League of Cities, 2017) have been developed from these models, especially the models of Von Thunen and Christaller (Planning Tank, 2016). Cologne in Germany is committed to 20 per cent reduction in GHG, 20 per cent increase in energy efficiency and an increase in the use of renewable energy by 20 per cent (EU Commission, 2017)

Max Weber, the German sociologist and philosopher, postulated that social hierarchies influenced settlement patterns based on wealth and status (Ahmed, 2004). In intelligent cities, social stratifications become difficult as planning zones are made rigid by planning authorities in order to ensure orderly development. This brings about inclusivity and closure in the income gap as urban entrepreneurs coach others in peer-to-peer learning.

Origin of Cities

According to the Merriam-Webster Dictionary, the word city is derived from the Latin words *cittade, civitas, civis* meaning fortified settlement, citizen of a state, and a chartered place which has powers to govern itself and it has a Cathedral. The other Latin word for city is *urbs* from which words such as urban and urbane are derived. The word urbane means refined and polished manners. Another word related to city from Latin is *polis* which means city. The Latin word *politeia* means the state whilst *polites* means citizen. Furthermore, there is the word municipal which has two Latin roots *munia* meaning an administrator or official and *capere* meaning to take; thus, municipal means taking control of local administration in a city (Amer, 2019).

Some of the oldest cities in the world are found in the Levant or Fertile Crescent namely, modern day Iran, Iraq, Israel, and Syria. Other places include Turkey, Greece, Italy, Egypt and India. Some ancient historical cities are Knossus (Crete), Damascus, Baghdad, Tehran, Istanbul, Byzantium, Rome, Carthage, Alexandria, Timbucktu, Jerusalem, Mecca, Athens, Harrapa-Mohendro Jaro, Syracuse, Beijing, Osaka, and Axum (Ethiopia) (ancientportsantiques.com). Rome was said to have been founded on River Tiber in 753 B.C during the time of the Etruscans by the twin brothers Romulus and Remus (Encyclopaedia Britannica).

Modern Cities

The rank-size of settlements starts from the smallest and lowest point at the bottom of the pyramid to the largest type of settlements. The smallest human settlement type is an isolated dwelling such as one held by a sage or a farmer, followed by a hamlet, then a village, small town, large town, city, and lastly a conurbation or mega-city such as Lagos in Nigeria or Mumbai in India or London in the UK.

Cities can be classified by their spatial functionality. London, New York and Johannesburg serve as centres of Finance, Transport hubs and Commercial capital cities, while Paris serves as the centre of fashion, culture and good taste. Rome and Mecca serve as religious and historical centres while Seattle and Chicago serve as centres of manufacturing and innovation (Mullan, 2019). The location or site of a city also adds to its functionality as some cities such as Le Havre, Manchester, Rotterdam, Hamburg, Singapore, Durban, Lagos, and Liverpool are cities famous for their merchandise, ports, and entreport trade (World Atlas, 2019). Some cities have become holiday resorts, tourist attractions, and centres of learning such as Oxford, Cambridge, Cape Town, Moscow, Boston, and Paris, among others (Hollands, 2008; Mumford, 1973).

Las Vegas in the USA is famous for its gambling and casinos (Pierceall, 2015). San Francisco and Los Angeles are famous for both Hollywood movies (Kruschewsky, 2018) as well as being part of the Silicon Valley for ICT innovations. Cities serve as consumption as well as production nodes in the concatenation of settlements as they discharge higher order functions to smaller dependent settlements (Hollands, 2008; Mumford, 1973). The Greek City States of the time of Pericles saw a boom in trade. The Peloponnesian Wars (431-405 B.C.) broke out among those cities such as Sparta, Thebes, Corinth, and Athens because of the struggle to gain control of trade in the Aegean Sea littoral (Encyclopaedia Britannica). In Europe, some cities bandied themselves together to form the Hanseatic League to promote regional trade during the period between the13th and the 15th centuries (cf. Encyclopaedia Britannica).

It has been suggested that intelligent city dwellers can use predictive analytics to make informed decisions concerning selection of courses to read at the university. Such an application can be extended to many functions such as shopping, selection of music to listen to, movies to watch, and also which novels to read, among other applications. Albino et al. (2013) observed that megacities of 20 million inhabitants or more are growing rapidly in Asia, Africa and Latin America and therefore there is urgent need for policy-makers in these parts of the world to become proactive by taking measures to grow intelligent cities. This is in order to reduce the negative effects of externalities created from consumption and production activities. This is where waste management, bio-chemical engineering and other sciences take front seat. Issues of air quality become prominent and many legislators make laws to regulate economic activities so as to preserve air quality (Airnow.gov, 2019). Already in Germany, diesel cars and fossil fuel powered vehicles are being phased out by 2025 (Reuters, 2018); similarly, in many other developed countries as well.

The UN ECOSOC Report (2016) lists issues to be addressed in connection with smart cities, with respect to skills gap, suitable governance model, inclusivity, and localisation of smart infrastructure. Many young people in cities in poor parts of the world lack employable skills. Hence, they need to be empowered with entrepreneurial skills, ICT coding and programming skills, financial management skills, and people management skills so that they can integrate into life and with people in intelligent cities. Many of these youth feel left out in governmental arrangements as they lack voice in decision making. There is also an attempt to train young leaders as youth ambassadors and peer-educators in order to bring about closure and inclusivity.

Smart city entrepreneurs engage the youth in mentorship programmes and they serve as business angels and venture capitalists in order to promote entrepreneurship among the youth. Many budding youth entrepreneurs lack financial muscle to bring their nascent ideas to fruition. Science, Innovation and Technology (SIT) can be propelled to new heights by introducing and promoting STEM (Science, Technology, Engineering, and Mathematics) education in schools (Live Science, 2019)

Intelligent cities are mindful of their carrying capacity of the land as they pay attention to issues such as underground water aquifers and regulation of water use for domestic and commercial purposes.

Intelligent Cities: Characteristics, Functionality and Historical Origins

The Greeks said that city air makes one free. The city is said to be a place of enlightenment, civilised behaviour, urbane manners, and a place to find the good life. Great geographers such as Smailes, Von Thunen, and Weber studied cities and their morphology, and came up with classification models of cities such as the functional city, primate city, and the millionaire city. A city is a large human settlement which is at the apex of all types of human settlements which ranges from the smallest in the form of a village/hamlet to a town, borough, district and conurbation (Amer, 2019 ; Mumford, 1973; Williams, 2018; Caragliu et al., 2011). Mumford (1973) cited in Amer (2019) states that cities are arenas of social drama and boiling pots of social engineering or *gestalt.* Cities are at the top of the settlement hierarchy and they perform functions for the lower order settlements in the form of providing higher order needs such as commerce, entertainment, medical care, education, superior administration and management, among others.

In 2014, the International Telecommunication Union (ITU) analysed 100 definitions of intelligent cities and concluded that an intelligent city uses ICTs and other technological solutions to enhance living in cities by improving the quality of life. Different definitions suit different stakeholders. An intelligent city also means conscious city dwellers that care about current and future needs and what is happening in their neighbourhoods or locale (Dijkstra & Poelman, 2012). It means outgoing socialites, who are eager to network, socialize and willing to engage, share knowledge, and use innovation creatively and expansively in their lives. Academics see intelligent cities as places where people engage themselves intellectually and cognitively in collective learning experiences, academic intercourse and semantics, which are reminiscent of Cicero's Rome or Solon's Athens, Pericles' and Socrates' Athens (Dijkstra & Poelman 2012).

Smart cities cannot self-destruct because they have Big Data Analytics which self-regulate city congestion, pollution, overpopulation, and use of spaces. Spain, UAE, Netherlands, Germany, UK, Japan, and South Africa, among others have made great strides in creating intelligent cities (UN 2016). The future of intelligent cities promises to be bright as city dwellers take to share riding, biking, housing accommodation, and many bonding activities in order to reduce slack time, idle capacity, down-time, and resource under-utilisation. Smart cities have far less commuters as travel within the city becomes auto-managed by super computers and AI. Travelling and decision making are not by impulse but by super-AI advisors which regulate city way of life.

Machine to human (M2H) interaction increases as intelligent cities emerge. According to McKinsey & Company (2018), travelling has contributed greatly to pollution in cities because of the need for people to collect parcels, access services such as medical care, beauty care, legal services, and access education, the need to visit relations, and attend social and public events. In an intelligent city, this trend is reduced and reversed in the IoT (Internet of Things) technology, as services and goods are taken to the abode of those who demand them. Mobility is greatly curtailed and leisure time increases. GHG is reduced by curtailing the amount of commuting or transportation which is done by intelligent city dwellers. Deliveries are done by drones, smart trains, super-transponders and cable-cars. Min et al. (2019) did research on sustainability and terminology connected with intelligent cities and they came up with a compendium on intelligent cities (cf. Harrison & Donnelly, 2011).

According to Min et al. (2019), the term smart city has been in use since there was realisation to have urban regeneration and revival from urban decay. In a study in 2017 by the European Commission, of more than 80 cities across Europe in 19 countries, best practices in intelligent cities were chronicled. The study examined issues such as renewable energy practices, waste recycling, smart street policing; street lighting, waste heat recovery, urban freight logistics, and electrical energy storage as the areas of focus. Bristol, London and Barcelona were among the cities in the empirical study. Lessons learnt from the study can be replicated in cities in third world countries in Africa, Asia, and Latin America.

Cities and human settlements, in general. are classified by their population size, functionality, and nodal connectivity by using Kansky's measurement (Raghav, 2019), and by their spread in terms of size of land area.

Some of the ancient and modern cities which have sprung up as port cities include Nineveh, Cape Town, Singapore, Le Havre, Tyre and Sidon (ancient ports antiques, 2019) Some famous historical cities include Rome, Boston, Tehran, London, Glasgow, Edinburgh, Cairo, Istanbul, and Baghdad (Encyclopaedia Britannica) Ancient cities of worship and pilgrimage include Mecca, Jerusalem, and Rome (Ancient Ports Antiques, 2019). Ancient centres of learning include Cambridge, Cordoba, Oxford, Paris, Bologna, Timbucktu, Alexandria, and Athens, among many others. Places of entertainment and tourism include Tel Aviv, Las Vegas, Addis Ababa, Barcelona, San Francisco, Istanbul, Belgrade, Bucharest, Prague, and Dubai; financial capitals include Hong Kong, London, New York, Johannesburg, and Frankfurt; transport hubs include Accra, Dubai, London, Addis Ababa, Amsterdam, Nairobi, and Mumbai; places of commerce and trade include Florence, Genoa, Venice, New York, Lagos, Tokyo, Beijing, and Shanghai; places of power and administration include Washington, Ottawa, Pretoria, New Delhi, Canberra, Wellington, and Manila; manufacturing hubs include Seoul, Chicago, Seattle, Pittsburgh, Osaka, Shanghai, and Hamburg, among others.

The Burgess or concentric settlement pattern which shows a series of different economic land use patterns as one moves away from the core towards the outer rings. Intelligent cities are planned to make room for social, economic and civil engineering activities. Intelligent cities require smart infrastructure such as eco-friendly bridges, eco-friendly roads and buildings, among others.

Figure 1 shows the hexagonal model of Christaller (Planning Tank, 2016) who used his theory to explain land use patterns in neighbourhoods where settlements coalesce to form interdependent clusters. The density gradient diminishes as one goes further away from the core (i.e. the central place) towards the periphery.

Convergence, Interoperability, and Efficiency of City Utilities

When electronic gadgets are interconnected on one platform, they are said to be interoperable and convergent. This means all the manufacturers agree to configure their products to be able to talk and communicate with one another (Technopedia, 2019). This helps consumers to cut down on cost of being the owners of many separate gadgets. The media of newspapers, radio, TV, and social media are accessed on one platform such as a smartphone. This brings efficiency in the use of energy and also it reduces

Figure 1. Hexagonal Model of Walter Christaller and Central Place Theory (CPT)

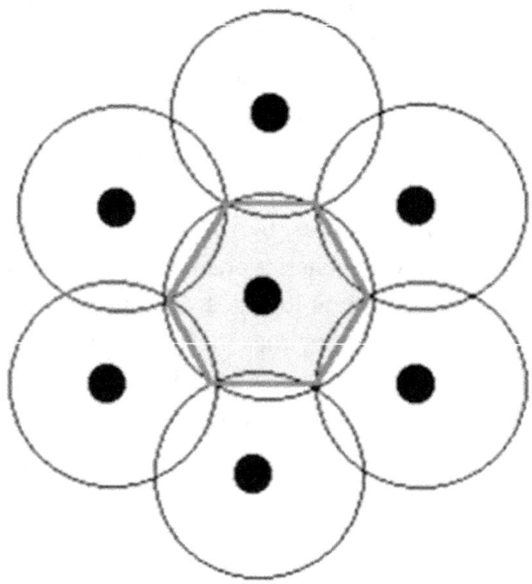

the amount of e-waste. Since all gadgets and products talk to one another, consumers and producers do not worry about stock or inventory planning as Artificial Intelligence (AI) and automated systems take care of these things.

Table 1 outlines some of the classical models of human settlements which are found in many information sources in Geography. All these models have relevance to modern intelligent cities and therefore they are worth studying in detail because intelligent cities have antecedents. They did not spring up from nowhere.

THEMATIC ISSUES AFFECTING INTELLIGENT CITIES

Housing

With the increase in global population, there is housing deficit around the world. Many cities now have reached their carrying capacity levels or threshold populations and they cannot accommodate more inhabitants because of strain on utilities such as water, electricity, and infrastructure such as schools, hospitals, roads and railways, among others. Cities such as Lagos in Nigeria, Kinshasa in Congo DR, and Cairo in Egypt have been projected to reach extremely high levels of population by 2030. There is therefore need to find ways to provide adequate decent housing for city dwellers. This requires looking for innovative and creative ways to design new buildings which are energy efficient, liveable, ecologically-friendly, and built with new materials which can reduce the amount of Greenhouse Gases emission (GHG) (Paskaleya, 2011).

Table 1. Models of Settlements

Serial Number	Author	Name of Model	Characteristics	Comments
1.	Johann Von Thunen (19th century) (German) World Atlas (2019)	4-Stage Model	4- Stage Concentric Circles Land Usage Patterns	Relevant to intelligent cities in terms of zoning land use
2.	Walter Christaller (German) in Planning Tank (2016)	Central Place Theory	Hexagonal Land Use Patterns	Relevant to intelligent cities in peri-urban relationships and planning dormitory towns
3.	Max Weber (1880) (German)	Social Stratification/Hierarchy	Isodapanes and land rents based on social ranking	Relevant to intelligent cities as income levels differ
4.	A. E. Smailes (English)	Urban Classification	Historical origins of English and Welsh cities and their functionality	Relevant to intelligent cities' configuration and archaeology
5.	Generic	Generic	Linear, Dispersed, Nucleated Radial Concentric/Burgess	Relevant to intelligent cities in terms of morphology and structural patterns
6.	Francois Perroux (1950s) Jacques Boudeville (born 1919)	Growth Pole	Core-Periphery Model where growth is uneven and from a core area growth spreads out to galvanize outlying areas	Relevant to intelligent cities urban planning
7.	Mark Jefferson (1939) Primate City	Primate City	Dominance of Primate City Population and Functionality	Relevant to planners and demographers in intelligent cities to achieve balance in population sizes
8.	Gunnar Myrdal(1956) (Swedish)	Cumulative Concentration Thesis	Core-Periphery Agglomeration Thesis of Economic Growth based on spiral and circular growth	Relevant to economists in intelligent cities in allocation of resources for urban regeneration
9.	Homer Hoyt (1939)	Radial or Sector Model	Growth at transport node	Combination of linear, nucleated and concentric settlements
10.	Ernest Burgess (1925) Zone Model	Concentric Rings in zones	Green belts at suburbs and brown belts in inner areas	

The need for intelligent cities requires moving away from the usual construction materials (cement, bricks mortar, iron rods, etc) towards a search for new eco-friendly materials that can contribute to the concept of intelligent cities. Many high-rise skyscrapers in cities are derogatorily referred to as 'concrete jungles' as they are not safe to have in terms of their ability to heat up and contribute to global warming and climate change. Currently, designers and architects do not incorporate green features into their designs. However, the high-rise buildings, in Singapore for example, reminds one of the legendary hanging gardens of ancient Babylon as designers there have incorporated horticulture into their designs.

Many of these high-rise buildings in Singapore are used by tenants to grow vegetables, crops and flowers in what is termed as truck farming and market gardening or horticulture. Currently, architects are testing new materials such as straw, mud, bamboo and graphite for the construction of eco-friendly city dwellings (Paslakeya, 2011; Gaur et al., 2015). The hope is to have buildings which help with community integration, inclusiveness, closure, and aesthetically appealing structures. The use of ICT 3D design and Computer Aided Design (CAD) help a lot on this score. Houses have to be designed to take care of weather patterns, earthquakes, and the direction of sunrise and sunset as well as the magnetic pole and true north-pole for those with superstitious beliefs, foibles, and idiosyncrasies (Gaur et al., 2015; Easy Render, 2019; and Paskaleya, 2011).

Cities, in general, are formed on the basis of their location, functionality, land use patterns, and accessibility (World Atlas, 2019). An edge city is one whose growth is constrained by a water body, thereby forcing it to have high rise skyscrapers in order to make maximum use of land and space. There is therefore vertical rather than horizontal expansion. Such high rise buildings pose risks in times of emergencies such as fire-outbreak, risk of falling from heights, and risk from natural disasters such as earthquakes. This requires having robust rescue systems in place including vibrant fire-fighting equipment, including robotic rescuers. Drones, sensors and early-warning systems can be put in place to avert many dangers, including fire outbreaks. With global warming and rising temperatures, cities near water-bodies will be prone to flooding, tsunamis and other hazards. Therefore, there is need to have heavy waterfront embankments such as the construction of dykes and breakwaters. The Dutch models of dykes and polders or reclaimed land comes to mind.

The following section focuses on city transport. Transport modes and routes are designed in line with the city housing configuration, topography, and engineering and architectural considerations.

Transport

In intelligent cities, traffic congestion is avoided by having functionally integrated transportation systems. Through machine learning and AI, an integrated transport system is designed whereby many students and workers commute easily to their destinations on time. However, with increase in the use of AI, robotics and ICT facilities, home based schooling and home working are gaining momentum and ascendancy in the city, thereby minimizing vehicular movement. People move minimally in order to receive medical care, to attend social events, and to participate in leisure activities. People are always alerted by e-communication systems to know when it is convenient to commute. Electric cars and motor bikes are self-driven such that all transport systems are highly synchronised, leading to fewer accidents. Intelligent city dwellers forget the past dystopia of living in unplanned and congested cities (UN, 2008; Psaltoglou, 2018).

A radial city is one whose functionality is based on a hub or node of transportation arteries which diverge or radiate in all directions from the centre to both near and distant places. Such an arrangement helps to connect the intelligent city to other satellite areas which act as suppliers as well as consumers in the supply chain. A hub city receives millions of tourist arrivals a day from all corners of the globe so there is a great need for vigilance in terms of disease control against infections from visitors who may act as disease vectors, transmitting dangerous diseases such as MERS, SARS, Ebola, Corona-Virus or Covoid ID 19, and TB (Tuberculosis). The next section examines health issues in intelligent cities.

Health

Sensors and smart bracelets which are worn by city dwellers can help give them health alerts and even diagnosis, prophylaxis. They may also receive information which recommends exercise regimen to be undertaken by them in order to stay fit. This two-way exchange of information alerts health centres and care-givers to despatch drones to deliver medical supplies to those in need. This scenario which has been painted shows that there is, therefore, need for fewer doctors and pharmacists. AI is used to conduct laboratory tests and instantly design new drugs from plants. Robots become pharmacopeia for storage of medical information. Newly invented AI based smart devices also help intelligent city dwellers to enjoy sports. In an intelligent city, there are fewer hospital visits and no need to visit health consultants as everything can be done remotely online. Health issues are closely related to sports so the next section looks at sports in a smart city.

Sports and Leisure

As a lot of sedentary work is done at home, intelligent city neighbourhoods become a hive of sporting activities and sports equipment and sports gear rise in demand. Sedentary lifestyle creates health problems of obesity, high blood pressure (HBP), and cardio-vascular complications. As many people work a shorter work week with plenty of time to enjoy leisure, they relapse into the indulgence of binge eating, debauched drinking lifestyle, and chain-smoking cigarettes, pipes, and shisha. However, the health-conscious people take to travelling abroad for sports, leisure, business and tourism. Human trainers are dispensed with and are replaced by apps on smart phones. Through infatuation with such health apps, many undiscovered athletes break world records. An intelligent city therefore has inhabitants who eat smart meals and exercise in a smart way in order to prolong life. The next section connects to this theme of smart lifestyle and sustainability.

Sustainability

At the core of the concept of intelligent cities is the concept of sustainability which refers to the ability to continue and sustain current activities over a longer period of time without jeopardizing the needs of future generations in terms of their accessibility to global commons or non-renewable resources (Easy Render, 2019; Gaur et al. 2015; UN, 2014). Smart cities are designed to achieve efficiency, effectiveness, openness and better economic environment (Elkington (2004) 3P model). Inhabitants of intelligent cities are subsumed to be creative, innovative and active participants in minimizing all lifestyle activities which threaten and leave heavy carbon footprint on the environment (Mori & Christodoulou, 2012; Morvaj et al., 2012; Monfared zadeh et al., 2015; Manville et al. 2014)

Intelligent city inhabitants therefore embrace the 17 goals of the UN Sustainable Development Goals which were launched in 2015 with target year of 2030 (Easy Render, 2019.) The SDG replaced the Millennium Development Goals (MDGs) which were revised in 2015. The MDGs were launched in the year 2000. Some of the 17 SDG Goals include paying attention to life below and above water as well as embracing healthy lifestyles, and converting from fossil fuels to cleaner and renewable energy sources such as solar, tidal, wind, and geothermal energies (UN SDGs, 2015). Intelligent cities engage in intelligent growth aided by Big Data planning and the use of ICT technologies as well as AI, Cloud Computing, and machine learning (Linthicum, 2010; Mahmood, 2013).

Figure 2 shows how Elkington came up with the Triple Bottom line model for people to pay attention to 3Ps, namely planet issues, profit for investors, and welfare of people (Elkington, 2004). Embracing these concepts helps intelligent cities to buy into the ideas of green cities. These ideas help intelligent cities to optimize the objective of satisfying different stakeholders. The next section on waste management shows how the concept of the 3Ps is put into practical use in order to reinforce the 3Rs of reduce, re-use, and recycle.

Figure 2. Elkington's 3P (Triple Bottom Line) Model

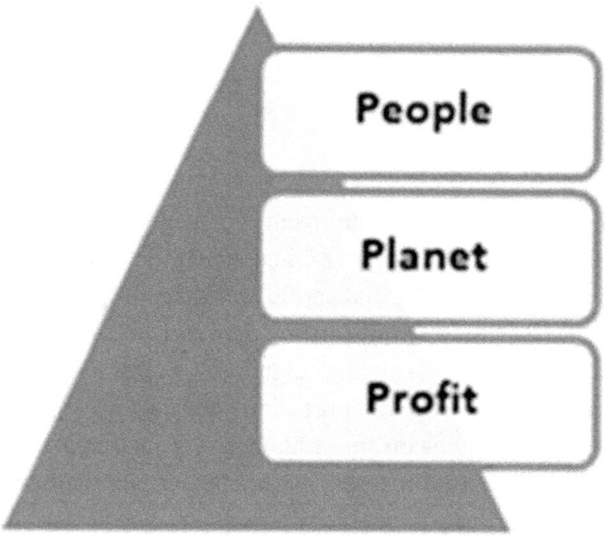

Waste Management

Cities generate large volumes of waste which contain large amounts of plastic and non-biodegradable materials as well which choke drainages and clutter waterways and the oceans, posing a threat to aquatic and marine life (EU, 2018; UN SDGs, 2015). Inefficient methods of waste disposal in cities are hazardous to the health of many city dwellers (WHO, 2019). However, waste has been identified as a source of bio-gas and heat which can be used to generate electricity (IBM, 2011; Smart Cities Dive, 2017). Waste bins can be fitted with sensors and they can be directed through trolley pathways such that they can self-travel to collection points, self-empty, and get back to their bases. There is no need for scavengers or bin collectors; as the bins can be made to compute their contents and classify the composition of waste so that estimates of waste contents can accurately be made in order to advise consumers about their consumption and litter disposal habits. There is no need to incinerate waste to produce smoke and toxic fumes to add to urban pollution. Empirical evidence suggests that the filthy areas of cities are also the areas where most criminals lurk and operate from. They are also some of the worst poverty-stricken areas (Witte, 1996). The next section examines urban crime and its detection.

Figure 3 shows the process of reducing carbon footprint by reducing consumption levels, recycling waste, and reusing scrap or residual of already manufactured items. This process can be maximised by using nanotechnology and microbes for mining to reduce dump-sites for mine waste. Microbes are efficient at recovering metals from ores (American Geo Sciences, 2019).

Figure 3. Resource Management Model

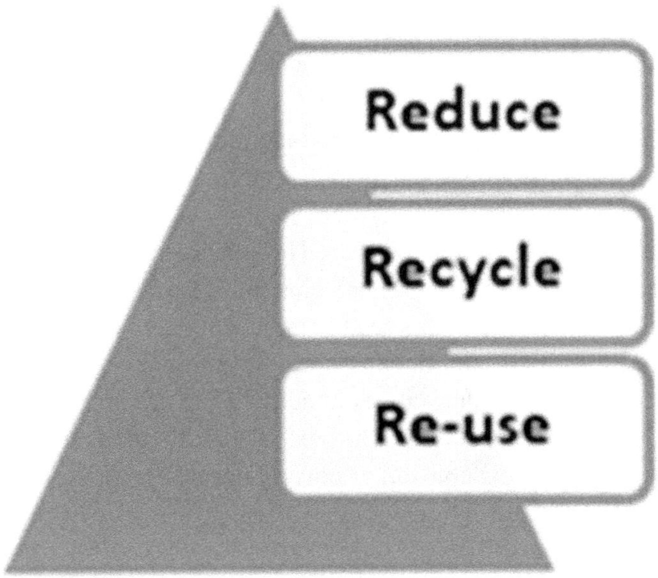

City Crime

City crimes raise questions of safety and security for city dwellers. Surveillance cameras and smart policing can deal with most crimes. It also requires crowd sourcing and cooperation of city-dwellers to provide information to security agents. Most city crimes in sin cities and hot spots in cities are motivated by factors such as gambling, prostitution, racism, drug addiction and drug peddling, gangsterism, terrorism, religious intolerance, civil wars, political machinations and manipulation, economic motives to dominate trade (taxi wars) and the menace of street kids from broken homes, ethnic wrangling, and the menace of organised armed robbers (Witte, 1996). In Accra in Ghana, for example, online fraudsters have set base to engage in cyber-crime by swindling unsuspecting people around the globe. They deceive their prey through clever and devious means (Darko, 2015). Most of the culprits are male school-leavers and school dropouts who cannot find decent jobs because of the high youth unemployment rate caused by disintermediation, use of capital-intensive methods of production, and dominance of the global economy by the rich and powerful nations. Urban poverty therefore is a fertile ground for brewing crime and criminals (Witte, 1996). Urban crime is therefore remotely and indirectly a result of global inequality in income distribution, and directly as a result of failed markets and governmental systems.

These young male online surfers and cyber-criminals and fraudsters are locally known in Ghana as *Sakawa Boys*. The BBC did a documentary on them (Darko, 2015). Similar setups are found in Lagos in Nigeria, Los Angeles in the USA, and Moscow in Russia. Some of these are computer hackers and geniuses who have misdirected their talents for nefarious and illegal online activities. Some of the cities in the world which are noted for high levels of crime in all forms are New York, Mexico City, Kingston, Rio de Janeiro, Bogota, Chicago, Johannesburg, Kabul, Mogadishu, Bangui, Medellin, Lagos, Nairobi, and Baghdad. It must be noted that there is no city without crime as crime is confined to certain suburbs of cities, especially the slums, ghettoes, shanty compounds, and old and decaying parts of cities which require urban regeneration and rebuilding. The UN Habitat for Human Settlements organisation is doing a fantastic job of working with governments around the world to provide decent accommodation for city dwellers in deprived areas

A case in point is the Ningo-Prampram Urban Housing Project in Accra in Ghana which is aimed at providing affordable accommodation in gated and serviced communities which are safe from crime. Another example is found in Kigali in Rwanda where state-of -the- art buildings have been constructed outside the city to provide excellent accommodation. The area has maximum security which is provided through the installation of modern surveillance cameras (Paskaleya, 2011).

Some of the mega cities mentioned in this section are nodal cities which radiate outwards following routes or passageways such as roads and railway lines. This follows the radial, sector, and nucleated settlement pattern. Land-use patterns show zones of activities in concentric circles as one moves away from the core towards the periphery or suburb. The next section discusses education which is a potential vehicle for emancipating many poor people out of poverty because its possession creates upward social mobility.

Education

In an intelligent city, conventional school and college buildings become redundant as school and college buildings have to be converted into other uses such as residential dormitories for workers and city dwellers. Most students learn from their homes by having e-teachers and e-tutors. Teachers are only employed to write course content and also to make videos for virtual classrooms. Informal schools are conducted in churches and town-halls where people gather for social functions. National school examinations become decentralised as students write exams from their homes at any time they are ready to do so (ACCA CBE, 2019)

This scenario eliminates examination malpractices such as leakage of question papers. Each student writes a different set of papers which is based on his or her continuous assessment results. This is made possible by continuous monitoring and assessment by using Big Data Analytics. E-books become tailor-made through Machine Learning for each online student. Students and pupils in online virtual classrooms are called by different names such as virtualites, e-learners, techno-studs, onliners, and techsophants or tech-sophists, among others.

In a smart city, students have no particular national school curriculum to follow as they are free to learn from any curricula in the world, depending upon their career choices, market demand and their abilities and inclinations. Students obtain hybrid education from elite Russell Group and Ivy League schools at much reduced rates by pursuing blended learning methods. Second chance education becomes galore online for school drop-outs. Andragogy or Adult-learners become spoilt for choice as they have many opportunities and invitations from elite online schools. Precocious children or gifted children are fast-tracked in virtual classrooms. All students are called upon to take a lot of studies on environmental studies in line with efforts to combat climate change and global warming. These issues are further discussed in the next section.

Environment Monitoring

To monitor environmental damage, drones and satellites are deployed to garner environmental data which is fed into City Databases (CDBs) for Data Analytics (Gaur et al, 2015). This data helps to inform city dwellers about the extent of weather changes and forecasts which are likely to affect their daily activities (UN Climate Change, 2016). Such information may trigger environmental activists to swing into action and also trigger creativity in fictional and non-fictional writers to redirect their activities towards environmental issues. Freak weather or exceptionally unexpected weather patterns have now become the norm due to global warming and climate change. The huge ice caps and permafrost on the Alps, Himalayas, Andes, Rockies, Antarctica and the Arctic are all breaking up and increasing sea levels. This poses a threat to low lying areas in coastal cities around the globe.

Changes in plate tectonics or isostasy deep inside the core of the earth are causing magmatic flows along fissures and cracks, causing violent eruptions of some dormant and active volcanoes. Underground seismic activities are causing tsunamis and flooding along the littoral in coastal areas. Exceptional dry weather conditions are causing uncontrollable bush fires in Australia, Europe, California and parts of Africa. All these events need close monitoring and evaluation in order to safeguard lives of those who live in intelligent cities such as Barcelona, Sydney, Manchester, Los Angeles, Jakarta, Lagos, and Honolulu. Exceptional heavy rains and storms have caused flooding in Italy, UK, France, Germany, and Mississippi State in the USA. The next section examines how the intelligent cities help city dwellers to have more leisure time and how this impacts on other environmental factors.

Entertainment

In ancient Greece and ancient Rome, the rich people, the patricians, the aristocrats, and the noble were given to indulgence in lavish entertainment. They also promoted the arts, drama, and humanities at the amphitheatres, and the Coliseum (refer to Encyclopaedia Britannica). Athletics was a big sport in Greece while wrestling was the favourite pastime of the Romans (refer to Encyclopaedia Britannica). In the era of intelligent cities, city dwellers enjoy more leisure time for sports and entertainment as work is mostly driven by AI and Machine Learning. People have a lot of time on their hands to read novels and drama books while these are being written by their authors through live streaming. Many unknown authors become instant celebrities as they are unearthed through the network. With more leisure comes more pleasure and indulgence as people do more sedentary activities in homes. This means that as there are fewer workers and more home-makers it leads to greater use of water as that leads to binge eating and increase in consumption which is good for business but bad for health. The use of algorithms by companies such as Netflix and Warner Brothers helps them hook their global clients to their streaming networks, thus maximizing profits. The next section examines water usage in intelligent cities.

Water Usage and Management

In this age of water scarcity, intelligent cities which are near the oceans, undertake desalination of water in order to obtain fresh water for drinking and for other domestic and industrial usage. Water in houses is regulated by automatic smart regulators which are mounted on taps (UNESCAP, 2019; UN SDGs, 2015). As population in cities increase rapidly and climatic conditions get worse, water becomes even more scarce. Scarcity of water brings about innovatory measures to conserve it and also to find new ways of producing it through scientific discoveries. The next section examines how intelligent cities manage the tenuous relationship between humans and their environment.

Ecology

The symbiosis between humans and plants requires careful balance in the exchange of gases. Cities need to be made green by reducing their carbon and energy footprints and undertaking the creation of greenbelts and wind breaks: Refer to goals 11 and 12 of UN SDGs (2015). City dwellers who practise truck farming and market gardening or horticulture can conserve water through practices such as drip irrigation, hydroponics, and dry farming. These activities are regulated and controlled by ICT based smart objects (UNESCAP, 2019).

City Councils employ Entomologists, Botanists, Scientists, Agriculturalists, Veterinarians, Hydrologists, Agronomists, Virologists, Immunologists, Soil Scientists, Geologists, Architects, Horticulturalists, and Civil Engineers to advise them on how to manage the environment. Citizens and city inhabitants come together by using their ICT based smart devices to help crowd source information for the efficient management of the environment. This Big Data which is fed into the City Data Repository enables interested parties to undertake data mining and cloud computing activities. Some of these activities are automated by advanced algorithms (Armhein &. Quint, 2009)

Results of environmental data analytics are shared on individual home dashboards in real time as up to the minute information. Citizens use such information to advise themselves on how to modify their behaviour in order to be eco-friendly. Of late, the issue of city air quality has grabbed the headlines in the media around the world. In highly polluted cities such as New Delhi in India, Beijing in China, and Seoul in South Korea, city dwellers have had to wear masks because of the poor air quality (Airnow.gov, 2019)

Major news media such as CNN, BBC, Al-Jazeera, Sky News, and RT do give daily updates on their networks of air quality indices of cities around the globe to inform city dwellers, travellers, policy makers and planners.

The following list presents some features of an intelligent cities. It shows how the EU Commission (2017) envisage how intelligent cities become good communities as people share pool of cars for going to work, for example, and shun owing their own personal cars. Electric cars are in vogue as they are noiseless and smokeless. People in intelligent cities only travel when it is necessary as they follow on-demand travel advisory services. Traffic control is also automated to avoid congestion in intelligent cities.

- Car-sharing pool
- Electric vehicles and Hybrid vehicles
- Automatic Traffic Control
- Travel Demand Management
- ICT and emerging technologies Planning Support

- Urban Freight Logistics- multi-mode transport for light goods

In the deployment of smart technologies, many moral and ethical issues also come to the fore. Some of these issues are examined in the next section.

Trust

Trust is built over time. City dwellers are in a melting pot of civilisation as cities are becoming cosmopolitan in nature with different nationalities, ethnic groups, races, and people with different sexual orientations, political persuasions, and religious inclinations. Despite a lot of things which divide them and make them unique, there are also many modern facets of technological advancement and social mobility which unite them in the intelligent city. Everyone in an intelligent city is a digital native, a web surfer and browser, a social media addict, and a virtual world statistic with a unique digital footprint.

Trust is what people look for before any social, commercial and political engagement or transaction. A trusted person is one who is reliable and can be depended upon to deliver something which they commit to deliver to another person. A person who is trusted will not betray another friend or cause any form of harm to them or their property or their connections. Trust is influenced by one's upbringing, one's personal values, one's religion, one's gender, one's level of education, and one's peer group (IGI Global, 2019). Some relevant questions are: can city dwellers trust city administrators with their personal data not to get into wrong hands? Or can they guarantee that their personal data will not be used to blackmail them or to harass and intimidate them? In this respect, safety of data becomes an equally important requirement.

Even though surveillance cameras that are mounted in city streets, shops and offices intrude on people's privacy, yet there is need to protect people for the greater good for the greater number (principle of utilitarianism of John Stuart Mills). Such surveillance cameras can be used proactively to apprehend criminals.

Table 2. Smart City of Things (SCoT)

1. Smart Water and Smart Energy
2. Smart Environment and Smart Living
3. Smart Housing and Smart Cooking
4. Smart Waste Management
5. Smart Learning and Smart Working
6. Smart Transport and Smart Commerce
7. Smart Commerce and Smart Shopping
8. Smart Health and Smart Leisure
9. Smart Security and Smart Policing
10. Smart Computing and Smart Infrastructure
11. Smart Governance and Smart City Planning
12. Smart Engagement and Interactivity
13. Smart Monitoring and Evaluation

Table 2 presents Smart City of Things, just a few of the important components of an intelligent city (UN ECOSOC, 2016). Clearly, if everything in the city becomes smart, as shown in Table 2, the negative impact of climate change will diminish and there will be utopia. Business will boom and competition will become keen as access to ICT by the majority of city dwellers will enable them to set up their businesses online, thus encouraging the rise of technopreneurs and netopreneurs (Pellicers, 2013).

Ethics and Security

Ethics is about right or wrong of a person's conduct. It is a big issue to decide who determines what is right or wrong conduct because right conduct is partially determined and prescribed by the law, sometimes by societal norms and unwritten social conventions, culture and customary usages. Besides, people differ in their views and emotions. However, whenever the rights of an individual clash with the rights of the state as enshrined in the constitution, the latter takes precedence (Archie & Archie, 2003). This is derived from the concept of sovereignty, and J.J. Rousseau's Social Contract (Archie & Archie, 2003). In an intelligent city, there is need to observe law and order in order to avoid anarchy or a Hobbesian state which was written about by Thomas Hobbes in his book, *Leviathan.* J.J. Rousseau wrote about the *Social Contract* and he pointed out the need for citizens to obey those they have given the mandate to govern by surrendering their will to them.

Whether it is the Nicomachean Ethics of Aristotle of seeking ultimate happiness and goodness or Platonic happiness in abstract concepts or the Socratian precept of equating happiness to having an enquiring spirit, one would state that what actuates happiness in people is neither time-bound nor spatially confined (Archie & Archie, 2003). Confucius taught the moral of teaching children good morals in order to have good and upright citizens. Therefore, in a free market world which is based on the assumption that humans are rational and they are the best judges of their own welfare, it is subsumed that humans themselves are the best judges to make their revealed preferences in the market place by their considered decisions, actions, utterances and behaviour. Therefore, it lies in the hands of individual citizens to make their own choices in order to maximize their welfare (Archie & Archie, 2003).

Designers, manufacturers, and policy-makers are all human beings who, one way or the other, subscribe to the observation that humans want to preserve their kind so they follow the Golden Rule of doing unto others what they would they did to them. It is a world of reciprocity. Therefore, city-dwellers in these Metropolis, Cosmopolis, Megalopolis, and Municipality are a community who engage one another in a courteous and polite manner in order to preserve order and cooperatively achieve both individual and collective goals. Intelligent city dwellers are better off by engaging in inclusive development rather than to seek exclusion of some part of the community (Wiig, 2015). That is the urbane and civilised life the ancient philosophers ascribed to city dwellers as people who are urbane, refined, cultured and astute in urban living, characterised by the desire for high taste.

Deontological ethics examines causes of actions and judges them by the propriety of such actions. On the other hand, teleological ethics examines results or effects of actions and judges them by the results. The latter is Machiavellian in nature since the end justifies the means (Archie & Archie, 2003).

The following list presents Some Features of Cognitive or Intelligent Cities:

- Street Lighting and Road Signs
- Surveillance Cameras
- Waste Management

- Self-driving cars as taxis
- Fire Detection Robots
- Energy Monitoring Drones
- Personal Body Cameras and Smart wearables
- Water and Waste Waste Monitoring
- Transport Congestion Sensors
- Parking Applications and Kiosks
- Community Engagement and Crowd Sourcing
- Broadband Infrastructure
- Vehicle Fleet Management
- Commuting and Logistics
- Bridges and Infrastructure Inspection

Some of the areas of focus of the National League of Cities (2017), as mentioned in the above list, envisage that new breakthroughs in crime detection make crime redundant and too transparent to commit such that crime becomes obsolete. Some of the intelligent systems also are designed to drive public utilities for maximum utilisation by citizens. For example, employees in the city receive alerts to go to job-demand places to do piece jobs involving either repair of public utilities or the installation of utilities. Many employees therefore engage in freelancing and moonlighting instead of being permanent employees of corporations and institutions.

With automation, majority of the labour force becomes self-employed and on the mailing list of several employers. Therefore, nobody is left unemployed except those who are heavily blacklisted for poor quality output in their previous assignments. These are automatically selected by the system for retraining online through computer simulation. There is also peer-to-peer learning through social media sites. The next section examines issues of personal privacy and confidentiality in intelligent cities where, it is feared that *Big Brother* is watching you.

Privacy

Privacy requires adherence to Data Protection Acts of not using personal data obtained through business transactions to hound clients or to pass such data to unlawful or unauthorised third parties (Bradley, 2019). This covers issues of plagiarism, copyright, intellectual property rights, and confidentiality. In many cities, plans are underway to use facial recognition in crime busting and for identification and processing at airports, banks, schools, and other places. However, what about look-alikes or identical twins? Many critics envisage that a lot of mistaken identities and collateral damage can be incurred by using these smart technologies in intelligent cities (Bradley, 2019). Voice-recognition methods are also being investigated.

All these issues border on clash between individual and state rights. The use of secret or hidden cameras may be considered unethical by some individuals. However, on the part of the state, prevention is better than cure. The state has a bigger stake in the welfare of its citizens than the citizens themselves (Mahmood, 2013a & 2013b; United Nations, 2008). This is the case of the whole being greater than the sum of its constituent parts. In this case, it is better the individual was naked than the whole state being naked. Revolutionaries may intone the saying that, ‘a revolution consumes its own people’. In both democratic and centralised political systems, it is recognised that the interests of the state supersede individual interests.

The following list highlights some of the issues which many city dwellers are worried about in intelligent cities. Some of these have been discussed above; some of the remaining of the following issues are further discussed in the upcoming sections.

- Human Rights and Safety
- Privacy, Security, and Trust
- Human Machine Conflict (HMC)
- Data Storage and Management
- Device Connectivity
- Future Directions in AI and Machine Learning

Human Rights

In 1948, there was the UN Declaration of Universal Human Rights which recognised inalienable rights of all humans to be free from want, to have freedom to life, liberty, and property, freedom of conscience, freedom of association, freedom of speech, freedom of movement, and other freedoms. In intelligent cities, these freedoms are extended by the use of Big Data Analytics, increased surveillance, greater convergence on platforms, and greater interconnectivity and interoperability (Bradley, 2019). City dwellers become safe because of efficient crime-busting techniques operated through smart policing, the use of monitors, sensors, and advanced surveillance cameras (Kourtit & Nijkamp, 2012; Psaltoglou, 2018).

Table 3 shows various areas of intelligent city applications (McKinsey & Company, 2018). The areas listed under various sub-groups indicate ICT-dependence activities such as tele-medicine, bike- sharing and citizen services, among others. In a smart city, citizens recognise the need to conserve fuel by accepting energy-saving arrangements such as car- sharing, bike-sharing, and peer-to-peer accommodation (Williams, 2018).

Table 4 lists some of the multidisciplinary areas which have interface with the operation of an intelligent city and they point also to the future of work and fields of future employment. These can be prioritised in colleges and universities in connection with admission of students into degree and diploma programmes.

Table 5 shows that within intelligent cities, people need to change their mind-sets, working habits, and set up functional institutions through reforms in order to ensure a bright future for all. Bureaucratic processes and procedures must become a thing of the past as AI speeds up processes, procedures, automation, and decision making. A new culture of speed, efficiency and accuracy emerges as many procedures become automated and decision making is done by machines through machine learning, use of advanced algorithms, and cloud computing (Erl, Mahmood, et al., 2013; Psaltoglou, 2018). Artificial Intelligence and Machine Learning help in dealing with complex phenomena such as meteorology, gaming or gambling, knowing the pharmacological properties of plants, and many other complex issues in medicine, engineering, and marketing.

Meteorology

As the global climate gets worse, there is the need to give precise information on weather patterns in order for citizens to make informed decisions. Weather information can be fed into City Databases and

Table 3. Digital Solutions for a more Liveable Future

Economic Development/Housing	Healthcare
1. Bike sharing	1. Water consumption tracking
2. Car sharing	2. Home energy auto systems
3. Digital payment in public transport	3. Home energy consumption tracking
4. Integrated Multimodal information	
5. Real time road navigation	Security
6. Smart deliveries	1. 1. Home security systems
7. Small parcel lockers	2. 2. Personal alert apps
8. Smart parking	
	Utilities
Mobility	1.Digital administration
1. 1. Lifestyle wearables	2. Citizen services
2. 2. Online care and scheduling	3. Online retraining (deskilling, retooling)
3. 3. Real time air quality information	4. Peer- to –peer accommodation platforms
4. 4. Telemedicine	

passed on to individual information platforms on an hourly basis as meteocasts, meteopods and meta-meteo-data which is disseminated to city dwellers on their dashboards. Such information assists farmers, traders, construction workers and the general public to plan outdoor activities (Gaur et al., 2015).

POPULATION DYNAMICS AND INTELLIGENT CITIES

A model is a simple representation of complex reality which has capacity to explain, forecast and help understand how a real phenomenon works (Science Learning, 2019). A model can be a mathematical representation or an assumption made in a series of logical propositions. In general, intelligent cities can be judged by eight determinant factors or criteria such as the following:

- Sustainability
- Efficiency
- Entrepreneurship
- Innovation
- Safety and Security
- Compliance
- Order
- Leisure

Sustainability criterion is used to judge practices which help intelligent cities survive environmental factors such as climate change and global warming, as well as those which help fit it into the external pressures introduced by globalisation (Komninos, 2009)

Table 4. Intelligent City - Areas of Study

Architecture	Engineering
Art & Culture	Finance
Accounting	Food & Nutrition
Agriculture	Geography
Banking	Geology
Beautification	History
Building & Construction	Horticulture
Commerce	Archaeology
Demography	Library Studies
Designing	Law
Education	Literature
Linguistics	Logistics
Interior Decoration	Medicine
Management	Planning
Natural Sciences	Land Economy
Meteorology	Marketing
Nursing	Media Studies
Politics/Public Administration	Public Relations
Police Science	Economics
Security; Cyber Security	Sociology
Transport Economics	Tourism & Hospitality
Statistics	Missiology
Psychology	Psychiatry
Horticulture	Waste Management
Veterinary Science	Fabrication
Chemical Engineering	Estate Management
Geodetic Engineering	Content Design; Film Making
Happiness Science	Leisure Management
Physiotherapy	Psychosocial Counselling
Data Science	Cloud Computing

Table 5. Intelligent Cities and effects of reforms

SMART CITIES 1	SMART CITIES 2	SMART CITIES 3
1. Work	People	Processes
2. Living	Institutions	Procedures
3. Productivity	Mind-set	Culture
QUALITY OF LIFE	FUTURE IS BRIGHT	PROGRESS

Efficiency measures the relative cost of doing business in the city in comparison to competitors around the world. This can further be judged by turnaround time and how tight or well-integrated the value and supply chains are. Furthermore, efficiency can be measured for example, by examining a simple logistic problem of deliveries to several destinations using the nearest neighbour principle to minimize journey costs by automating the schedules of deliveries (Raghav, 2019; CEADSEN, 2019).

Entrepreneurship is measured by the speed to market of new products as well as the number of born global companies registered online per year. This is linked to innovation and creativity. Israel has the highest rate of born global innovations in the world because of their huge investment in human capital.

Safety and security in intelligent cities are measured by the number of crimes that go undetected by the super system of sensors, drones, surveillance cameras, and crowd-sourcing. Safety, privacy and security are also linked to longevity and life expectancy which is measured by proxies of quality and quantity of medical care available to city dwellers (Gaur et al., 2015)

Compliance with the law is judged by how well citizens behave by complying with city laws and regulations relating to driving, city decongestion efforts, waste disposal efforts, greening the city efforts, following exercise regimen drawn up through e-governance messaging and other instructions sent out en mass through social media and other ICT and IoT based channels.

It is often said that it is divine to obey orders from above. Super-computers in cities are programmed by city authorities to churn out orders for the greater good for the greater number. Therefore, following orders becomes the order of the day in intelligent cities in order to sustain smart living; it is either collective doom or collective survival. Furthermore, intelligent cities are judged by the amount of leisure which city dwellers enjoy.

The intelligent city index (SCI) is a function of many variables such as the following (Mankiw, 2012; Lipsey & Chrystal, 2011):

SCI = f (Population, ICT, AI, ML, Global location, Legislation, GDP, Innovation, e-governance, e-commerce, PESTEL factors, Congestion, Mobility, Climate change)

SCI = k f (Pop. + ICT + AI + ML + GDP + Globalisation- Pollution- Culture- Political Culture)

SCI = k * Pop. + ICT+ AI+ML+GDP+ Globalisation/Pollution + Culture+ Political Culture

SCI = k* Pop. + GDP + Technology + Globalisation/Peoples' Behaviour

SCI = Capitalism/Socialism

$SCI = A\,K^{a}\,L^{b}\,(HC)/PB$ (100 to 1)

$SCI = a + b_1\,X_1 + b_2\,X_2 + b_3\,X_3 + b_4\,X_4 + b_5X_{5+}\,b_6\,X_6 \text{ -} b_7X_7 - b_8\,X_8$

where

SCI is smart city index
k is a constant of climate change
Pop. Is population size

ICT is information communication technology
AI is artificial intelligence
ML is machine learning
GDP is Gross Domestic Product
A is technological constant in Cobb-Douglas Production function
K is capital accumulation
L is labour force
a and b are growth factors or scales of both capital and labour which are homogenous and of order 1 (where $a + b = 1$)

It can be seen in the model above that the numerator represents capitalism or free enterprise as was written about in 1776 by Adam Smith in his *Wealth of Nations*. The denominator represents socialism as written about jointly by Karl Marx and Frederic Engels in their books *Das Kapital* and the *Communist Manifesto* in 1848, in which they advocated for collective state ownership of the means of production in a centrally-controlled or command regime.

Thus, in a smart city, people partially adopt capitalist entrepreneurship on the one hand and on the other hand behave in a collective herd mentality of obeying and complying with machine instructions in order to achieve maximum desired behaviour as mentioned in George Orwell's novel, 1984. Thus, capitalism in business allows individual initiatives for intelligent city survival, sustainability and efficiency, while homogenous behaviour which is similar to the one which exists in a socialist state allows the collective good of all to be attained. Thus, convergence of behaviour in a smart city becomes obligatory and a norm.

In an intelligent city, people accept some form of regimentation for the collective good by obeying orders issued by machines as well as being compliant with intelligent city green requirements. The smart city index (SCI) can therefore be represented as:

$SCI = A\, K^a L^b\, (HC)/PB$ (with Public Behaviour, PB, ranked from 1 to 100)

The numerator in the above formula shows the Cobb-Douglas production function which is discounted by PB (aggregate public behaviour) which is ranked from 1 to 100, with a score of 1 being the best behaviour and 100 being the worst behaviour (Lipsey & Chrystal, 2011; Mankiw, 2012). The Gravity Model states that the force of attraction (F) between two bodies, M_1 and M_2, is equal to a gravitational constant multiplied by the masses of the two bodies divided by the square of the distance between them. As per Anukoonwattaka (2016), F, the Force, is equal to:

$$F = g\, M_1\, M_2\, /d^2$$

The SCI and Gravity Model (GM) look similar in that the Smart City Index (SCI) is GDP divided by PB (aggregate public behaviour) while GM is the sum of masses of city populations divided by the square of the distance between them in order to ascertain the influence of one city on the other. After SCI has been calculated by using the formula above, the intelligent cities can be ranked from 1 to 100. The next section examines some basic economic models pertaining to intelligent cities.

Intelligent Cities: Population Growth and Equilibrium

A lot of assumptions have to be made about intelligent cities. It can be assumed that in a temporary short period of time, Aggregate Demand AD is greater than Aggregate Supply AS (i.e. AD > AS). This temporary disequilibrium state is like the pre-industrial subsistence stage of production. This disequilibrium in supply triggers local entrepreneurs to fill the market gap by being innovative and that stimulates output. In the short run (ranging from 1 to 5 years), AD = AS and this is the free market classical market equilibrium (Mankiw, 2012; Lipsey & Chrystal, 2011). However, in the long run, the intelligent city economy gets to the point where AD < AS and there is a surplus output which can be exported. This state is reached through specialisation, innovation, learning from past mistakes, and obtaining synergies and economies of scope and scale from linkages in the supply chain. A growing population provides the critical mass for market expansion.

The surplus state can arouse mercantilist sentiments of exporting more and importing less as it is happening in the USA now. All three situations combined trace an oscillating trade cycle which has depression, recovery, boom, and recession. This economic rhythm shows pendulum behaviour of the economy. Such oscillations in the economy trace out the long run Phillips Curve (LRPC) or the non-accelerating inflationary rate of unemployment or the long run Phillips Curve (NAIRU) which is fixed (Lipsey & Crystal, 2011, pp. 562-564; Mankiw, 2012, pp.794-795). The swings in the rhythm of the economy have implications for population growth, unemployment, and sustainability of the intelligent city. Population size determines and drives many decisions in the city. Therefore, the next section examines some population issues.

Population Models

In the developed economies, population growth is either stationary or it is negative. The latter is assumed here in view of the aging population and low birth rate in the developed countries. The formula below can be used to estimate the current population of a country given the last census estimate:

$$P_c = P_0 + B - D + I - E$$

where

P_c is current population size
P_0 is previous known population
B is number of births since last census
D is number of deaths since last census
I is number of immigrants allowed into the country since the last census
E is number of emigrants who have left the country since the last census

where the population is decreasing, the decay formula is given as:

$$A_1 = A_0 e^{-rt} \text{ or } P_1 = P_0 (1 - r)^n$$

Advanced countries are in the last stage of the Gompertz S Curve where growth has hit a ceiling. Some Intelligent cities in the developed countries have reached a point where they have to import labour. These include the UK, Spain, Italy, Switzerland, Sweden, and Germany, among others. The formulae for estimating future population size are given thus:

$$A_1 = A_0 e^{rt} \text{ or } P_1 = P_0 (1 + r)^n$$

The developing countries are in the middle stage of the Gompertz Curve which shows an elongated S shape. Intelligent cities have capacity to accurately forecast their optimum population with the help of Big Data and Data Analytics. Artificial Intelligence (AI) and Machine Learning (ML) can be used to plan population control as sustainability and efficiency have relevance only with regard to knowledge of population dynamics. AI and ML help women plan their births as well as decide when they should go on maternity leave. Knowledge of population dynamics is essential for policy makers in intelligent cities as well as for all intelligent city dwellers so that they can know the implications of their activities with regard to family planning.

The concept of growth poles was first suggested in the 1950s by two Frenchmen, Francois Perroux and Jacques Boudeville. A growth pole is a dominant economic area which could be a primate city or a conurbation or urban sprawling city that performs many functions for outlying areas. It serves as a hub within a cluster of settlements (Mumford, 1973).

Mullan (2019) provides a list of the top 10 intelligent cities in the world with three of them in Europe (London, Amsterdam, Paris), four in Asia (Singapore, Hong Kong, Tokyo, Seoul) and another three in North America (New York, Toronto, Reykjavik). The list is as follows:

- New York
- London
- Paris
- Tokyo
- Reykjavik
- Singapore
- Seoul
- Toronto
- Hong Kong
- Amsterdam

Tahir & Mulek (2016) state that the main criteria for designating a smart city are based on twin pillars of efficiency and sustainability. They list six criteria namely:

- smart environment
- smart mobility
- smart living
- smart people
- smart economy and
- smart government

A smart environment is where people cause minimum environmental damage to flora and fauna by undertaking measures to preserve and conserve the environment. This involves tree planting, afforestation, creating green belts, and having legislation in place to protect forests and animals. A smart environment has less pollution from toxic waste and effluent discharges. Fresh Water bodies are protected and conserved.

Smart mobility involves less city congestion, multi-mode mass movements in cities by coaches, use of bullet trains and overhead cable cars. It also means many city dwellers use bicycles and other means of transport, thus causing minimum pollution. For example, the use of water transport causes less pollution to the environment.

Smart living means healthy and comfortable living with emphasis on, perhaps, vegan diet. It also means having lots of exercise, using water wisely, using energy efficient means to cook such as using solar and gas stoves, for example. Smart living means also attending to cultural issues and being environmentally-friendly as well as being futuristic and conscious of the needs of future generations. Many people in poor areas in Africa use firewood and charcoal which is not smart living because trees are depleted rapidly, causing rapid deforestation and desert encroachment or desertification in the savannah and Sahel regions. People in intelligent cities are people who are ICT savvy, cognitive, caring, and au fait with the use of inter connected ICT based smart devices and the Internet of Things (IoT) vision, whereby their appliances are all interconnected remotely and they are all on the same page with developments in their micro and macro environments. People in intelligent cities are ICT-dependent and they are technophiles. They always seek ICT and high technology interventions and solutions.

In an intelligent city economy, business is conducted mostly online without cash except the use of e-money or bit-coins and crypto-currency which use secure block-chain technology. Transactions are carried out securely using smartcards or plastic money. There is high sense of innovation, entrepreneurship, integration, and interdependence. According to Tahir & Mulek (2016) a intelligent city invests in social well-being and human capital.

In an intelligent city, governance is carried out as e-governance whereby city councils hold electronic elections and crowd-source ideas from city-dwellers by using Big Data Analytics to collect and process data as well as also to micro-manage the lives of citizens. Democracy is therefore enhanced and made inclusive through the intermediation of ICT facilities. In intelligent cities, there is heavy surveillance of citizens in order to avert crime as the streets have hidden cameras and there is smart policing. City government aims at efficient service delivery and creation of economic opportunities for all through e-governance (Tahir & Mulek, 2016; Mahmood, 2013a & 2013b; United Nations, 2008).

CONCLUSION

This chapter in the book explored many multidisciplinary themes and applications of ICT and latest technologies in the era of development of intelligent cities. The chapter gave a background to the origin and functions of cities as well as the hierarchical position of cities in settlements. It also gave many examples of definitions of intelligent cities from authoritative sources. Characteristics of intelligent cities were given as well as establishing the rationale of adopting the concept of intelligent cities. Many examples were also given from around the globe about the efforts by many governments to turn cities into intelligent cities in response to the threats which are posed by global warming, climate change, and the need to embrace the changes thrust upon the world by rapid technological developments and rapid increases in population.

It turns out that demand exceeds supply in the short run in most cities as city utilities and services are stretched to their limits because of increased population size, increase in levels of poverty, and reduction in the stock of natural resources due to over-exploitation and improper use of those resources. The adoption of the concept of intelligent cities has been found as the panacea to most problems and challenges which modern cities face around the world. These problems include traffic congestion, spread of slums, and insufficiency of decent accommodation for city dwellers, choking of cities with filth and smog due to pollution from industries, rise in urban crime, rise in levels of urban unemployment, outbreak of communicable and infectious diseases such as SARS, MERS, Ebola, Cholera, and Corona Virus of 2020. There is also the problem of high levels of illiteracy and poverty among urban fringe populations, especially among city immigrants and those living in slum areas, ghettoes, and shanty compounds.

These challenges and problems require solutions and interventions. Policy makers need to understand the dynamics of population growth in order to plan ahead. In this vein, the chapter covers a section on models of population as well as a discussion on intelligent city models given by scholars such as Shapiro and Florida. Intelligent cities require smart policing and heavy surveillance in order to pre-empt and bust crimes. This leads to a whole raft of ethical dilemmas which border on issues of privacy, confidentiality, freedom, trust, protecting human rights, asserting the primacy of the state to unlimited access to information on citizens, among other issues.

Intelligent city management also covers other areas such as leisure, hospitality, health, education, commerce, transportation, tourism, waste management, city beautification, design and architecture of buildings and structures which should be safe, secure, aesthetically appealing, and technologically sound. This chapter included many examples of best practices from around the globe to make the discussion empirical. Finally, intelligent cities should be viewed from the perspective of how activities in them achieve convergence, integration, interoperability, synergy, sustainability, economy, efficiency, and effectiveness.

REFERENCES

ACCA CBE. (2019). *Platform Support* Retrieved from accaglobal.com/gb/en/student/exam-support-resources/professional-exams-study-resources/cbe-practice-platform/-support.html

Adepoju, A. (2004). *Trends in International Migration in and from Africa.* Retrieved from researchgate.net/publication/289568094-Trends-in-International-Migration-in-andaroundAfrica

Ahmed, A. I. M. U. (2004, January). Weber's Perspective on the City and Culture, Contemporary Urbanization and Bangladesh. *Bangladesh e-Journal of Sociology*, *1*(1), 1–13. Retrieved from https://pdfs.semanticscholar.org/b072/0ac93a6858f21ab2f3cf97468019deecf0e3.pdf

Airnow.gov. (2019). *Air Quality Index (AQI) Basics.* Retrieved from https://airnow.gov/index.cfm?action=aqibasics.aqi

Albino, V., Berardi, U., & Dangelico, R. M. (2013). Smart cities: definitions, dimensions, and performance. *Proceedings IFKAD 2013.*

Amer, S. B. (2019). *City Definitions*. Retrieved from smartcitiescentre.org/wp-content/uploads/18-sara-ben-amer-city-definitions-sarabenamer.pdf

American Geo Sciences (2019). *What is bio-mining?* Retrieved from americangeosciences.org/critical-issues/faq/what-biomining

Amrhein, G., & Quint, S. (2009). Cloud Computing for the Enterprise: Part 1: Capturing the Cloud. *DeveloperWorks*. Retrieved from www.ibm.com/developerworks/websphere/techjournal/0904_amrhein/0904_amrhein.html

Ancient Ports Antiques. (2019). *The Catalogue*. Retrieved from ancientportsantiques.com/the-catalogue/

Anukoonwattaka. (2016). *Introduction to the Basic Gravity Model*. Retrieved from unescap.org/sites/default/files/an/S2-Gravity-intro.pdf

Archie, L., & Archie, J.G. (2003). *Introduction to ethical studies- An Open Source Reader*. Ethicsbooks.

Bavel, J. V. (2013). *The World Population explosion: causes, backgrounds and projections for the future*. Retrieved from ncbi.nlm.nih.gov/pmc/articles/PMC3987379/

Bradley, R. (2019). *Data Privacy Concerns: An Overview for 2019*. Retrieved from medium.com@manifest/data-privacy-concerns

Caragliu, A., Del Bo, C., & Nijkamp, P. (2011). Smart Cities in Europe. *Journal of Urban Technology*, *18*(2), 65–82. doi:10.1080/10630732

Caragliu, A., Del Bo, C., & Nijkamp, P. (2011). *Smart Cities in Europe*. Retrieved from research.vu.nl/ws/portalfiles/portal/2474218/2009048.pdf

CEADSEN. (2019). *Nearest Neighbor Analysis*. Retrieved from ceadsen.nku.edu/longa/geomed/ppa/doc/NNA/NNA.htm

Darko, S. (2015). *Inside the world of Ghana's Internet fraudsters*. Retrieved from bbc.com/news/world-africa-32583161

Dijkstra, L., & Poelman, H. (2012). Cities in Europe: The new OECD-EC Definition. *RegionalFocusEurostat*. Retrieved from http://epp.eurostat.ec.europa.eu/portal/page/portal/nuts_nomenclature/local_administrative_units

Economist Intelligence Unit (EIU). (2013). *Hot spots 2025: Benchmarking the future Competitiveness of cities*. Academic Press.

Elkington, J. (2004). *Enter the Triple Bottom Line*. Retrieved from johnelkington.com/archive/TBL-elkington-chapter.pdf

Erl, T., Mahmood, Z., & Puttini, R. (2013). Cloud Computing: Concepts, Technology and Architecture. London: Prentice Hall.

EU. (2018). *Waste management in cities*. Retrieved from ec.europa.eu/easme/en/news/waste-management-in-cities

EU Commission. (2017). *(The) making of a smart city: best practices across Europe*. Retrieved from www.smartcities-infosystem.eu

Gaur, A., Scotney, B. Parr, G., McClean, S. (2015). *Smart City Architecture and its Applications based on IoT.* Retrieved from researchgate.net/publication/278726450_smart-city-architecture-and-its-application-based-on-IoT

Green Belt Movement. (2019). *Wangari Maathai*. Retrieved from https://www.greenbeltmovement.org/wangari-maathai

Guardian. (2018). *Why eating less meat is the best thing you can do for the planet in 2019*. Retrieved from https://www.the guardian.com/environment/2018/dec/21/lifestyle-change-eat-less-meat-climate-change

Harrison, C., & Donnelly, I. (2011). A Theory of Smart Cities. *Proceedings of the 55th Annual Meeting of the ISSS-2011*.

Hollands, R. (2008). Will the real smart city please stand up? Intelligent, progressive or entrepreneurial? *City: Analysis of Urban Trends, Culture, Theory, Policy, Action, 12*(3), 303-320. DOI: . doi:10.1080/13604810802479126

Hoyt Homer. (1939) Retrieved from https://planningtank.com/settlement-geography/hoyt-model-sector-model-land-use-1939-homer-hoyt

IBM (2011). *IBM's Smarter Cities Challenge: Philadelphia Report*. Retrieved from smartercitieschallenge.org/city_philadelphia_pa.html

IGI Global. (2019). *What is (sic) sources of Trust?* Retrieved from www.igi-global.com/dictionary/sources-trust-consumers-participation-permission/27807

Komninos, N. (2009). Intelligent cities: Towards interactive and global innovation environments. *International Journal of Innovative Research and Development, 1*(4), 337–355.

Kourtit, K., & Nijkamp, P. (2012). *Smart Cities in the Innovation Age. The European Journal of Social Science Research, 25*(2), 93–95. doi:10.1080/ 13511610.2012.660331

Kruschewsky, G. (2018). Los Angeles vs. San Francisco: Which city rules the West Coast? *Huffington Post*. Retrieved from huffpost.com/entry/los-angeles-vs-san-francisco

LeGates, T., & Stout, F. (2018). *Smart Cities: Digital Solutions for a more liveable future*. McKinsey Global Institute.

Linthicum, D. (2010). *Cloud Computing? Thank SOA*. Retrieved from https://www.thecloudtutorial.com/cloud-computing-soa.html

Lipsey, R., & Chrystal, A. (2011). *Economics*. Oxford: Oxford University Press.

Live Science. (2019). *What is STEM education?* Retrieved from livescience.com/43296-what-is-stem-education.html

Mahmood, Z. (2013a). *Cloud Computing: Methods and Practical Approaches*. London: Springer Verlag.

Mahmood, Z. (2013b). *E-Government Implementation and Practice in Developing Countries*. IGI Global Publishers.

Mankiw. (2012). *Principles of Economics*. London: Cengage Learning International.

Manville, C., Cochrane, G., Cave, J., Millard, J., Pederson, J., Thaarup, R., & Kotterink, B. (2014). *Mapping Smart Cities in the EU*. Brussels: European Union.

McKinsey & Company. (2018). Smart Cities: Digital Solutions for a more livable future June 2018. *McKinsey Global Institute*. Retrieved from https://www.mckinsey.com/~/media/mckinsey/industries/capital%20 projects%20and%20infrastructure/our%20insights/smart%20cities%20digital%20solutions%20for%2

Min, K., Yoon, M., & Furuya, K. (2019). Sustainability research terms used on smart cities. *Sustainability*, *11*, 3155. doi:10.3310u1113155

Monfaredzadeh, T., & Berardi, U. (2015). Beneath the smart city: Dichotomy between sustainability and competitiveness. *International Journal of Sustainable Building Technology and Urban Development*, *6*(3), 140–156.

Mori, K., & Christodoulou, A. (2012). Review of Sustainability Indices and Indicators: Towards a New City Sustainability Index (CSI). *Environmental Impact Assessment Review*, *32*(1), 94–106.

Morvaj, B., Lugaric, L., & Krajcar, S. (2012). Demonstrating Smart Buildings and Smart Grid features in a Smart Energy City. *Energetics (IYCE), Proceedings of the 2011 3rd International Youth Conference*.

Motta, R.C., de Oliveira, K.M., & Travassos, G.H. (2019). A conceptual perspective on interoperability and software systems operations. Retrieved from doi:10.1016/j.infsof.2019.07.001

Mullan, L. (2019). *Top 10 smart cities in the world*. Retrieved from gigabitmagazine/big-data/top-10-cities-world

Mumford, L. (1973). What is a city? In *The City Reader* (5th ed.). Academic Press.

National League of Cities (NLC). (2017). *Trends in Smart City Development*. Retrieved from https://www.nlc.org/sites/default/files/2017-01/Trends%20in%20Smart%20City%20Development.pdf

Nguyen, A. (2018). *10 typical tree planting projects in the world*. Retrieved from treedom.net/en/projects

Paskaleva, K. (2011). The smart city: A nexus for open innovation? *Intelligent Buildings International*, *3*(3), 153–171. doi:10.1080/17508975.2011.586672

Pellicer, S., Santa, G., Bleda, A., Maestre, R., Jara, A., & Skarmeta, G. (2013). A Global Perspective of Smart Cities: A Survey. *2013 Seventh International Conference on Innovative Mobile and Internet Services in Ubiquitous Computing (IMIS)*, 439–444. DOI:10.1109/IMIS.2013.79

Pierceall, K. (2015). *Group says New Las Vegas resort-casino to rise at stardust site by 2018*. Retrieved from washingtontimes.com/news/2015/may/5/group-says-new-las-vegas-resort-casino

Psaltoglou, A. (2018). *From Smart Cities to Cognitive Cities: Intelligence and Urban Utopias*. Retrieved from https://www.researchgate.net/figure/A-cognitive-system-for-transportation-infrastructure-Mostashari-2011_fig1_327395697

Saha, S. (2019). *Cumulative Causation Theory: Gunnar Myrdal*. Retrieved from https://bgc.org.in/pdf/open-educational-resources/geography/gunner-myrdals-theory_ug_II_ss_1.pdf

Science Learning. (2019). *Scientific Modelling*. Retrieved from sciencelearning.org.nz/resources/575-scientificmodelling

Smart Cities Dive. (2017). *Four Cities' Solutions to Sustainable Garbage Processing*. Retrieved from www.smartcitiesdive.com/ex/sustainablecitiescollective/Friday-fun-how-create-tomorrow-s-green-cities-today-s-garbage/1050616/

Tahir, Z., & Mulek, J. A. (2016). Main criteria in the development of smart cities determinants using analytical method. *Journal of Malaysian Institute of Planners*, *14*, 1–14. Retrieved from https://www.researchgate.net/publication/309958294_Main_criteria_in_the_development_of_smart_cities_determined_using_analytical_method

Tree Nation. (2019). *List of active projects*. Retrieved from www.tree.nation.com

UN. (2008). *United Smart Cities (USC)*. Retrieved from https://sustainabledevelopment.un.org/partnership/?p=10009

UN Climate Change. (2016). *The Paris Agreement*. Retrieved from unfccc.int/process-and-meetings/the-paris-agreement/the-paris-agreement

UN ECOSOC (Economic and Social Council) Report. (2016). *Smart Cities and Infrastructure Report 26 February 2016*. UN.

UN SDGs. (2015). *Envision 2030: 17 goals to transform the world for persons with disabilities*. Retrieved from https://www.un.org/development/desa/disabilities/envision2030.html

UNESCAP. (2019). *Water efficient infrastructure and water-smart cities (the case of Korea)*. Retrieved from unescap.org/sites/default/files/session%205-Yongdeok%20Cho-AWC.pdf

United Nations. (2014). *Cyber School Bus. Cities of Today, Cities of Tomorrow! Unit 1: What is a City?* Retrieved from https://www.un.org/cyberschoolbus/habitat/units/un01txt.asp

United Nations (UN). (2008). Benchmarking E-Government: A Global Perspective. Division of Public Economics and Public Administration Report United Nations, Department of Economic and Social Affairs, Population Division: World Urbanization Prospects, the 2011 Revision.

WHO. (2019). *Health and sustainable development: Improved urban waste management*. Retrieved from who.int/ sustainable-development-cities/strategies/urban-waste/en/

Wiig, A. (2015). The empty rhetoric of the smart city: From digital inclusion to economic promotion in Philadelphia. *Urban Geography*, *37*(4), 535–553. doi:10.1080/02723638.2015.1065686

Williams, H. (2018). *What is a smart city? How to define a smart city*. Retrieved from computerworld.com/article/3427835/what-is-a-smart-city-how-to-define- a smart-city.html

Witte, A. D. (1996). Urban Crime: Issues and Policies. *Housing Policy Debate*, *7*(4), 731–749.

World Atlas. (2019). *What is the von Thunen model?* Retrieved from worldatlas.com-what-is-the von-thunen-model.html

ADDITIONAL READING

Albino, V., Berardi, U., & Dangelico, R. M. (2015). Smart cities: Definitions, dimensions, performance, and initiatives. *Journal of Urban Technology*, *22*(1), 3–21. doi:10.1080/ 10630732.2014.942092

Eggers, W. D. (2005). *Government 2.0*. Rowman & Littlefield Publishers.

Giffinger, R., Fertner, C., Kramar, H., Pichler-Milanovic, N., & Meijers, E. (2013). Vienna University of Technology SRF Centre of Regional Science. Retrieved from http://www.smart-cities.eu/team_1.html

Government 2.0 Task Force. (2009), Engage: Getting on with Government 2.0, Report of Government 2.0 Task Force, March 2009. Retrieved from http://gov2.net.au/about/draftreport/, Accessed Dec 2014

O'Grady, M., & O'Hare, G. (2012). How Smart Is Your City? *Science*, *335*(3), 1581–1582. PMID:22461597

Paskaleva, K. (2009). Enabling the smart city: The progress of e-city governance in Europe. *International Journal of Innovative Research and Development*, *1*(4), 405–422.

Upendran, D., Chatterjee, S., Sindhumol, S., & Bijlani, K. (2016). Application of Predictive Analytics in Intelligent Course Recommendation ScienceDirect. *Procedia Computer Science*, *13*, 917–923.

KEY TERMS AND DEFINITIONS

Anthropocentric: Anything which is human-centred or meant for the good of humankind is said to be anthropocentric.

Architecture: This is the science of building design, and designing structures which are fit for purpose and which can stand the test of time.

Artificial Intelligence: This is machine recognition or thinking ability which is based on pre-determined data input or machine learning.

Big Data: This refers to large amounts of data from different sources; it is made up of all kinds of data (e.g., photos, music, text, figures and numerals, voice recording, videos, and a jumble of electronic data), which needs ordering in order to convey meaning to end-users.

Cloud Computing: Cloud computing refers to a remote server which is used to store large amounts of data which is safe and available at all time; the environment provides all kind of software, hardware and computing platforms that consumers can provision on a pay as you go basis.

Cognitive Cities: These, sometime also referred to intelligent cities, are modern cities which have many interconnected smart devices which communicate to each other to provide intelligent services to city residents and help them take intelligent decisions.

Crowd Sourcing: It is the ability to either carry out massive online survey or collect data from millions of people online at minimum cost.

Data Analytics: It is the analysis of big data to convey user-friendly information on trends embedded in the data.

Drones: These are mobile computers with artificial intelligence; generally flying objects to perform actions, either controlled from remote locations by human operators or they are automatically operated by using machine learning or artificial intelligence.

Green Cities: These are eco-friendly cities consisting of eco-friendly systems with minimum amount of carbon emissions.

Infrastructure: This refers to buildings, roads, rail, and other concrete structures that work together to aid in business transactions or other functional systems.

Intelligent Cities: These are smart cities, often referred to as cognitive cities or digital cities.

Internet of Things: This is a vision in which ICT based smart devices are connected to automatically perform the required operations to makes living easier; Intelligent cities and autonomous cars are just two examples of IoT.

Interoperability: It is the ability of electronic smart devices to communicate with each other in a seamless manner to perform operations to achieve common objectives.

Machine Learning: This is the ability of machines to learn and store knowledge that then enables them to act and think like humans in solving problems.

Megacities: These are large modern-day super cities which have millions of inhabitants and generally centres of commerce that have characteristics of intelligent digital cities.

Netiquette: Netiquette refers to etiquette or polite behaviour which is expected from online users towards other online users by avoiding vulgarity, irritation, racism, impiety and use of offensive language; it requires that online users are always courteous to other online users at all times.

Peri-Urban: The is the suburb or outlying area of a city, also called the urban fringe or city outskirt.

Primate Cities: This is a city which is twice as big as the next largest city and which is predominant a commercial hub.

Prosumers: Here, the terms 'producers' and 'consumers' are combined into one to refer to both.

Smart City: Also known as intelligent cities, this refers to a new vision for urban development, using IoT-based technologies and tools, to integrate and effectively manage a city's resources and processes (e.g., transportation, healthcare, commerce, utilities, city's e-government functions), to improve the quality of life of citizens.

Sustainability: This is where an on-going activity is capable of being continued for a longer period of time into the future.

Technophiles: These are people who like technology-based devices and embrace technology in their daily life to solve problems and make life easier and more comfortable.

Technophobes: These are people who fear to embrace new technology.

Truck Farming: This is where small scale farmers in heavily populated cities cultivate in urban fringes and suburbs, providing for city dwellers; they use large vehicles to ferry their produce to markets as they usually have perishable goods such as vegetables and fruits, among others.

Urban Implosion: When two big urban settlements or agglomerations come together to form a sprawling settlement such as Lagos in Nigeria or Mumbai in India, this process of coming together is referred to as a conurbation or urban implosion.

Urbanisation: It is the process whereby towns and villages grow to reach the rank of cities in terms of population size, area size, and complexity of the functions performed for their outlying areas and inhabitants.

Chapter 10
Internet of Things Application for Intelligent Cities:
Security Risk Assessment Challenges

Wissam Abbass
https://orcid.org/0000-0001-5370-7470
INPT, Rabat, Morocco

Amine Baina
INPT, Rabat, Morocco

Mostafa Bellafkih
INPT, Rabat, Morocco

ABSTRACT

The rapid growth of the world's population is placing a huge strain on the existing infrastructures. As a quest for accommodating this growth, interest is turned to the internet of things (IoT). In fact, the IoT is significantly improving today's quality of life by innovating the provided services and enhancing communication and interaction. Furthermore, it has also empowered real-time decision making by introducing dynamic services for innovative traffic handling, energy-efficient infrastructure saving, and public safety ensuring. However, IoT applications for smart cities is still a major issue as it lacks assuring privacy and security within provided services. In this chapter, the authors pinpoint IoT's security risk assessment challenges and examine its critical influence on smart cities. Additionally, they highlight the key aspects characterizing a smart city which also represent the critical assets requiring security risk assessment. Moreover, they discuss the resulting issues and their related countermeasures.

DOI: 10.4018/978-1-7998-5062-5.ch010

INTRODUCTION

The Internet of Things (IoT) is the latest paradigm spanning various technological fields, from data discovery and processing to networking, device connectivity, and data analysis (Losavio, et al., 2018). It is used in many applications ranging from home security and factory automation to smart living and health care. The IoT offers many benefits and applications to intelligent cities, also called smart cities, including:

- Optimizing city's standard services
- Improving situational awareness
- Detection and correction of faults
- Management of infrastructure
- Better work practices
- Monitoring air quality
- Managing city resources
- Energy monitoring and saving
- Waste management
- Better communication capabilities
- Preventing traffic congestion.

The IoT represents a large set of diverse smart 'things' that are equipped with sensors and processing capabilities, connected via the Internet. The number of devices connected to the Internet already exceeds the number of people on earth and is expected to reach 50 to 200 billion devices by 2020 (Bujari and Furini, 2018). Connected devices generate and transmit huge amounts of data that allow citizens to better understand the state and evolution of their cities (Ismagilova, et al., 2019). Additionally, the collected data allows generation of significant insights which later lead to better decision making. Indeed, intelligent transport, smart energy management systems and digital environmental monitoring are all examples of the IoT applications for intelligent cities, also known as digital cities. Furthermore, relying on the IoT tolerates a smart management of the cities' services. Thus, making tomorrow's cities smarter and more sustainable by accommodating other devices able to be shared and scalable (Harmon, et al., 2015). Indeed, by conceiving smart cities, citizens are constructively engaged and influence the city management (Cardullo, Paolo, et al., 2019).

As portrayed in Figure 1, an intelligent city designates the incorporation of compound systems within its urban infrastructure allowing smarter living, architecture, mobility, health, environment and government (Ahvenniemi and Huovila, 2017). For instance:

- The ability to communicate using a smartphone via video in order to allow doctors to check on their patients at home more frequently. Indeed, the patient's data can be put to use as the benefits can range from an increased use of digital medical monitoring to genomic sequencing to obtain completely personalized treatments.
- Intelligent waste collection management systems that make it is possible to know the level of waste in real time. Thus, making it possible to optimize the collection routes and reduce the number of kilometers traveled for collection.

- Better control of energy expenditure where smart cities can also have real-time control of their energy consumption (using smart energy meters) and thus achieve positive savings both for their finances and for the environment.
- Direct and flexible communication between citizens and their community via technologies such as social media and Web 2.0.
- An improved public transport network allowing better management of urban traffic and optimizing the real-time flow of infrastructure.

Figure 1. The benefits of IoT applications for intelligent cities

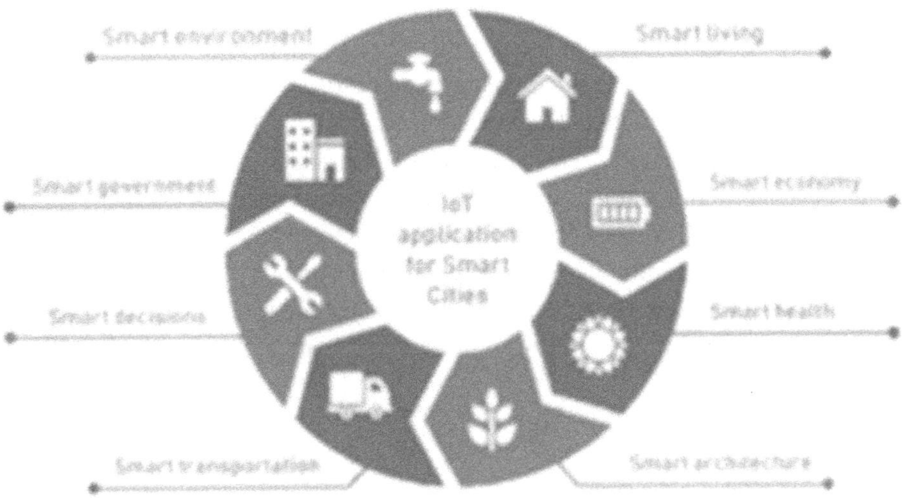

However, considering the rapid growth of cities in scope and capabilities, it consequently places the related infrastructures and exchanged data at a huge strain (Braun and Trevor, et al., 2018). In fact, the smart city technologies are relatively young which gives rise to bugs and security holes that attackers can easily target. Technical limitations can also prevent security. For example, some connected objects are designed to consume as few resources as possible (Soyata, 2019). Therefore, they are not powerful enough to support encryption technology that would secure their communication and interaction. Moreover, a smart city features a large set of devices sharing data which undoubtedly need to be confidential and secured (Singh, et al., 2019). Different risk scenarios can suddenly surface as data is compromised which can result in unacceptable situations, for example road congestion possibilities, disruption in supply of water or energy, etc. (Elmaghraby, et al., 2014). Actually, it is important for cities to become intelligent only if they are ready to face vulnerabilities. Among the most likely risks, we can have:

- Denial of service on critical gates,
- Intrusion on vital IT systems,
- Illegal management of certain devices,
- Corruption of collection and analysis of sensitive data,

- Illegal use of an urban service;
- Dissemination of false information by taking control of the communication server,
- Mass leak of confidential data.

These risks have already taken strike as back in 2017, when hackers manage to comprise Dallas warning system, causing panic to citizens by triggering 156 emergency sirens (Tweneboah-Koduah, et al., 2017). The same year, hackers used a ransomware to lock all the door of a hotel in the Alps, trapping all guests into their rooms and demanding ransom for freeing them.

Finding an appropriate response to these risks is a crucial issue (Yaqoob and Ibrar, 2017). In fact, IoT security issues limit its widespread adoption within smart cities. Security constraints are swiftly heightened due to today's modern and ubiquitous nature of intelligent cities. Moreover, the deployment of new technologies at an unprecedented pace gives the attackers a major advantage; and thus requiring urgent need to put in place a coherent and sustainable Security Risk Assessment strategy. An efficient strategy is the key to leveraging IoT applications for smart cities.

This chapter reviews the key findings from existing research on the Security Risk Assessment challenges related to intelligent cities from an IoT perspective. Its organization is as follows: The next section provides the basic background of our research work; The following section details IoT technologies and opportunities for intelligent cities. The section after this articulates the Security Risk Assessment challenges followed by a last section introducing an experimental study of Morocco's smart city Casablanca. Finally, we summarize findings and provide research ideas for the future.

In this contribution, we will use the following terms interchangeably: intelligent cities, smart cities, and digital cities.

INTELLIGENT CITIES OVERVIEW

From Traditional Cities to Intelligent Cities

As interest towards innovative urbanization and optimized performance grows intensively, the concept of an intelligent (or smart) city is widely promoted. Such a city aims at developing the citizens' urban environment by providing high standard of Quality of Services (QoS) through the use of innovative distributed computing technologies. An intelligent city is conceived as a modern and innovative digital city using Information and Communication Technologies (ICT) in order to improve the quality of life and everyday operations' sustainability and availability (Bifulco, Francesco, et al. 2016). Indeed, ICT and sensor technologies have a key role to play in building intelligent cities and smart spaces. Smart urban infrastructure incorporates a variety of factors: ICT, the IoT, public-private partnerships, social and human capital. The main goal of an intelligent city is to accommodate the optimum balance between demand and supply within its different services including health, transportation, energy, education, business, leisure, etc. (Batty, MIchael, et al., 2012). Nevertheless, considering the current emergence of smart devices connecting them via the existing networks is highly recommended. Indeed, the IoT is the key integration of different technological advancements as it concerns Wireless Sensor Networks (WSN), Ubiquitous Computing (UC), Machine Learning, and Machine-to-Machine (M2M) communication (Silva, et al., 2018). In fact, a Smart City is fundamentally an application of the IoT which encompasses its key operations: data collection, transmission, storage, processing and management.

The IoT allows connecting the city's various devices to the Internet for effective monitoring and management of daily activities. In the area of urban development, ICTs and IoT are important building blocks for creating a smart city to administer, serve and support the ever-growing urban population (Zavratnik, et al. 2018). IoT application for smart cities enhances the services and capabilities, such as the following:

- Transport, traffic, parking and mobility,
- Communication, interaction
- Services, security, privacy
- Control and maintenance of public spaces,
- Buildings and workplaces,
- Administrative transactions in the city,
- Buildings performance,
- Energy consumption,
- Utilities, transportation
- Waste management.

The beneficiaries of these mechanisms are consumers, citizens, visitors, government institutions and private companies. In addition, the different data collected by connected devices allow citizens to better understand the state, available services and evolution of the city (Hancke, et al., 2013). Indeed, intelligent transportation, intelligent energy management and smart environmental monitoring are all examples of applications of the IoT for intelligent cities (Angelidou, 2014).

Intelligent Cities Concept

The need of intelligent (or smart) cities matches the needs of any IoT system which are as follows:

- Connectivity and Telecommunications: for example, in a local network within a geographical proximity, the ability to monitor street lights of a same street via Bluetooth), or in a global network, to integrate data from social networks to manage traffic flows.
- Analysis and Decision: this refers to acting semantically on the received data for an alarm center and the use of Machine Learning predictive approaches in order to obtain accurate data.
- Temporal need: as the pervasiveness features are a requirement, there is a need to monitor real-time data. For example, managing the temperature of a device, handling a massive data storage and supervising a user behavior.

More and more cities around the world are becoming smart cities. These cities integrate sensors and various smart devices through a network, usually the Internet, in order to optimize the functioning of the city and the efficiency of services while improving their interface with citizens directly. The data collected is analyzed to manage many municipal assets such as power plants, water and sewer systems, traffic and transportation systems and many other community services. In general, IoT based technologies helps to enhance functionality of smart cities in a number of ways, including the following:

- *Data Acquisition*: Data collection and transfer is independent of the devices where operations are conducted.

- *Visualization and control*: Systematic control is applied through an understanding of the complex services semantic.
- *Analysis and reporting*: Real-time calculations and estimation are performed on the archived databases using advanced analysis, artificial intelligence and machine learning.
- *Design*: The use of automation solutions in order to facilitate different projects of the city.
- *Data management*: This comparing and organizing data in real-time in order to make efficient decisions.
- *Traffic supervision*: Dynamic and intelligent solutions are used in order to handle mobility and maintain sensible timing and energy.

The potential of smart cities is virtually unlimited and their growth is anticipated to speed up greatly in the coming years.

IOT AS ENABLER FOR INTELLIGENT CITIES

The ideal concept of a smart city has become possible because of the IoT paradigm and IoT enabled technologies (such as wireless networks, connectivity protocols, Cloud computing, etc.) which halp to provide a number of diverse services in sectors such as transport, energy, waste, safety, health, health, commerce, etc. (Maksimovic, 2018). From the quality of air in the atmosphere to the flow of traffic on the roads to the connectivity between people via social media, and through the management of energy and waste, the Smart City features cover all the government-related and citizen-related operations of a city. However, it fundamentally requires further understanding of the opportunities and technologies that the latest IoT technologies offer in order to be manage all these efficiently, safely, and securely.

Promising Opportunities

Considering the growing advancement of technologies, smart cities extend the IoT vision in order to adopt all its capabilities. In fact, smart cities rely on the IoT in order to comprehensively control, monitor and manage their utilities and services. IoT paradigm introduces a high interoperability and scalability in dealing with the city's urban development. Accordingly, IoT-enabled smart cities offer a modern digital era and improve public safety. IoT applications for smart cities enhance effectively our everyday activities by significantly improving today's quality of life. It allows cities to better handle traffic congestion, to constructively improve decision making and to better manage and use the existing infrastructures (Fortino, et al., 2014). Indeed, smart cities rely on the IoT in order to efficiently tackle inefficiencies which further allows creating sustainable communities (Mohanty, et al., 2016). IoT applications for smart cities advocate incorporating sensors within city's assets. adoptability is classified according to the type of the network, scalability, flexibility, heterogeneity, and the citizen involvement.

IoT applications for smart cities are significantly diverse. However, these applications must follow a common is implementation approach as these identify the basic foundation for tomorrow's cities much improved and safer operations. Therefore, a basic smart city platform would acknowledge the state of required change; and then upgrade its existing architecture and tools rather than wasting time in rebuilding – emphasis is on reuse. Table 1 depicts the basic fundamentals of a generic smart city platform for intelligent cities.

Table 1. IoT enabled generic smart city platform

The plateform fundamentals	Main artefacts
Scalability	In order to be able to scale, a basic architecture is founded. It would highlight the imminent enhancements and designate the allowed changes. The architecture comprises four main artefacts: · Smart city as an IoT system, · Gateways connecting the IoT system to the network devices and the cloud domain, · Data Lake storing the transmitted data between the different artefacts, · Big Data in order to extract a sufficient output from the Data Storage.
Semantic analysis	Considering the Big Data added value of data analytics, a devices monitoring is achieved. In fact, data is semantically visualized and analysed.
Deep analytics	Thanks to Big Data use and the semantic analysis, an inclusion of Deep Learning benefits is possible. In fact, it leads to an identification of hidden patterns and correlations within IoT sensors which furtherly allow conceiving predictive models. Predictive models allow setting rules to the IoT actuators.
Smart control	Considering the inclusion of the Deep Learning advanced tools, a smart control can significantly improve IoT devices automation. Indeed, as IoT actuators become smart, rules are swiftly applied.
Dynamic interaction	Regardless of the automated control, a user's action on IoT devices is mandatory. The platform grants the user to control and monitor the devices.
Connectivity	IoT applications for smart cities focus on achieving a smart platform able to endorse new services, tools and sensors.

Smart city platform, as outlined in Table 1, leverages communication within and between cities. Deep Learning and Big Data are the drivers behind the smart cities' initiative. Around the world, to emphasis their infrastructure, cities are enriched by real digital architectures composed of sensors (to detect presence, movement, temperature, pollution, etc.) and computer networks (Serrano and Bajo, 2019). The result is generation of huge amount of data where Deep Learning methods are used in order to analyze, predict and then help decision-making, especially within the security context. Since deep learning offers different algorithms, it makes it possible to analyze abnormal or deviant behavior compared to a normal situation. Alerts are sounded as required within the real-time execution and procedures are launched in order to keep the resources safe and available. Furthermore, Big Data contributes highly to the sustainable development, especially in the context of smart cities (Cronemberger, Felippe, and Gil-Garcia, 2019). Big Data allows collecting data in order to identify needs and accordingly satisfy the cities' citizens with increased adaptability. Big Data analysis can provide information to help reduce pollution. Moreover, by incorporating sensors in smart roads, it is possible to measure total traffic at different times of the day, as well as to ward drivers of possible delays etc. ahead. This data can be sent to a center responsible for coordinating regulatory officers. Traffic can thus also be efficiently managed to reduce carbon dioxide emissions. By incorporating sensors into cars, it is possible to guide vehicles to the nearest parking stations. Thus, by combining Big Data and Deep Learning, predictive analyses can be further improved (Mohammadi, 2018). This leads consequently to predict the different risk scenarios based on historical and geographical data. Subsequently, data is a powerful tool as devices' sensors the cities' infrastructures are linked and can easily interact via an intelligent network such as Internet in order to better manage production and consumption flows.

Innovative Technologies

Intelligent Cities and more generally the IoT are based on a large set of diverse technologies. As detailed in Table 2, IoT related technologies for smart cities include mainly the RFID identifiers and sensors embedded within the smart devices within the wireless networks.

Table 2. IoT-based main technologies for smart cities

Technology	Specification	Description
Smart sensing	Radio-Frequency Identification (RFID)	Allows to automatically identify and track tags related to devices. It transmits data from the tag to a reader which is eventually transmitted to an RFID computer program in order to be incorporated in the network.
Smart connectivity	Wireless Sensor Networks (WSN)	Encompasses dedicated sensors for monitoring and tracking physical requirements of the network. It allows mainly measuring environmental conditions. Often combined with RFIDs in order to track devices' changes.
	Addressing	Internet addressing allows different nodes to communicate over the Internet. Accordingly, IoT allows uniquely addressing the large scale of a smart city devices via the Internet.
Smart computing	Middleware	Middleware application main objective is to decrease devices' heterogeneity, constrained storage and restricted performance. It ensures a suitable interconnection of the interconnected devices to the application layer.

The IoT is the most important basic infrastructure for smart cities as it optimizes the access to the public resources and grants transparency within its devices' management. Relying on the IoT is the means to achieving communication and collaboration between every IoT based artefacts. As a result, it embraces a digital equity, induces a vigorous knowledge with a consideration and an inclusion of the user experience. As a result, the main pillars emerging from the IoT application for smart cities introduce the following:

- An interconnexion of devices where physical and virtual nodes are integrated into the global networks.
- An interconnexion of Services combining various applications into compatible services and using Big Data and Deep Learning as tools to understand and process related data.
- An interconnexion of People using, connecting and interacting with the devices in order to transmit and exchange data, and monitor operations.

SECURITY RISK ASSESSMENT

Current Trade Challenges

Equipped with sensors to collect data, relying on the cloud to optimize services, and favoring a growing use of digital entities, smart cities are becoming increasingly connected. In order to guarantee security and privacy, a smart city must encompass different constraints upon data management and data exploitation. As a matter of a fact, the related security risks are rapidly increasing. Thus, it is paramount to preserve the city's functioning and its data private and confidential. In fact, although there are the above-mentioned advantages, the IoT encompasses a myriad of security risks. Securing IoT applications for smart cities is a major challenge as it highly affects the sustainability and confidentiality of our everyday life activities. Conducting Security Risk Assessment relies largely on assessing risks and deciding the appropriate countermeasures to withstand their impacts. This section highlights how a threat may exploit IoT devices related vulnerabilities. Li et al. (2017) state that smart cities' security and trustworthiness is its weakest link. In fact, attackers often seek to gain advantage by affecting the reliability of the IoT based services by causing disruption to the interconnected devices. Security Risk Assessment determines the suitable countermeasures for mitigating risks occurrence (Abbass, 2019). It focuses on 4 main concepts as detailed in Table 3.

Table 3. Security Risk Assessment main concepts

Security risk assessment concepts	General description	IoT designation
Assets	Resources that are considered critical and of greater value requiring imminent protection.	Designate the sensors related data such as temperature or humidity, also security cameras, smartwatches Smart TVs, and every device to the Internet.
Threats	Represent the potential events that would cause negative impacts.	Interception of communication, compromising sensitive data, third party intrusion, forced authentication and extraction of confidential credentials.
Vulnerabilities	Flaws and weaknesses likely to be exploited by threats.	Weak authentication, unencrypted communications between devices and web user interfaces susceptible to Cross-Site-Scripting (XSS) and SQL injections.
Impacts	Consequences of a successful attack.	Installing botnets within the devices, Eavesdropping and confidentiality integrity and availability compromise.

In an intelligent city, security risks are increased due to the continuous interconnection between the connected devices which are the pillars of the innovative services offered by the city. In fact, these services are carried by separate but interconnected devices in order to connect traffic lights, or energy distribution, etc. Furthermore, sharing confidential data within the interconnected objects leads consequently to serious consequences which impact consequently business activities and reputation. In fact, different risk scenarios exist; the most common ones consider scanning the device's IP address in order to discover open ports. In case of a successful attempt, it further cracks authentication' credentials, implements the malware and further ties to infect the other peers. Indeed, the main risks related to the smart city security are due to the following:

- The large scope of vulnerabilities considering the huge number of citizens' smartphones, cars, homes and other interconnected objects,
- The myriad of threats due to complexity and heterogeneity of interconnected objects,
- The low maturity level of connected objects security,
- The unsuitable wireless networks including no authentication or encryption and therefore sensitive to attacks,
- The laborious data control due to use of the Cloud,
- The deficiency of standard and reference architecture between different subsystems, which makes the development of security solutions very difficult,
- The lack of strong government regulation to take into account the Security Risk assessment issue.

Moreover, as nearly every city device is connected to the Internet, it rises its odds of being attacked and consequently influencing and spreading the attack to its peered devices. Also, due to near permanent Internet connection, the attack can be remotely launched using either Zigbee protocols or standard TCP/IP. Additionally, considering the devices' low security levels and use of default generic passwords, attackers can easily launch successful attacks. Accordingly, back in 2016, a cyber-attack has launched DDoS attacks as IoT appliances were connected to a botnet (De Donno, et al., 2018). As an example: the malware Mirai breaking unsecure authentication of IoT devices and creating links to a botnet (Antonakakis, 2017) and Persirai botnet targeting IP cameras by gaining access via TCP Port (Martin, et al., 2017). As displayed in Figure 2, risks are always inherent as these can be the issue of a citizen's behavior resulting subsequently in intentional or accidental threats.

Figure 2. The landscape of Security Risk Assessment within a smart city

Smart cities are emerging into a highly complex and interdependent network of devices and users. Therefore, a Security Risk Assessment strategy should not be limited only to a single device but should embody end-to-end security countermeasures. It must be paradoxical by focusing on data integrity and security while supporting the infrastructure's dynamic nature. The strategy should primarily embody innovation engineering capable of supporting smart cities in their action and development (Baig, 2017). Nevertheless, due to the fact that a smart city is perceived as a public entity incorporating heterogenous devices, attention should be given to the different required levels of security (Vitunskaite, 2019).

Table 4 portrays the most prominent risks targeting IoT based application for intelligent cities with their related mitigation countermeasures.

Table 4. IoT most prominent security risks and their related countermeasures

IoT technology	Security risks	Mitigation countermeasures
Smart sensing	Abuse of RFID tags, jamming, DoS, spoofing, buffer overflows and worms injected into the RFID system in order to harm its operation.	Selective blocking, cryptography and encryption, RFID tags' management, data coding and transmission supervision.
Smart connectivity	DoS, DDoS, Bandwidth deprivation, eavesdropping, Software attacks, mishandling of sensitive data.	Using PKI, cryptography and encryption, digital signatures and secure routing protocols for WSNs Network.
Addressing	Man in the Middle, side-channel attacks, spoofing, TCP session hijacking and Ad hoc security issues.	IEEE standard solutions
Smart computing	Malicious applications, botnets, social engineering threats, spywares, malwares.	Firewalls, secure APIs, Proxy, IDS/IPS.

Smart cities are increasingly dependent on sensors for security which is insufficient. While industry needs to improve their systems, community leaders need to be more involved in security risk assessment. Without securing a city's different objects, the global security can easily be damaged (Lam, et al., 2019). In fact, one single risk can drastically disrupt the operation of an entire city.

Today, since 80% of the population live in cities and a massive urbanization process is underway worldwide. Accordingly, the aim is to increase the resilience level. Security Risk Assessment needs to be more effective upstream (to thwart threats) and downstream (for optimal management). Indeed, it is ultimately to induce balance between human actions and the innovative technology to a city's services (Caragliu, Andrea, et al., 2019).

Suggested Solutions

Driven by increasing urbanization and fueled by the IoT, intelligent cities are poised to grow exponentially. Actually, smart cities aim at improving public safety through a better data sharing and collaboration between its different actors within its urban areas. In the years to come, more than 600 cities are getting ready to become smart. The trend is international and is driven by the desire to improve quality of life. Moreover, in order to fully exploit their potential, smart cities must actively involve their citizens. Initiatives to make cities smarter must aim to encourage data sharing with and within citizens as much as possible, in order to integrate it into the decision-making process. These initiatives must seek to impose technologies and innovations on citizens in order to solve their daily activities. By effectively organizing, viewing and sharing data, cities management understand the needs and concerns of their citizens much better, and thus develop policies that precisely match their expectations.

The transformation of cities into smart cities has led to a complete transformation of information systems that support this vision. For a very long time, it remained very centralized and based on internal services. Technological innovations have led to the creation of open and interconnected autonomous systems. This openness, with the distribution of the data that it generates, multiplies the potential security flaws, especially since many of these subsystems are outsourced to partners and therefore very difficult to control. Finally, the growing technological complexity of these subsystems, due to the artificial intelligence component that they sometimes integrate with, and their interdependencies at all levels (physical, economic, institutional and social), may further increase their vulnerabilities to attacks. The safety of the communities in which the population live, work and grow is an essential driver of a smart city. Therefore, it is essential to do the following:

- Collaborate to improve overall security: With the emergence of IoT, the security systems modernization and their switchover over IP connectivity between all stakeholders in urban life, should extend through reliable sensors and systems. By creating secure gateways, collaborating and sharing data and resources, all actors should increase their resilience to feared events.
- Deploy dedicated tools preserving data confidentiality and privacy: Protecting citizens sensitive data should be a prime priority. A good approach would be to embed data protection aspects from the design stage, which would incorporate privacy protection into the relevant processes, procedures, and activities from the beginning, rather than to address security and privacy issues at a later stage (Li, Zhen, et al., 2018).
- Educate and engage the Community: In order for a smart city to successfully exist and thrive, it is required to involve citizens to actively engage with the novel technologies. Therefore, every project should include an educational process that would instruct and inform citizens with relevant guidelines. However, citizens must be motivated in order to better use the proposed technologies to avail the relevant services provided by the cities (Kumar, et al., 2018).
- Tighten encryption mechanisms: This requires using testbed models in order to better deal with cyber-security schemes. Indeed, major constraints should be applied in order to verify identities and guarantee secure access to the valuable data; for example, an RFID tag identification algorithm guarantying the accuracy of different RFID tags simultaneously and lowering the devices computational power (Butt and Talal, et al., 2019).
- Plan for resiliency of projects: it is essential to guarantee the survival of the city in the event of a major disaster affecting its various services. The aim is to restart a paused or broken activity as quickly as possible with the minimum loss of data and time.

The main goal of Security Risk Assessment solutions is to principally guarantee data availability, integrity, correctness, privacy, trust and confidentiality at all levels.

AN EXPERIMENTAL STUDY: CASABLANCA SMART CITY

Presently, urbanism is making its digital transformation by putting IoT based connected devices and information systems to develop more efficient and smarter cities. This digitization of the urban space has a real impact on the development of territories, as promised within and by the City of Casablanca in Morocco. Indeed, as part of the Moroccan initiative for the development of its cities, the objective was to start with enhancing Casablanca's public space around four main goals and axes as described in the following sub-sections.

Making Casablanca a Smart or Intelligent City is the objective agreed by the management of the metropolis since the hosting of the second edition of the Smart City Expo for a transformation in the management of the city and well-being of its citizens. The main pillars of the project are to implement a Smart Grid, an urban creativity and an intelligent mobility. The proposed approach is based upon and implemented through the city's information systems, which is the key to support all the additional and value-added services. This approach has the merit of facilitating risk analyses and identifying solutions to improve the security of the city of Casablanca.

The economical capital of Morocco has been chosen by the Smart City Expo World Congress, an organization supported and approved by the UN and the World Bank, to be the host city and representative of the congress in Africa. To gain the qualification of smart city, the white city launched a new challenge by wanting to become the African model of the smart city. the idea is to develop means and solutions capable of meeting the needs of the population while preserving resources and the environment. The main objective is to adopt a global digital strategy bringing intelligence, performance and sustainability to its public, as well as social and environmental services. Indeed, the digital transformation of the economy, society and public services to anchor Morocco into a global movement of digital transformation is an ambition set to the horizon of 2020.

In the context of globalization, social tensions, economic crises and post-industrial transitions, Morocco is looking for new urban models capable of meeting its cities' expectations, hence the value of flexibility of the business services management within the world of uncertainty.

Main Goals

In 2013, Morocco took the initiative to implement a numeric strategy for developing a smart city. It put forward the initiatives at national and local level for a cluster initiative e-Madina, Casablanca smart city. The initiative is a lever for economic, social and environmental progress which constitute a first step towards the transformation of the city of Casablanca, first for a better quality of life, and secondly for the renewal of its activities and attractiveness, and mainly for relocating the city on an international scale to put on a map among the world's major cities.

The e-Madina numeric strategy focuses mainly on:

- Social improvement,
- Energy management for public buildings,
- Street lighting management,
- Implementing public services by advocating the digital citizenship,
- Traffic flows safety and improvement,
- Easy integration into existing systems and infrastructures,

- Advanced reporting and predictive analytics,
- Scalability for projects and systems,
- Flexibility with new operations,
- Secured access to information,
- Management of urban waste collection,
- Management of green spaces,
- Forest fire detection and avoidance,
- Digitalizing small and medium-sized organization.

In fact, e-Madina main goal is to make the city of Casablanca more attractive, more efficient and more competitive for organizations, citizens and tourists. It promotes developing a smart city ecosystem enhancing the city's initiatives through the use of the digital latest technologies. As depicted in Figure 3, e-Madina supports a 4P model focusing on Public, Private, People and Partnership which takes into account all the actors of the urban ecosystem.

Figure 3. Casablanca smart city cluster

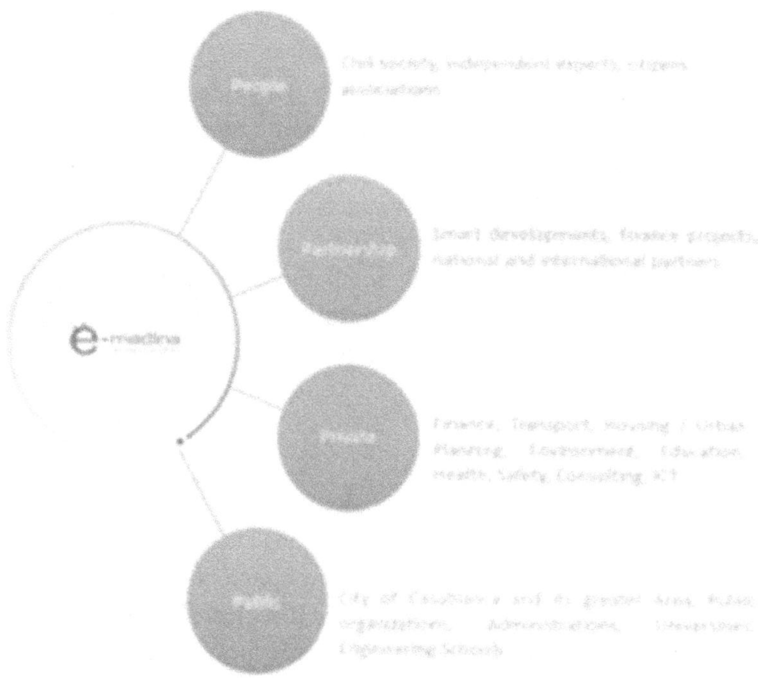

Currently, a number of different projects have already been implemented, for instance:

- A Social Sustainable Solar smart city,
- A virtual museum of the city,
- Development of urban Video Surveillance.

The Casablanca example indicates that medium-sized cities have the opportunity to develop and transform, incrementally. This is because these cities can use ICTs and other distributed technologies to contribute to addressing the 'soft' domains of education, culture, social cohesion, participatory governance and welfare, as well.

Governance Model

The governance of Casablanca lies in the hands of the Government. In fact, its model enables the city to communicate effectively with its citizens in order to deliver the public services in an optimal way. This communication is based on websites, mobile applications and digital communication platforms of all kinds, with the main goal of bringing the administration closer to the end user, streamlining the exchanges, shortening the waiting times and improving the social, commercial and communication environment for entrepreneurship, research and investment.

Inspired by feedback from smart cities in developed countries, the strategy proposes a bottom-up approach based on the use of already existing infrastructures and technologies such as smartphones, other mobile terminals, and Internet to develop applications and services. This would help to formulate the pillars of a smart city that will reinforce the other achievements to create an interconnected smart city environment.

The need for effective connectivity grants for the citizens in order to enable them with appropriate data and available services. The model's main features include:

- A social innovation meeting the needs of citizens,
- An improvement of the well-being of citizens as well as implications for the development of the city,
- A reduction of deployment and maintenance costs by adopting innovative economic devices and models,
- Sustainable approaches for natural resources exploitation,
- An evolutionary digital transformation of society and the economy,
- An exploitation of the ubiquitous mobile networks,
- Adoption of intelligent models: crowdsourcing and crowdfunding

Furthermore, the governance model major goal is to improve the living conditions of the citizens, strengthen mobility, promote economic attractiveness and advance the business climate to make it a hub nationally, continentally and internationally. Accordingly, the governance model aims at providing the before-mentioned features at four main bases:

- *Transport*: To provide additional and sufficient means to serve most of the major neighborhoods of the metropolis known for their high density, as well as the implementation of novel road infrastructure. Latest IoT based technologies and novel infrastructures can include video surveillance, smart lightning and smart ecological means.
- *Social*: To improve living conditions of residents specifically people with special needs. The goal is to be able to offer services that meet their special needs.

- *Economical*: To promote the economic attractiveness of Casablanca through the restructuring of the economic activity areas, the development of new industrial and commercial districts and further improvement of the business environment.
- *Entertainment*: This base aims to anchor the region's positioning as a national and international shopping and leisure destination. The goal is to better manage the public space by promoting sports complex, rehabilitation of green spaces, as well as enhancing the urban and architectural heritage of the city of Casablanca.

The e-Madina initiative is setting up indicators that maintain Casablanca's activities under effective monitoring.

CONCLUSION

The IoT is an attractive paradigm and smart city development is an application of the IoT. In fact, it grants a city a swift management of its infrastructure and public services, as the resulting applications allow constructively optimizing the use of resources and the infrastructure installations and the energy costs. IoT enable cities responding to the citizens demands in a more significant, secure and fast manner.

Smart cities are emerging innovation, not only for the government but also for the industry as well as for the research domain. However, notwithstanding the numerous benefits of IoT and intelligent cities vision, there are also security concerns. As part of a Smart City initiative, the consequences and impact of a security breach can be severe. Smart Cities Security needs to be a collaborative project involving many partners - sensor and actuator manufacturers, gateway providers, standards organizations, security software, security experts, city managers, and even Smart City infrastructure operating system developers. All of these actors are an essential part of a complete, vital and connected IoT security ecosystem.

Moving towards smart cities development, we face a diverse set of technological, social and ethical challenges. Securing the applications, based on the IoT for smart cities, is a major challenge as it highly affects its common sustainability, privacy and confidentiality. In fact, the use of traditional security risk assessment models are unsuitable due to the heterogenous nature of networked devices and the different interconnected infrastructures with varying communication protocols. This chapter has reviewed the key findings from existing research on the Security Risk Assessment challenges related to smart cities from an IoT perspective; a case study of the Casablanca intelligent city is also presented. The main goal has been to assess IoT related security risks within the development and management of intelligent cities, which is a new vision of digital urbanization.

REFERENCES

Abbass, W., Bakraouy, Z., Baina, A., & Bellafkih, M. (2019). Assessing the Internet of Things Security Risks. *Journal of Communication*, *14*(10), 958–964. doi:10.12720/jcm.14.10.958-964

Ahvenniemi, H., Huovila, A., Pinto-Seppä, I., & Airaksinen, M. (2017). What Are the Differences between Sustainable and Smart Cities? *Cities (London, England)*, *60*, 234–245. doi:10.1016/j.cities.2016.09.009

Angelidou, M. (2014). Smart City Policies: A Spatial Approach. *Cities (London, England), 41*, S3–S11. doi:10.1016/j.cities.2014.06.007

Antonakakis, M. (2017). Understanding the Mirai Botnet. *26th USENIX Security Symposium*, 1093–1110.

Baig, Z. A., Szewczyk, P., Valli, C., Rabadia, P., Hannay, P., Chernyshev, M., ... Peacock, M. (2017). Future Challenges for Smart Cities: Cyber-Security and Digital Forensics. *Digital Investigation, 22*, 3–13. doi:10.1016/j.diin.2017.06.015

Batty, M., Axhausen, K. W., Giannotti, F., Pozdnoukhov, A., Bazzani, A., Wachowicz, M., ... Portugali, Y. (2012). Smart Cities of the Future. *The European Physical Journal. Special Topics, 214*(1), 481–518. doi:10.1140/epjst/e2012-01703-3

Bifulco, F., Tregua, M., Amitrano, C. C., & D'Auria, A. (2016). ICT and Sustainability in Smart Cities Management. *International Journal of Public Sector Management, 29*(2), 132–147. doi:10.1108/IJPSM-07-2015-0132

Braun, T., Fung, B. C. M., Iqbal, F., & Shah, B. (2018). Security and Privacy Challenges in Smart Cities. *Sustainable Cities and Society, 39*, 499–507. doi:10.1016/j.scs.2018.02.039

Bujari, A., Furini, M., Mandreoli, F., Martoglia, R., Montangero, M., & Ronzani, D. (2018). Standards, Security and Business Models: Key Challenges for the IoT Scenario. *Mobile Networks and Applications, 23*(1), 147–154. doi:10.100711036-017-0835-8

Butt, T. A., & Afzaal, M. (2019). Security and Privacy in Smart Cities: Issues and Current Solutions. In A. Al-Masri & K. Curran (Eds.), *Smart Technologies and Innovation for a Sustainable Future* (pp. 317–323). Cham: Springer International Publishing. doi:10.1007/978-3-030-01659-3_37

Caragliu, A., & Del Bo, C. F. (2019). Smart Innovative Cities: The Impact of Smart City Policies on Urban Innovation. *Technological Forecasting and Social Change, 142*, 373–383. doi:10.1016/j.techfore.2018.07.022

Cardullo, P., & Kitchin, R. (2019). Being a 'Citizen'in the Smart City: Up and down the Scaffold of Smart Citizen Participation in Dublin, Ireland. *GeoJournal, 84*(1), 1–13. doi:10.100710708-018-9845-8

Cronemberger, F., & Ramon Gil-Garcia, J. (2019). Big Data and Analytics as Strategies to Generate Public Value in Smart Cities: Proposing an Integrative Framework. In Setting Foundations for the Creation of Public Value in Smart Cities. Springer.

De Donno, M., Dragoni, N., Giaretta, A., & Spognardi, A. (2018). DDoS- Capable IoT Malwares: Comparative Analysis and Mirai Investigation. *Security and Communication Networks, 2018*, 2018. doi:10.1155/2018/7178164

Elmaghraby, A. S., & Losavio, M. M. (2014). Cyber Security Challenges in Smart Cities: Safety, Security and Privacy. *Journal of Advanced Research, 5*(4), 491–497. doi:10.1016/j.jare.2014.02.006 PMID:25685517

Fortino, G., & Trunfio, P. (2014). *Internet of Things Based on Smart Objects: Technology, Middleware and Applications*. Springer. doi:10.1007/978-3-319-00491-4

Hancke, G. P., & Hancke, G. P. Jr. (2013). The Role of Advanced Sensing in Smart Cities. *Sensors (Basel), 13*(1), 393–425. doi:10.3390130100393 PMID:23271603

Harmon, R. R., Castro-Leon, E. G., & Bhide, S. (2015). Smart Cities and the Internet of Things. *Portland International Conference on Management of Engineering and Technology (PICMET)*, 485–94. 10.1109/PICMET.2015.7273174

Ismagilova, E., Hughes, L., Dwivedi, Y. K., & Ravi Raman, K. (2019). Smart Cities: Advances in Research—An Information Systems Perspective. *International Journal of Information Management, 47*, 88–100. doi:10.1016/j.ijinfomgt.2019.01.004

Kumar, & Singh, Gupta, & Madaan. (2018, Apr.). Moving towards Smart Cities: Solutions That Lead to the Smart City Transformation Framework. *Technological Forecasting and Social Change.*

Lam & Ma. (2019). Potential Pitfalls in the Development of Smart Cities and Mitigation Measures: An Exploratory Study. *Cities, 91*, 146–56.

Li, W., Song, H., & Zeng, F. (2017). Policy-Based Secure and Trustworthy Sensing for Internet of Things in Smart Cities. *IEEE Internet of Things Journal, 5*(2), 716–723. doi:10.1109/JIOT.2017.2720635

Li, Z., & Qi, L. (2018). Economic Solutions to Improve Cybersecurity of Governments and Smart Cities via Vulnerability Markets. *Government Information Quarterly, 35*(1), 151–160. doi:10.1016/j.giq.2017.10.006

Losavio, M. M., Chow, K. P., Koltay, A., & James, J. (2018). The Internet of Things and the Smart City: Legal Challenges with Digital Forensics, Privacy, and Security. *Security and Privacy, 1*(3), e23. doi:10.1002py2.23

Maksimovic, M. (2018). Greening the Future: Green Internet of Things (G-IoT) as a Key Technological Enabler of Sustainable Development. In *Internet of Things and Big Data Analytics toward Next-Generation Intelligence* (pp. 283–313). Springer. doi:10.1007/978-3-319-60435-0_12

Martin, V., Cao, Q., & Benson, T. (2017). Fending off IoT-Hunting Attacks at Home Networks. *Proceedings of the 2nd Workshop on Cloud-Assisted Networking*, 67–72. 10.1145/3155921.3160640

Mohammadi, M., & Al-Fuqaha, A. (2018). Enabling cognitive smart cities using big data and machine learning: Approaches and challenges. *IEEE Communications Magazine, 56*(2), 94–101. doi:10.1109/MCOM.2018.1700298

Mohanty, S. P., Choppali, U., & Kougianos, E. (2016). Everything You Wanted to Know about Smart Cities: The Internet of Things Is the Backbone. *IEEE Consumer Electronics Magazine, 5*(3), 60–70. doi:10.1109/MCE.2016.2556879

Serrano, E., & Bajo, J. (2019). Deep neural network architectures for social services diagnosis in smart cities. *Future Generation Computer Systems, 100*, 122–131. doi:10.1016/j.future.2019.05.034

Silva, B. N., Khan, M., & Han, K. (2018). Towards Sustainable Smart Cities: A Review of Trends, Architectures, Components, and Open Challenges in Smart Cities. *Sustainable Cities and Society, 38*(April), 697–713. doi:10.1016/j.scs.2018.01.053

Singh, Kumar, & Prakash. (2019). *Importance and Needs of IoT in Developing Smart Cities*. Ved.

Soyata, T., Habibzadeh, H., Ekenna, C., Nussbaum, B., & Lozano, J. (2019). Smart City in Crisis: Technology and Policy Concerns. *Sustainable Cities and Society*, *50*, 101566. doi:10.1016/j.scs.2019.101566

Tweneboah-Koduah, S., Skouby, K. E., & Tadayoni, R. (2017). Cyber Security Threats to IoT Applications and Service Domains. *Wireless Personal Communications*, *95*(1), 169–185. doi:10.100711277-017-4434-6

Vitunskaite, M., He, Y., Brandstetter, T., & Janicke, H. (2019). Smart Cities and Cyber Security: Are We There yet? A Comparative Study on the Role of Standards, Third Party Risk Management and Security Ownership. *Computers & Security*, *83*, 313–331. doi:10.1016/j.cose.2019.02.009

Yaqoob, I., Ahmed, E., Rehman, M. H., Ahmed, A. I. A., Al-garadi, M. A., Imran, M., & Guizani, M. (2017). The Rise of Ransomware and Emerging Security Challenges in the Internet of Things. *Computer Networks*, *129*, 444–458. doi:10.1016/j.comnet.2017.09.003

Zavratnik, V., Kos, A., & Duh, E. S. (2018). Smart Villages: Comprehensive Review of Initiatives and Practices. *Sustainability*, *10*(7), 2559. doi:10.3390u10072559

ADDITIONAL READING

Ahmed, S. H., & Rani, S. (2018). A Hybrid Approach, Smart Street Use Case and Future Aspects for Internet of Things in Smart Cities. *Future Generation Computer Systems*, *79*, 941–951. doi:10.1016/j.future.2017.08.054

Cunningham, T. (2015), A Cyber-Threat Intelligence Program – How to develop one and why it matters (Unpublished doctoral dissertation). Luleå University of Technology

Eom, J. (2014). Roles and Responsibilities of Cyber Intelligence for Cyber Operations in Cyberspace. *International Journal of Software Engineering and Its Applications*, *8*(9), 137–146.

ITU. (2012). Internet of thing global initiative, Recommendations ITU-T Y. 2060, June.

Komninos, N., Bratsas, C., Kakderi, C., & Tsarchopoulos, P. (2019). Smart City Ontologies: Improving the Effectiveness of Smart City Applications. *Journal of Smart Cities*, *1*(1), 31–46.

Köseli, M. (2015). Bilimsel Araştırma Yöntemlerinin İstihbarat Analizinde Kullanılması. *Journal of Security Strategies*, *21*(11), 87–117.

Maple, C. (2017). Security and privacy in the Internet of things. *Journal of Cyber Policy*, *2*(2), 155–184. doi:10.1080/23738871.2017.1366536

Miorandi, D., Sicari, S., De Pellegrini, F., & Clamatac, I. (2012). Internet of things: Vision, applications and research challenges. *Ad Hoc Networks*, *10*(7), 1497–1516. doi:10.1016/j.adhoc.2012.02.016

Mohbey, K. K. (2019). An Efficient Framework for Smart City Using Big Data Technologies and Internet of Things. In *Progress in Advanced Computing and Intelligent Engineering* (pp. 319–328). Springer. doi:10.1007/978-981-13-0224-4_29

Zanilla, A., Bui, N., Castellani, A., Vangelista, L., & Zorzi, M. (2014). Internet of things for smart cities. *IEEE Internet of Things Journal*, *1*(1), 22–32. doi:10.1109/JIOT.2014.2306328

KEY TERMS AND DEFINITIONS

Attacker: An entity, a person, or device that attempts to cause harm to an information system or results in a denial of existing service.

Authentication: The process or action of verifying the identity of a user or process, e.g. verification based on username and password credentials.

Big Data: This refers to extremely large data sets that are analyzed computationally to reveal patterns, trends, and associations, especially relating to human behavior and interactions.

Cloud: Remote server and distributed computing environment used to store data and provision computing related services as and when needed on a pay-as-you-go basis.

Deep Learning: An artificial intelligence function that imitates the workings of the human brain in processing data and creating patterns for use in decision making.

Denial of Service (DoS): An attack shutting down running of a service or network in order to render it inaccessible to its users (whether human person or a processing device).

Internet of Things (IoT): This refers to a system of inter-connected computing and smart devices, that are provided with unique identifiers and the ability to transfer data over a network without requiring human interaction.

Interoperability: Characteristic of a product or system, whose interfaces are completely understood, to work with other products or systems.

Malware: Software that is specifically designed to disrupt, damage, or gain unauthorized access to a computer system.

RFID Identifiers: Stands for radio frequency identification – a method for tracking goods by means of tags which transmit a radio signal.

Risk Assessment: A systematic process of evaluation and measurement of potential risks that may be involved in a projected activity or undertaking.

Scalability: Property of a system to handle a growing/reducing amount of work by adding or removing resources to the system.

Sensor: A device that can sense and record physical stimulus (e.g., temperature, speed, pressure, etc.) from physical environments.

Situational Awareness: This is the perception of environmental elements and events with respect to time or space, and the projection of their future status.

Smart or Intelligent Cities: Cities where citizens conserve resources by relying on ICT solutions for maximum efficiency and transparency.

TCP/IP: A set of networking protocols allowing two or more computers to communicate.

Urbanization: This refers to the population shift from rural to urban residency, and the technological ways in which each society of people adapts to the technological changes.

Zigbee: An IEEE 802.15.4-based specification for a suite of high-level communication protocols used to create personal area networks.

Chapter 11
A Machine Learning-Based Cyber Defence System for an Intelligent City

Jared Piconi
La Trobe University, Australia

Omaru Maruatona
Aiculus, Australia

Alex Ng
https://orcid.org/0000-0002-5809-2150
La Trobe University, Australia

A. S. M. Kayes
La Trobe University, Australia

Paul A. Watters
La Trobe University, Australia

ABSTRACT

In 2017, a Price Waterhouse report on intelligent cities advised that technology would be a key enabler for efficient management of resources for overpopulated cities. The increased reliance on technology to drive daily lives of people is the main reason why many believe that smart technology is a major part of intelligent cities. By design, intelligent cities are heavily reliant on connected devices. Consequently, reliance on technology poses the risk that any disruption of technology-based services may put human lives in danger. In this context, attacks on technological frameworks can mean attacks on the entire city. Unfortunately, there is no existing technology that is able to successfully defend against such cyberattacks. In this research, the authors review the concepts, principles, and technologies, and propose an approach for a cyber defence system, based on artificial intelligence and machine learning, to detect and prevent an integrated cyberattack.

DOI: 10.4018/978-1-7998-5062-5.ch011

INTRODUCTION

With the world population rapidly growing and the majority residing in urban areas, an increasing demand is being placed on the infrastructure, systems and technologies of cities in order to best serve their people. As an indicator of our increasing reliance on technology, statistics show that the number of smartphone users in the world is expected to pass the five billion mark by the end of 2019 (PWC, 2018). This population growth coupled with modern technological advancements has coincided with the emergence of something known as a smart city. The objective of an intelligent city, also known as smart city or digital city, is to best address the complex needs of modern business organisations, government departments, technological infrastructures, security and privacy of citizens, and of course the citizens themselves (Khatoun & Zeadally, 2016). It has been widely reported that the intelligent city market is expected to exceed $2.5 trillion by 2025 (PWC, 2018).

An intelligent city, also known as smart city, can be defined as "an environment of diverse and interconnected components constantly exchanging data through the use of intelligent technologies and platforms including the Internet of Things (IoT) and the Cloud based technologies (Zubair A. Baig, 2017). Similarly, the work of Ivezic (2016) defines an intelligent city as "an urbanized area where multiple sectors cooperate to achieve sustainable outcomes through the analysis of contextual real-time information shared among sector-specific information and operational technology systems". Within the available literature, it is evident that for a city to be considered *intelligent* or *smart*, it is critical for the information processing and data exchange to be constantly occurring in real-time for quick decision making. Many aspects of modern life such as healthcare, commerce, manufacturing, education, provision of utilities, and transportation are now incorporating smart technologies. Even within homes, smart technology is increasingly present in everyday devices such as televisions, refrigerators, security cameras and thermostats (Elmaghraby & Losavio, 2014).

When we consider the vast amount of data produced and stored by intelligent city technologies, one of the key aspects to address within intelligent city development and management is the necessity to appropriately manage cyber-security and mitigate cyber-crimes. Cyber-attacks are becoming increasingly prevalent as technology manifests itself as indispensable in a modern society. According to Dilek et al. (2015), cyber-crime can be defined as "any crime that is facilitated or committed using a computer, network, or hardware device" and can consist of offenses such as computer intrusions, misuse of intellectual property rights, theft of data, denial of an electronic service, economic espionage, online extortion, international money laundering, non-delivery of goods or services, as well as many other related threats and offenses.

With the number of cyber-attacks and cyber-criminals increasing exponentially, it is becoming more and more difficult to safeguard personal, business related, and even government departmental technologies and the subsequent data contained in technological smart devices. It can be argued that most cybercrimes simply represent the migration of real-world crimes to cyberspace, providing criminals with new ways to commit old crimes (Brenner, 2010). Additionally, it is even more apparent that unlike traditional crime, human intervention is inadequate to address these cyber-criminal activities and to avoid the damage associated with such offences.

Furthermore, as new forms of cyber-crime develop, almost daily, the legislation associated with the crimes fail to keep up, making it even more difficult to develop appropriate jurisdiction and hold criminals accountable (Clough, 2015). Additionally, such advanced information technology has facilitated the globalization of cyber-crimes by eliminating country borders and making it much harder to monitor, detect, prevent or capture cyber criminals (Dilek et al., 2015).

When considering an intelligent city, implementing cyber security mechanisms is paramount as attackers exploiting inherent vulnerabilities can, and as has been seen in the past, cause significant disruption to service delivery and ultimately damage a city's ability to function optimally. This could have widespread impact on essential services including access to electricity, water and gas or could have far-reaching effects on human safety by targeting transportation infrastructure and technologies. There are multiple examples in the media and within the literature on smart cities already being confronted with cyber-crime including: identity spoofing, data tampering, eavesdropping, malicious code, and lack of e-services availability (Khatoun & Zeadally, 2016; Zubair A. Baig, 2017; Kimani, Oduol, & Kibet, 2019).

This research aims to explore the available literature regarding the current cyber threats that intelligent cities face, address the current frameworks and systems utilised to help solve cyber security concerns and ultimately the potential for artificial intelligence to be utilised in creating adequate cyber defence systems for intelligent cities.

Currently, companies are tasked with managing their own cyber defence systems and this is usually an afterthought to functionality and the timely completion of a product or service. The goal of this chapter is to find out if there is a need for a harmonised cyber defence system based on Artificial Intelligence (IA) approach that enables protection of the critical infrastructure within a smart city. If the outcome correlates to the need for such a system, it is proposed that this contribution will assist in future research in creating an AI based system that will be able to detect, alert and predict attack anomalies in areas such as power provision infrastructure, water and sanitary systems, telecommunications, banking, transportation and any other sector that will receive and transmit the generated data to the system. By utilising AI, the data collected and provided will be used to train the machine learning algorithms to not only detect known attacks but also learn and alert to any new attacks that have not been discovered yet or even to alert in advance any possible persistent threats.

The organisation of this chapter is as follows:

- The first part discusses current cyber threats and vulnerabilities in intelligent cities. This section analyses recent vulnerabilities, security concerns and cyber incidents targeting various elements of a city's critical infrastructure. The second part reviews some current alternatives in addressing cyber security for intelligent cities. Also, in this part, a brief discussion on potential approaches in addressing cyber threats in intelligent cities is laid out. This part also reviews different attack types affecting intelligent cities and their impact.
- This research proposes cyber monitoring system for an intelligent city. This system does not need to be completely built from scratch. Instead, many existing cyber securities threats monitoring and defence approaches, frameworks and products, can be leveraged to come up with a new system. In part three of this chapter, a review of potentially useful cyber defence frameworks, including the NIST Smart City Model (NIST, 2018), CISA (National Critical Functions Strategic Risk Management Process and Security) Secure framework for future smart city, are briefly reviewed. Given the need for aggregation, correlation, alerting and reporting capabilities, the Security Information and Event Management (SIEM) is also discussed as a potential component of the intelligent city cyber monitoring system.
- Compared to traditional systems, intelligent frameworks have features that make them relevant and desirable for today's era; mainly for the reason that they are adaptive, require less rules main-

tenance, and have no significant issues with system brittleness, etc. (Maruatona, 2013). In this context, part four compares Traditional cyber systems against Artificial Intelligence based cyber defence systems. In part 5, a detailed discussion of our proposition for an Artificial Intelligence based cyber defence system for intelligent cities is given.

In this chapter, we use the smart city and intelligent city terms interchangeably.

CYBER THREATS AND VULNERABILITIES IN INTELLIGENT CITIES

The world of interconnected devices is developing rapidly, with the estimated number of IoT based connected devices doubling by the year 2020 (Drizik, 2018). This rapid increase in the connectivity of devices that historically were non-smart devices adds further complexity to securing and protecting them. Ultimately, this causes an increasing concern with the security of devices, software and data as countermeasures are either wholly neglected or implemented as an afterthought rather than being engrained in the deployment and development of these new technologies and applications (Drizik, 2018; Dagle, 2009).

To have a better understanding of the aspects that could be compromised, it is important to understand what infrastructure and services a smart city comprises of. IoT based and Industrial Control Systems (ICS) devices play an essential role in managing the critical infrastructure of an intelligent city by collecting and processing data from multiple distributed devices spread throughout the city to implement smart services to critical infrastructure systems (Zubair A. Baig, 2017). Some of the relevant elements of an intelligent city include the following:

- Traffic control and monitoring systems
- Environmental monitoring systems
- Sewerage and waste management systems
- Power and electrical grids infrastructure
- Water and sanitation plants
- Mobile and telecommunication systems
- Medical and structural health systems
- Intelligent lighting and smart homes
- Surveillance and security systems

It is widely understood that by increasing the number of devices and creating an entirely interconnected city, the threat landscape accordingly increases dramatically (Zubair A. Baig, 2017; Ivezic, 2016). A smart city is one of the most complex cyber-physical systems (CPS) ever built (Yan & He, 2017) and the intelligent grid architecture focused on in this chapter is taken from the NIST smart city framework (NIST, 2018). But when creating one of the most complex CPS, one notices cause for concern over security of IoT based systems, privacy of data, and safety of people. One such situation of significant interest is not only the compromise of a single system but the compromise of multiple systems within a smart city which could equate to a cascading failure of the cities systems and infrastructure (Brown, Carlyle, & Salmeron, 2016; Pacheco & Hariri, 2016; Adrien, et al., 2011; Zhang, et al., 2017).

To further understand the security concerns of a smart city, it is imperative to understand the type of hardware and software that power intelligent cities. These cities are built upon the Internet of Things

(IoT) paradigm, and collaborated with critical infrastructure through industrial control systems (Clotet, Moyano, & Leon, 2018). IoT based devices are designed to work in a variety of areas and are categorized into three main groups viz: enterprise, consumer and industry (Kimani, Oduol, & Kibet, 2019). The available literature details many examples of security flaws in connection with IoT based smart devices, with one consumer example being GPS and the consequent risks to personal safety and personal security of movement. Many individuals set the GPS location of their vehicle, or location of where they live on their smartphone. If this information is obtained by attackers, this could be used to identify where the vehicle is or when the home is unattended so they could perform theft or burglary or to monitor or stalk the individual (Elmaghraby & Losavio, 2014). Although consumer/personal security is paramount, the most damaging threats to the functionality of intelligent cities, and in turn the persons living within the city, is to enterprise related and industrial devices. These areas comprise major businesses and infrastructure critical to a city's resources, such as the power grid or water sources. Unfortunately, as a recent study, conducted by Dilek et al. (2015), discovered, the vulnerabilities are not only numerous, but the manufacturers are also lacking in their ability and skills to appropriately secure them.

An example of the far-reaching consequences of cyber-physical system vulnerabilities leading to failures, is the attack on Ukraine's power grid in 2015. Another example is the cyber-attack that caused a six-hour blackout to hundreds of thousands of people in and around the capital city of Ukraine (Dmitriy & Sullivan, 2017). The vulnerability was so severe that workers at the power station watched helplessly, unable to stop the hackers as they took control of their systems and input the codes to cause widespread outages, bringing over thirty substations down. Not only did this well-devised cyber-attack bring down the substations, they also disabled backup power supplies that sabotaged the operators' workstations. They also implemented malware on the system to wipe out essential system files. Additionally, they flooded call centres with fake calls resulting in Denial of Service (DoS) such that legitimate calls could not be made to the power companies (Dmitriy & Sullivan, 2017; Whiting, 2018; Whitehead, Owens, Gammel, & Smith, 2017). This attack highlights the far-reaching threat landscape and the ways in which cyber-attacks can not only shut down a system but can also use multiple techniques to affect many other connected systems. On a recent study conducted by IBM, there were over 17 zero-day vulnerabilities found in various controls and sensors in intelligent cities around the world (IBM, 2018). This figure further proves the point of just how vulnerable intelligent cities are.

It can be discovered by reviewing existing literature that there are many key areas that contribute to the vulnerabilities within smart devices and systems embedded in intelligent cities (Khatoun & Zeadally, 2016; Zubair A. Baig, 2017 ; Dmitriy & Sullivan, 2017). These aspects being as follows:

- Lack of or insufficient cyber response plans and teams
- Poor cybersecurity testing and vulnerability assessments
- Poor implementation of security i.e. encryption protocols with poor and outdated algorithms
- Increasingly large and more sophisticated attack surfaces
- Lack of patching and patch deployment issues
- Utilisation of insecure legacy systems

In addition to these security flaws, it has also been noted that many companies fail to disclose vulnerabilities of IoT products within their company and systems. Research from Mansfield (2018) has demonstrated that, of the 331 companies examined, only 32 had some form of online disclosure scheme and following this only 3 had a hard deadline of a 90-day fix for any reported issues. This point further emphasises the case that companies cannot be trusted to provide timely fixes for vulnerabilities; and utilising other forms of cyber defences is imperative in keeping smart connected cities safe and secured.

Variety of Attacks and Their Impact

Intelligent cities are extremely susceptible to a range of cyberattack techniques which range from remote code execution, signal interruptions and jamming, malware infection, data manipulation and denial of service attacks (Zubair A. Baig, 2017; Kimani, Oduol, & Kibet, 2019; Dilek et al., 2015; Clotet et al., 2018). There are numerous repercussions from these different attack types being carried out but what is most alarming, as previously mentioned, is a coordinated attack on critical infrastructure systems. It is evident that critical infrastructure needs rigorous and proven defences to protect them from attacks that have the potential to cripple or worse, severely damage a city's ability to correctly function (Khatoun & Zeadally, 2016).

As it has been researched and documented, the current attack surface for an intelligent city is wide open (Khatoun & Zeadally, 2016; Drizik, 2018; Mansfield-Devine, 2018). With such little defence systems in place and a wide-open threat landscape, exploiting the vulnerabilities within the industrial control systems (which include SCADA devices, Programmable Logic Controllers, Distributed Control Systems), which are now becoming internet-facing has the possibility of bringing an entire city down (Kimani et al., 2019). As presented by Kimani et al. (2019), the critical infrastructure targets, that are currently the target of cyber-attacks, show energy as the main attack area which equates for 54% of attacks.

These types of attacks have the possibility to cause economic loss, power outages, contamination of water, shutdown of transport systems, mass panic and even injury or loss of life. This is evident in a recent attack on smart city infrastructure which did not eventuate in a loss of life but very easily could have done so (Baker, 2008).

The following attack is a very real threat and although a hypothetical one, it has the possibility to be carried out by hackers. An attacker might be able to connect to the electric power system to gain access to the network and alter public transportation to potentially paralyse intelligent transportation systems, with thousands of passengers on board at rush hour. The attacker could also launch false alarms and modify traffic lights and controllers with the potential for collisions, (Khatoun & Zeadally, 2016; Zubair A. Baig, 2017).

It is imperative to understand that finding workable solutions to the current security threats faced in smart cities is necessary, otherwise the public simply will not trust the intelligent city deployments and reject the deployment of new technologies (Efthymiopoulos, 2016). With reference to the literature sources already discussed and all the findings within, it is clear that intelligent cities in their current state require additional layers of security with one of these being in the form of cyber defence systems.

As part of our research, we have developed a threat matrix after the analysis of the key literature resources. Based on this, the top 10 threats have been listed below and plotted in Table 1. The numbers listed within the table corresponding to a specific attack type – and there is no specific order or importance attached to the mentioned threats.

1. Targeting IoT Components with known vulnerabilities
2. Security Misconfigurations
3. Using disgruntled Employees
4. Distributed Denial of Service (DDoS)

5. SQL Injection
6. Data leaks
7. Broken Access Control
8. Social Engineering attacks
9. Phishing attacks
10. Insufficient Logging and monitoring

FRAMEWORKS AND SYSTEMS FOR CYBER DEFENCE

Table 1. Intelligent city vulnerability risk matrix

	Negligible	Minor	Moderate	Significant	Severe
Very Likely	LOW MED	MEDIUM	MED HI 7,10	HIGH	HIGH 1,2,5
Likely	LOW	LOW MED	MEDIUM	MED HI 4,9	HIGH
Possible	LOW	LOW MED	MEDIUM	MED HI	MED HI 8
Unlikely	LOW	LOW MED 6	LOW MED	MEDIUM 3, 5	MED HI
Very Unlikely	LOW	LOW	LOW MED	MEDIUM	MEDIUM

There is currently a multitude of cyber defence systems, frameworks, software and devices in use today, but it is imperative to understand whether they are suitable for a much wider context within a smart city scenario. Although some of the cyber security tools, systems and frameworks could be perfectly applicable when applied in a smart environment, where they log data and monitor feeds from a variety of inter-connected smart devices and sub-systems, they have never been used in this way. This is mainly because such a smart city cyber monitoring system will require the input of systems and devices managed by different government organisations and private companies, with diverse communication protocols. Although collating such data would be massively useful for the purpose of defending the city's critical infrastructure against a devastating cyber-attack, most organisations are likely to resist sharing their security data and feeds with other organisations.

If the difficulty of collecting relevant data from different public and private organisations can be overcome, designing and building an integrated cyber threat monitoring and defence system for a smart city would not require any new specific toolsets. Instead, it would leverage existing frameworks and specialised products especially architected and applied in an environment where data is coming from several disparate toolsets and organisations. An important part of such a system is its architecture, which if done correctly will scale, would have the required near real-time capability and be highly accurate. The following are a sample of cyber defence frameworks and available systems that could potentially be combined with other frameworks and systems to build an effective intelligent city integrated cyber threat monitoring and defence system.

Cyber Defence Frameworks

In this section, we briefly discuss a number of available frameworks that can be adopted for cyber defence of a large smart environment.

NIST Smart City Model

The NIST (2018) Smart City Model is a framework that has been developed to facilitate the deployment of smart city technologies. It is one of the most widely adopted reference models (Khatoun & Zeadally, 2016) and has three main components, viz: 1) the discussion and use cases, 2) spreadsheets describing and comparing various applications, and 3) the smart city application framework tool. The framework tool provides the means for stakeholders to evaluate the needs of a city and assess whether, from a technical perspective, the city is technically ready, through determining the requirements for specific smart city applications. As can be seen in the NIST IES city assessment tool (Khatoun & Zeadally, 2016), the user can input information and then select "elaborate", which will then provide them more detailed requirements, readiness parameters and benefits for that specific smart city application.

The framework, as briefly mentioned above, does not provide specific technological solutions to certain applications but merely provides stakeholders with a guidance tool. The framework can be seen as a valuable starting point in the development of a smart city development plan by establishing some recommended requirements for specific applications. The NIST (2018) publication notes its limitations, stating that "once a smart city project is engaged, the more detailed analytical tools may be appropriate to aid in the specification and implementation process". Ultimately the NIST (2018) Smart City Model is a great framework on how to design and implement processes for specific areas within a smart city but fails to highlight the current threats that a smart city can potentially be facing and how to protect against them.

CISA Strategic Risk Management Process

The CISA Strategic Risk management Process for national critical functions is designed to protect the functions which consist of internet, electricity, supply chains, transport, law enforcement, emergency management and access to medical records of citizens, to name but a few within the government and the private sectors. The framework was developed out of a need to protect these areas, if they experience any disruptions. As stated by the Department of Homeland Security (2018), any disruptions would have a debilitating effect on economic security, citizen security, public health and safety or any combination of these. The process outlined in this framework to manage risks in critical infrastructure is in the form of the following functions: identify, analyse, prioritize and manage.

This framework is a solid base for providing a risk management framework that holistically captures risks and associated dependencies and mitigates avoidable risks to ultimately harden the systems implemented across the critical infrastructures. CISA is also creating a risk register which ultimately provides dependency modelling and consequence modelling which are critical services that can help determine the negative impact of critical services being interrupted. While the strategic risk management process for critical functions is a necessity, it does not mitigate and alert to the cyber-attacks discussed previously in regard to intelligent cities and therefore is a good starting point for risk management in an intelligent city. The risk register could be beneficial to help train an AI system using those data sets. However, it is not adequate in itself to protect the city.

SEFSCITY Secure Framework

The SEFSCITY secure framework for future intelligent cities is designed with the end user in mind and is concerned with keeping their data safe and devices secure. The framework is based around the utilisation of cloud computing and big data technologies to process the mass amounts of data that comes with an enabled intelligent city. This security framework utilises a form of message authentication code alongside the elliptical curve discrete logarithm problem. While this is a promising option to secure data and ensure the confidentiality and integrity of the transference phase of the data, it misses a key point; if an attacker utilises a vulnerability on the IoT device and gains access, encryption will not help (Foresti, Yung, & Martinelli, 2012).

The process outlined in this framework is suitable to mitigate the interception of data on the condition that the IoT devices or ICS have the system resources and processing power to encrypt the data it is sending out which generally is not the case (Alromaihi & Balakrishna, 2018). One key takeaway from this framework is the cloud computing system designed to deal with the large amounts of data associated with a smart city environment. The data management framework utilises: 1) Apache Flume to ingest log data of management systems, 2) Hadoop HDFS and Hive which stores the data collected from the plethora of connected devices, and 3) Spark services to distribute log data into clusters and then implement parallel computing for analysis. It is a well-designed system to deal with the issue of utilising and collecting massive amounts of data generated from inter-connected smart objects (Djigal, Jun, & Jiamin, 2017; Peng, Li, Li, & Zhu, 2016; Tankard, 2011). However, this system does not directly monitor intelligent city key data to detect coordinated threats targeting the city's critical infrastructure.

Cyber Defence Systems

A system, whose functionality involves scraping data from a host of IoT devices and other IT and OT systems from different organisations providing various intelligent city related systems, will require robust aggregation, correlation, alerting and reporting capabilities. In terms of today's cyber security products, a system that usually provides these capabilities is a SIEM.

Security Information and Event Management (SIEM) software products and services combine both security information management and security event management (Hanlon & O'Connor, 2014). Deploying a SIEM is currently one of the best methods of providing real-time analysis of security alerts that are generated by applications and network hardware. These alerts allow the users to view the logs to understand, detect, block or stop an attack. This is currently the most popular way of protecting a network and most commonly used in corporate environments.

Relating to intelligent cities, these systems have not been adapted to intake the massive amount of raw data that comes from a large number of devices consistently sending data in real-time. Additionally, they generally utilise rule-based alerting to detect intrusion which may not be useful for a smart city environment (Swift, 2019). The capabilities of a SIEM product were hard to find as these systems are usually patented. However, the authors have summarised the generic capabilities of a SIEM and it is important to note these, to assist with further research in this area. Refer to Figure 1.

Figure 1. Capabilities of a SIEM

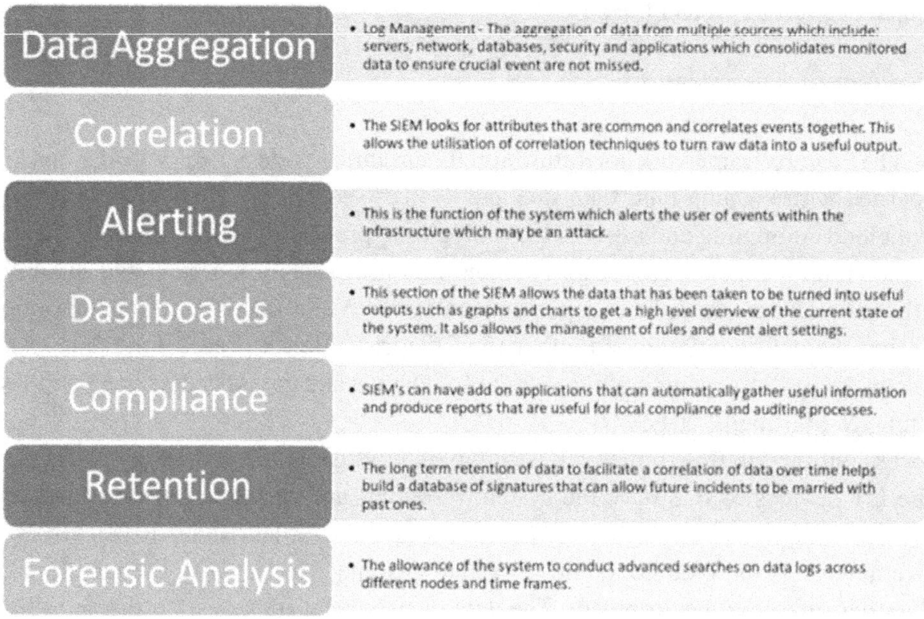

There are numerous weaknesses when utilising an existing system to manage the security of a smart city. Some of the main issues are with respect to the mass amount of data generated by a variety of devices within a smart city, that is required to be analysed. Ability of these systems to be scaled is another issue (Van Zoonen, 2016). As outlined by Nigon et al. (2016), to further support the previous statement, with a large amount of data and ever-changing multi-level dynamics, intelligent cities are far too complex for a top-down approach.

With regard to the types of attacks intelligent cities face, the threat landscape is too large; and the attacks are increasing in complexity. Because of the difficulty to detect them with the historical approach, there is a need for a new breed of detection systems (Allam & Dhunny, 2019). Managing the alerts with the current systems with a human approach is a severely ineffective means of dealing with any adverse event scenario. The intelligent city alerts would be far too many and too frequently for a security operation centre to manage effectively and this is where the use of an artificial intelligence systems come in. While there are no specific AI systems to manage intelligent city security, there are some multiple systems and frameworks, as discussed above, to manage cyber defences which have the possibility to be adapted and implemented in a intelligent city scenario.

TRADITIONAL VS ARTIFICIAL INTELLIGENCE CYBER DEFENCE

Cyber defence systems need to be designed to be adaptable, robust, flexible and able to detect a wide variety of threats. These should then be able to make intelligent decisions in real-time (Dilek et al., 2015; Chen & Wang, 2005; Dasgupta, 2006). It has become increasingly apparent that the current advances in attack types and the increasing predominance of cybercriminals, human interaction and monitoring is no longer sufficient for a timely response to an attack (Dilek et al., 2015).

Network-specific attacks, as opposed to physical attacks, are carried out by intelligent malware. Thus, it is necessary to resolve them with an artificial intelligence defence system that is not only able to detect, evaluate and respond to threats but is also able to learn from past attacks to be more effective (Stytz, Lichtblau, & Banks, 2005). As previously mentioned, the current defence systems have the possibility of detecting and alerting to current threats that are known, such as a known signature detection. However, they are lacking in the detection of advanced persistent threats that are becoming more commonplace and provide new security challenges for critical infrastructures (Tankard, 2011; Huang & Zhu, 2019).

AI-Based Cyber Defence System for Intelligent Cities

Figure 2 illustrates a typical cyber lifecycle, with the proposed benefits of an artificial intelligence system which has been developed in the present research which indicates when and where a functioning and useful AI system would intervene. As seen from the weaknesses of the previous systems and frameworks which have been discussed above, there is a need for a new type of defence system.

Figure 2. Cyber Defence System Life Cycle Intervention

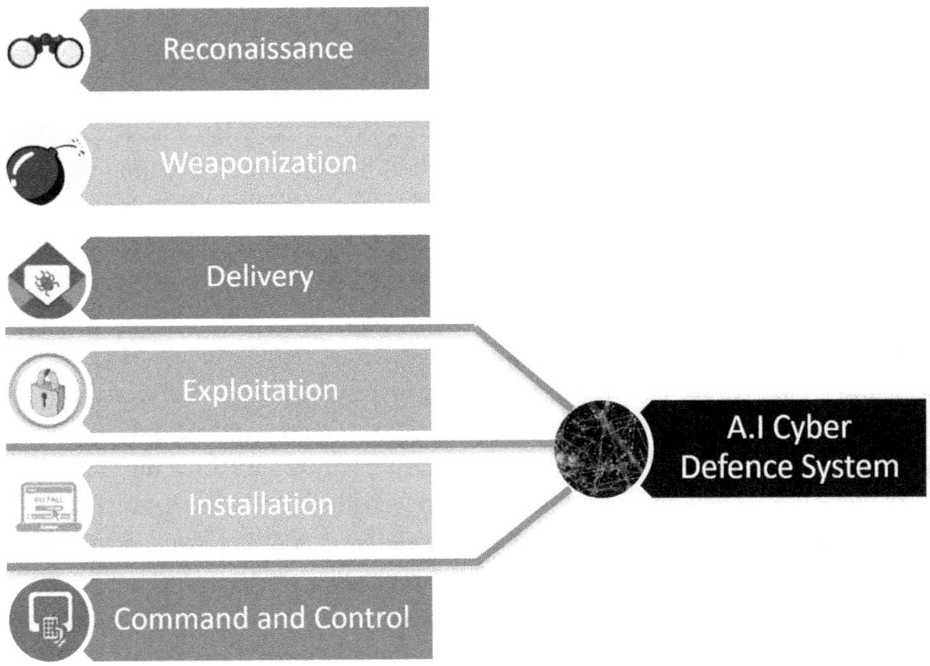

The proposed defence system uses artificial intelligence so that the system can learn from complicated data sets that are aggregated over time and fed into a sophisticated algorithm that then alerts and possibly mitigate cyber-attacks. Not only is this system able to protect against the more standard cyber-attack, it will also, theoretically, be prepared to alert against advanced persistent threats and even with the possibility of warnings against new types of attacks.

As outlined by Tyugu (2013), humans are unable to effectively handle the amount of data to be used while defending cyberspace without considerable automation. With the current set of growing intelligence and malware coupled with the increasingly sophisticated nature of cyber-attacks, it is unavoidable that we develop intelligent cyber defence methods (Allam & Dhunny, 2019; Tyugu, 2013; Trifonov, Nakov, & Mladenov, 2017). Although it is evident, there is further research that needs to be conducted on how to improve this artificial intelligence cyber defence system. For instance, whether neural networks, deep learning, expert systems, fuzzy logic or another entirely different method would be the most appropriate approach.

PROPOSED AI/ML-BASED CYBER DEFENCE SYSTEM

With the rapid increase in technology and implementation of Internet of Things based devices, there is a fast-growing panic in the world of cybersecurity. As previously mentioned, the risks are currently far too significant for a successful or widespread transition of cities into intelligent cities. This provides support for the hypothesis that intelligent cities are not adequately prepared to mitigate and protect against the many attack vectors which they currently face.

The aim of this study is to develop a system that can adequately track, manage, predict and mitigate the numerous vulnerabilities and potential threats to ensure the safety and security of an intelligent city. This is achieved through the proposal of utilising artificial intelligence, in real-time, to achieve all the aforementioned goals. This can be done through the utilisation of machine learning algorithms within Python and Jupyter Notebooks. A prototype can be built utilising machine learning algorithms of Support Vector Machine, Random Forest Algorithm and K-Nearest Neighbour. The hypothesised outcome is that at least one of these algorithms will be able to accurately predict anomalous behaviour over the IoT network with well performing recall and precision scores.

As previously explained, an intelligent city has an extremely large amount of data being transmitted at any one time (Komamizu, Amagasa, Shiokawa, & Kitagawa, 2016; Elshawi, Sakr, Talia, & Trunfio, 2018). This creates a complex scenario for anyone trying to log the data in a traditional way. Utilising the current methods are far too difficult for any security operations centre to adequately manage the process. The implementation and utilisation of machine learning is predicted to solve this issue. This research utilises a synthetic dataset based on a smart home to produce the Correlation heatmap for complex data visualisation and evaluate the performance using three different modelling mechanisms: Support Vector Machine (SVM), Random Forest (RF) and K-Nearest Neighbour (KNN).

To materialise this system, multiple challenges are anticipated given the variety of technologies, data and feeds and the cooperation of government and private organisations. One of the biggest challenges is finding an adequate data set with enough data to train the algorithm against (Elshawi et al., 2018). Without an adequate data set, it is extremely difficult to achieve the desired outcome. There are two options for this study: either create a synthetic dataset or utilise a real data set that has already been compiled. It is important to note that simple machine learning algorithms can perform almost identically to complex algorithms if given enough data (Banko & Brill, 2001; Halevy, Norvig, & Pereira, 2009).

A key challenge is acquiring clean data that is correct and appropriate, and free of misrepresentations or noise. In real life, that is likely be an issue of concern. Another issue relates to the insufficiency of quality of training data. This inevitably make it harder for the system to perform well and in turn dramatically increase the number of false positives and false negatives. The solution to these potential occurrences is to identify what is an outlier or noise and discard these or manually fix the errors in the data. Utilising data of poor quality will make it increasingly more difficult for the proposed system to correctly detect underlying patterns and anomalies. Cleaning up the training data will greatly increase the performance of the system and produce a better outcome.

Some datasets contain what is known as nonrepresentative training data. These outliers can cause issues when training against the dataset and skew the outcomes. Ultimately, this will lead to a model that is unlikely to predict specific scenarios accurately. A solution to this scenario is to fill in the missing data manually or remove this data from the set. This is also known as sampling bias.

Given the amount of data that a smart city produces in real time, one issue that may occur is overfitting the data. When the data set is acquired and algorithm trained, having too many vectors can cause issues that would make the model overly complex. Some possible solutions to this are to utilise a model with fewer parameters, reducing the number of attributes in the training data or constraining the model. Removing outliers will also be a sound solution if this case arises. One solution which is going to be explored in more detail is the implementation of a principal component analysis which will reduce the dimensionality of the data to a specific number of components within a given parameter (DeZyre, 2018).

Acquiring the Dataset

Selecting the relevant data is an important process when training the dataset for machine learning algorithms. This process proved to be highly difficult as acquiring a dataset of an entire intelligent city is nearly impossible. The first reason is that finding a city that is converted into a reasonably fully functioning smart city is quite difficult. Secondly, the cities that are classified as intelligent cities are incredibly reluctant to share all the sensitive data that their city produces. Privacy and confidentiality of such data is the main reason for not allowing it to be available for research. The relevant ethical and legal issues cannot be resolved that easily. Thus, a viable option is to create a synthetic dataset of a smart city, however, this also presents difficulties, due to the sheer amount of data that would need to be synthetically created. Without a real dataset to base the synthetic data on, it would be challenging to predict what data various IoT devices would send and receive, the time and frequency with which it would be sent or received.

Notwithstanding the issues mentioned above, the option that this project has considered is to obtain and also partly develop synthetic dataset based on a smart home environment. This utilised synthetic dataset has been previously created and has the added benefit of including anomalous factors into the dataset based on real-world events. This dataset increased the workflow of the entire project which ensured its validity. The open-source dataset was obtained from Kaggle (2019).

Preparing the Environment

To begin the process, Anaconda was utilised. Anaconda is an open source distribution which allowed the implementation of machine learning to develop, test and train the synthetic dataset. Multiple Python libraries were installed within the Anaconda distribution. The following Python libraries and installation code were utilised and installed as follows:

- Jupyter Notebook: This is an open-source web application that this project has used to prepare the synthetic data, train algorithms, various visualisations and confusion matrixes. Jupyter notebook comes with the Anaconda distribution.
- Pandas: Python data analysis library was used for standardisation, manipulation and analysis, e.g.: *conda install -c anaconda pandas*
- Numpy: This was installed and used for preprocessing and data definitions. Numpy was installed as follows: *conda install -c anaconda numpy*
- Matplotlib: Matplotlib was utilised for data visualisation, correlations as well as outputting confusion matrices. It was installed as follows: *conda install -c anaconda matplotlib*
- Scikit-learn: Scikit-learn is an open-source machine learning software utilising Python which will be utilised for the machine learning and other processes to predict any anomalous behaviours within the synthetic dataset. Sci-kit learn is built upon NumPy and matplotlib which was previously installed. Scikit-learn is installed as follows: *conda install -c scikit-learn*
- Seaborn: Seaborn was utilised for complex data visualisation with the ability to output high quality statistical graphs. It was installed as follows: *conda install -c anaconda seaborn*
- Python Confusion Matrix: PyCM was utilised as a final point to be able to create a confusion matrix to get an accurate outcome of where the machine learning algorithm does and does not do well. It was installed as follows: *conda install -c sepandhaghughu pycm*

Preparing Data for Machine Learning Algorithms

Pre-processing the data is an important step where values within the dataset are cleansed. This function takes the DataFrame of the raw data from the .csv file and removes all rows which contain NaN, true, false, twenty, non-related (for example: org.ds2os.vsl.core.utils.Address Parameters@2c82b391) and none values from the *value* column. The data type of the *value* column is then changed from string values to floats. All columns aside from *value* are then encoded and then all columns except *normality* are standardized. Finally, the standardized encoded DataFrame is returned along with the non-encoded *normality* column.

The preprocessing module of the Sklearn library is used to encode the new DataFrame features to numeric values. X is used as an ndarray of 335479, 11 (which does not include the *value* column) which is the rows and columns of data within the dataset. It is then transposed to reflect 11, 335479. This is to transpose the data to reflect 11 rows of data instead of 3354769 for ease of processing. Once each index contains the encoded data of one feature, a loop is run 11 times to store the encoded data.

Following this, the normality column is removed as it contains the normal or anomalous label and should not be standardised. The remaining columns are standardised and then the normality column is appended to the new standardized DataFrame.

Selecting a Training Model

When selecting a training model, it was evident that selecting a model that has performed similar tasks was important. Viewing similar work of anomaly-based detection, the area of fraud detection was used as a starting point. From reviewing numerous pieces of work, one of the more interesting and well-performing algorithms that presented itself was Random Forest Algorithm as presented in (Kumar, Soundarya, Kavitha, Keerthika, & Aswini, 2019) (Xuan, et al., 2018) (Bauder & Taghi, 2018) (Mishra

& Ghorpade, 2018) (Varmedja, Karanovic, Arsenovic, & Anderla, 2019). To compare and contrast against the Random Forest Algorithm, two other supervised algorithms of Support Vector Machine and K-Nearest Neighbour were also used.

Exploratory Data Analysis

Exploratory data analysis provides insight about data which will help determine the features that correlate with one another, features that are are important to consider and their relationship to the target label. A correlation heatmap is used to analyse various relationships among different features. From the correlation heatmap as shown in Figure 2, it is evident that the following features are highly correlated. A higher correlation can cause confusion for the algorithm when trying to differentiate certain classes from each other but conversely can assist the algorithm in detecting anomalous behaviour in certain classes. From Figure 2, we note that the following features are correlated.

- sourceID and sourceType (0.9)
- sourceID and operation (0.51)
- sourceAddress and destinationServiceAddress (0.98)
- sourceAddress and accessNodeAddress (0.98)
- sourceLocation and destinationLocation (0.95)
- destinationServiceAddress and accessNodeAddress (1)
- destinationServiceType and accessNodeType (0.98)

From the previously mentioned Principle Component Analysis, a PCA visualisation is created. This assists in visualising the spread of data in a two-dimensional principal component space. Figure 3 b shows the distribution of anomalies in the data.

To understand the two-component principal component analysis, a Bi-Plot visualisation was created to visualise the data with all the labels and direction of maximum variance in the data indicated by the arrowhead. The name at the end of the arrowhead represents the feature in that particular direction. From the biplot in Figure 4, we can see that the following features are important.

- sourceID
- sourceType
- sourceLocation
- operation
- destinationLocation
- accessNodeType
- destinationServiceType

Model Training and Validation

For the model training and validation, the principal components were passed to the algorithms. When training the data, it is important to define a training testing split of the data. Defining the function splits

Figure 3. Correlation Heatmap

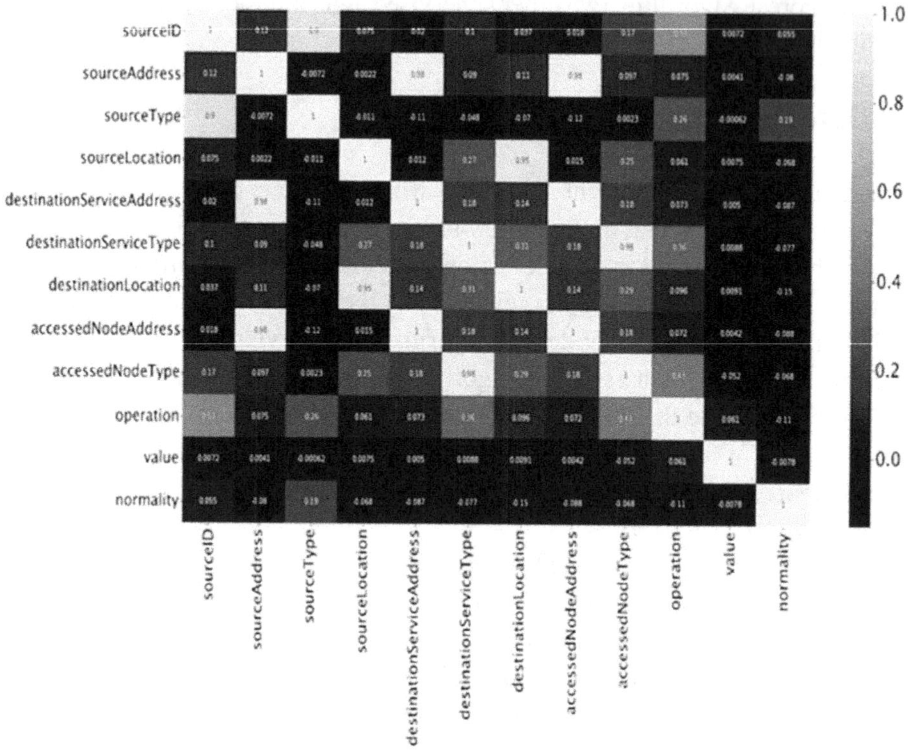

the total data into two parts: one for training purposes and the other for testing purposes. For this data, the test to training data split was 80:20.

Using the built-in function from sklearn.model_selection train_test_split was imported. The standardised and encoded data frame of normal and anomalous data the testsize was passed which is 80%. This looks as follows:

```
def split_data(x,y,testsize):
    x_train,x_test, y_train, y_test =train_test_split(x, y, test_
size=testsize,random_state=10, shuffle=True)
    return x_train, x_test, y_train, y_test
```

Once the training and testing data had been split, the functions had to be created to train the classifiers. The first classifier chosen was Random Forest, that was already imported from Sklearn. Random forest is an ensemble classifier which uses a number of smaller classifiers and aggregates them to give a strong prediction with regard to the data. In this scenario, the number of trees to be used in the forest = 100 with a randomiser value for random_state = 40. The function appeared as follows:

```
def train_RandomForest(x_train,y_train):
clf = RandomForestClassifier(n_estimators=100,random_state=40)
```

Figure 4. Two Component Principal Component Analysis

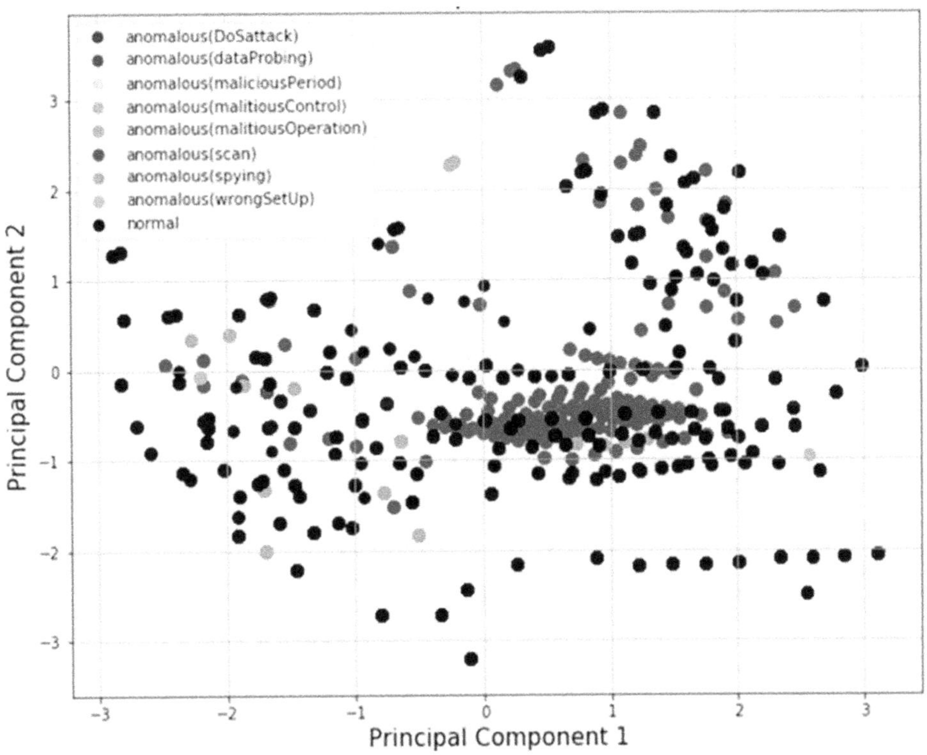


```
clf.fit(x_train, y_train)
return clf
```

Support vector machine was the next classifier to train against the data. Support vector machines are a non-probabilistic binary linear classifier. The gamma parameter in this formula defines how far the influence of a single training example will reach. This value affects the decision boundary and for this process, the gamma value was auto adjusted to allow the algorithm to decide on the value of the decision boundary to allow it to become more linear. The function appeared as follows:

```
def train_svm(x_train,y_train):
    clf = svm.SVC(gamma='auto')
    clf.fit(x_train, y_train)
    return clf
```

The third and final machine learning algorithm that was chosen was K-Nearest Neighbour (KNN). This learns by the parameter of n_neighbours and whenever there is new data the algorithm looks to the data and the nearest data to it. The value of n_neighbours results in drastic differences in training values; it is considered a high variance classifier. KNN looks for the nearest neighbours that has been defined and assigns those neighbours to the data point. For this function, we defined n_numbers as 3. The formula was as follows:

Figure 5. PCA Bi-Plot Visualisation (Principal Component Analysis)

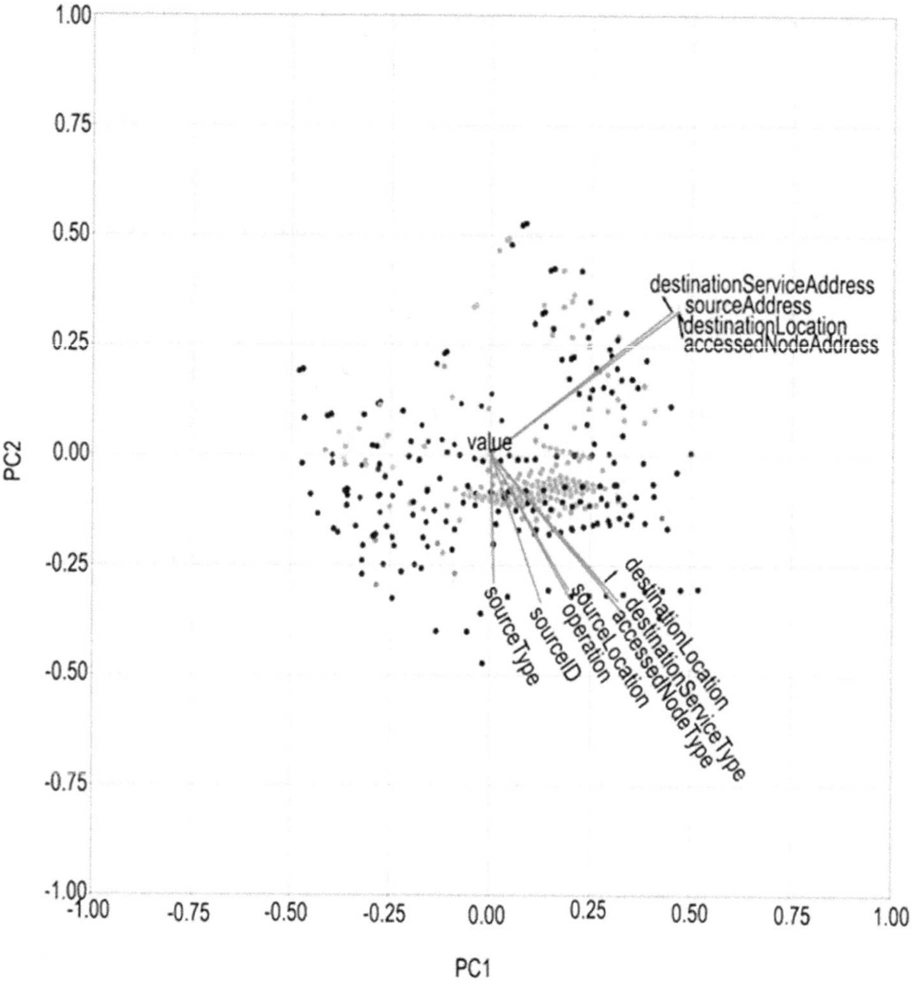


```
def train_KNN(x_train,y_train):
    clf = KNeighborsClassifier(n_neighbors=3)
    clf.fit(x_train,y_train)
    return clf
```

After running all three of the machine learning algorithms, the results was returned with a validation accuracy score that equates to a percentage of correct predictions. This is now discussed below in section on Results and analysis.

RESULTS AND ANALYSIS

This section presents results split between the three different algorithms to ultimately determine which algorithm performed the best through the means of validation accuracy score, F1 scores and confusion matricies.

Support Vector Machines - Results

From training and running the support vector machines algorithm, it resulted in a validation accuracy of 0.9697895552640992. This is a correct class predictor of 96.9790% accuracy on whether the data is anomalous or normal. To further explore this, a confusion matrix was implemented to gauge true positive, true negative, false positive and false negative. In this scenario: 1) a true positive is when the algorithm manages to correctly classify anomalous behaviour; 2) true negative is when the algorithm correctly classifies normal behaviour; 3) false positive is when the algorithm classifies the data as an anomaly when it in fact is normal; and 4) false negative is when the algorithm classifies the data as a normality when it is actually anomalous. These points are plotted in Figure 5. Diagonally from the top left to the bottom right of Figure 6, these are all correct predictions of the SVM model. The left side of the figure defines the true labels and when moving horizontally if there is a number; and if does not align vertically with the same label, it is a false positive or incorrect prediction. Looking vertically, if a prediction does not align horizontally with the same label, it is classified as a false negative. From the confusion matrix of the SVM model (Figure 6), it is clearly evident that it had difficulty correctly predicting anomalous data, aside from the anomalous(DoSattack) scenario. This is evident through looking at the normal predicted and moving vertically there is numerous anomalies that are false negatives.

Figure 6. Confusion Matrix for SVM (Y-axix frefers to True Labels and X-axix to Predicted Labels)

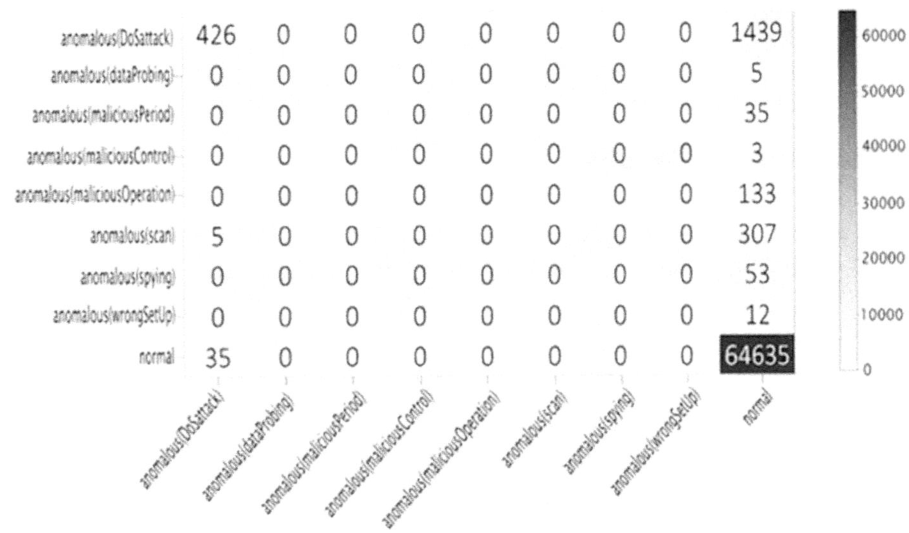

Random Forest – Results

From training and running the random forest algorithm, it resulted in a validation accuracy of 0.9697895552640992. This equates to a correct class predictor of 97.7763% accuracy on the data. To understand the accuracy of anomaly detection, Figure 7 outlines how random forest performs with the actual label and the predicted label. This is seen through the number of false negatives that have been predicted. 1439 anomalous(DoSattack) have been predicted as normalities. This means that the 1439 denial of service attacks would not have been picked up and would have been allowed to proceed through the network. Looking at the true positive and true negatives that have been correctly predicted, it is evident that RF has performed better than SVM. This can be seen, as SVM only predicted 8 anomalous(spying) and 426 anomalous(DoS) attacks correctly. RF managed to predict 12 anomalous(wrongSetUp) correctly, 61 anomalous(Spying), 312 anomalous(Scan), 133 anomalous(maliciousOperation), 3 anomalous(malicionPeriod), 5 anomalous(dataProbing) and 426 anomalous(DoSattack) correctly.

Figure 7. Random Forest Confusion Matrix (Y-axix frefers to True Labels and X-axix to Predicted Labels)

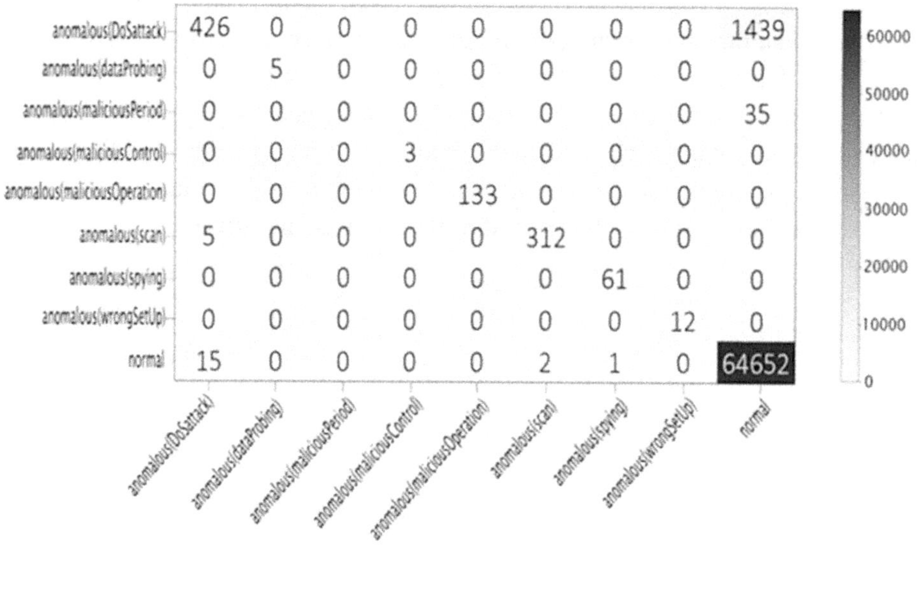

K Nearest Neighbour - Results

The validation accuracy of KNN resulted as 0.97795695719566. This equates to a correct predictor of 97.7957%. K-Nearest Neighbour results will be further discussed in the analysis section below. Refer to Figure 8.

Analysis

While the results of the three algorithms resulting in validation accuracies of 96.9790% for SVM, 97.7763% for RF and 97.7957% would in fact seem like a very positive outcome, it is important to

Figure 8. K-NN Confusion Matrix (The Y-axix frefers to True Labels and X-axix Refers to Predicted Labels)

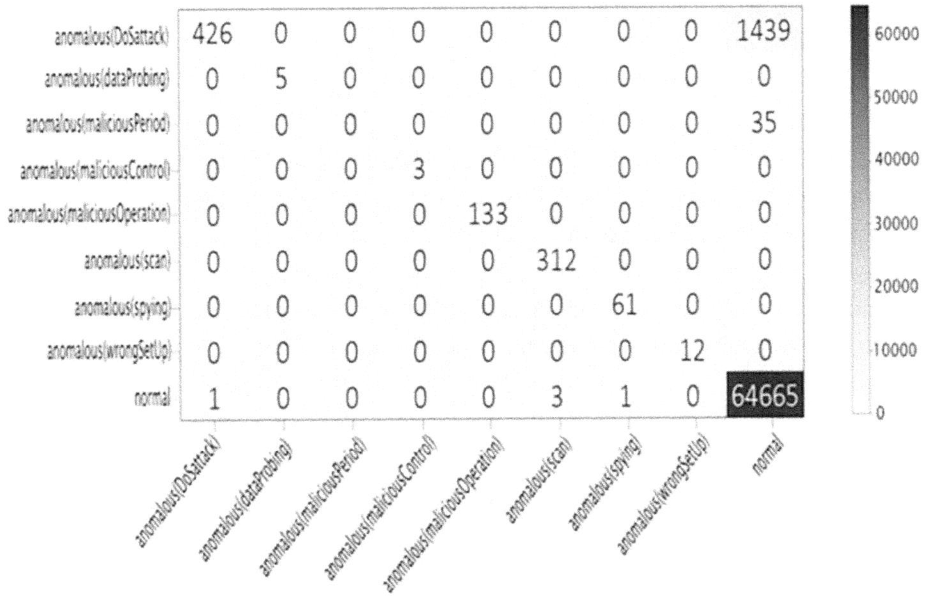

remember the amount of data within the datasets. The datasets managed to accurately predict nearly all the normal behaviours over the network. This can equate to a high validation accuracy over all three models respectively. Reviewing the false positives and false negatives is the best way to get an accurate insight into how well anomalous behaviour is being predicted. SVM predicted most of the anomalous behaviour as a false negative and this model, as it stands, would not be a sound solution.

The RF algorithm had lower false negatives than SVM, it performed better, overall. It accurately managed to predict most of the anomalous behaviour within the dataset as previously mentioned. Overall KNN was the most accurate and best predictor of the three machine learning algorithms. It managed to perform slightly better than RF. Whereas RF predicted normal behaviour as anomalous, KNN was able to classify this correctly as normal aside from one piece of data. The only area that KNN was worse than RF was in classifying one piece of normal data as an anomalous(scan). From all three datasets, it is evident that all struggled with accurately predicting anomalous(DoSattack) as all three wrongly classified 1439 pieces of data and 426 correctly.

To accurately analyse the outcome, precision and recall scores were calculated on the highest performing algorithm of KNN. To analyse this, the recall and precision scores were calculated with the following formulae:

Recall = TP / (TP + FN) and
Precision = TP / (TP + FP).

When looking at the KNN algorithm, the total number for the relevant sections of the above formula are:

True Positive = TP = 952
False Positive = FP = 5

True Negative = TN = 64665
False Negative = FN = 1474
Recall = 952 / (952 + 1474) = 0.3924
Precision = 952 / (952 + 5) = 0.9947

The recall score of 0.3924 indicates that the algorithms ability to predict anomalous behaviours on the network was in fact subpar. Where this outcome has likely manifested is in predicting 1439 anomalous(DoSattack) as normal behaviour within the dataset. This shows that the algorithm struggles in this area.

The precision score of 0.9947 is an extremely high score. This shows that the algorithm is extremely good at predicting normal behaviour over the network. There were only five occurrences over the entire dataset for which it indicated a false positive. Overall, to see the total performance metric of the algorithm, the F-Score equation was implemented. The F-Score was as follows:

F-Score/Measure = 2*((precision*recall) /(precision+recall))
F-Score = 2*((.9947*.3924) / (0.9947+0.3924))
F-Score = 0.5628

The F-Score conveys the balance in comparison between the recall and precision. The F-Score can also be broken down to its each individual class to see how well the algorithm performed on each. The results are provided in Table 2.

Table 2. K-NN F-Score of Each Class

Class	F1 Score
anomalous(DoSattack)	0.3655083655083655
anomalous(dataProbing)	1.0
anomalous(maliciousPeriod)	0.0
anomalous(malitiousControl)	1.0
anomalous(malitiousOperation)	1.0
anomalous(scan)	0.9836065573770492
anomalous(spying)	0.9836065573770492
anomalous(wrongSetUp)	1.0
normal	0.9845992139658166

While the outcome of the overall F-Score is a respectable one, it is evident that the recall and precision values are too far apart. Ultimately, the F-Score would need to even be out by increasing recall score and of course in turn, the precision score lowers due to the algorithm's changes.

Defense of an Integrated Intelligent City Cyber Attack - Example

The end goal of this research is to implement the prototype in order to prevent an integrated intelligent city cyber attack. While this is a static dataset, we can compare it to real-world events. One possible cyberattack that can occur within a fully integrated intelligent city is an attack on multiple systems which combine to cripple an entire city. In this scenario, it is hypothesised that the data is from various critical infrastructures of a smart city. The critical infrastructures that the data is hypothesised to be from, are the traffic control systems, emergency department control system, 4G/5G network and the power grid. If the algorithm for KNN was implemented and was able to be run on dynamic data in real-time, the outcome of the attack would be as follows.

The attacker would begin by scanning the networks of the critical infrastructure and of the scans, the algorithm managed to pick up all 312 occurrences. This is excellent as probing networks is a method by which attackers managed to penetrate the network through scenarios such as conducting syn scans through programs like Nmap to map out and determine the state of the network. If a scan were successful, the attacker would then set up a way to spy on the network to search for flaws. Assuming the attacker managed to scan and find weaknesses in the network to begin spying on it. The algorithm managed to predict all cases of spying over the networks. Next, the attacker would attempt to initiate a malicious operation over the network and again the algorithm managed to predict all 133 cases of this. If the malicious operation was successful, the attacker would then move onto malicious control of the critical system but again the algorithm managed to accurately identify this behaviour. When the malicious period has begun, the malicious period would be continued over the network but in this instance, the algorithm failed to pick up the 35 cases of data that fall under this category.

Comparing these results back to the critical systems previously mentioned, it is evident that an attacker would fail to make a successful cyber attack on the infrastructure initially. The attacker would possibly then aim to disrupt the systems through a denial of service attack. From the algorithms results, it would only be able to detect 29% of the denial of service attacks. This denial of service attack could have far-reaching consequences. Traffic control systems could be disrupted which could cause significant traffic problems for the city or potential car accidents. With the disruption reaching into the 4G/5G network, any emergency calls would not be able to go out, and any emergencies would not be able to be tended to which could potentially result in a loss of life. Disruptions to the emergency department control system would further enhance the potential increase in injury and loss of life as critical response teams would not be able to be deployed. Finally, a disruption to the power grid could cause panic in the city with all businesses not being able to operate, emergency departments being forced onto backup power which can only sustain operations for a finite time. Ultimately, following the reinstatement of all the critical systems, the public uncertainty and distrust of the systems which improve their quality of life drastically would be extremely high.

CONCLUSION

Intelligent cities are currently an extremely complex environment and throughout this chapter, it was emphasized that security within a smart city is generally an afterthought and not something that is considered during the deployment of IoT based connected devices and software. Furthermore, the literature provided evidence that there was an ongoing increase in threat agents, and attack vectors threatening the confidentiality, integrity and availability of connected systems within a smart city. After thoroughly researching the available literature, it was noted that there was no one single system which was able to satisfy the requirement of protecting a smart city. There was also a realisation that a system utilising artificial intelligence or machine learning technique within a smart city environment was not currently available. This suggested that the creation of a harmonised, integrated cyber defence system was a critical and urgent requirement, when the transition of cities was currently undergoing into becoming smarter cities. Having a system that can monitor all the functional sections of a smart city, with a minimal increase in overhead to the connected devices via the IoT, and being a post-deployment system is the solution to ensuring the cybersecurity of these cities.

In this contribution, we have proposed a machine learning based cyber defence system. Proposing this solution solves the issue of the rapid increase in the deployment of IoT based smart devices, providing appropriate security. This also addresses the primary concern of monitoring and mitigating cyber-attacks on critical infrastructure within a city to avoid a catastrophic multi-faceted attack which has the potential to cause significant economic loss and possible loss of life within an integrated intelligent city.

The discussion presented in this chapter also brings to light the need for further research into creating a system specific to intelligent cities, that will further harmonise connectivity of smart objects and devices and be able to learn from complicated data sets to ultimately provide alerts and mitigate against cyber-attacks to critical services in an intelligent city.

REFERENCES

Adrien, B., Juan, H.-S., Maarc, S., Mischa, D., Apostolos, K., & Dominique, B. (2011). Security and Privacy in your Smart City. *Centre Tecnologic de Telecomunicacions de Catalunya*, *3*(1), 1–6.

Allam, Z., & Dhunny, Z. (2019). On big data, artificial intelligence and smart cities. *Cities (London, England)*, *89*, 80–91. doi:10.1016/j.cities.2019.01.032

Alromaihi, S., & Balakrishna, C. (2018). Cyber Security Challenges of Deploying IoT in Smart Cities for Healthcare Applications. *2018 6th International Conference on Future Internet of Things and Cloud Workshops (FiCloudW)*.

Baker, G. (2008, January 11). Schoolboy hacks into city's tram system. *Weekly Telegraph*, p. 34.

Banko, M., & Brill, E. (2001). Scaling to Very Very Large Corpora for Natural Language Disambiguation. *ACL '01 Proceedings of the 39th Annual Meeting on Association for Computational Linguistics*.

Bauder, R., & Taghi, K. (2018). Medicare Fraud Detection using Random Forest with Class Imbalanced Big Data. *2018 IEEE International Conference on Information Reuse and Integration (IRI)*. 10.1109/IRI.2018.00019

Brenner, S. (2010). *Cybercrime: Criminal Threats from Cyberspace*. Dayton: Praeger.

Brown, G., Carlyle, M., & Salmeron, J. (2016). Defending Critical Infrastructure. *Homeland Security: Operations Research Initiatives and Applications*, *36*(6), 530–544.

Chen, H., & Wang, F. Y. (2005). Introduction: Artificial Intelligence for Homeland Security. *IEEE Intelligent Systems*, *20*(5), 12–16. doi:10.1109/MIS.2005.88

Clotet, X., Moyano, J., & Leon, G. (2018). A real-time anomaly-based IDS for cyber-attack detection at the industrial process level of Critical Infrastructures. *International Journal of Critical Infrastructure Protection, 32*, 11–20. doi:10.1016/j.ijcip.2018.08.002

Clough, J. (2015). *Principles of cybercrime*. Cambridge: Cambridge University Press. doi:10.1017/CBO9781139540803

Cyber security challenges for IoT-based smart grid networks. (2019). *International Journal of Critical Infrastructure Protection, 25*(16), 36-49.

Dagle, J. E. (2009). Cyber security of the electric power grid. IEEE/PES Power Systems Conference and Exposition.

Dasgupta, D. (2006). Computational Intelligence in Cyber security. *IEEE International Conference on Computtional Intelligence for Homeland Security and Personal Safety*.

Department of Homeland Security. (2018). *National Critical Functions*. Retrieved May 7, 2019, from https://www.dhs.gov/cisa/national-critical-functions

DeZyre. (2018). *Principal Component Analysis*. Retrieved August 16, 2019, from https://www.dezyre.com/data-science-in-python-tutorial/principal-component-analysis-tutorial

Dilek, S., Cakir, H., & Aydin, M. (2015). Applications of artificial intelligence techniques to combat cyber crime: A review. *International Journal of Artificail Intelligene & Applications*, *6*, 1–19.

Djigal, H., Jun, F., & Jiamin, L. (2017). Secure Framework for Future Smart City. *2017 IEEE 4th International Conference on Cyber Security and Cloud Computing*.

Dmitriy, K., & Sullivan, J. E. (2017). How cyber-attacks in Ukraine show the vulnerability of the U.S. power grid. *The Electricity Journal*, *30*(3), 30–35. doi:10.1016/j.tej.2017.02.006

Drizik, J. (2018, January 17). *Cyber risk is a growing challenge. So how can we prepare?* Marsh LLC.

Efthymiopoulos, M.-P. (2016). Cyber-security in smart cities: The case of Dubai. *Journal of Innovation and Entrepreneurship*, *5*(11), 1–16. doi:10.118613731-016-0036-x

Elmaghraby, A. S., & Losavio, M. M. (2014). Cyber security challenges in Smart Cities: Safety, security and privacy. *Journal of Advanced Research*, *1*(5), 491–497. doi:10.1016/j.jare.2014.02.006 PMID:25685517

Elshawi, R., Sakr, S., Talia, D., & Trunfio, P. (2018). Big Data Systems Meet Machine Learning Challenges: Towards Big Data Science as a Service. *Big Data Research*, *14*(1), 1–11. doi:10.1016/j.bdr.2018.04.004

Foresti, S., Yung, M., & Martinelli, F. (2012). *Computer Security - ESORICS 2012*. Pisa: Springer. doi:10.1007/978-3-642-33167-1

Halevy, A., Norvig, P., & Pereira, F. (2009). The Unreasonable Effectivenesss of Data. *IEEE Intelligent Systems*, *24*(2), 8–12. doi:10.1109/MIS.2009.36

Hanlon, J., & O'Connor, K. (2014, April). SIEM. *Security Information and Event Management*, p. 36.

Huang, L., & Zhu, Q. (2019). Adaptive Strategic Cyber Defense for Advanced Persisten Threats in Critical Infrastructure Networks. *Performance Evaluation Review*, *46*(2), 52–56. doi:10.1145/3305218.3305239

IBM. (2018). *The Dangers of Smart City Hacking*. New York: IBM.

Ivezic, M. (2016). *Cyber Security – A necessary pillar of Smart Cities*. London: EY.

Kaggle. (2019). *DS2OS traffic traces - IoT traffic traces gathered in a the DS2OS IoT environment*. Retrieved August 10, 2019, from https://www.kaggle.com/francoisxa/ds2ostraffictraces

Khatoun, R., & Zeadally, S. (2016, August 12). Smart Cities: Concepts, Architectures, Research Opportunities. *Communications of the ACM*, *59*(8), 46–57. doi:10.1145/2858789

Kimani, K., Oduol, V., & Kibet, L. (2019). Cyber security challenges for IoT-based smart grid networks. *International Journal of Critical Infrastructure Protection, 25*, 36-49.

Komamizu, T., Amagasa, T., Shiokawa, H., & Kitagawa, H. (2016). Towards Real-time Analysis of Smart City Data: A Case Study on City Facility Utilizations. *2016 IEEE 18th International Conference on High Performance Computing and Communications*.

Kumar, M., Soundarya, V., Kavitha, S., Keerthika, S., & Aswini, E. (2019). Credit Card Fraud Detection Using Random Forest Algorithm. *2019 3rd International Conference on Computing and Communications Technologies (ICCCT)*.

Mansfield-Devine, S. (2018). Firms failing to disclose IoT vulnerabilities. *Network Security*, *2018*(12), 1–2. doi:10.1016/S1353-4858(18)30118-1

Maruatona, O. (2013). *Internet Banking Fraud Detection Using Prudent Analysis*. University of Ballarat.

Mishra, A., & Ghorpade, C. (2018). Credit Card Fraud Detection on the Skewed Data Using Various Classification and Ensemble Techniques. *2018 IEEE International Students' Conference on Electrical, Electronics and Computer Science (SCEECS)*.

Nigon, J., Glize, E., Dupas, D., Crasnier, F., & Boes, J. (2016). Use Cases of Pervasive Artificial Intelligence for Smart Cities Challenges. *2016 Intl IEEE Conferences on Ubiquitous Intelligence & Computing, Advanced and Trusted Computing, Scalable Computing*.

NIST. (2018, September 30). *A Consensus Framework for Smart City Architectures*. Retrieved April 22, 2019, from https://s3.amazonaws.com/nist-sgcps/smartcityframework/files/ies-city_framework/IES-CityFramework_Version_1_0_20180930.pdf

Pacheco, J., & Hariri, S. (2016). *IoT Security Framework for Smart Cyber Infrastructures. 2016 IEEE 1st International Workshops on Foundations and Applications of Self* Systems*.

Peng, W., Li, Y., Li, B., & Zhu, X. (2016). An Analysis Platform of Road Traffic Management System Log Data Based on Distributed Storage and Parallel Computing Techniques. *2016 IEEE International Conferences on Big Data and Cloud Computing (BDCloud), Social Computing and Networking (SocialCom), Sustainable Computing and Communications*.

PWC. (2018). *Creating the smart cities of the future*. Geneva: World Economic Forum.

Stytz, M. R., Lichtblau, D. E., & Banks, S. B. (2005). *Toward using intelligent agents to detect assess, and counter cyberattacks in a network-centric environment*. Alexandria, VA: Belvoir Defense Technical Information Center. doi:10.21236/ADA464134

Swift, D. (2019). Successful SIEM and Log. *SANS Institute - Information Security Reading Room, 11*(1), 1-40.

Tankard, C. (2011). Advanced Persistent threats and how to monitor and deter them. *Network Security, 2011*(8), 16–19. doi:10.1016/S1353-4858(11)70086-1

Trifonov, R., Nakov, O., & Mladenov, V. (2017). *Artificial Intelligence in Cyber Threat Intelligence.* Sofia: Technical University of Sofia.

Tyugu, E. (2013). Artificial Intelligence in Cyber Defense. *International Conference on Cyber Conflict.*

Van Zoonen, L. (2016). Privacy concerns in smart cities. *Government Information Quarterly, 33*(1), 472–480. doi:10.1016/j.giq.2016.06.004

Varmedja, D., Karanovic, S., Arsenovic, M., & Anderla, A. (2019). Credit Card Fraud Detection - Machine Learning methods. *2019 18th International Symposium INFOTEH-JAHORINA (INFOTEH).*

Whitehead, D. E., Owens, K., Gammel, D., & Smith, J. (2017). *Ukraine Cyber-Indued Power Outage: Analysis and Practical Mitigation Strategies*. Houston: IEEE.

Whiting, N. (2018). Cyberspace Triggers a New Kind of Arms Race. *Signal, 72*(6), 38–40.

Xuan, S., Liu, G., Li, Z., Zheng, L., Wang, S., & Jiang, C. (2018). Random Forest for Credit Card Fraud Detection. *2018 IEEE 15th International Conference on Networking, Sensing and Control (ICNSC).*

Yan, J., & He, H. (2017). Cyber-physical attacks and defences in the smart grid: A survey. *IET Cyber-Physical Systems: Theory & Applications, 1*(1), 13–27.

Zhang, K., Ni, J., Yang, K., Liang, X., Ren, J., & Shen, S. (2017, January). Security and Privacy in Smart City Applications: Challenges and Solutions. *Enabling Mobile and Wireless Technologies for Smart Cities: Part 1*, 122-129.

Zubair, A. (2017). Future challenges for smart cities: Cyber-security and digital forensics. *Digital Investigation, 22*, 3–13. doi:10.1016/j.diin.2017.06.015

ADDITIONAL READING

Amit, I., Matherly, J., Hewlett, W., Xu, Z., Meshi, Y., & Weinberger, Y. (2018). Machine Learning in Cyber-Security - Problems, Challenges and Data Sets. The AAAI-19 Workshop on Engineering Dependable and Secure Machine Learning Systems, 2019, https://sites.google.com/view/edsmls2019/home

Buczak, A. L., & Guven, E. (2016). A Survey of Data Mining and Machine Learning Methods for Cyber Security Intrusion Detection. *IEEE Communications Surveys and Tutorials, 18*(2), 1153–1176. doi:10.1109/COMST.2015.2494502

Chen, X. Huang, Xinyi, & Zhang, Jun. (2019). Machine Learning for Cyber Security, Second International Conference, ML4CS 2019, Xi'an, China, September 19-21, 2019

Das, R., & Morris, T. (2017). Machine Learning and Cyber Security. *2017 International Conference on Computer, Electrical & Communication Engineering (ICCECE)*, 1-7.

Dilek, S., Huseyin, C., & Aydin, M. (2015). Applications of artificial intelligene techniques to combating cyber crimes: A review. *International Journal of Artificial Intelligence & Applications*, *6*(1), 21–39. doi:10.5121/ijaia.2015.6102

Dong, J., Hannon, C., Zhiyi, L., Pablo, C., Srinivasan, R., Burgess, P., & Nathan, B. (2016). Smart street lighting system: A platform for innovative smart city applications and a new frontier for cyber-security. *The Electricity Journal*, 28–35.

Rotuna, C., Cirnu, C. E., Smada, D., & Gheorghita, A. (2017). Smart City Applications Built on Big Data Technolog and Secure IoT. *Ecoforum*, *6*(3), 13–19.

Varshney, K., & Alemzadeh, H. (2016). *On the Safety of Machine Learning: Cyber-Physical Systems*. Decision Sciences, and Data Products.

Wong, S. (2015). How to take out a city: Smart cities' vulnerability and hacking Technology. *New Scientist*, *227*(3033), 18–19. doi:10.1016/S0262-4079(15)30924-6

KEY TERMS AND DEFINITIONS

Artificial Intelligence Cyber Defence System: AI systems being applied to detect vulnerabilities and execute response actions toward cyberattacks. Features may include self-healing, cyber defence, decision-support, risk management, pattern detection, cyber situation consciousness, projection, malware recognition, and data correlation. The goal is to strengthen the communication and information systems security by providing AI-based network resilience, prevention, and protection against cyber threats.

Cyber Attack: A cyberattack is a malicious action taken deliberately to exploit a computer system, networks and technology-dependent enterprises to cause damage or disrupt a computer network or system which may involve the attempt to access files and infiltrate, steal confidential data or compromise an IT related system. Result may be data theft, denial of service, attempt to access files and compromise security.

Cyber Defence Framework: A set of cybersecurity mechanisms that aim at detecting cyber threats and resolution of such threats. Frameworks also provide detailed guidance to an organisation to align its cybersecurity activities with its business requirements, risk tolerances, and resources.

Cyber Threat: A cyber threat refers to the possibility of a cyber attack on IT related systems and devices. Threat may result in a malicious attempt, or develop into a security attack, through cyberspace with the purpose to cause damage or disrupt a computer network or system, Examples are data theft, denial of service, attempt to access files or steal confidential data.

Internet of Things (IoT): This refers to a network of connected devices for generation, storage and transmission of data. Devices may be computing equipment, sensor-based objects, smart devices, mobile phones, etc. This is the latest networking paradigm: applications include smart cities, internet of vehicles, industrial internet of things, etc.

Smart City: A smart or intelligent city refer to new breed of technology -cities. Its elements include smart homes, smart manufacturing, intelligent transportation, etc. Development and management of such cities is technology based. The goal is to enable people living more comfortably in smart homes and work in more desirable environments.

Vulnerability: A vulnerability is a weakness or flaw that exists in a system due to design, operation procedures or implementation fault. A vulnerability may result in cyber-attack to compromise a system. Result may be data theft, denial of service, or an attempt to access and modify files.

Chapter 12
Hybrid Intrusion Detection System for Smart Home Applications

Fida Hussain
Canterbury Christ Church University, UK

Abhaya Induruwa
Canterbury Christ Church University, UK

Man Qi
Canterbury Christ Church University, UK

ABSTRACT

Smart homes, which incorporate IoT technologies to provide home security, efficient environmental services, conveniences, and improved living standards, are becoming the centre of smart urban developments. With the increased inter-connectivity of smart objects and sensors, there is now, also, an increased level of cyber threats, which can compromise privacy and security. These threats either modify packets of information or inject modified packets into the networks. This chapter examines current intrusion detection systems (IDSs) and presents a unique solution to overcome intrusion detection challenges. It discusses the implementation of smart home IDS (SHIDS), using a machine learning based signature and anomaly intrusion detection scheme to detect network intrusions in the smart home. Suggested mechanism is based on naïve Bayes technique to improve the detection performance. The performance of SHIDS has been tested with network intrusions resulting from DoS, probe, remote-to-local (R2L), and user-to-root (U2R) attacks.

DOI: 10.4018/978-1-7998-5062-5.ch012

INTRODUCTION

Cities play a significant role in the developing environmental, social, and economic aspects nationally and globally (Ismagilova, et al., 2019). More than half of the world's population currently lives in cities and it is predicted that 2.5 billion more will move to cities. Its is estimated that new inhabitants would make it about 60% of the world population to live in cities by 2050. This is supported by the United Nations' (UN) prediction that 6.5 billion people will live in cities by 2050 (Ismagilova, et al., 2019). A consequence of this rapid urbanisation is that cities are facing many challenges as their assets and foundations are brought under huge pressure. An emerging trend to efficiently run cities and to decrease the impact of these challenges is the deployment of digital and communication technologies to build an integrated infrastructure (combining sensors and actuators, intelligent objects such as smart phones, networking, storage resources and physical computing) giving rise to the concept, now commonly known as smart or intelligent or digital cities.

Ismagilova, et al. (2019) report that many countries have already developed smart cities by utilising these technologies to refine different aspects of the city's performance, monitoring and the management of critical services relating to provision of water, transport, traffic, health, environment, public and private sector, homes, quality of life for residents and many more. Many cities are becoming smarter than ever before, due to revolutionary developments in the area of Internet of Things (IoT). IoT has been steadily incorporated into some of the main operational and service aspects of a city such as Smart Energy, Smart Buildings (including Smart Homes), Smart Mobility, Smart Healthcare, Smart Security, Smart Governance, Smart Citizens, Smart Infrastructure (Arasteh, et al., 2016). Hui, et al. (2017) argue that "SH [Smart Home] is the basic building block of smart cities" and setting up of smart cities is the key for manageable fast urbanisation.

The ubiquity and popularity of the Internet, and low cost of IoT based devices, are the contributing factors to the rapid development of Smart Homes in intelligent cities. In contemporary Smart Home systems, there is a plethora of sensors, actuators, and relevant digital devices, all inter-connected wirelessly. Most of the time, they operate behind the scenes monitoring, collecting, storing, communicating, and processing data. Easy access to these wireless networks and devices on them is an essential prerequisite of the legitimate users of these Smart Homes, but that is also the very reason that has brought about new security and privacy challenges.

One of the main objectives in building Smart Homes within Intelligent Cities equipped with smart devices and smart appliances with Internet connectivity is to provide comfort to its user and efficiency of processing of devices and operation of systems within the homes such as security. To its owners, a Smart Home is a complex digital ecosystem that can be accessed, like any other Internet device or service, from anywhere in the world. However, the very same conveniences a Smart Home offers its owner also allow, unfortunately, an intruder to launch attacks on it remotely or sitting within its wireless network range. From the intruder's viewpoint, attacking a Smart Home either remotely or in the wireless range is more attractive than a physical break-in (Jose, 2017). Unlike other wireless sensor networks, the Smart Home wireless networks pose new security challenges. The attacker can easily fulfil their objective by targeting and compromising a less secure smart device in the home and make it to obey their orders. Smart Home is appealing to the intruder to target, as most of the existing defensive strategies are focusing on the attacks that happen in the application layer and the transmission layer and not so much in the physical network layer.

One way to improve the security of Smart Homes is to refuse access through the Internet, but that would defeat the very purpose of making a home "smart" and it would also kill the idea of Home and City Automation. Therefore, it is very important that access to Smart Homes over the Internet is provided securely and all in-house wireless networks with their connected devices are secured as much as possible. One way to achieve this is by narrowing access to the home over the Internet, and by restricting access only to a few selected people to the trusted fixed number of devices in the home. To accomplish this, it is necessary to identify *a priori* the number of devices and the users who need access to them.

A smarter and better way to improve the security of Smart Homes and to defend them from intruders is to identify authorised homeowner and other users, and identify the legitimate smart devices connected in the Smart Home. It is very challenging to analyse and understand the behaviour of a Smart Home all the time not only because of the uncertain behaviour and nature of human occupants, but also because the home could be occupied by guests and untrusted people. It is generally accepted that security in the IoT is a primary concern specially in detecting vulnerabilities or possible intrusions in the home network. Bertino & Islam (2017) propose that IoT devices with high computation power make them an easy and attractive target for attackers seeking to compromise these devices and use them to create large-scale botnets. Therefore, pinpointing the IoT devices used at home and accessing them securely is the next logical step towards a secure smart home. This makes the use of effective machine learning techniques for intrusion detection in an IoT environment an essential feature of an Intrusion Detection Systems (IDS).

Most advanced Smart Home protection schemes are helpful today in securing Smart Homes from remote attacks launched through the Internet but give very little importance to other types of network attacks. Considering this and the weaknesses of the defensive techniques used in a smart home environment, Smart Home Intrusion Detection System (SHIDS), a new hybrid approach to defend against activities on both the local and the wide area networks has been developed and tested. SHIDS is an integrated security framework built on a machine learning-based signature and anomaly intrusion detection scheme which improves the detection of network intrusions in Smart Homes. Specifically, the use of machine learning based naïve Bayes techniques improve the performance of the detection technique by analysing and identifying false positives accurately, thereby drastically reducing the false-positive ratio.

The IoT paradigm is gaining prominence around the world in terms of academic research, as well as its many useful applications for the society at large e.g. in the case of connected autonomous vehicles. It is also getting implemented in the real world in a big way, helping to control multiple devices from remote locations. Several IoT devices like smart watches, smart meters, Smart Home appliances, etc., already improve the quality of life of people. Smart watches and smart wearables contain sensors that can monitor the health of the wearer and upload data on to cloud storage regularly. Similarly, other devices can be controlled from remote locations and can be switched on and off whenever necessary (Romdhani, 2017). These devices have gained popularity with the reduction in the price of sensors, an increase in the popularity of cloud storage, and big data. This has enabled more access to these resources, thereby strengthening the IoT network.

This has led to the creation of a dedicated industry that participates in the security and updating of IoT networks. There is a general concern and a low level of confidence about the data obtained from these devices and how these data are used. However, there is a lack of research regarding the application of machine learning in the context of IoT by the concentration of detecting the existing intrusions. This was the main contribution of (Costa, et al., 2019). Recently, the use of machine learning for helping security and detecting has become necessary in the IoT network for facing the previously known challenges (Mehmood & Rais, 2016), (Bhuyan & Kalita, 2014). However, there are very few studies that have used machine learning in the context of security challenges in IoT networks (Costa, et al., 2019). Traditional machine learning algorithms like Bayesian Belief Networks (BBN) (Mascaro, et al., 2014), Support Vector Machines (SVM) (Mascaro et al., 2014) (Zhang, et al., 2013) and others (Buczak & Guven, 2016) (Hurley, et al., 2016) were used for cybersecurity. But, since the amount of data generated in the network is very high, a deep learning-based technique is proposed in Yavuz (2018). This can perform better with larger data and can adapt to a range of attack scenarios.

This chapter describes the design of an efficient security framework using a machine learning-based signature and anomaly intrusion detection scheme which detects the network intrusions in the Smart Homes. Specifically, machine learning-based naïve Bayes technique is used to improve the detection process which can analyse and identify false positive ratio accurately, giving a drastic reduction in the rate of false-positives.

The contribution in this chapter aims to do the following:

- examine present IDS evaluation techniques concerning detection, prevention, type of attack, challenges and applications and provide a unique solution to overcome the network security issues.
- create an optimised intrusion detection scheme for a wireless Smart Home network to gather information, identify and log the attacks, and alert an administrator.
- detect and prevent intrusions by using signature and intrusion-based hybrid intrusion detection system
- minimise the incorrectly classified instances, mean absolute error (MAE), and Root Mean Squared Error (RMSE) in a cloud-based IDS by incorporating effective clustering and classification technique
- test and report the performance of the proposed technique with network attacks like DoS, Probe, R2L, and U2R, and then validate the performance with a traditional approach.

RELATED WORK

Zarpelao, et al. (2017) present a review of different cybersecurity issues in the IoT platform. In addition, they propose a taxonomy algorithm toward classifying the articles concerning different attributes, namely, Intrusion Detection System (IDS) placement strategy, validation strategy, security threat, and intrusion detection. However, their approach does not cover the wide range of attacks and IoT technologies. Moreover, it does not adequately deal with different placement and detection approaches in IoT platforms. Hence, there is a need for a more complete and effective approach for attack and intrusion detection in IoT platforms.

Costa, et al. (2019) survey optimised security protocols which attain reasonable protection and keep data secure while preserving moderate or low energy consumption. They also discuss the intelligent approach that was applied in the context of intrusion detection in computer networks. The suggested methods seek to obtain a better average recognition rate. However, the false-positive ratio remains a complex issue. Few of the researchers have suggested intrusion detection methods that minimised the false positive ratio but there is an increase in training time and classification time. Some of them perform an inverse process that stabilises the false-positive ratio but at a high computational burden for both testing and training data. Hence, there is a need for an efficient method to minimise false-positive ratios but at the same time enhance the system performance where detection in real-time is a relevant factor.

Modi, et al. (2013) classify cloud-based Intrusion Detection System (IDS) into two types, namely signature and anomaly-based systems. The signature-based intrusion detection method is used where it frequently updates the database of signatures. However, the signature-based method is not successful for new attacks (Oktay & Sahingoz, 2013); (Iqbal, et al., 2016). For detecting unknown attacks, the anomaly-based IDS is utilised. However, when compared to the signature-based method, the anomaly technique is prone to high false-positive ratios (Fung & Zhu, 2016), (Abdelouahab, et al., 2015). On the other hand, an anomaly-based detection approach cannot detect any new kind of attack. Rajendran, et al. (2015) point out that Hybrid Intrusion Detection would be the solution to the above problem. More specifically, to enhance the detection accuracy, the hybrid combination will be better suited (Abusitta, et al., n.d.), (Fung & Zhu, 2016).

Research by Arrington et al. (2016) presents a host-based IDS that utilises a machine learning technique for anomaly-based intrusion detection. This suggested framework, which is based on artificial immune systems, increases the system complexity with the growth of the IoT platform and leads to degrading system performance while increasing resource consumption. A study by Raza, et al. (2013) presents a hybrid signature and anomaly-based IDS for intrusion detection on IoT networks. However, due to the centralised design, the suggested model is particularly oriented towards detecting routing protocol attacks in IoT. As a result, it cannot monitor the traffic amongst local nodes. To address this issue, Liu, et al. (2018) propose an IDS by combining Principal Component Analysis (PCA) algorithms and suppressed fuzzy clustering methods. The results obtained show better detection efficiency when relating to the neural network and Bayesian algorithms but the accuracy and efficiency of the IDS decreases when there is an increase in data volume. Hence, there is a necessity to improve the performance of the detection model. It is also reported in the literature that the use of traditional machine learning approaches result in less scalability and low accuracy when used for the detection of attacks on large-scale IoT networks (Thamilarasu & Chawla, 2019).

Bertino & Islam (2017) argue that, due to the diversity of devices, communication protocols and media, and software in the IoT, it is essential to extend all existing security mechanisms to fit the IoT ecosystem. In that context intrusion detection techniques and systems coupled with good antivirus software become even more important. However, it must be noted that when new types of security mechanisms are created against specified security threats, the attackers quickly conquer the situation by inventing new types of attack methods (Bertino & Islam, 2017). For instance, in the recent Distributed Denial of Service (DDoS) amplification attack, the locations became untraceable by defenders when the attackers utilised spoofed source IP (Internet Protocol) addresses. This means more destructive attacks of high complexity, when compared to Mirai, can be anticipated due to the vulnerabilities of IoT systems. Hence, there is a greater need for an effective system for protecting IoT systems (Kolias, et al., 2017), (Shakeri, et al., 2019). Though, there is a high level of exiting research activity, use of machine learning for anomaly detection in IoT environments is still in its infancy (Thamilarasu & Chawla, 2019). The work reported in this chapter aims to address this and reports the effective use of machine learning techniques for intrusion detection in an IoT environment.

Jose (2017) has worked on Intelligent Home Automation Security System (IHASS), and Sahoo, et al. (2012) mention how wireless communication is insecure in the face of internal and external attacks. A network that is deployed in a home is vulnerable to different type of attacks such as Sinkhole attack, Cloned ID attack, Sybil attack, and Selective Forwarding attack. The SVELTE IDS proposed by Reza, et al. (2013) use a hybrid intrusion detection approach consisting of both anomaly and signature-based filters to detect attacks such as routing and network-based attacks. SVELTE IDS which is based on a distributed architecture, may not be suitable for home automation networks due to the fact that on a home network, most devices are limited to a specific function and have limited processing power and, therefore, a distributed defensive approach may not be suitable at all times.

In the present-day scenario, the Internet and computer networks are playing a major role in modern society in terms of computational power, affordability, scalable storage, and resource availability. Specifically, the amount of data generated by day-to-day activities and stored in the cloud is also increasing exponentially (Costa, et al., 2019). Not only managing and accessing this huge collection of data become complex but it also gives rise to critical threats, ranging from viruses to network intrusions, causing significant data losses in numerous platforms. To combat this, some researchers suggest the use of intelligent frameworks and intrusion detection tools (KarsligEl, et al., 2017), (Bamakan, et al., 2016) which involve data encryption, firewalls, and intrusion detection systems for protecting networks from attacks and intrusions. However, with all the advancements in research, network security remains a crucial task. More specifically, a primary concern arises in intrusions detection in IoT with the broader acceptance of Internet Protocol version 6 (IPv6). The cooperative synergy among IPv6 and the IoT model, enabling free and easy access to the Internet by various gadgets, for example, cognitive buildings, clothing, house hold devices, wearable devices, microwaves, Smart Home (Wu, et al., 2019) (Pasini, et al., 2016) become responsible for making system security more complex. This makes the search for intrusion detection approach for IoT, and the creation of a security framework for the Smart Home, to become highly essential (Costa, et al., 2019).

SMART HYBRID INTRUSION DETECTION SYSTEM (SHIDS)

In the foregoing, it was argued that a signature-based detection system is weak in detecting novel attack types whereas anomaly detection is more attractive in such situations. Therefore, a hybrid intrusion detection system comprising both signature and anomaly-based techniques is better suited for effectively detecting intrusions in a smart home network. This section briefly describes such a Smart Hybrid IDS (SHIDS) system developed and tested by the authors.

SHIDS uses Particle Swarm Optimisation (PSO) and Simulated Annealing (SA) for feature selection and extraction. The process, very briefly, is as follows. The results are then optimised by using naive Bayes and Decision Tree Classification techniques. KDD CUP 99 data (Tavallaee, et al., 2009) which is the most widely used dataset for the evaluation of such intrusion detection systems, have been used to test the efficacy of SHIDS. Tavallaee, et al. (2009) believe that KDD CUP 99 can be an effective benchmark data set to help researchers compare different intrusion detection methods. The performance of SHIDS when run with KDD CUP 99 data is compared against the results of a study on the application of Machine Learning Techniques to the Network Intrusion Detection System published by Seraphim, et al., (2019).

The SHIDS can be described in four steps as follows:

1. Use a Hybrid Intrusion Detection System (HIDS) which is anomaly-based intrusion detection system to detect intrusions from a local area network. Anomaly detection is performed by assigning anomaly's score to an entire packet which will be analysed to create behaviour profile by packet analysis. This score depends on the measure of the event between the source and destination port.
2. Use the KDD CUP 99 dataset which comprises 4GB of compressed raw (binary) tcpdump data. About 5 million linking records, each about 100 bytes, can be extracted from this dataset. KDD

training data set consists of approximately 4,900,000 single connection vectors each of which contains 41 features and is labelled as either normal or an attack. These features selected and extracted by PSO and SA approach are used for network attack detection and categorisation.

3. After applying optimisation technique on feature selection, pass the selected features onto two different classifier techniques namely: naive Bayes and Decision Tree Classification to optimise such parameters as several clusters (K), confidence and max_runs. Specifically, Classification and Regression Trees (CART) based decision tree has been used, whereas the classification constructed through the binary splitting of the attribute will select the splitting attribute by utilising the Gini Index. To predict a dependent variable for a given set of predictor variables over a given period, the regression feature is utilised. It is observed that the suggested technique performs better with large data volumes and is adaptable to different attack scenarios in IoT.
4. To increase the overall efficiency of the IDS, a drop in the number of applicable traffic features without incurring any adverse effects on classification accuracy is required. To achieve this, the combination of naive Bayes and Decision Tree Classification techniques are utilised.

The naïve Bayes classifier operates on the probability of one attribute that does not affect the probability of another. First, the prior probability for the given intrusion data set (estimated by counting how often each class occurs in the dataset) is found and then the class conditional probability for the data set (estimated by counting how often each attribute value occurs in the class in the dataset) is evaluated. After that, the highest classifier probability is obtained and based on that the detection rate and false positive rate for the intrusion data set are found. To enhance the system classification performance, the naïve Bayes classifier that combines the Bayes' model with decision rules like the hypothesis which is the most probable outcomes is used. The architecture of the SHIDS is shown in Figure 1 and the algorithm used to optimise and classify the dataset is shown in Box 1.

For the algorithm in Box 1, various variables mean as follows:

D_{INP} INPut Data
F Features
X_i location
v_i Velocity
f_i Frequency at different iteration level
X_{best} optimal solution
D Entropy

Some of the processes relevant to SHIDS are now described in the following sections.

Data Pre-Processing

Pre-processing of data is needed for processing the text in the dataset. In order to remove any noise present and to correct the data, a statistical transform tool is necessary. In this work, missing values and outliers are compensated for by making the distribution more normal via the outlier detection technique. However, the missing values depend on individual features. While some features may have zero as a missing value, others have zero as part of its value where binary data is considered. After this, the features are extracted for efficient prediction. Since there are many features, only the necessary features must be

Figure 1. System Architecture

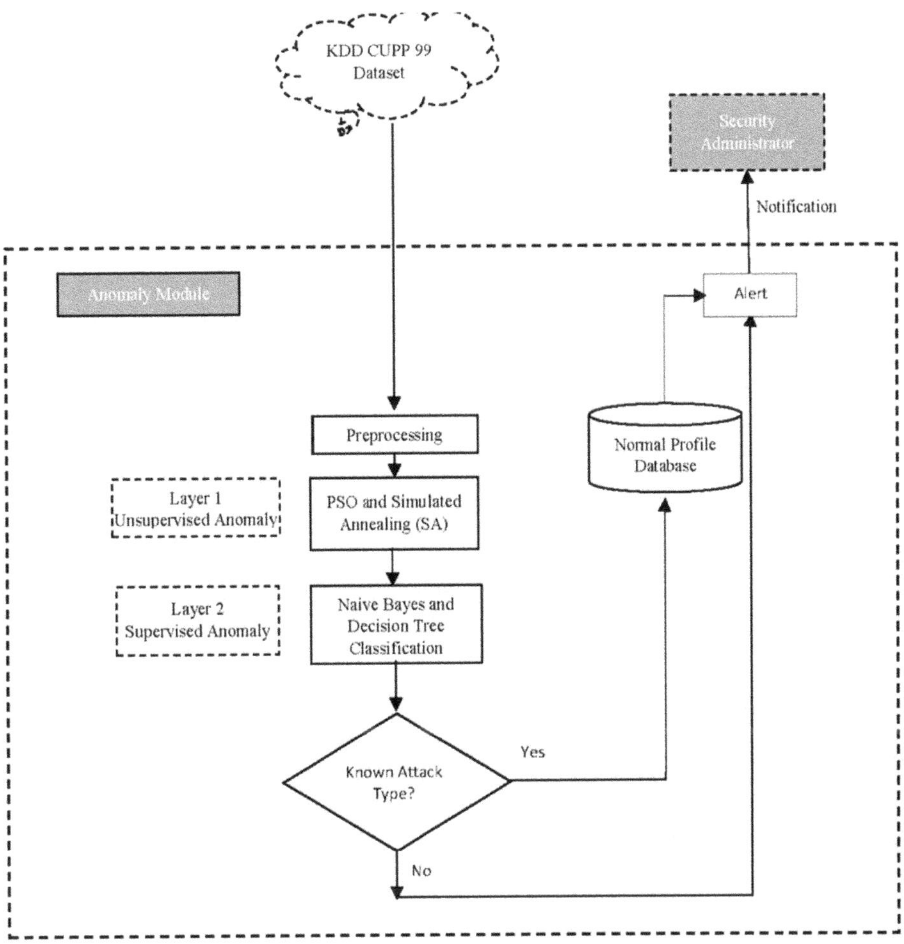

extracted. Hence, a combination of Particle Swarm Optimisation (PSO) and Simulated Annealing (SA) is used for determining and extracting the most suitable features.

With respect to the Particle Swarm Optimisation (PSO), the performance of IDS depends upon the correct dataset in Wireless Sensor Networks (WSNs). The dataset has a sample which is also known as features. Generally, these features are not known before, therefore it may affect the categorisation of performance of WSNs. PSO was proposed by Kennedy and Eberhart in 1995 (Chen, et al., 2014) and it is said that it was inspired by birds flocking to look for food in a specific area, where none of the birds knows where exactly the food is. The idea is that with each reiteration, birds know they get near to food. The best way to get food is to follow the birds that are near to the food. In PSO terminology, the group of birds is a swarm and a single bird is a particle.

In general, the use of PSO can be divided into three different categories (Chen, et al., 2014). The first category is the modification of PSO implementation itself both mathematics and topology wise. The second one is combining with another optimisation technique such as Genetic Algorithms (GA) (Shi, et al., 2005), (Robinson, et al., 2002). The third one is taking advantages of its chaotic behaviour such

Box 1. Algorithm for hybrid optimisation and classifier technique

Input: Dataset D_{INP}
Output: Reduced features F, Normal and/or Attack Data

Do
Initialise: $D_{INP} = \{F\}$
Take Class CL from D_{INP}
For each attribute value
Remove noise from D_{INP}
Calculate prior probability for each class in D_{INP}.
End For
For each attribute value
Calculate class conditional probabilities for each attribute value in D_{INP}
while $1 \leq t \leq$ Max_no._of_iterations do

for i=1 to n do
Generate new f_i (Frequency at iteration t)
Update X_i and v_i (Location and velocity)
Select a X_i from X_{best}
Generate a new X_{new} // generate the new value
Calculate fit (X_{new})
if $fit(X_i) \leq fit(X_{new})$ // calculate the new location value
$fit_{temp}(i) \leftarrow fit(X_{new})$
$X_{temp}(i) \leftarrow X_{new}$
$X_{best} \leftarrow X_{new}$ // To calculate the best solution
//Invoke Naïve Bayes classifier on a dataset with F features
S=S-{F}
F=S // Set F with reduced features
//For every feature 'F', calculate gain ratio by dividing the
//information gain of an attribute with a splitting value of
//that attribute.

$$\text{Gain (D,A)= Entropy(D)-} \sum_{j=1}^{v} \frac{|Dj|}{|D|} \text{ Entropy } D_j$$

Find the attribute with the highest gain ratio
Construct a decision node that divides the dataset on the attribute 'X_{best}'.
End if
End For
end while
End Do

as reliability, probability and stochastic property (Alatas, et al., 2009). The Chaos-PSO (COSPSO) is a hybrid of Chaos with PSO that was applied to the test function (Yang, et al., 2012).

In this implementation, the hybrid pre-processing technique PSO-AS, based on Particle Swam Optimisation and Simulated Annealing has been used. To find local and global optima it is important to use a hybrid of both PSO and AS to get combined advantages of both. Figure 2 shows the optimisation technique using PSO-AS algorithms.

Intrusion feature selection (also called attribute selection or feature extraction) recognises and selects an effective attribute subset to describe an effective mode in a relatively large data set that generally contains redundant and irrelevant attributes. During feature subset selection, the algorithm generally selects the minimal effective attribute set using the principle of classification accuracy, and the classification distribution will be maintained concord as much as possible before and after the intrusion feature selection.

Figure 2. System Flow Diagram for PSO-SA Optimisation Technique

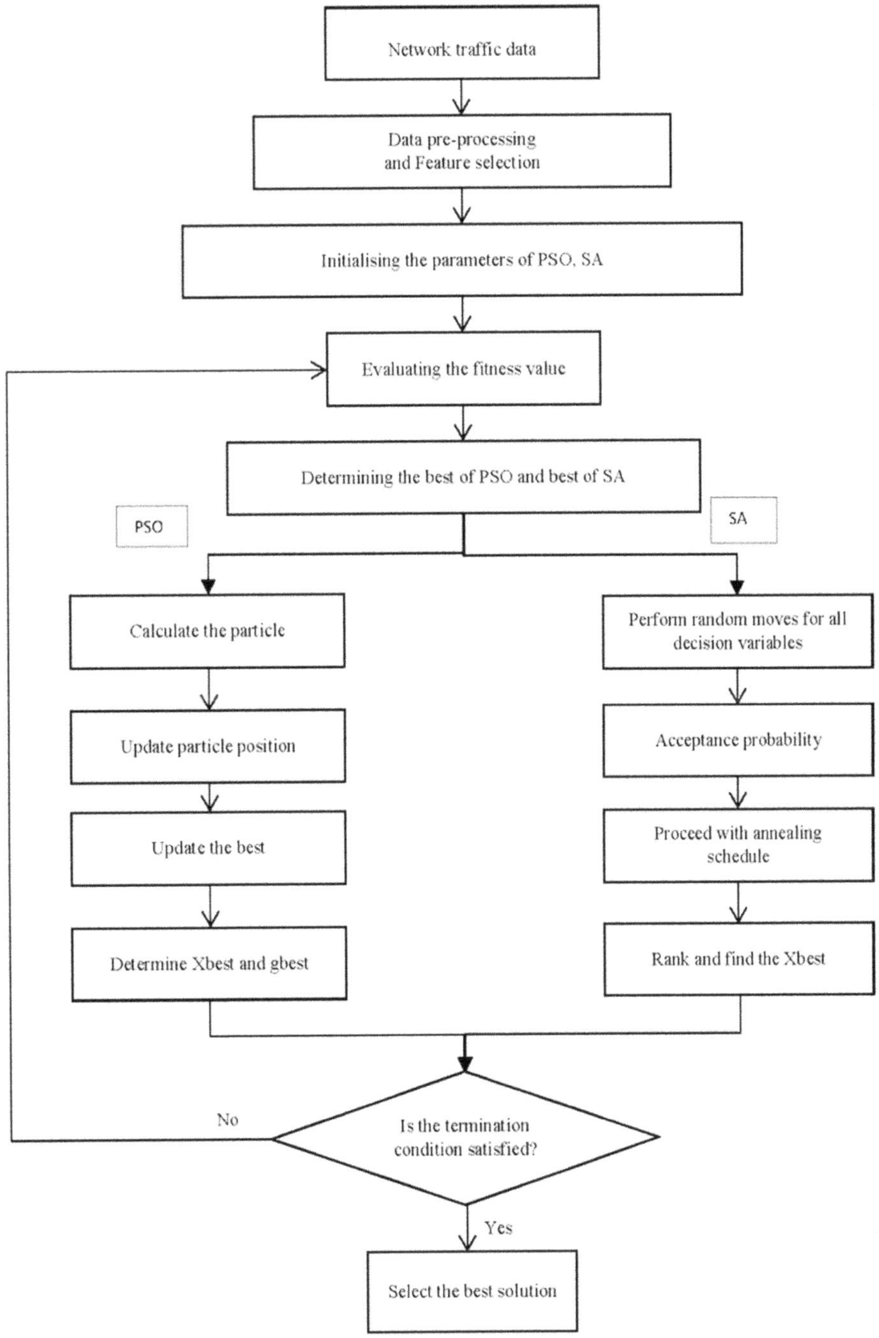

The feature selection depends on the below equation (Jabeena, et al., 2013).

$$V_i^{k+1} = wV_i^k + c_1 \, rand_1 (\ldots) \times (pbest_i - s_i^k) + c_2 \, rand_2(\ldots) \times (gbest - s_i^k)$$

where

V_i^k: direction of IDS agent i at iteration k,
w: weighting function,
c_j: weighting factor,
rand: uniformly distributed random number between 0 and 1,
s_i^k: current position of IDS agent i at iteration k,
$pbest_i$: pbest of IDS agent i,
gbest: gbest of the group.

In the above equation, the weighting function w is obtained from:

w = wMax-[(wMax-wMin) x iter]/maxIter

where

wMax= initial weight,
maxIter = maximum iteration number,
wMin = final weight,
iter = current iteration number

The number of clusters k is initialised to n + 1 where n is the number of categories that make up the problem: k = n + 1.

The best optimal feature is selected by:

Figure 3.

$$P = \left(exp \frac{\left(-(f_i - f_{i-1})\right)}{T_i} \right)$$

where f represents sum of Euclidean distances between each centroid and its objects and T is an iteration value.

Machine-Learning Implementation

Regarding what happens after feature extraction depends on the condition of the data traffic, it will go either to training phase or to detection phase. During the training phase, as it is learning the process, it will collect all the relevant information which has been asked about the packet behaviour and create a profile about the current data traffic. This is sent and stored in the Normal Profile Database (NPD) (see figure 1). If the IDS is in active mode and packets are in the detection phase, it will be sent to Anomaly Detection (AD). Anomaly Detection will check the behaviour and properties of the packets. An algorithm based on Bayes' theorem is used for further investigation and making sure the results produced by Anomaly Detection minimise false positive alarms. If the properties of the packet do not match the normal profile which is stored in the database (NPD), anomaly detection will send a message to alarm generation to trigger an alarm and inform the security management. If security management think that the alert received relates to normal data traffic and not a threat to the security system, then it will/can change the data label appropriately and store in the normal profile database (NPD) to avoid triggering an alarm next time. Figure 4 represents the system flow for SHIDS.

Figure 4. SHIDS Flow Diagram

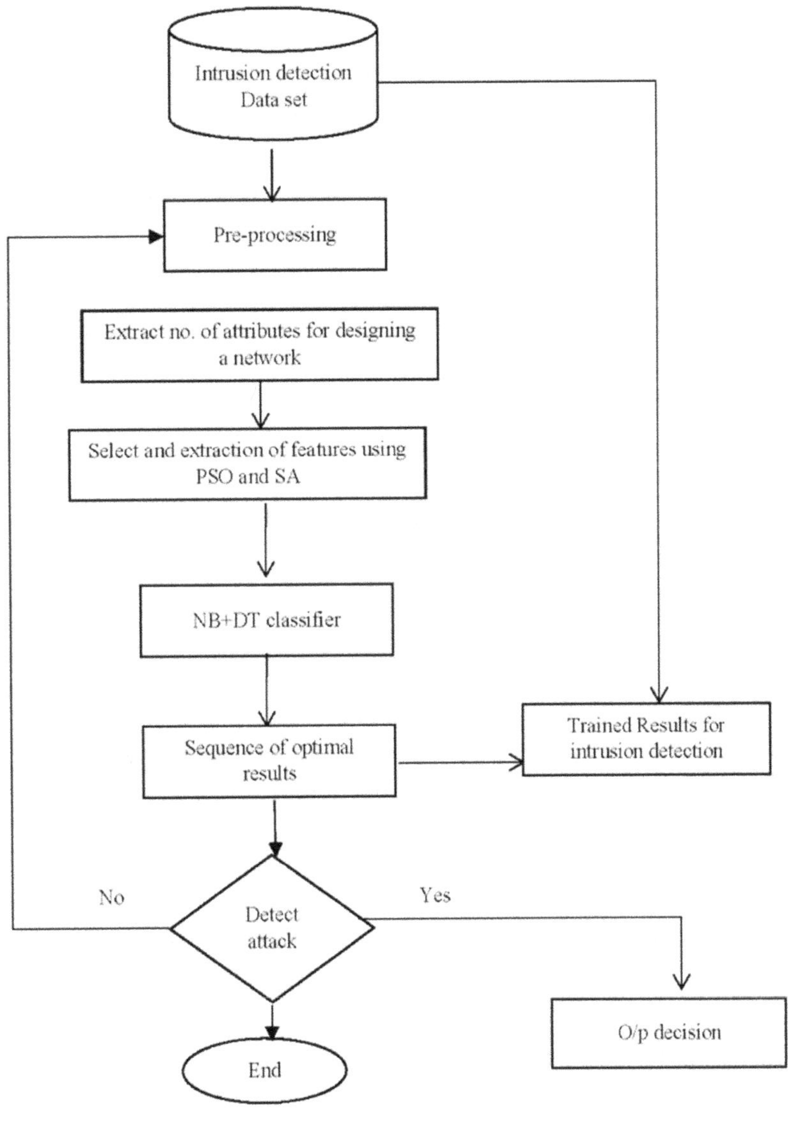

Simulation of the Proposed System

To evaluate the performance of the proposed detection system against various attacks on IoT networks, simulations of Denial of Service (DoS) attack, Probing (Probe) attack, Remote to Local (R2L) attack, and User to Root (U2R) attack have been carried out. The following is a brief description of each of the four types of attacks.

1. DoS attack: It is a type of attack that stops an authorised user from accessing services. By carrying out a DoS attack, it is possible to shut down one machine or even the whole network to make it out of reach to the legitimate user (Cosimo, et al., 2018). The outcome of DoS attacks is simulated by flooding target device with traffic, or letting it crash after receiving the attack information.
2. Probe attack: This is an attack where the attacker attempts to gather information about the target machine or the network, to map out the network. Information about target may reveal useful information such as open ports, its IP address, hostname, and operating system (Khamphakdee & Saiyod, 2015). This information can be used later on to launch attacks on the target.
3. R2L attack: It is a type of attack that is used by an attacker to try to gain access to target machine without the owner's permission (Jeya & Ravichandran, 2012).
4. U2R attack: This is a variety of attack where the attacker already has access to the victim's machine and is used to gain admin privileges by escalating permission to the root (Cosimo, et al., 2018).

Different types of popular attacks in each of the above categories are shown in Table 1.

Table 1.Different types of DoS, Probe, R2L and U2R attacks

Attack	Attack types
DoS	back, land, Neptune, pod, smurf, teardrop
Probe	ipsweep, nmap, portsweep and satan
R2L	ftp, write, guess_password, imap, multihop, phf, spy, warezmaster
U2R	buffer_overflow, loadmodule, perl, rootkit

Hybrid intrusion detection takes advantage of both anomaly and intrusion-based detection systems. Snort (2020), an open source intrusion prevention system, is a signature-based detection system that has a strong community following, and has been available for personal and commercial use for over a decade. There are different types of subscriptions e.g. free or paid. Snort works on signature-based rules, which are divided into header and option (Park & Kang, 2017). The header is a very important part that contains Rule Action, Protocol Type, Source IP Address, Source Port, Traffic Transmission Direction, Destination IP Address, and Destination Port. Rule Action plays a vital role because it must decide whether the packet goes through when the packet matches the rule. Snort also has other rule actions like alert, pass, and log but in many cases, it will alert an admin. Snort has multiple components which work together to detect different type of signature-based attacks.

Snort detects and monitors network activities and compares with set signatures to protect the network from threats. It triggers an alarm in case of detection of an anomaly. This approach is simple to use but not very useful for detecting new intrusions because the anomaly detection system is trained for a certain time to define normal behaviour and analyse the event on the base of the pre-defined normal behaviour. The normal profile is created during the training period and this is a time-consuming process. Even though it is effective in identifying new intrusions, it may also generate false positive alarms because not every unusual activity needs to be an intrusion (Xie, et al., 2011).

An IDS can be designed either as centralised or distributed. In a centralised approach, data collection, audit and control are all central, whereas in the disturbing approach data collection, audit and control are distributed. An example of a distributed IDS is the mobile agent-based IDS. The anomaly-based IDS, SHIDS presented here is a centralised IDS where data collection, audit and control processing is done centrally.

In the Smart Hybrid Intrusion Detection System (SHIDS) presented in this chapter, Snort network-based IDS, a tried and tested solution, is used to give protection to Smart Home from external threats coming from the Wide Area Network (WAN). Snort has two types of IDS, one is host-based IDS, and the other is network-based IDS. In order to provide strong security, both host-based and network-based IDS are needed. Snort can monitor both hardware and software and it does have an alarm system to notify the admin. It can detect a single session attack (making it much easier to detect an inexperienced hacker) as well as multiple session attacks by a determined experienced hacker who will somehow find a way to hack into the network.

Hardware Implementation

A Smart Home consists of various smart devices that are connected directly to the network. Hence, they are vulnerable to network attacks arising from the Internet. These connected devices must be protected by preventing intrusions. In order to do this, different modules are utilised in the implementation. The traffic flow must be monitored continuously by the classifier for identifying any sign of intrusion in the network. Two modules are used here, namely processing module and a power supply module. Since the voltage and current ratings of electronic components vary drastically, it is necessary to keep a separate module to distribute the power.

Raspberry Pi 4B is used in this work as the processor to classify the data accordingly. It is a powerful unit with a high processing power when compared to other controllers in a similar price range. Tensor Flow, a library package exclusive for AI is installed in the Raspberry Pi module to give hybrid artificial intelligence capability. The trained data model is uploaded onto the Raspberry Pi module and the module is connected to the network. The data acquisition module is used to continuously scan the individual packets in the network. The incoming packets are classified by the AI module based on the previously trained data into a legitimate packet or an anomaly. In case that an anomaly is detected, the process must block the packet from entering the system. Only the intrusion free packets will be allowed to enter the smart devices. By using this setup, smart devices can be protected from vulnerabilities arising from network attacks.

PERFORMANCE EVALUATION OF SHIDS

The performance of suggested IDS framework has been measured using four factors namely, true negatives (δ), false negatives (β), true positives (α) and false positives (γ) (Elrawy et al., 2018). During the prediction of anomaly class, a correct classification which represents an intrusion is known as true positive (αA) and incorrect classification of an intrusion, i.e. there is no intrusion, is known as false positive (γA). Subsequently, a correct classification that is designated as no intrusion is known as true negative (δA) and incorrect classification which shows no intrusion when there is an intrusion is known as false negative (βA). The probability of detecting intrusions, that means True Positive Rate (TPR) is mathematically written as:

$$TPR = \frac{\alpha_A}{\alpha_A + \beta_A}$$

Similarly, the probability of incorrectly identifying normal behaviour as an intrusion is known as a False Positive Rate (FPR) which is mathematically written as:

$$FPR = \frac{\gamma_A}{\gamma_A + \delta_A}$$

The fraction of total relevant records in a database that is retrieved by searching is known as Recall (R), which is the same as the measure of TPR. The fraction of relevant records among the records retrieved is known as Precision (P) which is mathematically written as:

$$P = \frac{\alpha_A}{\alpha_A + \gamma_A}$$

The stability amongst precision and recall is known as F-score which is mathematically written as:

$$F = \frac{2 * P * R}{P + R}$$

The overall success rate is measured based on the fraction of correct classifications which is mathematically written as:

ErrorRate = 1 - SuccessRate

Based on the above parameters, the performance was validated and compared in terms of how the proposed framework gives better performance compared to the traditional methods.

Experimental Results

In order to assess the effectiveness of the proposed SHIDS framework, it was evaluated by simulating different types of network traffic using KDD CUP 99 data comprising 4GB of compressed raw (binary) tcpdump data. Figure 5 shows the frequency of occurrence of each traffic class classified as Dos, Probe, R2L, U2R attacks in the data set.

Figure 5. Frequency of Occurrence of Attack Classes

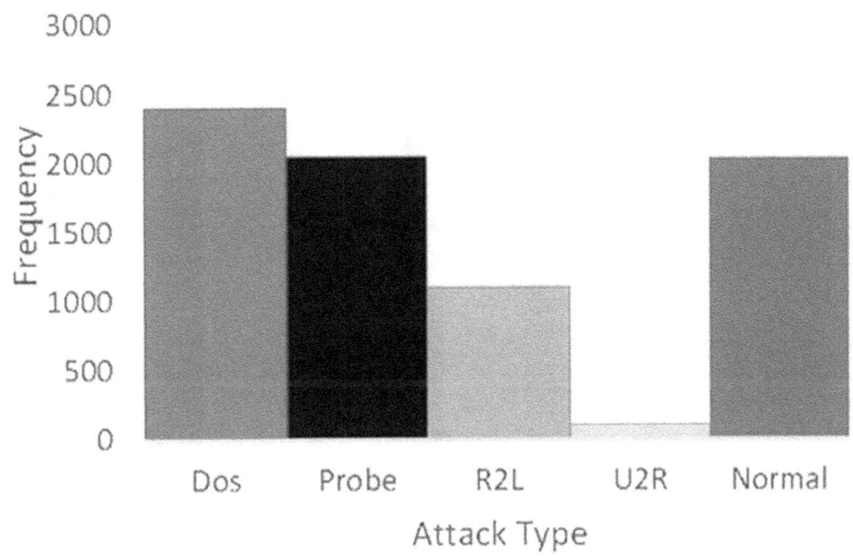

The Accuracy, False Positive Rate, Precision, Recall, F-score measured based on the below formulae, are shown in Figure 6.

$$Accuracy = \frac{|TP| + |TN|}{|TP| + |FP| + |TN| + |FN|}$$

$$Recall = \frac{|TP|}{|TP| + |FN|}$$

$$Precision = \frac{|TP|}{|TP| + |FP|}$$

$$Success\_rate = (TP + TN / \left(TP + TN + FP + FN\right)$$

where
TP- True positive, TN- True Negative, FP- False Positive, FN- False negative.

ErrorRate = 1- SuccessRate

F1 Score = 2*(Recall * Precision) / (Recall + Precision)

Figure 6. Experiment Results

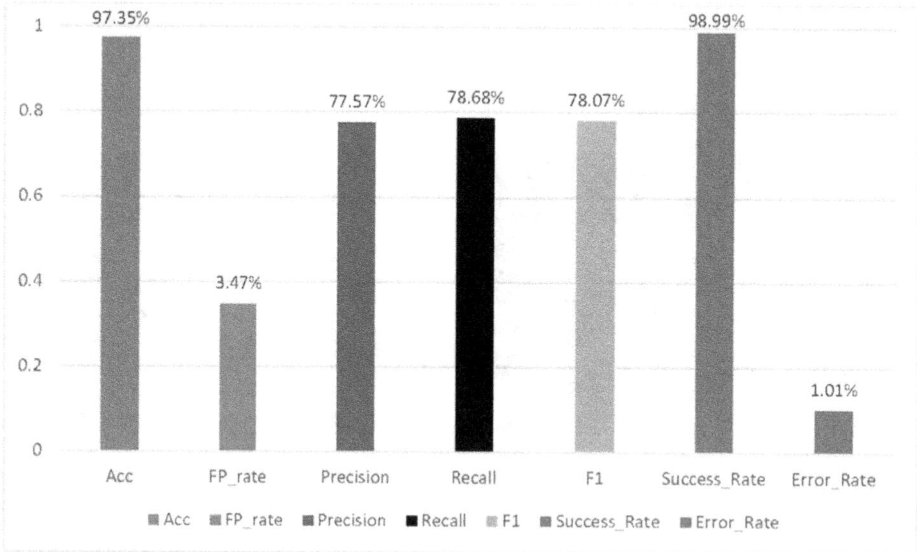

Comparison of Results

In order to check the validity of SHIDS against other known methods, the results were compared with the results published by Seraphim et al. (2019) which have been obtained using Decision Tree Model, KNeighbour Classifier, and Logistic Regression Methods. The result of the comparison is shown in Table 2. As can be seen, the accuracy obtained using SHIDS at 97.48% is higher than the traditional IDS

Table 2. Experimental Results Comparison

Author	Method	Accuracy	Precision	Re-call	F-score
Research by (Seraphim, et al., 2019)	Logistic regression	0.85	0.85	0.85	0.85
	KNeighbour classifier	0.87	0.88	0.87	0.87
	Decision Tree	0.95	0.86	0.91	0.88
SHIDS model		0.9748	0.7772	0.7862	0.7812

methods whereas the precision, re-call, and F-score show a slightly lower performance. This is work in progress and further studies will be carried out to see if these parameters can be improved.

CONCLUSION

In the SHIDS framework presented in this chapter, machine learning-based naïve Bayes technique is used to improve the detection performance which can analyse and identify false ratio accurately. It can be seen from experiment results presented in Figure 6 that the false-positive rate at 3.47% and error rate of 1.01% are indeed very low. High accuracy is most important for any IDS and the ability to detect attacks depends upon higher percentage accuracy. This in turn leads to least number of false alarms (Wahyudi & Kalamullah, 2016). In this case, feature engineering would be useful to reduce the amount of data by selecting only useful features leading to improved accuracy. Doing so will result in a further reduction of meaningless calculations on useless features. Misleading or unrelated features can also be removed as a measure towards improving accuracy further.

REFERENCES

Abdelouahab, J., Rodrigues's, D., De Melo's, L., & Abdelouahab's, Z. (2015). EICIDS-elastic and internal cloud-based detection system. *International Journal of Communication Networks and Information Security*, *7*, 34–49.

Abusitta, A., Bellaiche, M., & Dagenais, M. (2019, October). Multi-cloud cooperative intrusion detection system: Trust and fairness assurance. *Springer Nature*, *74*(9-10), 637–653. doi:10.100712243-019-00724-8

Alatas, B., Akin, E., & Ozer, A. B. (2009). Chaos embedded particle swarm optimization algorithms. *Chaos, Solitons, and Fractals*, *4*(4), 1715–1734. doi:10.1016/j.chaos.2007.09.063

Arasteh, H., Hosseinnezhad, V., Loia, V., Tommasetti, A., Troisi, O., Shafie-khah, M., & Siano, P. (2016). Iot-based smart cities: A survey. *IEEE 16th International Conference on Environment and Electrical Engineering,* 1-6.

Arrington, B., Barnett, L., Rufus, R., & Esterline, A. (2016). Behavioral Modeling Intrusion Detection System (BMIDS) (BMIDS) Using Internet of Things (IoT) Behavior-Based Anomaly Detection via Immunity-Inspired Algorithms. *International Conference on Computer Communication and Networks,* 1-6. 10.1109/ICCCN.2016.7568495

Bamakan, S. M. H., Wang, H., Yingjie, T., & Shi, Y. (2016). An effective intrusion detection framework based on MCLP/SVM optimized by time-varying chaos particle swarm optimization. *Neurocomputing, 16*(7), 90-102.

Bertino, E., & Islam, N. (2017). Botnets and Internet of Things Security. *Computer, 50*, 76-79.

Bhuyan, M. H., & Kalita, D. K. B. J. K. (2014). Network Anomaly Detection: Methods, Systems and Tools. *IEEE Communications Surveys and Tutorials*, *16*(1), 303–336. doi:10.1109/SURV.2013.052213.00046

Buczak, A. L., & Guven, E. (2016). A Survey of Data Mining and Machine Learning Methods for Cyber Security Intrusion Detection. *IEEE Communications Surveys and Tutorials*, *18*(2), 1153–1176. doi:10.1109/COMST.2015.2494502

Chen, S., Xu, Z., Tang, Y., & Liu, S. (2014). An Improved Particle Swarm Optimization Algorithm Based on Centroid and Exponential Inertia Weight. *Mathematical Problems in Engineering*, *07*, 1–14. doi:10.1155/2014/976486

Cosimo, I., Ahsan, A., Mandar, G., & Kia, D. (2018). Statistical Analysis Driven Optimized Deep Learning System for Intrusion Detection. Academic Press.

da Costa, K. A., Papa, J. P., Lisboa, C. O., Munoz, R., & de Albuquerque, V. H. C. (2019). Internet of Things: A survey on machine learning-based intrusion detection approaches. *Computer Networks*, *151*, 147–157. doi:10.1016/j.comnet.2019.01.023

Elrawy, M. F., Awad, A. I., & Hamed, H. F. A., 2018. Intrusion detection systems for IoT-based smart environments: a survey. *Journal of Cloud Computing: Advances, Systems and Applications.*

Fung, C. J., & Zhu, Q. (2016). FACID: A trust-based collaborative decision framework for intrusion detection networks. *Ad Hoc Networks*, *53*, 17–31. doi:10.1016/j.adhoc.2016.08.014

Hui, T. K., Sherratt, R. S., & Sánchez, D. D. (2017). Major requirements for building Smart Homes in Smart Cities based on Internet of Things technologies. *ScienceDirect*, *76*, 358–369. doi:10.1016/j.future.2016.10.026

Hurley, T., Perdomo, J. E., & Perez-Pons, A. (2016). HMM-based Intrusion Detection System for Software Defined Networking. *15th IEEE International Conference on Machine Learning and Applications (ICMLA),* 617-621. 10.1109/ICMLA.2016.0108

Iqbal, S., Kiah, L. M., Dhaghighi, B., Hussain, M., Khan, S., Khan, M. K., & Choo, K. R. (2016). On cloud security attacks: A taxonomy and intrusion detection and prevention as a service. *Journal of Network and Computer Applications*, *74*, 98–120. doi:10.1016/j.jnca.2016.08.016

Ismagilova, E., Hughes, L., Dwivedi, Y. K., & Raman, K. R. (2019). Smart cities: Advances in research—An information systems perspective. *ScienceDirect*, *47*, 88–100. doi:10.1016/j.ijinfomgt.2019.01.004

Jabeena, A., Jayabarathi, T., Singh, G. S., & Kumar, A. A. (2013). PSO Based Optimization Technique for Wireless Optical Communication. *Asian Journal of Computer and Information Systems*, *1*(1), 11–16.

Jeya, P. G., & Ravichandran, M. R. a. C. S. (2012). Efficient Classifier for R2L and U2R Attacks. *International Journal of Computer Applications, 45*(21), 28-32.

Jose, A. C. (2017). Intelligent Home Automation Security System based on Novel Logical Sensing and Behaviour Prediction. In Intelligent home automation security system based on novel logical sensing and behaviour prediction. University of Pretoria.

KarsligEl, M. E., Yavuz, A. G., Güvensan, M. A., & Bank, K. H. a. H. (2017). Network intrusion detection using machine learning anomaly detection algorithms. *2017 25th Signal Processing and Communications Applications Conference (SIU),* 1-4.

Khamphakdee, N., & Saiyod, N. B. S. (2015). Improving Intrusion Detection System Based on Snort Rules for Network Probe Attacks Detection with Association Rules Technique of Data Mining. *J. ICT Res. Appl*, *8*(3), 234–250. doi:10.5614/itbj.ict.res.appl.2015.8.3.4

Kolias, C., Kambourakis, G., Stavrou, A., & Voas, J. (2017). DDoS in the IoT. *Mirai and Other Botnets. in Computer*, *50*(7), 80–84.

Liu, L., Xu, B., Zhang, X., & Wu, X. (2018). An intrusion detection method for internet of things based on suppressed fuzzy clustering. *EURASIP Journal on Wireless Communications and Networking*, *2018*(1), 1–7. doi:10.118613638-018-1128-z

Mascaro, S., Nicholso, A. E., & Korb, K. B. (2014). Anomaly detection in vessel tracks using Bayesian networks. *International Journal of Approximate Reasoning*, *55*(1), 84–98. doi:10.1016/j.ijar.2013.03.012

Mehmood, T., & Rais, H. B. M. (2016). Machine learning algorithms in context of intrusion detection. *3rd International Conference on Computer and Information Sciences (ICCOINS),* 369-373. 10.1109/ICCOINS.2016.7783243

Modi, C., Patel, D., Patel, H., Borisaniya, B., Patel, A., & Rajarajan, M. (2013). A survey of intrusion detection techniques in Cloud. *Journal of Network and Computer*, *36*(1), 42–57. doi:10.1016/j.jnca.2012.05.003

Oktay, U., & Sahingoz, O. K. (2013). Proxy Network Intrusion Detection System for Cloud Computing. *2013 The International Conference on Technological Advances in Electrical, Electronics and Computer Engineering (TAEECE),* 98-104. 10.1109/TAEECE.2013.6557203

Park, W., & Kang, B. H. (2017). *An Enhancement of Optimized Detection Rule of Security Monitoring and Control for Detection of Cyberthreat in Location-Based Mobile System*. doi:10.1155/2017/8501976

Pasini, D., Ventura, S. M., Rinaldi, S., Bellagente, P., Flammini, A., & Ciribini, A. L. C. (2016). Exploiting Internet of Things and building information modeling framework for management of cognitive buildings. *IEEE International Smart Cities Conference (ISC2),* 1-6. 10.1109/ISC2.2016.7580817

Rajendran, P. K., Rajesh, M., & Abhilash, R. (2015). Hybrid Intrusion Detection Algorithm for Private Cloud. *Indian Journal of Science and Technology, 12*(8).

Raza, S., Wallgren, L., & Voigt, T. (2013). SVELTE: Real-time intrusion detection in the Internet of Things. *Ad Hoc Networks*, *11*(08), 2661–2674. doi:10.1016/j.adhoc.2013.04.014

Robinson, J., Sinton, S., & Samii, Y. R. (2002). Particle swarm, genetic algorithm, and their hybrids: optimization of a profiled corrugated horn antenna. *Proceedings of the IEEE International Symposium in Antennas and Propagation Society, 1*, 314–317.

Romdhani, I. (2017). Confidentiality and Security for IoT Based Healthcare. In Securing the internet of things. Syngress. doi:10.1016/B978-0-12-804458-2.00009-3

Sahoo, M. S., Mishra, M. P. K., & Satpathy, P. N. (2012). Secure Routing in Wireless Sensor Networks. *International Journal of Computer Science, 9*(2).

Seraphim, I., Palit, S., Srivastava, K., & Poovammal, E. (2019). Implementation of Machine Learning Techniques applied to the Network Intrusion Detection System. *International Journal of Engineering and Advanced Technology*, *8*(5), 2721-2726.

Shakeri, R., Al-Garadi, M. A., Badawy, A., Mohamed, A., Khattab, T., Al-Ali, A., ... Guizani, M. (2019). Design Challenges of Multi-UAV Systems in Cyber-Physical Applications: A Comprehensive Survey, and Future Directions. *IEEE Communications Surveys and Tutorials*, *21*(4), 1–50. doi:10.1109/COMST.2019.2924143

Shi, X., Liang, Y. C., Lee, H. P., Lu, C., & Wang, L. M. (2005). An improved GA and a novel PSO-GA-based hybrid algorithm. *Information Processing Letters*, *93*(5), 255–261. doi:10.1016/j.ipl.2004.11.003

Snort. (2020). https://www.snort.org/

Tavallaee, M., Bagheri, E., Lu, W., & Ghorbani, A. A. (2009). A detailed analysis of the KDD CUP 99 data set. *2009 IEEE Symposium on Computational Intelligence for Security and Defense Applications*, 1-6. 10.1109/CISDA.2009.5356528

Thamilarasu, G., & Chawla, S. (2019). Towards Deep-Learning-Driven Intrusion Detection for the Internet of Things. *Sensors (Basel)*, *19*(09), 1977. doi:10.339019091977 PMID:31035611

Wahyudi, M. B., & Kalamullah, R. (2016). *Improving Intrusion Detection System Detection Accuracy and Reducing Learning Time by Combining Selected Features Selection and Parameters Optimization*. IEEE.

Wu, W., Pirbhulal, S., Zhang, H., & Mukhopadhyay, S. C. (2019). Quantitative Assessment for Self-Tracking of Acute Stress Based on Triangulation Principle in a Wearable Sensor System. *IEEE Journal of Biomedical and Health Informatics, 23*(2), 703-713.

Yang, C. H., Tsai, S. W., Chuang, L. Y., & Yang, A. C. H. (2012). An improved particle swarm optimization with double-bottom chaotic maps for numerical optimization. *Applied Mathematics and Computation, 219*(1), 260–279. doi:10.1016/j.amc.2012.06.015

Yavuz, F. Y. (2018). *Deep Learning in Cyber Security for Internet of Things*. Graduate School of Natural and Applied Sciences.

Zarpelao, B. B., Miani, R. S., Kawakani, C. T., & de Alvarenga, S. C. (2017). A survey of intrusion detection in Internet of Things. *ACM Digital Library*, *84*(C), 25–35. doi:10.1016/j.jnca.2017.02.009

Zhang, Y., Meratnia, N., & Havinga, P. J. (2013). Distributed online outlier detection in wireless sensor networks using ellipsoidal support vector machine. *Ad Hoc Networks*, *11*(3), 1062–1074. doi:10.1016/j.adhoc.2012.11.001

ADDITIONAL READING

Airehrour, D., Gutierrez, J., & Ray, S. K. (2016). Secure routing for internet of things: A survey. *Journal of Network and Computer Applications*, *66*, 198–213. doi:10.1016/j.jnca.2016.03.006

Costa, D. G., Ivanovitch, S., Luiz, G. A., & Paulo, V. F. P. (2014). ability issues in wireless visual sensor networks. *Sensors (Basel)*, *14*(2), 2795–2821. doi:10.3390140202795 PMID:24526301

Jose, A. C., & Malekian, R. (2015). Smart Home Automation Security: A Literature. *Smart Computing Review*, *5*(4), 269–285.

Thamilarasu, G., & Chawla, S. (2019). Towards Deep-Learning-Driven Intrusion Detection for the Internet of Things. *Sensors (Basel)*, *19*(09), 1977. doi:10.339019091977 PMID:31035611

Zekrifa, D. M. S. (2014). *Hybrid Intrusion Detection System*. The University of South Australia.

KEY TERMS AND DEFINITIONS

Anomaly Detection: This refers to the act of identifying rare events in network operation that could be the result of network intrusions. Anomalies raise suspicion because they differ significantly from the data observed during normal operation.

Best Solution (in PSO): This is the optimum solution for the particle's position obtained by applying the cost optimisation function to randomly moving particles.

Classification and Regression Trees (CART): CART refers to decision tree algorithms used in machine learning for the purpose of classification and regression. It belongs to supervised machine learning algorithm category. The classification tree part of the CART algorithm is used to identify the class within which the target variable is most likely to fall whereas the regression tree is used to predict the value of the variable.

Euclidean Distance: Euclidean distance is used in Particle Swarm Optimisation (PSO) to measure the distance between individual particles in Euclidean space.

Gini Index: Gini index gives the probability of incorrectly classifying labels when randomly selected from data. It is a measure of the impurity of the dataset and is used as a metric in CART to create binary splits.

Internet of Things (IoT): This refers to a collection of physical objects such as home appliances, personal digital devices, wearables, security systems, smart buildings, smart vehicles, national critical infrastructure, etc., embedded with sensors, actuators, microcontrollers, storage, and software and connected to other similar devices and systems over the Internet.

Logistic Regression Methods: It refers to statistical techniques used in machine learning for solving binary classification problems in which there is one or more independent variables that determines the dependent variable (outcome).

Machine Learning: It refers to a technique that automates building analytical models using algorithms to analyse large amounts of data. It is based on the idea that computer systems can learn by identifying patterns within data and make decisions with minimal human intervention.

Mirai: This refers to a malware that takes advantage from flaws or security holes in IoT devices, especially those running Linux, and turns them into remotely controlled bots. It infects smart home devices and change them into bots or zombies that can be controlled remotely.

Naïve Bayes: Naïve Bayes is an algorithm used in supervised machine learning for predictive modelling. It is based on the assumption that the features of a measurement are statistically independent.

Network Intrusion: This refers to any unauthorised activity on a computer network with the aim of either stealing valuable network resources or jeopardising the security of the network and data.

Particle: In an optimisation problem where there is a population of candidate solutions, the candidate solution that is chosen as the one to optimise over the search space is called a particle.

Particle Swarm Optimisation (PSO): This refers to an optimisation technique that optimises a problem by iteratively improving a candidate solution a.k.a. particle. Simple mathematical formulae are used to move these particles in search space in search of the optimised solution.

Principal Component Analysis (PCA): It refers to a statistical technique that transforms a set of possibly correlated variables into a set of linearly correlated variables called principal components.

Simulated Annealing (SA): This refers to the unsupervised machine learning algorithm based on features optimisation to extract different features from the given data with the objective of finding a global optimum.

Compilation of References

Aarts, E., & Encarnação, J. L. (2005). Into Ambient Intelligence. In True Visions: Tales on the Realization of Ambient Intelligence. Springer Verlag.

Abbasi, M. H., Majidi, B., & Manzuri, M. T. (2018). Deep cross altitude visual interpretation for service robotic agents in smart city. In *2018 6th Iranian Joint Congress on Fuzzy and Intelligent Systems (CFIS)* (pp. 79-82). IEEE.

Abbass, W., Bakraouy, Z., Baina, A., & Bellafkih, M. (2019). Assessing the Internet of Things Security Risks. *Journal of Communication*, *14*(10), 958–964. doi:10.12720/jcm.14.10.958-964

Abdelouahab, J., Rodrigues's, D., De Melo's, L., & Abdelouahab's, Z. (2015). EICIDS-elastic and internal cloud-based detection system. *International Journal of Communication Networks and Information Security*, *7*, 34–49.

Aboueleta, M. (2019). *Atmospheric water harvesting: Our one environmental chance against a water crisis*. Retrieved from https://www.quantumrun.com/article/atmospheric-water-harvesting-our-one-environmental-chance-against-water-crisis

Abu-Matar, M., & Mizouni, R. (2018). Variability Modeling for Smart City Reference Architectures. In *The Fourth IEEE Annual International Smart Cities Conference (ISC2 2018)*. Kansas City, MO: IEEE. 10.1109/ISC2.2018.8656967

Abusitta, A., Bellaiche, M., & Dagenais, M. (2019, October). Multi-cloud cooperative intrusion detection system: Trust and fairness assurance. *Springer Nature*, *74*(9-10), 637–653. doi:10.100712243-019-00724-8

ACCA CBE. (2019). *Platform Support* Retrieved from accaglobal.com/gb/en/student/exam-support-resources/professional-exams-study-resources/cbe-practice-platform/-support.html

Adepoju, A. (2004). *Trends in International Migration in and from Africa.* Retrieved from researchgate.net/publication/289568094-Trends-in-International-Migration-in-andaroundAfrica

Adrien, B., Juan, H.-S., Maarc, S., Mischa, D., Apostolos, K., & Dominique, B. (2011). Security and Privacy in your Smart City. *Centre Tecnologic de Telecomunicacions de Catalunya*, *3*(1), 1–6.

Africa, P. (2019a). *Ghana. Energy sector overview.* Retrieved from https://www.usaid.gov/sites/default/files/documents/1860/Ghana_-_November_2018_Country_Fact_Sheet.pdf

Africa, P. (2019b). *Two out of three people in Sub-Saharan Africa lack access to electricity*. Retrieved from https://www.usaid.gov/powerafrica

African Union. (2019). *Our aspirations for the Africa we want*. Retrieved from https://au.int/en/agenda2063/aspirations

Ahmed, A. I. M. U. (2004, January). Weber's Perspective on the City and Culture, Contemporary Urbanization and Bangladesh. *Bangladesh e-Journal of Sociology*, *1*(1), 1–13. Retrieved from https://pdfs.semanticscholar.org/b072/0ac93a6858f21ab2f3cf97468019deecf0e3.pdf

Ahvenniemi, H., Huovila, A., Pinto-Seppä, I., & Airaksinen, M. (2017). What Are the Differences between Sustainable and Smart Cities? *Cities (London, England)*, *60*, 234–245. doi:10.1016/j.cities.2016.09.009

Airnow.gov. (2019). *Air Quality Index (AQI) Basics*. Retrieved from https://airnow.gov/index.cfm?action=aqibasics.aqi

Alam, M., Tehranipoor, M. M., & Guin, U. (2017). Tsensors vision, infrastructure and security challenges in trillion sensor era. *Journal of Hardware and Systems Security*, *1*(4), 311–327.

Alatas, B., Akin, E., & Ozer, A. B. (2009). Chaos embedded particle swarm optimization algorithms. *Chaos, Solitons, and Fractals*, *4*(4), 1715–1734. doi:10.1016/j.chaos.2007.09.063

Albino, V., Berardi, U., & Dangelico, R. M. (2013). Smart cities: definitions, dimensions, and performance. *Proceedings IFKAD 2013*.

Albino, V., Berardi, U., & Dangelico, R. M. (2015). Smart cities: Definitions, dimensions, performance, and initiatives. *Journal of Urban Technology*, *22*(1), 1–19. doi:10.1080/10630732.2014.942092

All3Dp. (2019). *About Us*. Retrieved from https://all3dp.com/about/

Allam, Z., & Dhunny, Z. (2019). On big data, artificial intelligence and smart cities. *Cities (London, England)*, *89*, 80–91. doi:10.1016/j.cities.2019.01.032

Alonso, I. G. (2011). Service robotics. In *Service robotics within the digital home* (pp. 89–114). Dordrecht: Springer.

Alromaihi, S., & Balakrishna, C. (2018). Cyber Security Challenges of Deploying IoT in Smart Cities for Healthcare Applications. *2018 6th International Conference on Future Internet of Things and Cloud Workshops (FiCloudW)*.

Alsouda, Pllana, & Kurti. (2018). *A Machine Learning Driven IoT Solution for Noise Classification in Smart Cities*. Academic Press.

AmazonGo. (2019). https://www.amazon.com/b?ie=UTF8&node=16008589011

Amer, S. B. (2019). *City Definitions*. Retrieved from smartcitiescentre.org/wp-content/uploads/18-sara-ben-amer-city-definitions-sarabenamer.pdf

American Geo Sciences (2019). *What is bio-mining?* Retrieved from americangeosciences.org/critical-issues/faq/what-biomining

Amrhein, G., & Quint, S. (2009). Cloud Computing for the Enterprise: Part 1: Capturing the Cloud. *DeveloperWorks*. Retrieved from www.ibm.com/developerworks/websphere/techjournal/0904_amrhein/0904_amrhein.html

Ancient Ports Antiques. (2019). *The Catalogue*. Retrieved from ancientportsantiques.com/the-catalogue/

Andoni, M., Robu, V., Flynn, D., Abram, S., Geach, G., Jenkins, D., ... Peacock, A. (2019). Blockchain technology in the energy sector: A systematic review of challenges and opportunities. *Renewable & Sustainable Energy Reviews*, *100*, 143–174. doi:10.1016/j.rser.2018.10.014

Andročec, D., Novak, M., & Oreški, D. (2018). Using Semantic Web for Internet of Things Interoperability: A Systematic Review. *International Journal on Semantic Web and Information Systems*, *14*(4), 147–171.

Angelidou, M. (2014). Smart City Policies: A Spatial Approach. *Cities (London, England)*, *41*, S3–S11. doi:10.1016/j.cities.2014.06.007

Anthopoulos, L., Janssen, M., & Weerakkody, V. (2016). A unified smart city model (USCM) for smart city conceptualization and benchmarking. In Smart Cities and Smart Spaces: Concepts, Methodologies, Tools, and Applications (pp. 247–264). IGI Global.

Anthopoulos, L. (2015a). Defining Smart City Architecture for Sustainability. *Proceedings of the 14th IFIP Electronic Government (EGOV) and 7th Electronic Participation (ePart) Conference 2015.*

Anthopoulos, L. (2015b). Understanding the smart city domain: A literature review. In *Transforming city governments for successful smart cities* (pp. 9–21). Springer. doi:10.1007/978-3-319-03167-5_2

Antonakakis, M. (2017). Understanding the Mirai Botnet. *26th USENIX Security Symposium*, 1093–1110.

Anukoonwattaka. (2016). *Introduction to the Basic Gravity Model.* Retrieved from unescap.org/sites/default/files/an/S2-Gravity-intro.pdf

Arasteh, H., Hosseinnezhad, V., Loia, V., Tommasetti, A., Troisi, O., Shafie-khah, M., & Siano, P. (2016). Iot-based smart cities: A survey. *IEEE 16th International Conference on Environment and Electrical Engineering,* 1-6.

ArchDaily. (2019). *Shore connection for berthed ships: SIHARBOR.* https://new.siemens.com/global/en/products/energy/medium-voltage/solutions/siharbor.html,Siemens

Archie, L., & Archie, J.G. (2003). *Introduction to ethical studies- An Open Source Reader.* Ethicsbooks.

Aries, P. (1974). *Western attitudes toward death: From the Middle Ages to the present.* Johns Hopkins University Press.

Arrington, B., Barnett, L., Rufus, R., & Esterline, A. (2016). Behavioral Modeling Intrusion Detection System (BMIDS) (BMIDS) Using Internet of Things (IoT) Behavior-Based Anomaly Detection via Immunity-Inspired Algorithms. *International Conference on Computer Communication and Networks,* 1-6. 10.1109/ICCCN.2016.7568495

Asante, F. A., & Amuakwa-Mensah, F. (2015). Climate Change and Variability in Ghana: Stocktaking. *Climate (Basel), 3*(1), 78–99. doi:10.3390/cli3010078

Ashwini, A. (2017). *What is the technology stack used in IOT product (Internet of Things)?* Available at: https://www.quora.com/What-is-the-technology-stack-used-in-IOT-product-Internet-of-Things

Augusto, J. C. (2007). *Ambient Intelligence: The Confluence of Ubiquitous/Pervasive Computing and Artificial Intelligence. In Intelligent Computing Everywhere* (pp. 213–234). Springer.

Augusto, J. C., & McCullagh, P. (2007). Ambient intelligence: Concepts and Applications. *Computer Science and Information Systems, 4*(1), 1–27.

Auto Connected Car News. (2017). *Vehicle to Pedestrian (V2P) and Bicycle Demoed.* Available at: https://www.autoconnectedcar.com/2017/07/vehicle-to-pedestrian-v2p-and-bicycle-demoed/

Badii, C., Bellini, P., Cenni, D., Difino, A., Nesi, P., & Paolucci, M. (2017). Analysis and assessment of a knowledge-based smart city architecture providing service APIs. *Future Generation Computer Systems, 75,* 14–29. doi:10.1016/j.future.2017.05.001

Bagdasarov, Z., Martin, A. A., & Buckley, M. R. (2018). Working with robots: Organizational considerations. *Organizational Dynamics.* doi:10.1016/j.orgdyn.2018.09.002

Baig, Z. A., Szewczyk, P., Valli, C., Rabadia, P., Hannay, P., Chernyshev, M., ... Peacock, M. (2017). Future Challenges for Smart Cities: Cyber-Security and Digital Forensics. *Digital Investigation, 22,* 3–13. doi:10.1016/j.diin.2017.06.015

Baker, G. (2008, January 11). Schoolboy hacks into city's tram system. *Weekly Telegraph,* p. 34.

Bamakan, S. M. H., Wang, H., Yingjie, T., & Shi, Y. (2016). An effective intrusion detection framework based on MCLP/SVM optimized by time-varying chaos particle swarm optimization. *Neurocomputing, 16*(7), 90-102.

Banh, A., Rea, D. J., Young, J. E., & Sharlin, E. (2015). Inspector baxter: The social aspects of integrating a robot as a quality inspector in an assembly line. In *Proceedings of the 3rd international conference on human-agent interaction* (pp. 19-26). ACM.

Banko, M., & Brill, E. (2001). Scaling to Very Very Large Corpora for Natural Language Disambiguation. *ACL '01 Proceedings of the 39th Annual Meeting on Association for Computational Linguistics.*

Barrionuevo, J. M., Berrone, P., & Ricart, J. E. (2012). Smart cities, sustainable progress. *IESE Insight, 14*(14), 50–57. doi:10.15581/002.ART-2152

Bartlett, B., Estivill-Castro, V., & Seymon, S. (2004). Dogs or robots: why do children see them as robotic pets rather than canine machines? In *Proceedings of the fifth conference on Australasian user interface-Volume 28* (pp. 7-14). Australian Computer Society, Inc.

Bastidas, V., Helfert, M., & Bezbradica, M. (2018). A Requirements Framework for the Design of Smart City Reference Architectures. *Proceedings of the 51st Hawaii International Conference on System Sciences.* 10.24251/HICSS.2018.317

Batty, M., Axhausen, K. W., Giannotti, F., Pozdnoukhov, A., Bazzani, A., Wachowicz, M., ... Portugali, Y. (2012). Smart cities of the future. *The European Physical Journal. Special Topics, 214*(1), 481–518. doi:10.1140/epjst/e2012-01703-3

Bauder, R., & Taghi, K. (2018). Medicare Fraud Detection using Random Forest with Class Imbalanced Big Data. *2018 IEEE International Conference on Information Reuse and Integration (IRI).* 10.1109/IRI.2018.00019

Bauman, Z. (1991). *Modernity and the Holocaust.* Polity Press.

Bavel, J. V. (2013). *The World Population explosion: causes, backgrounds and projections for the future.* Retrieved from ncbi.nlm.nih.gov/pmc/articles/PMC3987379/

Beegle, K., Christiaensen, L., Dabalen, A., & Gaddis, I. (2016). *Poverty in a rising Africa.* Retrieved from https://www.un.org/africarenewal/sites/www.un.org.africarenewal/files/Poverty%20in%20a%20Rising%20Africa%20Overview.pdf

Bellini, P., Benigni, M., Billero, R., Nesi, P., & Rauch, N. (2014). Km4City ontology building vs data harvesting and cleaning for smart-city services. *Journal of Visual Languages and Computing, 25*(6), 827–839. doi:10.1016/j.jvlc.2014.10.023

Bertino, E., & Islam, N. (2017). Botnets and Internet of Things Security. *Computer, 50,* 76-79.

Bettencourt, L. (2014). The Uses of big data in cities. *Big Data, 2*(1), 12–22. doi:10.1089/big.2013.0042 PMID:27447307

Bevölkerung und Bevölkerungsvorgänge der hessischen Gemeinden am 30. Juni 2019. (2019). [Statistical]. Hessisches Statistisches Landesamt.

Bhavithra, J., & Saradha, A. (2018). Personalized web page recommendation using case-based clustering and weighted association rule mining. *Cluster Computing,* 1–12.

Bhorat, H., & Naidoo, K. (2014). *Drivers of inequality in the context of the growth-poverty-inequality nexus in Africa: An overview of key issues.* Retrieved from https://www.undp.org/content/dam/rba/docs/Reports/undp-rba_Income%20Inequality%20in%20SSA_Chapter%203.pdf

Bhuyan, M. H., & Kalita, D. K. B. J. K. (2014). Network Anomaly Detection: Methods, Systems and Tools. *IEEE Communications Surveys and Tutorials, 16*(1), 303–336. doi:10.1109/SURV.2013.052213.00046

Bifulco, F., Tregua, M., Amitrano, C. C., & D'Auria, A. (2016). ICT and Sustainability in Smart Cities Management. *International Journal of Public Sector Management, 29*(2), 132–147. doi:10.1108/IJPSM-07-2015-0132

Bill and Melinda Gates Foundation. (2013). *Water, sanitation & hygiene: reinvent the toilet challenge fact sheet.* Retrieved from https://docs.gatesfoundation.org/Documents/Fact_Sheet_Reinvent_the_Toilet_Challenge.pdf

Blumer, H. (2009). *Symbolic Interactionism: Perspective and Method.* University of California Press.

Bonomi, F. (2013). The Smart and Connected Vehicle and the Internet of Things, Synchronization. *Telecommunication Systems*, *2013*, 1–53.

Bradley, R. (2019). *Data Privacy Concerns: An Overview for 2019.* Retrieved from medium.com@manifest/data-privacy-concerns

Braun, T., Fung, B. C. M., Iqbal, F., & Shah, B. (2018). Security and Privacy Challenges in Smart Cities. *Sustainable Cities and Society*, *39*, 499–507. doi:10.1016/j.scs.2018.02.039

Brenner, S. (2010). *Cybercrime: Criminal Threats from Cyberspace.* Dayton: Praeger.

Briggs, G., & Scheutz, M. (2014). How robots can affect human behavior: Investigating the effects of robotic displays of protest and distress. *International Journal of Social Robotics*, *6*(3), 343–355.

Brown University. (2019). *Favelas in Rio de Janeiro, past and present.* Retrieved from https://library.brown.edu/create/fivecenturiesofchange/chapters/chapter-9/favelas-in-rio-de-janeiro-past-and-present/

Brown, T. L. (2019). Ambient Intelligence Changes the Office Environment. In Advances in the Technology of Managing People: Contemporary Issues in Business (pp. 149-160). Emerald Publishing Limited.

Brown, G., Carlyle, M., & Salmeron, J. (2016). Defending Critical Infrastructure. *Homeland Security: Operations Research Initiatives and Applications*, *36*(6), 530–544.

Buczak, A. L., & Guven, E. (2016). A Survey of Data Mining and Machine Learning Methods for Cyber Security Intrusion Detection. *IEEE Communications Surveys and Tutorials*, *18*(2), 1153–1176. doi:10.1109/COMST.2015.2494502

Buhalis, D. (2019). Technology in tourism-from information communication technologies to eTourism and smart tourism towards ambient intelligence tourism: A perspective article. *Tourism Review.*

Bujari, A., Furini, M., Mandreoli, F., Martoglia, R., Montangero, M., & Ronzani, D. (2018). Standards, Security and Business Models: Key Challenges for the IoT Scenario. *Mobile Networks and Applications*, *23*(1), 147–154. doi:10.100711036-017-0835-8

Bürgerservice: Blinde und Sehbehinderte. (2019). [Citiy Administration]. Leben in Marburg. https://www.marburg.de/portal/seiten/blinde-und-sehbehinderte-900000287-23001.html

Butt, T. A., & Afzaal, M. (2019). Security and Privacy in Smart Cities: Issues and Current Solutions. In A. Al-Masri & K. Curran (Eds.), *Smart Technologies and Innovation for a Sustainable Future* (pp. 317–323). Cham: Springer International Publishing. doi:10.1007/978-3-030-01659-3_37

C40.org. (2019). *Why cities: Ending Climate change begins in the cities.* https://www.c40.org/ending-climate-change-begins-in-the-city,C40Cities

Cagliero, L., Garza, P., Pasini, A., & Baralis, E. M. (2018). Additional reviewer assignment by means of weighted association rules. *IEEE Transactions on Emerging Topics in Computing.*

Calzada, I., & Cobo, C. (2015). Unplugging: Deconstructing the Smart City. *Journal of Urban Technology*, *22*(1), 23–43. doi:10.1080/10630732.2014.971535

Campbell, P. (2018). *Trucks Headed for a Driverless Future*. https://www.ft.com/content/7686ea3e-e0dd-11e7-a0d4-0944c5f49e46

Campbell, F. A. K. (2001). Inciting Legal Fictions: Disability's Date with Ontology and the Ableist Body of the Law. *Griffith Law Review*, *10*, 42–62.

Cao, B. Y., Ma, S. Q., & Cao, H. H. (Eds.). (2014). *Ecosystem assessment and fuzzy systems management*. Springer International Publishing.

Caragliu, A., Del Bo, C., & Nijkamp, P. (2009). Smart Cities in Europe 49-59. *Proceedings of the 3rd Central European Conference in Regional Science—CERS 2009.*

Caragliu, A., Del Bo, C., & Nijkamp, P. (2011). *Smart Cities in Europe*. Retrieved from research.vu.nl/ws/portalfiles/portal/2474218/2009048.pdf

Caragliu, A., & Del Bo, C. F. (2019). Smart Innovative Cities: The Impact of Smart City Policies on Urban Innovation. *Technological Forecasting and Social Change*, *142*, 373–383. doi:10.1016/j.techfore.2018.07.022

Caragliu, A., Del Bo, C., & Nijkamp, P. (2011). Smart Cities in Europe. *Journal of Urban Technology*, *18*(2), 65–82. doi:10.1080/10630732

Cardullo, P., & Kitchin, R. (2019). Being a 'Citizen'in the Smart City: Up and down the Scaffold of Smart Citizen Participation in Dublin, Ireland. *GeoJournal*, *84*(1), 1–13. doi:10.100710708-018-9845-8

CEADSEN. (2019). *Nearest Neighbor Analysis*. Retrieved from ceadsen.nku.edu/longa/geomed/ppa/doc/NNA/NNA.htm

Chamoso, P., González-Briones, A., Rodríguez, S., & Corchado, J. M. (2018). Tendencies of Technologies and Platforms in Smart Cities: A State-of-the-Art Review. Wireless Communications and Mobile Computing.

Chayes, S. (2015). *Thieves of State*. New York, NY: W. W. Norton & Company.

Cheng, B. (2016). *Number of 'cage home' residents rises to nearly 200,000*. https://www.hongkongfp.com/2016/03/20/hkfp-lens-number-of-cage-home-residents-rises-to-nearly-200000/,HKFPLens

Cheng, B., Longo, S., Cirillo, F., Bauer, M., & Kovacs, E. (2015). Building a big data platform for smart cities: Experience and lessons from santander. In *Big Data (BigData Congress), 2015 IEEE International Congress on* (pp. 592–599). IEEE.

Chen, H., & Wang, F. Y. (2005). Introduction: Artificial Intelligence for Homeland Security. *IEEE Intelligent Systems*, *20*(5), 12–16. doi:10.1109/MIS.2005.88

Chen, S., Xu, Z., Tang, Y., & Liu, S. (2014). An Improved Particle Swarm Optimization Algorithm Based on Centroid and Exponential Inertia Weight. *Mathematical Problems in Engineering*, *07*, 1–14. doi:10.1155/2014/976486

Chourabi, H., Nam, T., Walker, S., Gil-Garcia, J. R., Mellouli, S., & Nahon, K. (2012). Understanding Smart Cities: An Integrative Framework. In *2012 45th Hawaii International Conference on System Sciences*. IEEE Explore.

Chourabi, H., Nam, T., Walker, S., Gil-Garcia, J. R., Mellouli, S., & Nahon, K., … Scholl, H. J. (2011). Understanding smart cities: An integrative framework. Proceedings of the Annual Hawaii International Conference on System Sciences, 2289–2297. 10.1109/HICSS.2012.615

Chu, J. (2019). Breaching a "carbon threshold" could lead to mass extinction. *MIT News*. http://news.mit.edu/2019/carbon-threshold-mass-extinction-0708

Circle, S. (2014). *Five principles for a smarter city*. https://www.smart-circle.org/smartcity/blog/five-principles-smarter-city/

Clement, S. J., McKee, D. W., & Xu, J. (2017). Service-oriented reference architecture for smart cities. In *Service-Oriented System Engineering (SOSE), 2017 IEEE Symposium on* (pp. 81–85). IEEE. 10.1109/SOSE.2017.29

Clotet, X., Moyano, J., & Leon, G. (2018). A real-time anomaly-based IDS for cyber-attack detection at the industrial process level of Critical Infrastructures. *International Journal of Critical Infrastructure Protection, 32*, 11–20. doi:10.1016/j.ijcip.2018.08.002

Clough, J. (2015). *Principles of cybercrime*. Cambridge: Cambridge University Press. doi:10.1017/CBO9781139540803

Cloutier, R., Muller, G., Verma, D., Nilchiani, R., Hole, E., & Bone, M. (2010). The concept of reference architectures. *Systems Engineering, 13*(1), 14–27.

CMSWire. (2019). *Connected Car Experiences in 2019*. Available at: https://www.cmswire.com/digital-experience/connected-car-experiences-in-2019-exploring-the-possibilities/

CNBC. (2018). *New Story is a nonprofit that hopes to use 3-D printers to build homes in poor communities*. Retrieved from https://www.cnbc.com/2018/03/15/new-story-nonprofit-will-3-d-print-homes-in-poor-communities.html

Cohen, B. (2015). *The 3 Generations of Smart Cities*. https://www.fastcompany.com/3047795/the-3-generations-of-smart-cities

Consultancy.eu. (2019). *Europe best prepared for autonomous driving, Netherlands leads*. https://www.consultancy.eu/news/2609/europe-best-prepared-for-autonomous-driving-netherlands-leads

Cooke, E., Hague, S., & McKay, A. (2016). *The Ghana Poverty and Inequality Report*. Retrieved from https://www.unicef.org/ghana/media/531/file/The%20Ghana%20Poverty%20and%20Inequality%20Report.pdf

Coordes, G. (2016b, July 5). Marburg, das "Mekka der Blinden." *Ärzte Zeitung*. https://www.aerztezeitung.de/panorama/article/914262/marburg-mekka-blinden.html

Coordes, G. (Ed.). (2016a). *Blindenstadt Marburg*. Magistrat der Universitätsstadt Marburg. https://www.marburg.de/downloads/datei/

Cor, A. (2019b). *Top five teams win a share of $100,000 in virtual modeling stage of NASA's 3D-Printed Habitat Competition*. Retrieved from https://www.nasa.gov/directorates/spacetech/centennial_challenges/3DPHab/five-teams-win-a-share-of-100000-in-virtual-modeling-stage

CorA. (2019a). *Home*. Retrieved from https://www.apis-cor.com

Cormier, D., Newman, G., Nakane, M., Young, J. E., & Durocher, S. (2013). Would you do as a robot commands? An obedience study for human-robot interaction. *International Conference on Human-Agent Interaction*.

Cosimo, I., Ahsan, A., Mandar, G., & Kia, D. (2018). Statistical Analysis Driven Optimized Deep Learning System for Intrusion Detection. Academic Press.

Cox, A., Parslow, P., De Lathouwer, B., Klien, E., Kempen, B., & Lonien, J. (2016). *D4.2 - Definition of Smart City Reference Architecture, ESPRESSO systEmic Standardisation apPRoach to Empower Smart citieS and cOmmunities*. Academic Press.

Cree.com. (2014). *Cree LED Street Lights Transform California City's Streets, Cut Energy Use by 65 Percent*. https://www.cree.com/news-media/news/article/cree-led-street-lights-transform-california-city-s-streets-cut-energy-use-by-65-percent

Cronemberger, F., & Ramon Gil-Garcia, J. (2019). Big Data and Analytics as Strategies to Generate Public Value in Smart Cities: Proposing an Integrative Framework. In Setting Foundations for the Creation of Public Value in Smart Cities. Springer.

Cyber security challenges for IoT-based smart grid networks. (2019). *International Journal of Critical Infrastructure Protection, 25*(16), 36-49.

da Costa, K. A., Papa, J. P., Lisboa, C. O., Munoz, R., & de Albuquerque, V. H. C. (2019). Internet of Things: A survey on machine learning-based intrusion detection approaches. *Computer Networks, 151*, 147–157. doi:10.1016/j.comnet.2019.01.023

da Silva, W. M., Alvaro, A., Tomas, G. H. R. P., Afonso, R. a., Dias, K. L., & Garcia, V. C. (2013). Smart cities software architectures - a survey. *Proceedings of the 28th Annual ACM Symposium on Applied Computing - SAC '13*, 1722. 10.1145/2480362.2480688

Dagle, J. E. (2009). Cyber security of the electric power grid. IEEE/PES Power Systems Conference and Exposition.

Darko, S. (2015). *Inside the world of Ghana's Internet fraudsters.* Retrieved from bbc.com/news/world-africa-32583161

Dasgupta, D. (2006). Computational Intelligence in Cyber security. *IEEE International Conference on Computtional Intelligence for Homeland Security and Personal Safety.*

De Donno, M., Dragoni, N., Giaretta, A., & Spognardi, A. (2018). DDoS- Capable IoT Malwares: Comparative Analysis and Mirai Investigation. *Security and Communication Networks, 2018*, 2018. doi:10.1155/2018/7178164

De Oliveira Fernandes, E., Meeus, L., Leal, V., Azevedo, I., Delarue, E., & Glachant, J.-M. (2011). *Smart Cities Initiative: how to foster a quick transition towards local sustainable energy systems.* European University Institute. doi:10.2870/34539

Deakin, M., & Al Waer, H. (2011). From Intelligent to Smart Cities. *Journal of Intelligent Buildings International: From Intelligent Cities to Smart Cities., 3*(3), 140–152. doi:10.1080/17508975.2011.586671

Demir, K. A. (2019). *Human-Robot Co-working and Roboethics* (Master's Thesis). Gebze Technical University, Izmit, Turkey.

Demir, K. A., Caymaz, E., & Elçi, M. (2017). *Issues in integrating robots into organizations.* In The 12th International Scientific Conference "Defence Resources Management in the 21st Century", Brazov, Romania.

Demir, K. A. (2017). Research Questions in Roboethics. *Mugla Journal of Science and Technology, 3*(2), 160–165.

Demir, K. A., & Cicibas, H. (2017). *Industry 5.0 and a Critique of Industry 4.0.* In *4th International Management Information Systems Conference*, Istanbul, Turkey.

Demir, K. A., & Cicibaş, H. (2018). *The Next Industrial Revolution: Industry 5.0 and Discussions on Industry 4.0. In Industry 4.0 from the Management Information Systems Perspectives.* Peter Lang Publishing Group.

Demir, K. A., Döven, G., & Sezen, B. (2019). Industry 5.0 and Human-Robot Co-working. *Procedia Computer Science, 158*, 688–695.

Demir, K. A., Turan, B., Onel, T., Ekin, T., & Demir, S. (2019). Ambient Intelligence in Business Environments and Internet of Things Transformation Guidelines. In *Guide to Ambient Intelligence in the IoT Environment* (pp. 39–67). Cham: Springer.

Denscombe, M. (2007). *The good research guide for small-scale social research projects* (3rd ed.). McGraw-Hill.

Department of Homeland Security. (2018). *National Critical Functions.* Retrieved May 7, 2019, from https://www.dhs.gov/cisa/national-critical-functions

Desouza, K. C., & Flanery, T. H. (2013). Designing, planning, and managing resilient cities: A conceptual framework. *Cities (London, England), 35*, 89–99. doi:10.1016/j.cities.2013.06.003

DeZyre. (2018). *Principal Component Analysis.* Retrieved August 16, 2019, from https://www.dezyre.com/data-science-in-python-tutorial/principal-component-analysis-tutorial

Diamandis, P. H., & Kotler, S. (2012). *Abundance: The future is better than you think.* Simon and Schuster.

Digest of UK Energy Statistics (DUKES): Electricity. (2019). *DUKES chapter 5: statistics on electricity from generation through to sales.* Department for Business, Energy and Industrial Strategy.

Dijkstra, L., & Poelman, H. (2012). Cities in Europe: The new OECD-EC Definition. *RegionalFocusEurostat.* Retrieved from http://epp.eurostat.ec.europa.eu/portal/page/portal/nuts_nomenclature/local_administrative_units

Dilek, S., Cakir, H., & Aydin, M. (2015). Applications of artificial intelligence techniques to combat cyber crime: A review. *International Journal of Artificail Intelligene & Applications, 6*, 1–19.

DIN. (2017). *DIN SPEC 91357, Reference Architecture Model Open Urban Platform (OUP).* Berlin, Germany: Deutsches Institut für Normung.

Dinakaran, R. (2018). *Deep Learning Based Pedestrian Detection at Distance in Smart Cities.* doi:10.1007/978-3-030-29513-4_43

Ding, Y., Chen, X., Lamata, L., Solano, E., & Sanz, M. (2019). *Logistic Network Design with a D-Wave Quantum Annealer.* arXiv:1906.10074

Ding, M., & Pérez, D. L. (2018). Promises and caveats of uplink IoT ultra-dense networks. In *2018 IEEE Wireless Communications and Networking Conference (WCNC)* (pp. 1-6). IEEE.

Djigal, H., Jun, F., & Jiamin, L. (2017). Secure Framework for Future Smart City. *2017 IEEE 4th International Conference on Cyber Security and Cloud Computing.*

Dmitriy, K., & Sullivan, J. E. (2017). How cyber-attacks in Ukraine show the vulnerability of the U.S. power grid. *The Electricity Journal, 30*(3), 30–35. doi:10.1016/j.tej.2017.02.006

domo.com. (2019). *Data Never Sleeps 7.0.* https://www.domo.com/learn/data-never-sleeps-7

Donnelly, E. (2019). *Future success: megacities of more than 10 million set to provide big business opportunities.* https://www.independent.ie/business/world/future-success-megacities-of-more-than-10-million-set-to-provide-big-business-opportunities-37961796.html

Drizik, J. (2018, January 17). *Cyber risk is a growing challenge. So how can we prepare?* Marsh LLC.

Economist Intelligence Unit (EIU). (2013). *Hot spots 2025: Benchmarking the future Competitiveness of cities.* Academic Press.

Edvinsson, L. (2006). Aspects on the city as a knowledge tool. *Journal of Knowledge Management, 10*(5), 6–13. doi:10.1108/13673270610691134

Efthymiopoulos, M.-P. (2016). Cyber-security in smart cities: The case of Dubai. *Journal of Innovation and Entrepreneurship, 5*(11), 1–16. doi:10.118613731-016-0036-x

Eger, J. M. (2009). Smart growth, smart cities, and the crisis at the pump a worldwide phenomenon. *I-WAYS-The Journal of E-Government Policy and Regulation, 32*(1), 47–53. doi:10.3233/IWA-2009-0164

EIP_SCC Urban Platform. (2016). *Requirements Specification For Urban Platforms v2.2 (consultation stage)* (Vol. 2). Academic Press.

Elhoseny, H., Elhoseny, M., Abdelrazek, S., Bakry, H., & Riad, A. (2016). Utilizing service oriented architecture (SOA) in smart cities. *International Journal of Advancements in Computing Technology*, *8*(3), 77–84.

Elkington, J. (2004). *Enter the Triple Bottom Line*. Retrieved from johnelkington.com/archive/TBL-elkington-chapter.pdf

Elmaghraby, A. S., & Losavio, M. M. (2014). Cyber Security Challenges in Smart Cities: Safety, Security and Privacy. *Journal of Advanced Research*, *5*(4), 491–497. doi:10.1016/j.jare.2014.02.006 PMID:25685517

Elrawy, M. F., Awad, A. I., & Hamed, H. F. A., 2018. Intrusion detection systems for IoT-based smart environments: a survey. *Journal of Cloud Computing: Advances, Systems and Applications.*

Elshawi, R., Sakr, S., Talia, D., & Trunfio, P. (2018). Big Data Systems Meet Machine Learning Challenges: Towards Big Data Science as a Service. *Big Data Research*, *14*(1), 1–11. doi:10.1016/j.bdr.2018.04.004

Emory. (2016). *Should public toilets be part of urban sanitation solutions for poor families living in slums?* School of Public Health. Retrieved from http://sanipath.org/wp-content/uploads/2016/04/SaniPath-Public-Latrin-Policy-Brief.pdf

Engstrom, R., Pavelesku, D., Tanaka, T., & Wambile, A. (2019). Mapping poverty and slums using multiple methodologies in Accra, Ghana. *2015 Joint Urban Remote Sensing Event. JURSE, 2019.* doi:10.1109/JURSE.2019.8809052

Engstrom, R., Sandborn, A., Yu, Q., Burgdorfer, J., Stow, D., Weeks, J., & Graesser, J. (2015). Mapping slums using spatial features in Accra, Ghana. *2015 Joint Urban Remote Sensing Event. JURSE, 2015.* doi:10.1109/JURSE.2015.7120494

EPA. (2019). *Facts and Figures about Materials, Waste and Recycling*. Retrieved from https://www.epa.gov/facts-and-figures-about-materials-waste-and-recycling/national-overview-facts-and-figures-materials

Eregha, P. B. (2015). Foreign direct investment inflow, volatility, and domestic investment in West Africa. *Journal of Developing Areas*, *49*(2), 273–294. doi:10.1353/jda.2015.0018

Ericsson. (2019). https://www.ericsson.com/4a45a8/assets/local/digital-services/trending/scalable-network/executive_guide_network_slicing.pdf

Erl, T., Mahmood, Z., & Puttini, R. (2013). Cloud Computing: Concepts, Technology and Architecture. London: Prentice Hall.

Essig, S., Allebé, C., Remo, T., Geisz, J. F., Steiner, M. A., Horowitz, K., ... Young, D. L. (2017). Raising the one-sun conversion efficiency of III–V/Si solar cells to 32.8% for two junctions and 35.9% for three junctions. *Nature Energy*, *2*(9), 17144.

EU Commission. (2017). *(The) making of a smart city: best practices across Europe*. Retrieved from www.smartcities-infosystem.eu

EU. (2018). *Waste management in cities*. Retrieved from ec.europa.eu/easme/en/news/waste-management-in-cities

Falconer, G., & Mitchell, S. (2012, Sept.). Smart City Framework: A Systematic Process for Enabling Smart+Connected Communities. *Point of View*.

Farms, S. (2019). *Innovation. Sundrop is the future*. Retrieved from https://www.sundropfarms.com/innovation/

Feld. (2019). A Hybrid Solution Method for the Capacitated Vehicle Routing Problem Using a Quantum Annealer. *Frontiers in ICT, 6*.

Felsman, C. (2014). *Schools in Slums: A surprising number of the poorest kids are enrolled in private preschools*. Retrieved from https://educationinnovations.org/blog/schools-slums-surprising-number-poorest-kids-are-enrolled-private-preschools

Fischer, G., & Herrmann, T. (2011). Socio-technical systems: A meta-design perspective. *International Journal of Sociotechnology and Knowledge Development*, *3*(1), 1–33. doi:10.4018/jskd.2011010101

Fittkau, L. (2017, October 19). Hauptstadt der Blinden. In *Länderreport*. Deutschlandfunk Kultur. https://www.deutschlandfunkkultur.de/marburg-hauptstadt-der-blinden.1001.de.html?dram:article_id=398614

Fiware. (2019). *Fiware Wiki*. Retrieved from https://forge.fiware.org/plugins/mediawiki/wiki/fiware/index.php

Fogelin, R. J. (1971). Three Platonic Analogies. *The Philosophical Review*, *80*(3), 371. doi:10.2307/2184102

Foresti, S., Yung, M., & Martinelli, F. (2012). *Computer Security - ESORICS 2012*. Pisa: Springer. doi:10.1007/978-3-642-33167-1

Forlizzi, J. (2007). How robotic products become social products: an ethnographic study of cleaning in the home. In *Proceedings of the ACM/IEEE international conference on Human-robot interaction* (pp. 129-136). ACM.

Forlizzi, J., & DiSalvo, C. (2006). Service robots in the domestic environment: a study of the roomba vacuum in the home. In *Proceedings of the 1st ACM SIGCHI/SIGART Conference on Human-Robot Interaction* (pp. 258-265). ACM.

Fortino, G., & Trunfio, P. (2014). *Internet of Things Based on Smart Objects: Technology, Middleware and Applications*. Springer. doi:10.1007/978-3-319-00491-4

Foucault, M. (1973). *The Birth of the Clinic: An Archaeology of Medical Perception*. Pantheon Books.

Fraunhofer FOKUS OpenMTC. (2019). *OpenMTC*. Retrieved from https://www.open-mtc.org

Frey, J. H., & Fontana, A. (1991). Article. *The Group Interview in Social Research.*, *28*(2), 13.

Fromhold-Eisebith, M. (2017). Cyber-Physical Systems in Smart Cities - Mastering Technological, Economic, and Social Challenges. *Smart Cities: Foundations, Principles, and Applications*, 1–22.

Fuchs, C. (2019). Beyond Big Data Capitalism, Towards Dialectical Digital Modernity: In C. Fuchs & D. Chandler (Eds.), Digital Objects, Digital Subjects (pp. 43–52). University of Westminster Press.

Fung, C. J., & Zhu, Q. (2016). FACID: A trust-based collaborative decision framework for intrusion detection networks. *Ad Hoc Networks*, *53*, 17–31. doi:10.1016/j.adhoc.2016.08.014

Gafni, R., & Geri, N. (2013). Generation Y versus Generation X: Differences in Smartphone Adaptation. *Proceedings of the Chais Conference on Instructional Technologies Research*, 6.

Galster, M., & Avgeriou, P. (2011). Empirically-grounded reference architectures: a proposal. In *Proceedings of the joint ACM SIGSOFT conference--QoSA and ACM SIGSOFT symposium--ISARCS on Quality of software architectures--QoSA and architecting critical systems--ISARCS* (pp. 153–158). ACM. 10.1145/2000259.2000285

Gams, M., Gu, I. Y. H., Härmä, A., Muñoz, A., & Tam, V. (2019). Artificial intelligence and ambient intelligence. *Journal of Ambient Intelligence and Smart Environments*, *11*(1), 71–86.

Ganchev, I., & Ji, Z. (2014). *Generic IoT Architecture for Smart Cities*. Telecommunications Research Centre (TRC) University of Limerick.

Gandotra, P., Kumar Jha, R., & Jain, S. (2017). A survey on device-to-device (D2D) communication: Architecture and security issues. *Journal of Network and Computer Applications*, *78*, 9–29. doi:10.1016/j.jnca.2016.11.002

Garrido-Marijuan, A., Pargova, Y., & Wilson, C. (2017). The Making of a Smart City: Best Practices across Europe, Belgium 2017. Academic Press.

Gartner. (2017). *Gartner Says 8.4 Billion Connected "Things" Will Be in Use in 2017, Up 31 Percent From 2016.* https://www.gartner.com/en/newsroom/press-releases/2017-02-07-gartner-says-8-billion-connected-things-will-be-in-use-in-2017-up-31-percent-from-2016,Gartner

Gaur, A., Scotney, B. Parr, G., McClean, S. (2015). *Smart City Architecture and its Applications based on IoT.* Retrieved from researchgate.net/publication/278726450_smart-city-architecture-and-its-application-based-on-IoT

Geiskkovitch, D. Y., Cormier, D., Seo, S. H., & Young, J. E. (2016). Please continue, we need more data: An exploration of obedience to robots. *Journal of Human-Robot Interaction, 5*(1), 82–99.

Geoportal. (2019). https://www.marburg-biedenkopf.de/dienste_und_leistungen/auto_und_mobilitaet/_Auto-Mobilitaet.php

Ghana Statistical Service. (2019a). *Population by regions - Greater Accra.* Retrieved from http://www.statsghana.gov.gh/regionalpopulation.php?population=MTM0NTk2MjQzOS4yMDE1&&Greater%20Accra&regid=3

Ghana Statistical Service. (2019b). *Population by regions - Greater Accra.* Retrieved from https://www.statsghana.gov.gh/regionalpopulation.php?population=MTI5MzE3OTU5OC40NDg1&&Ashanti&regid=1

GhanaPost. (2019). *Ghana's official digital addressing system.* Retrieved from https://ghanapostgps.com/mapview.html

Ghanaweb. (2019a). *Akufo-Addo wants vibrant Ghana aluminium industry.* Retrieved from https://www.ghanaweb.com/GhanaHomePage/business/Akufo-Addo-wants-vibrant-Ghana-aluminium-industry-727527

Ghanaweb. (2019b). *Ghana's housing deficit hits two million – Housing Minister.* Retrieved from https://www.ghanaweb.com/GhanaHomePage/business/Ghana-s-housing-deficit-hits-two-million-Housing-Minister-739235

Giffinger, R. (2007). Smart Cities: Ranking of European medium-sized cities. Centre of Regional Science, Vienna UT.

Giffinger, R., Fertner, C., Kramar, H., Kalasek, R., Pichler-Milanovic, N., Meijers, E., & Studies, M. (2007). Smart cities Ranking of European medium-sized cities. *October, 16*, 13–18.

Giffinger, R., Pichler-Milanović, N., Fertner, C., Kramar, H., Kalasek, R., & Meijers, E. (2007). *Smart Cities: Ranking of European Medium-Sized Cities.* Centre of Regional Science, Vienna University of Technology. www.smart-cities.eu

Glebova, I. S., Yasnitskaya, Y. S., & Maklakova, N. V. (2014). Assessment of Cities in Russia According to the Concept of" Smart City" in the Context of the Application of Information and Communication Technologies. *Mediterranean Journal of Social Sciences, 5*(18), 55. doi:10.5901/mjss.2014.v5n18p55

Glenn, G. (2014). *In Ghana, school absenteeism and poverty go hand-in-hand.* Retrieved from http://news.wypr.org/post/ghana-school-absenteeism-and-poverty-go-hand-hand#stream/0

Goasduff, L. (2019). *Top Trends on the Gartner Hype Cycle for Artificial Intelligence, 2019.* https://www.gartner.com/smarterwithgartner/top-trends-on-the-gartner-hype-cycle-for-artificial-intelligence-2019/

Goddard, D. (2015). *Daredevil (TV Series).* Netflix.

Golledge, R. G. (1993). Geography and the disabled: A survey with special reference to vision impaired and blind populations. *Transactions of the Institute of British Geographers, 18*(1), 63–85. doi:10.2307/623069

Green Belt Movement. (2019). *Wangari Maathai.* Retrieved from https://www.greenbeltmovement.org/wangari-maathai

Grewal-Carr, V., & Marshall, S. (2016). *Blockchain: Enigma. Paradox. Opportunity.* Deloitte. Retrieved from https://www2.deloitte.com/content/dam/Deloitte/uk/Documents/Innovation/deloitte-uk-blockchain-full-report.pdf

Griffor, E. R., Greer, C., Wollman, D. A., & Burns, M. J. (2017). *Framework for Cyber-Physical Systems: Volume 1, Overview*. Academic Press.

Grigorescu, S. D., Argatu, F. C., Paturca, S. V., Cepisca, C., Seritan, G. C., Adochiei, F. C., & Enache, B. (2019). Robotic Platform with Medical Applications in the Smart City Environment. In *2019 11th International Symposium on Advanced Topics in Electrical Engineering (ATEE)* (pp. 1-6). IEEE.

Grönwall, J., & Oduro-Kwarteng, S. (2017). Groundwater as a strategic resource for improved resilience: A case study from peri-urban Accra. *Environmental Earth Sciences*, *77*. doi:10.100712665-017-7181-9

Guallart, V. (2015). *The 8 principles on How to Make a Smart City*. Barcelona: Architect Institute for Advanced Architecture of Catalonia; https://smartnet.niua.org/content/3e4c6fcf-f273-4722-a531-e9decc8bf5de

Guardian. (2018). *Why eating less meat is the best thing you can do for the planet in 2019*. Retrieved from https://www.the guardian.com/environment/2018/dec/21/lifestyle-change-eat-less-meat-climate-change

Gyenesei, A. (2000, September). Mining weighted association rules for fuzzy quantitative items. In *European Conference on Principles of Data Mining and Knowledge Discovery* (pp. 416-423). Springer. 10.1007/3-540-45372-5_45

Hahn, H. (1994). The minority group model of disability: Implications for medical sociology. *Research in the Sociology of Health Care*, *11*, 3–24. doi:10.4135/9781412939645.n69

Halevy, A., Norvig, P., & Pereira, F. (2009). The Unreasonable Effectivenesss of Data. *IEEE Intelligent Systems*, *24*(2), 8–12. doi:10.1109/MIS.2009.36

Han, J., Jo, M., Park, S., & Kim, S. (2005). The educational use of home robots for children. In *Robot and Human Interactive Communication, 2005. ROMAN 2005. IEEE International Workshop on* (pp. 378-383). IEEE.

Hancke, G. P., Hancke, G. P. Jr, & ... (2012). The role of advanced sensing in smart cities. *Sensors (Basel)*, *13*(1), 393–425. doi:10.3390130100393 PMID:23271603

Han, J., Pei, J., & Kamber, M. (2011). *Data mining: concepts and techniques*. Elsevier.

Hanlon, J., & O'Connor, K. (2014, April). SIEM. *Security Information and Event Management*, p. 36.

Hansen, R., & King, D. (2013). Sterilized by the State: Eugenics, Race, and the Population Scare in Twentieth-Century North America. Cambridge University Press. doi:10.1017/CBO9781139507554

Harmon, R. R., Castro-Leon, E. G., & Bhide, S. (2015). Smart Cities and the Internet of Things. *Portland International Conference on Management of Engineering and Technology (PICMET)*, 485–94. 10.1109/PICMET.2015.7273174

Harrison, C., & Donnelly, I. (2011). A Theory of Smart Cities. *Proceedings of the 55th Annual Meeting of the ISSS-2011*.

Hasse, F., von Perfall, A., Hillebrand, T., Smole, E., Lay, L., & Charlet, M. (2016). *PwC global power & utilities, Blockchain - an opportunity for energy producers and consumers?* Retrieved from https://www.pwc.com/gx/en/industries/assets/pwc-blockchain-opportunity-for-energy-producers-and-consumers.pdf

He, W., Goodkind, D., & Kowal, P. (2016). *International Population Reports*. P95/16-1, An aging world: 2015. US Census Bureau. https://www.census.gov/content/dam/Census/library/publications/2016/demo/p95-16-1.pdf

Heidegger, M. (1962). *Being and Time*. Harper & Row.

He, Y., Zhu, C., He, Z., Gu, C., & Cui, J. (2017). Big data oriented root cause identification approach based on Axiomatic domain mapping and weighted association rule mining for product infant failure. *Computers & Industrial Engineering*, *109*, 253–265. doi:10.1016/j.cie.2017.05.012

Höjer, M., & Wangel, J. (2014). Smart Sustainable Cities: Definition and Challenges. In L. M. Hilty & B. Aebischer (Eds.), *ICT Innovations for Sustainability. Advances in Intelligent Systems and Computing 310.* Springer International Publishing.

Hollands, R. (2008). Will the real smart city please stand up? Intelligent, progressive or entrepreneurial? *City: Analysis of Urban Trends, Culture, Theory, Policy, Action, 12*(3), 303-320. DOI: . doi:10.1080/13604810802479126

Holmes, G., Donkin, A., & Witten, I. H. (1994). *Weka: A machine learning workbench.* Academic Press.

Honeywill, R. (2019). *What is an Intelligent City?* Available at: https://rosshoneywill.com/articles/what-is-an-intelligent-city/

Hoyt Homer. (1939) Retrieved from https://planningtank.com/settlement-geography/hoyt-model-sector-model-land-use-1939-homer-hoyt

Huang, L., & Zhu, Q. (2019). Adaptive Strategic Cyber Defense for Advanced Persisten Threats in Critical Infrastructure Networks. *Performance Evaluation Review, 46*(2), 52–56. doi:10.1145/3305218.3305239

Hughes, B. (1999). The Constitution of Impairment: Modernity and the aesthetic of oppression. *Disability & Society, 14*(2), 155–172. doi:10.1080/09687599926244

Hughes, B. (2007). Being disabled: Towards a critical social ontology for disability studies. *Disability & Society, 22*(7), 673–684. doi:10.1080/09687590701659527

Hui, K., Chen, C., & Zhan, Y. (2019, May). A Novel Approach to Classify Video Scenes via Mining Weighted Association Rules of Actions in Temporal Sequence. In *Proceedings of the 2019 4th International Conference on Multimedia Systems and Signal Processing* (pp. 145-152). ACM.

Hui, T. K., Sherratt, R. S., & Sánchez, D. D. (2017). Major requirements for building Smart Homes in Smart Cities based on Internet of Things technologies. *ScienceDirect, 76*, 358–369. doi:10.1016/j.future.2016.10.026

HumanDrive. (2020). *Nissan Leaf Completes the UK's Longest and most Complex Autonomous Journey.* Available at: https://humandrive.co.uk/nissan-leaf-completes-the-uks-longest-and-most-complex-autonomous-car-journey/

Hurley, T., Perdomo, J. E., & Perez-Pons, A. (2016). HMM-based Intrusion Detection System for Software Defined Networking. *15th IEEE International Conference on Machine Learning and Applications (ICMLA),* 617-621. 10.1109/ICMLA.2016.0108

IBM (2011). *IBM's Smarter Cities Challenge: Philadelphia Report.* Retrieved from smartercitieschallenge.org/city_philadelphia_pa.html

IBM. (2018). *The Dangers of Smart City Hacking.* New York: IBM.

Ibrahim, Victor, & Nor, Adewole, Ibrar, Abdullah, … Haruna. (2016). The Role of Big Data in Smart City. *International Journal of Information Management, 36.* doi:10.1016/j.ijinfomgt.2016.05.002

IDC. (2018). *IDC Forecasts Smart Cities Spending to Reach $158 Billion in 2022, with Singapore, Tokyo, and New York City Among Top Spenders.* https://www.idc.com/getdoc.jsp?containerId=prUS44159418

IEEE Standards. (2019). https://standards.ieee.org/initiatives/iot/stds.html

IES-City PWG. (2018). *IES-City Framework - A Consensus Framework for Smart City Architectures v1.0.* Academic Press.

IGI Global. (2019). *What is (sic) sources of Trust?* Retrieved from www.igi-global.com/dictionary/sources-trust-consumers-participation-permission/27807

Ihsanullah, I. (2018). Carbon nanotube membranes for water purification: Developments, challenges, and prospects for the future. *Separation and Purification Technology*, *209*, 307–337. doi:10.1016/j.seppur.2018.07.043

Illich, I. (1985). *Tools for conviviality*. Marion Boyars.

Illich, I. (1995a). Guarding the Eye in the Age of Show. *Anthropology and Aesthetics*, *28*(Autumn), 47–61. doi:10.1086/RESv28n1ms20166929

Illich, I. (1995b). *The Scopic Past and the Ethics of the Gaze: A Plea for the Historical Study of Ocular Perception. Working Papers No. 6*. University Park, PA: Science, Technology, and Society Studies. http://www.davidtinapple.com/illich/1998_scopic_past.PDF

Imrie, R. (1996). Ableist Geographies, Disablist Spaces: Towards a Reconstruction of Golledge's "Geography and the Disabled.". *Transactions of the Institute of British Geographers*, *21*(2), 397. doi:10.2307/622489

Indhumathy, M., Nabhan, A. R., & Arumugam, S. (2018). A weighted association rule mining method for predicting HCV-human protein interactions. *Current Bioinformatics*, *13*(1), 73–84. doi:10.2174/1574893611666161123142425

Indu, S. K. (2019). Internet of Vehicles (IoV): Evolution, Architecture, Security Issues and Trust Aspects. *Int J Recent Technology and Engineering*, *7*(6), 2019.

Insider, B. (2019). *Automotive Industry Trends: IoT Connected Smart Cars and Vehicles*. Available at: https://www.businessinsider.com/internet-of-things-connected-smart-cars-2016-10?r=US&IR=T

Institute for Security Studies. (2019). *Reducing poverty in Africa: huge potential, but more realistic goals needed.* Retrieved from https://issafrica.org/about-us/press-releases/reducing-poverty-in-africa-huge-potential-but-more-realistic-goals-needed

Intelligent Community Forum. (2006). *What is an Intelligent Community*. Available at: https://www.intelligentcommunity.org/what_is_an_intelligent_community

International Telecommunication Union Telecommunication Standardization Sector Focus Group on Smart Sustainable Cities (ITU-T FG-SSC). (2014). *Smart sustainable cities: An analysis of definitions. Technical report*. ITU.

International Telecommunication Union. (2014). https://www.itu.int/en/ITU-T/focusgroups/ssc/Pages/default.aspx

Investopedia. (2017). *V2X (Vehicle-to-Vehicle or Vehicle-to-Infrastructure)*. Available at: https://www.investopedia.com/terms/v/v2x-vehicletovehicle-or-vehicletoinfrastructure.asp

Iqbal, S., Kiah, L. M., Dhaghighi, B., Hussain, M., Khan, S., Khan, M. K., & Choo, K. R. (2016). On cloud security attacks: A taxonomy and intrusion detection and prevention as a service. *Journal of Network and Computer Applications*, *74*, 98–120. doi:10.1016/j.jnca.2016.08.016

Isikli, E., Ustundag, A., & Cevikcan, E. (2015). The effects of environmental risk factors on city life cycle: A link analysis. *Human and Ecological Risk Assessment*, *21*(5), 1379–1394. doi:10.1080/10807039.2014.957953

Ismagilova, E., Hughes, L., Dwivedi, Y. K., & Ravi Raman, K. (2019). Smart Cities: Advances in Research—An Information Systems Perspective. *International Journal of Information Management*, *47*, 88–100. doi:10.1016/j.ijinfomgt.2019.01.004

Ismail, N. (2017). *The Present and Future pf Connected Car Data*. Available at: https://www.information-age.com/present-future-connected-car-data-123466309/

ISO 15704. (2000). *ISO 15704, Industrial automation systems - Requirements for enterprise-reference architectures and methodologies*. Geneva, Switzerland: International Organization for Standardization.

ISO. (2014). ISO 37120:2014, Sustainable development of communities - Indicators for city services and quality of life (Vol. 2014). Geneva, Switzerland: ISO.

ISO/IEC JTC 1. (2014). *Smart Cities - Preliminary Report 2014*. Geneva, Switzerland: International Organization for Standardization/International Electrotechnical Commission.

ISO/IEC. (2016). *ISO/IEC AWI 30145-2, Information technology - Smart city ICT reference framework - Part2 - Smart City Knowledge Management Framework*. Geneva, Switzerland: International Organization for Standardization/International Electrotechnical Commission.

ISO/IEC/IEEE 42010. (2011). ISO/IEC/IEEE 42010:2011 - Systems and software engineering - Architecture description (Vol. 2011). Geneva, Switzerland: International Organization for Standardization/International Electrotechnical Commission. doi:10.1109/IEEESTD.2011.6129467

IST Advisory Group. (2003). *Ambient Intelligence: From Vision to Reality*. European Commission.

ITS. (2019). *Connected Vehicles: Vehicles to pedestrian communication*. Available at: https://www.its.dot.gov/factsheets/pdf/CV_V2Pcomms.pdf

ITU-T FG-SCC. (2014a). *Smart sustainable cities: An analysis of definitions*. International Telecommunication Union.

ITU-T FG-SCC. (2014b). *Technical Report on "Overview of key performance indicators in smart sustainable cities."*. Geneva, Switzerland: International Telecommunication Union.

ITU-T FG-SCC. (2016). *ITU-T Y.4400 series – Smart sustainable cities – Setting the framework for an ICT architecture - Supplement 27*. Geneva, Switzerland: ITU.

ITU-T. (2012). *Recommendation ITU-T Y.2060 - Overview of the Internet of things*. Geneva, Switzerland: International Telecommunication Union.

ITU-T. (2014). *ITU-T Y.2066 - Common requirements of the Internet of things*. International Telecommunication Union.

Ivezic, M. (2016). *Cyber Security – A necessary pillar of Smart Cities*. London: EY.

Jabeena, A., Jayabarathi, T., Singh, G. S., & Kumar, A. A. (2013). PSO Based Optimization Technique for Wireless Optical Communication. *Asian Journal of Computer and Information Systems*, *1*(1), 11–16.

Jee, C. (2018). *The world's first autonomous tram just launched in Germany*. https://www.technologyreview.com/f/612186/the-worlds-first-autonomous-tram-just-launched-in-germany/

Jena, A. K., Biswas, P., & Saha, H. (2017). Aquaponics- a step towards urban agriculture. Innovative Farming. *International Journal of Agriculture.*, *1*, 163–167. Retrieved from https://www.researchgate.net/publication/316191741_Advanced_Farming_Systems_in_Aquaculture_Strategies_to_Enhance_the_Production

Jeya, P. G., & Ravichandran, M. R. a. C. S. (2012). Efficient Classifier for R2L and U2R Attacks. *International Journal of Computer Applications, 45*(21), 28-32.

Jose, A. C. (2017). Intelligent Home Automation Security System based on Novel Logical Sensing and Behaviour Prediction. In Intelligent home automation security system based on novel logical sensing and behaviour prediction. University of Pretoria.

Kachouie, R., Sedighadeli, S., Khosla, R., & Chu, M. T. (2014). Socially assistive robots in elderly care: A mixed-method systematic literature review. *International Journal of Human-Computer Interaction*, *30*(5), 369–393.

Kafer, A. (2013). *Feminist, Queer, Crip*. Indiana University Press.

Kagermann, H., Wahlster, W., & Johannes, H. (2013). *Recommendations for implementing the strategic initiative INDUSTRIE 4.0: Final report of the Industrie 4.0 working group*. Frankfurt: Acatech.

Kaggle. (2019). *DS2OS traffic traces - IoT traffic traces gathered in a the DS2OS IoT environment.* Retrieved August 10, 2019, from https://www.kaggle.com/francoisxa/ds2ostraffictraces

Kakarontzas, G., Anthopoulos, L., Chatzakou, D., & Vakali, A. (2014). A conceptual enterprise architecture framework for smart cities: A survey based approach. *11th International Conference on E-Business, ICE-B 2014 - Part of 11th International Joint Conference on e-Business and Telecommunications, ICETE 2014*, 47–54. 10.5220/0005021400470054

Kakderi, C. (2013). *Smart City Design Principles.* https://www.urenio.org/2013/10/09/smart-city-design-principles/

KarsligEl, M. E., Yavuz, A. G., Güvensan, M. A., & Bank, K. H. a. H. (2017). Network intrusion detection using machine learning anomaly detection algorithms. *2017 25th Signal Processing and Communications Applications Conference (SIU),* 1-4.

Kautzleben, V. (2010, October 15). Eine Stadt mit vielen weißen Stöcken. *Oberhessische Presse.* https://www.op-marburg.de/UNIversum/Das-pralle-Leben/Eine-Stadt-mit-vielen-weissen-Stoecken

Kedir, A., Elhiraika, A., Chinzara, Z., & Sandjong, D. (2017). Growth and development finance required for achieving sustainable development goals (SDGs) in Africa. *African Development Review, 29*(S1), 15–26. doi:10.1111/1467-8268.12230

Kevrekidis, I. G., Gear, C. W., & Hummer, G. (2004). Equation-free: The computer-aided analysis of complex multiscale systems. *AIChE Journal. American Institute of Chemical Engineers, 50*(7), 1346–1355. doi:10.1002/aic.10106

Khamphakdee, N., & Saiyod, N. B. S. (2015). Improving Intrusion Detection System Based on Snort Rules for Network Probe Attacks Detection with Association Rules Technique of Data Mining. *J. ICT Res. Appl, 8*(3), 234–250. doi:10.5614/itbj.ict.res.appl.2015.8.3.4

Khan, M. S., Muyeba, M., & Coenen, F. (2008, September). A weighted utility framework for mining association rules. In *2008 Second UKSIM European Symposium on Computer Modeling and Simulation* (pp. 87-92). IEEE.

Khatoun, R., & Zeadally, S. (2016, August 12). Smart Cities: Concepts, Architectures, Research Opportunities. *Communications of the ACM, 59*(8), 46–57. doi:10.1145/2858789

Khatri, Y. (2019). *China launches blockchain-based identification system for Smart Cities.* https://www.theblockcrypto.com/linked/45654/china-launches-blockchain-based-identification-system-for-smart-cities

Khramtsov, D. (2018). *Augmented reality applications in smart cities.* https://medium.com/coinmonks/augmented-reality-applications-in-smart-cities-8f511fe5895

Kimani, K., Oduol, V., & Kibet, L. (2019). Cyber security challenges for IoT-based smart grid networks. *International Journal of Critical Infrastructure Protection, 25*, 36-49.

Kingsbury, C. (2019). *In the Dark.* CBS Television Distribution.

Kolias, C., Kambourakis, G., Stavrou, A., & Voas, J. (2017). DDoS in the IoT. *Mirai and Other Botnets. in Computer, 50*(7), 80–84.

Komamizu, T., Amagasa, T., Shiokawa, H., & Kitagawa, H. (2016). Towards Real-time Analysis of Smart City Data: A Case Study on City Facility Utilizations. *2016 IEEE 18th International Conference on High Performance Computing and Communications.*

Komninos, N. (2009). Intelligent cities: Towards interactive and global innovation environments. *International Journal of Innovative Research and Development, 1*(4), 337–355.

Kotoulas, S., Lopez, V., Lloyd, R., Sbodio, M. L., Lecue, F., Stephenson, M., ... Mac Aonghusa, P. (2014). SPUD—Semantic processing of urban data. *Journal of Web Semantics, 24*, 11–17. doi:10.1016/j.websem.2013.12.003

Kourtit, K., & Nijkamp, P. (2012). Smart cities in the innovation age. *Innovation (Abingdon), 25*(2), 93–95. doi:10.1080/13511610.2012.660331

Kourtit, K., & Nijkamp, P. (2012). *Smart Cities in the Innovation Age. The European Journal of Social Science Research, 25*(2), 93–95. doi:10.1080/ 13511610.2012.660331

Kresin, F. (2016). 22. Smart cities value their smart citizens. In G. Marini (Ed.), *Urban Europe* (pp. 181–186). Amsterdam University Press. doi:10.1515/9789048535811-023

Kruschewsky, G. (2018). Los Angeles vs. San Francisco: Which city rules the West Coast? *Huffington Post*. Retrieved from huffpost.com/entry/los-angeles-vs-san-francisco

Kumar, M., Soundarya, V., Kavitha, S., Keerthika, S., & Aswini, E. (2019). Credit Card Fraud Detection Using Random Forest Algorithm. *2019 3rd International Conference on Computing and Communications Technologies (ICCCT)*.

Kumar, & Singh, Gupta, & Madaan. (2018, Apr.). Moving towards Smart Cities: Solutions That Lead to the Smart City Transformation Framework. *Technological Forecasting and Social Change*.

Kumi, E. N. (2017). *The electricity situation in Ghana: Challenges and opportunities*. CGD Policy Paper. Washington, DC: Center for Global Development. Retrieved from https://www.cgdev.org/sites/default/files/electricity-situation-ghana-challenges-and-opportunities.pdf

Lakshmi, K. S., & Vadivu, G. (2019). A novel approach for disease comorbidity prediction using weighted association rule mining. *Journal of Ambient Intelligence and Humanized Computing*, 1–8. doi:10.100712652-019-01217-1

Lam & Ma. (2019). Potential Pitfalls in the Development of Smart Cities and Mitigation Measures: An Exploratory Study. *Cities, 91*, 146–56.

Lazarina. (2015). *Describing a Smart City, Case Study: City of Stockholm*. Politechnico Di Milano, School of Architecture and Society.

Lea, R. J. (2017). *Smart cities: An Overview of the technology trends driving smart cities*. Academic Press.

Lee, E. A. (2008). Cyber physical systems: Design challenges. In *2008 11th IEEE International Symposium on Object and Component-Oriented Real-Time Distributed Computing (ISORC)* (pp. 363–369). IEEE.

Lee, J. H., Hancock, M. G., & Hu, M.-C. (2014). Towards an effective framework for building smart cities: Lessons from Seoul and San Francisco. *Technological Forecasting and Social Change, 89*, 80–99. doi:10.1016/j.techfore.2013.08.033

Lee, S., Bae, M., & Kim, H. (2017). Future of IoT networks: A survey. *Applied Sciences (Basel, Switzerland), 7*(10), 1072.

LeGates, T., & Stout, F. (2018). *Smart Cities: Digital Solutions for a more liveable future*. McKinsey Global Institute.

Lehne, M., Luijten, S., Vom Felde, G. I. P., & Thun, S. (2019). The Use of FHIR in Digital Health-A Review of the Scientific Literature. *Studies in Health Technology and Informatics, 267*, 52–58. PMID:31483254

Leydesdorff, L., & Deakin, M. (2011). The triple helix model of smart cities: A neo-evolutionary perspective. *Journal of Urban Technology, 18*(2), 53–63. doi:10.1080/10630732.2011.601111

Libelium. (2019). *Top 50 Sensor Applications for A Smarter World*. http://www.libelium.com/resources/top_50_iot_sensor_applications_ranking

Lin, S.-W., Miller, B., Durand, J., Joshi, R., Didier, P., Chigani, A., … Witten, B. (2017). Industrial Internet Reference Architecture 1.8 - Volume G1: Reference Architecture. Industrial Internet Consortium.

Lin, P., Abney, K., & Bekey, G. A. (2011). *Robot ethics: the ethical and social implications of robotics*. MIT Press.

Linthicum, D. (2010). *Cloud Computing? Thank SOA*. Retrieved from https://www.thecloudtutorial.com/cloud-computing-soa.html

Lipsey, R., & Chrystal, A. (2011). *Economics*. Oxford: Oxford University Press.

LiSA geht im Jägertunnel in den Testbetrieb. (2018). https://www.marburg.de/portal/meldungen/-lisa-geht-im-jaegertunnel-in-den-testbetrieb-900004262-23001.html

Li, T., Li, Y., An, D., Han, Y., Xu, S., Lu, Z., & Crittenden, J. (2019). Mining of the association rules between industrialization level and air quality to inform high-quality development in China. *Journal of Environmental Management*, *246*, 564–574. doi:10.1016/j.jenvman.2019.06.022 PMID:31202021

Liu, J., Xiong, K., Fan, P., & Zhong, Z. (2017). RF energy harvesting wireless powered sensor networks for smart cities. *IEEE Access: Practical Innovations, Open Solutions*, *5*, 9348–9358.

Liu, L., Xu, B., Zhang, X., & Wu, X. (2018). An intrusion detection method for internet of things based on suppressed fuzzy clustering. *EURASIP Journal on Wireless Communications and Networking*, *2018*(1), 1–7. doi:10.118613638-018-1128-z

Liu, Y., Wei, J., & Rodriguez, A. F. C. (2014). Development of a strategic value assessment model for smart city. *International Journal of Mobile Communications*, *12*(4), 346–359. doi:10.1504/IJMC.2014.063652

Live Science. (2019). *What is STEM education?* Retrieved from livescience.com/43296-what-is-stem-education.html

Li, W., Song, H., & Zeng, F. (2017). Policy-Based Secure and Trustworthy Sensing for Internet of Things in Smart Cities. *IEEE Internet of Things Journal*, *5*(2), 716–723. doi:10.1109/JIOT.2017.2720635

Li, Y. (2005, September). A user-guided association rules mining method and its application. In *The Fifth International Conference on Computer and Information Technology (CIT'05)* (pp. 150-156). IEEE.

Li, Z., & Qi, L. (2018). Economic Solutions to Improve Cybersecurity of Governments and Smart Cities via Vulnerability Markets. *Government Information Quarterly*, *35*(1), 151–160. doi:10.1016/j.giq.2017.10.006

Lombardi, P., Giordano, S., Farouh, H., & Yousef, W. (2012). Modelling the smart city performance. *Innovation (Abingdon)*, *25*(2), 137–149. doi:10.1080/13511610.2012.660325

Losavio, M. M., Chow, K. P., Koltay, A., & James, J. (2018). The Internet of Things and the Smart City: Legal Challenges with Digital Forensics, Privacy, and Security. *Security and Privacy*, *1*(3), e23. doi:10.1002py2.23

Lu, J., & Qian, Z. (1999). Mining Boolean and General Fuzzy Weighted Association Rules database. *Mathematica Applicata*, *12*, 38–43.

Maddox, T. (2018). *Smart Cities: A cheat sheet*. https://www.techrepublic.com/article/smart-cities-the-smart-persons-guide/

Mahabir, R., Crooks, A., Croitoru, A., & Agouris, P. (2016). The study of slums as social and physical constructs: Challenges and emerging research opportunities. *Regional Studies. Journal of Regional Science*, *3*(1), 399–419. doi:10.1080/21681376.2016.1229130

Mahizhnan, A. (1999). Smart cities: The Singapore case. *Cities (London, England)*, *16*(1), 13–18. doi:10.1016/S0264-2751(98)00050-X

Mahmood, A., Butler, B., Sheng, Q. Z., Zhang, W. E., & Jennings, B. (2019). Need of Ambient Intelligence for Next-Generation Connected and Autonomous Vehicles. In *Guide to Ambient Intelligence in the IoT Environment* (pp. 133–151). Cham: Springer.

Mahmood, Z. (2013a). *Cloud Computing: Methods and Practical Approaches*. London: Springer Verlag.

Mahmood, Z. (2013b). *E-Government Implementation and Practice in Developing Countries*. IGI Global Publishers.

Mahmood, Z. (2020). Connected Vehicles in the IoV: Concepts, Technologies, and Architectures. In Z. Mahmood (Ed.), *Connected Vehicles in the Internet of Things. Springer*. doi:10.1007/978-3-030-36167-9_1

Maier, A., & Student, D. (2015). Industrie 4.0 – Der Große Selbstbetrug. *Manager Magazine*. https://www.manager-magazin.de/magazin/artikel/digitale-revolutionindustrie-4-0-ueberfordert-deutschen-mittelstand-a-1015724.html

Maier, M. W. (1998). Architecting principles for systems-of-systems. *Systems Engineering*, *1*(4), 267–284. doi:10.1002/(SICI)1520-6858(1998)1:4<267::AID-SYS3>3.0.CO;2-D

Maksimovic, M. (2018). Greening the Future: Green Internet of Things (G-IoT) as a Key Technological Enabler of Sustainable Development. In *Internet of Things and Big Data Analytics toward Next-Generation Intelligence* (pp. 283–313). Springer. doi:10.1007/978-3-319-60435-0_12

Mankiw. (2012). *Principles of Economics*. London: Cengage Learning International.

Mansfield-Devine, S. (2018). Firms failing to disclose IoT vulnerabilities. *Network Security*, *2018*(12), 1–2. doi:10.1016/S1353-4858(18)30118-1

Manville, C., Cochrane, G., Cave, J., Millard, J., Pederson, J., Thaarup, R., & Kotterink, B. (2014). *Mapping Smart Cities in the EU*. Brussels: European Union.

Markets and Markets. (2019). *Edge Computing Market by Component (Hardware, Platform, and Services), Application (Smart Cities, IIoT, Content Delivery, Remote Monitoring, AR and VR), Organization Size (SMEs and Large Enterprises), Vertical, and Region - Global Forecast to 2024*. https://www.marketsandmarkets.com/Market-Reports/edge-computing-market-133384090.html

Markopoulos, P. (2016). Ambient intelligence: Vision, research, and life. *Journal of Ambient Intelligence and Smart Environments*, *8*(5), 491–499.

Martinidis, G. (2017). *Three Generations of Evolving Smart Cities*. https://www.urenio.org/2017/07/19/three-generations-evolving-smart-cities/

Martin, V., Cao, Q., & Benson, T. (2017). Fending off IoT-Hunting Attacks at Home Networks. *Proceedings of the 2nd Workshop on Cloud-Assisted Networking*, 67–72. 10.1145/3155921.3160640

Maruatona, O. (2013). *Internet Banking Fraud Detection Using Prudent Analysis*. University of Ballarat.

Mascaro, S., Nicholso, A. E., & Korb, K. B. (2014). Anomaly detection in vessel tracks using Bayesian networks. *International Journal of Approximate Reasoning*, *55*(1), 84–98. doi:10.1016/j.ijar.2013.03.012

Maslin Nir, S. (2019). *In Ghana, free high school brings opportunity and grumbling*. Retrieved from https://www.nytimes.com/2019/06/15/world/africa/ghana-free-senior-high-school-brings-chaos.html

Mattila, J., Seppälä, T., Naucler, C., Stahl, R., Tikkanen, M., Bådenlid, A., & Seppälä, J. (2016). *Industrial Blockchain Platforms: An Exercise in Use Case Development in the Energy Industry*. ETLA Working Papers 43, The Research Institute of the Finnish Economy. Retrieved from https://ideas.repec.org/p/rif/wpaper/43.html

McKee, D. W., Clement, S. J., Almutairi, J., & Xu, J. (2017). Massive-scale automation in cyber-physical systems: Vision & challenges. In *Autonomous Decentralized System (ISADS), 2017 IEEE 13th International Symposium on* (pp. 5–11). IEEE.

McKinsey & Company. (2018). Smart Cities: Digital Solutions for a more livable future June 2018. *McKinsey Global Institute.* Retrieved from https://www.mckinsey.com/~/media/mckinsey/industries/capital%20projects%20and%20infrastructure/our%20insights/smart%20cities%20digital%20solutions%20for%2

Mehdi, M. (2018). Path Planning in Support of Smart Mobility Applications using Generative Adversarial Networks. *2018 IEEE International Conference on Internet of Things (iThings) and IEEE Green Computing and Communications (GreenCom) and IEEE Cyber, Physical and Social Computing (CPSCom) and IEEE Smart Data (SmartData).*

Mehmood, T., & Rais, H. B. M. (2016). Machine learning algorithms in context of intrusion detection. *3rd International Conference on Computer and Information Sciences (ICCOINS),* 369-373. 10.1109/ICCOINS.2016.7783243

Menouar, H., Guvenc, I., Akkaya, K., Uluagac, A. S., Kadri, A., & Tuncer, A. (2017). UAV-enabled intelligent transportation systems for the smart city: Applications and challenges. *IEEE Communications Magazine*, *55*(3), 22–28.

Ministry of Energy. (2019). *Energy sector recovery program.* Retrieved from http://energycom.gov.gh/files/2019%20 1111%20ESRP%20ESTF_Clean_v3.0redacted%20final.pdf

Ministry of Finance. (2019). *Mid-year fiscal policy review of the 2019 budget statement and economic policy & supplementary estimates of the government of Ghana for the 2019 financial year.* Retrieved from https://drive.google.com/file/d/10bSdWq1woHs-IjK2HAz-UYKs7SlDjTym/view

Min, K., Yoon, M., & Furuya, K. (2019). Sustainability research terms used on smart cities. *Sustainability*, *11*, 3155. doi:10.3310u1113155

Mishra, A., & Ghorpade, C. (2018). Credit Card Fraud Detection on the Skewed Data Using Various Classification and Ensemble Techniques. *2018 IEEE International Students' Conference on Electrical, Electronics and Computer Science (SCEECS).*

Mission, S. C. (2017). *Smart City Features.* http://smartcities.gov.in/content/innerpage/smart-city-features.php

Modi, C., Patel, D., Patel, H., Borisaniya, B., Patel, A., & Rajarajan, M. (2013). A survey of intrusion detection techniques in Cloud. *Journal of Network and Computer*, *36*(1), 42–57. doi:10.1016/j.jnca.2012.05.003

Mohammadi, M., & Al-Fuqaha, A. (2018). Enabling cognitive smart cities using big data and machine learning: Approaches and challenges. *IEEE Communications Magazine*, *56*(2), 94–101. doi:10.1109/MCOM.2018.1700298

Mohanty, S. P., Choppali, U., & Kougianos, E. (2016). Everything you wanted to know about smart cities: The internet of things is the backbone. *IEEE Consumer Electronics Magazine*, *5*(3), 60–70. doi:10.1109/MCE.2016.2556879

Mol, Khan, Aalders, & Schouten. (2015). *Report: A spotlight on Smart City Eindhoven.* Venturespring B.V.

Monfaredzadeh, T., & Berardi, U. (2015). Beneath the smart city: Dichotomy between sustainability and competitiveness. *International Journal of Sustainable Building Technology and Urban Development*, *6*(3), 140–156.

Moore, S. K. (2018). Self-powered image sensor could watch you forever. *IEEE Spectrum.*

Mori, K., & Christodoulou, A. (2012). Review of Sustainability Indices and Indicators: Towards a New City Sustainability Index (CSI*). Environmental Impact Assessment Review*, *32*(1), 94–106.

Morvaj, B., Lugaric, L., & Krajcar, S. (2012). Demonstrating Smart Buildings and Smart Grid features in a Smart Energy City. *Energetics (IYCE), Proceedings of the 2011 3rd International Youth Conference.*

Motta, R.C., de Oliveira, K.M., & Travassos, G.H. (2019). A conceptual perspective on interoperability and software systems operations. Retrieved from doi:10.1016/j.infsof.2019.07.001

Mueller, T. (2017). *Redefining the Smart City Concept: A New Smart City Definition.* https://hub.beesmart.city/strategy/en/towards-a-new-smart-city-definition

Mullan, L. (2019). *Top 10 smart cities in the world.* Retrieved from gigabitmagazine/big-data/top-10-cities-world

Muller, G., & van de Laar, P. (2010). Researching reference architectures. In *Views on Evolvability of Embedded Systems* (pp. 107–119). Springer. doi:10.1007/978-90-481-9849-8_7

Mulvey, L. (1975). Visual Pleasure and Narrative Cinema. *Screen, 16*(3), 6–18. doi:10.1093creen/16.3.6

Mulvey, L. (2013). *Fetishism and Curiosity: Cinema and the Mind's Eye.* Palgrave Macmillan.

Mumford, L. (1973). What is a city? In *The City Reader* (5th ed.). Academic Press.

Murphy, J. (2018). *7 amazing body parts that can now be 3D printed.* MDLinx. Retrieved from https://www.mdlinx.com/internal-medicine/article/2668

Muyeba, M., Khan, M. S., & Coenen, F. (2008, May). Fuzzy weighted association rule mining with weighted support and confidence framework. In *Pacific-Asia Conference on Knowledge Discovery and Data Mining* (pp. 49-61). Springer.

Mylrea, M. E., & Gourisetti, S. N. G. (2017). Blockchain for smart grid resilience: Exchanging distributed energy at speed, scale and security. *Resilience Week (RWS)*, 18-23. doi:10.1109/RWEEK.2017.8088642

Naab, F. Z., Abubakari, Z., & Ahmed, A. (2019). The role of climate services in agricultural productivity in Ghana: The perspectives of farmers and institutions. *Climate Services, 13*, 24–32. doi:10.1016/j.cliser.2019.01.007

Nakagawa, E. Y., Antonino, P. O., & Becker, M. (2011). Reference architecture and product line architecture: A subtle but critical difference. In *European Conference on Software Architecture* (pp. 207–211). Springer. 10.1007/978-3-642-23798-0_22

Nam, T., & Pardo, T. A. (2011). Conceptualizing smart city with dimensions of technology, people, and institutions. In *Proceedings of the 12th annual international digital government research conference: digital government innovation in challenging times* (pp. 282–291). 10.1145/2037556.2037602

Nanjie, L. (2011). *Internet of Vehicles: Your next connection.* Huawei WinWin.

Naphade, M., Banavar, G., Harrison, C., Paraszczak, J., & Morris, R. (2011). Smarter Cities and Their Innovation Challenges. *Computer, 44*(6), 32–39. doi:10.1109/MC.2011.187

National League of Cities (NLC). (2017). *Trends in Smart City Development.* Retrieved from https://www.nlc.org/sites/default/files/2017-01/Trends%20in%20Smart%20City%20Development.pdf

Ndehedehe, C. E., Awange, J. L., Kuhn, M., Agutu, N. O., & Fukuda, Y. (2017). Analysis of hydrological variability over the Volta river basin using in-situ data and satellite observations. *Journal of Hydrology. Regional Studies, 12*, 88–110. doi:10.1016/j.ejrh.2017.04.005

Neirotti, P., De Marco, A., Cagliano, A. C., Mangano, G., & Scorrano, F. (2014). Current trends in Smart City initiatives: Some stylised facts. *Cities (London, England), 38*, 25–36. doi:10.1016/j.cities.2013.12.010

Nelson and Pade. (2019a). *A non-gmo fish food for aquaponics.* Retrieved from https://aquaponics.com/product-category/aquaponic-growing-supplies/fish-food/

Nelson and Pade. (2019b). *Aquaponics.* Retrieved from https://aquaponics.com/

Neukart, Compostella, G., Seidel, C., von Dollen, D., Yarkoni, S., & Parney, B. (2017). Traffic Flow Optimizer using a Quantum Annealer. *Frontiers in ICT*, *4*, 29. doi:10.3389/fict.2017.00029

Nguyen, A. (2018). *10 typical tree planting projects in the world*. Retrieved from treedom.net/en/projects

Nia, E. M., Zawawi, N. A. W. A., & Singh, B. S. M. (2017). A review of walking energy harvesting using piezoelectric materials. *IOP Conference Series. Materials Science and Engineering*, *291*(1), 012026.

Nigon, J., Glize, E., Dupas, D., Crasnier, F., & Boes, J. (2016). Use Cases of Pervasive Artificial Intelligence for Smart Cities Challenges. *2016 Intl IEEE Conferences on Ubiquitous Intelligence & Computing, Advanced and Trusted Computing, Scalable Computing*.

Niras.dk. (2019). *Copenhagen Smart City*. https://www.niras.dk/media/1585/kimspiegelbergsteltzer.pdf

NIST. (2016). *Current standards landscape for smart manufacturing systems. National Institute of Standards and Technology, NISTIR* (Vol. 8107). U.S. Department of Commerce, National Institute of Standards and Technology.

NIST. (2018, September 30). *A Consensus Framework for Smart City Architectures*. Retrieved April 22, 2019, from https://s3.amazonaws.com/nist-sgcps/smartcityframework/files/ies-city_framework/IES-CityFramework_Version_1_0_20180930.pdf

Norman, D. (1998). *The Invisible Computer*. The MIT Press.

Nuclear Energy by the Numbers. (2019). Nuclear Energy Institute.

O'grady, M., & O'hare, G. (2012). How smart is your city? *Science*, *335*(6076), 1581–1582. PMID:22461597

O'Neill, P. H. (2019). *Ransomware hackers hit nearly two dozen Texas cities*. https://www.technologyreview.com/f/614200/ransomware-hackers-hit-nearly-two-dozen-texas-cities/

OASIS RM-SOA TC. (2019). *OASIS SOA Reference Model (SOA-RM) Technical Committee FAQ*. Retrieved October 17, 2019, from https://www.oasis-open.org/committees/soa-rm/faq.php

Object Management Group (OMG). (2005). https://www.omg.org/robotics/robotic_systems_rfi.htm

Oktay, U., & Sahingoz, O. K. (2013). Proxy Network Intrusion Detection System for Cloud Computing. *2013 The International Conference on Technological Advances in Electrical, Electronics and Computer Engineering (TAEECE)*, 98-104. 10.1109/TAEECE.2013.6557203

Othman, Z. A., Ismail, N., Bakar, A. A., Latif, M. T., & Abdullah, S. M. S. (2018). Rules discovery of high ozone in Klang areas using data mining approach. *International Journal on Advanced Science. Engineering and Information Technology*, *8*(6), 2683–2689.

Owusu-Ansah, F. E., Tagbor, H., & Togbe, M. A. (2016). Access to health in city slum dwellers: The case of Sodom and Gomorrah in Accra, Ghana. *African Journal of Primary Health Care & Family Medicine*, *8*(1), e1–e7. doi:10.4102/phcfm.v8i1.822 PMID:27247151

Pacheco, J., & Hariri, S. (2016). *IoT Security Framework for Smart Cyber Infrastructures. 2016 IEEE 1st International Workshops on Foundations and Applications of Self* Systems.*

Park, W., & Kang, B. H. (2017). *An Enhancement of Optimized Detection Rule of Security Monitoring and Control for Detection of Cyberthreat in Location-Based Mobile System*. doi:10.1155/2017/8501976

Parks, Mclaren, Toumi, & Rivett. (2019). *Experiences and lessons in managing water from Cape Town*. Grantham Institute Briefing paper No 29.

Park, S. Y., Lee, K., Song, H., & Yoon, E. (2018). Simultaneous Imaging and Energy Harvesting in CMOS Image Sensor Pixels. *IEEE Electron Device Letters*, *39*(4), 532–535.

Pasini, D., Ventura, S. M., Rinaldi, S., Bellagente, P., Flammini, A., & Ciribini, A. L. C. (2016). Exploiting Internet of Things and building information modeling framework for management of cognitive buildings. *IEEE International Smart Cities Conference (ISC2),* 1-6. 10.1109/ISC2.2016.7580817

Paskaleva, K. (2011). The smart city: A nexus for open innovation? *Intelligent Buildings International*, *3*(3), 153–171. doi:10.1080/17508975.2011.586672

Patel, N. (2018). *Figure of the week: Understanding poverty in Africa*. Retrieved from https://www.brookings.edu/blog/africa-in-focus/2018/11/21/figure-of-the-week-understanding-poverty-in-africa/

Pellicer, S., Santa, G., Bleda, A., Maestre, R., Jara, A., & Skarmeta, G. (2013). A Global Perspective of Smart Cities: A Survey. *2013 Seventh International Conference on Innovative Mobile and Internet Services in Ubiquitous Computing (IMIS)*, 439–444. DOI:10.1109/IMIS.2013.79

Peng, W., Li, Y., Li, B., & Zhu, X. (2016). An Analysis Platform of Road Traffic Management System Log Data Based on Distributed Storage and Parallel Computing Techniques. *2016 IEEE International Conferences on Big Data and Cloud Computing (BDCloud), Social Computing and Networking (SocialCom), Sustainable Computing and Communications.*

Pesce, N. L. (2019). *Commuters in this city waste 119 hours sitting in traffic every year*. https://www.marketwatch.com/story/commuters-in-this-city-waste-119-hours-sitting-in-traffic-every-year-2019-08-22

Pierceall, K. (2015). *Group says New Las Vegas resort-casino to rise at stardust site by 2018*. Retrieved from washingtontimes.com/news/2015/may/5/group-says-new-las-vegas-resort-casino

Pozdniakova, A. (2018). Smart city strategies "London-Stockholm-Vienna-Kyiv": in search of common ground and best practices. *Acta Innovations*. doi:10.32933/ActaInnovations.27.4

Proctor, R. N. (2002). *Racial hygiene: Medicine under the Nazis*. Harvard University Press.

Psaltoglou, A. (2018). *From Smart Cities to Cognitive Cities: Intelligence and Urban Utopias*. Retrieved from https://www.researchgate.net/figure/A-cognitive-system-for-transportation-infrastructure-Mostashari-2011_fig1_327395697

Pucher, J., & Buehler, R. (2008). Making Cycling Irresistible: Lessons from The Netherlands, Denmark and Germany. *Transport Reviews*, *28*(4), 495–528. doi:10.1080/01441640701806612

PWC. (2018). *Creating the smart cities of the future*. Geneva: World Economic Forum.

Qin, S., Liu, F., Wang, C., Song, Y., & Qu, J. (2015). Spatial-temporal analysis and projection of extreme particulate matter (PM10 and PM2. 5) levels using association rules: A case study of the Jing-Jin-Ji region, China. *Atmospheric Environment*, *120*, 339–350. doi:10.1016/j.atmosenv.2015.09.006

Qiu, S., Feng, F., Li, Z., Yang, G., Xu, P., & Li, Z. (2019, April). Data mining based framework to identify rule based operation strategies for buildings with power metering system. []. Tsinghua University Press.]. *Building Simulation*, *12*(2), 195–205. doi:10.100712273-018-0472-6

Rahimi, Z., Noferesti, S., & Shamsfard, M. (2019). Applying data mining and machine learning techniques for sentiment shifter identification. *Language Resources and Evaluation*, *53*(2), 279–302. doi:10.100710579-018-9432-0

Raj, D. (2016). Fog. *IEEE 8th Int. Conf. Cloud Comput. Technol. Sci.*, 90–93.

Rajendran, P. K., Rajesh, M., & Abhilash, R. (2015). Hybrid Intrusion Detection Algorithm for Private Cloud. *Indian Journal of Science and Technology, 12*(8).

Ramos, C., Augusto, J. C., & Shapiro, D. (2008). Ambient intelligence—The next step for artificial intelligence. *IEEE Intelligent Systems*, *23*(2), 15–18.

Ran, P., Wang, Y., Liu, P., & Li, Z. (2016). A Fuzzy Association Rule Mining Based Approach to Identifying SO 2 Emission Concentration Patterns of Coal-fired Boilers. *Chemical Engineering Transactions*, *52*, 631–636.

Ravikumar, G. V. V. (2019). Advanced Robotics: Autonomous Solutions Take on Dull, Dirty, and Dangerous Work. *Infosys Insights*. https://www.infosys.com/insights/ai-automation/advanced-robotics.html

Ray, C., Mondada, F., & Siegwart, R. (2008). What do people expect from robots? In *Intelligent Robots and Systems, 2008. IROS 2008. IEEE/RSJ International Conference on* (pp. 3816-3821). IEEE.

Raza, S., Wallgren, L., & Voigt, T. (2013). SVELTE: Real-time intrusion detection in the Internet of Things. *Ad Hoc Networks*, *11*(08), 2661–2674. doi:10.1016/j.adhoc.2013.04.014

Remagnino, P., & Foresti, G. L. (2005). Ambient Intelligence: A New Multidisciplinary Paradigm. *IEEE Transactions on Systems, Man, and Cybernetics. Part A, Systems and Humans*, *35*(1), 1–6.

Report, W. C. (2016). *Urbanization and development: Emerging Futures*. Retrieved from http://wcr.unhabitat.org/

Robinson, J., Sinton, S., & Samii, Y. R. (2002). Particle swarm, genetic algorithm, and their hybrids: optimization of a profiled corrugated horn antenna. *Proceedings of the IEEE International Symposium in Antennas and Propagation Society*, *1*, 314–317.

Romdhani, I. (2017). Confidentiality and Security for IoT Based Healthcare. In Securing the internet of things. Syngress. doi:10.1016/B978-0-12-804458-2.00009-3

Sachsenmeier, P. (2016). Industry 5.0—The Relevance and Implications of Bionics and Synthetic Biology. *Engineering*, *2*, 225–229.

Saha, S. (2019). *Cumulative Causation Theory: Gunnar Myrdal*. Retrieved from https://bgc.org.in/pdf/open-educational-resources/geography/gunner-myrdals-theory_ug_II_ss_1.pdf

Sahoo, M. S., Mishra, M. P. K., & Satpathy, P. N. (2012). Secure Routing in Wireless Sensor Networks. *International Journal of Computer Science*, *9*(2).

Saito, N., Ruhanen, L., Noakes, S., & Axelsen, M. (2018). Community engagement in pro-poor tourism initiatives: Fact or fallacy? Insights from the inside. *Tourism Recreation Research*, *43*(2), 175–185. doi:10.1080/02508281.2017.1406566

Samani, H., & Zhu, R. (2016). Robotic automated external defibrillator ambulance for emergency medical service in smart cities. *IEEE Access: Practical Innovations, Open Solutions*, *4*, 268–283.

Santana, E. F. Z., Chaves, A. P., Gerosa, M. A., Kon, F., & Milojicic, D. S. (2017). Software platforms for smart cities: Concepts, requirements, challenges, and a unified reference architecture. *ACM Computing Surveys*, *50*(6), 78.

Saraswat, D., Dev, V., & Singh, P. (2018, September). Analyzing the performance of the Indian Cricket Team using Weighted Association Rule Mining. In *2018 International Conference on Computing, Power and Communication Technologies (GUCON)* (pp. 161-164). IEEE. 10.1109/GUCON.2018.8675115

Schelin, D., Batmaz, B., Carvalho, M., Ivanic, S., & Flach, J. (2019, November 15). *Group Interview: The Relation Between the City of Marburg and its Blind/Visually Impaired Citizens* (D. Schelin, Interviewer) [Audio recording]. Academic Press.

Schelin, D. (2019). *Vision and Blindness in Film*. Büchner Verlag.

Science Learning. (2019). *Scientific Modelling*. Retrieved from sciencelearning.org.nz/resources/575-scientificmodelling

Seo, Y., Alam, M. T., & Tang, W. (2018). *Gasification of Municipal Solid Waste*. doi:10.5772/intechopen.73685

Seraphim, I., Palit, S., Srivastava, K., & Poovammal, E. (2019). Implementation of Machine Learning Techniques applied to the Network Intrusion Detection System. *International Journal of Engineering and Advanced Technology*, *8*(5), 2721-2726.

Serholt, S., Basedow, C. A., Barendregt, W., & Obaid, M. (2014). Comparing a humanoid tutor to a human tutor delivering an instructional task to children. In *2014 IEEE-RAS International Conference on Humanoid Robots* (pp. 1134-1141). IEEE.

Serrano, E., & Bajo, J. (2019). Deep neural network architectures for social services diagnosis in smart cities. *Future Generation Computer Systems*, *100*, 122–131. doi:10.1016/j.future.2019.05.034

Shakeri, R., Al-Garadi, M. A., Badawy, A., Mohamed, A., Khattab, T., Al-Ali, A., ... Guizani, M. (2019). Design Challenges of Multi-UAV Systems in Cyber-Physical Applications: A Comprehensive Survey, and Future Directions. *IEEE Communications Surveys and Tutorials*, *21*(4), 1–50. doi:10.1109/COMST.2019.2924143

Shapiro, J. M. (2006). Smart Cities: Quality of Life, Productivity, and the Growth Effects of Human Capital. *The Review of Economics and Statistics*, *88*(2), 324–335. doi:10.1162/rest.88.2.324

Sherazi, H. H. R., & Khan, Z. A. (2019). A Heterogeneous IoV Architecture for Data Forwarding in Vehicle to Infrastructure Communication, Mobile. *Information Systems*, *2019*, 3101276.

Sheridan, T. B., & Verplank, W. L. (1978). *Human and computer control of undersea teleoperators*. Man-Machine Systems Laboratory Report, MIT. www.dtic.mil/cgi-bin/GetTRDoc?AD=ADA057655,

Shi, X., Liang, Y. C., Lee, H. P., Lu, C., & Wang, L. M. (2005). An improved GA and a novel PSO-GA-based hybrid algorithm. *Information Processing Letters*, *93*(5), 255–261. doi:10.1016/j.ipl.2004.11.003

Siebers, T. (2005). Disability Aesthetics. *PMLA*, *120*(2), 542–546.

Silva, B. N., Khan, M., & Han, K. (2018). Towards Sustainable Smart Cities: A Review of Trends, Architectures, Components, and Open Challenges in Smart Cities. *Sustainable Cities and Society*, *38*, 697–713.

Silva, B. N., Khan, M., & Han, K. (2018). Towards sustainable smart cities: A review of trends, architectures, components, and open challenges in smart cities. *Sustainable Cities and Society*, *38*, 697–713. doi:10.1016/j.scs.2018.01.053

Simon, Y. R. (1986). *The Definition of Moral Virtue* (V. Kuic, Ed.). Fordham University Press.

Singh, Kumar, & Prakash. (2019). *Importance and Needs of IoT in Developing Smart Cities*. Ved.

Skouby, K. E., Kivimäki, A., Haukiputo, L., Lynggaard, P., & Windekilde, I. M. (2014). Smart cities and the ageing population. *The 32nd Meeting of WWRF*.

Smart Cities Dive. (2017). *Four Cities' Solutions to Sustainable Garbage Processing*. Retrieved from www.smartcitiesdive.com/ex/sustainablecitiescollective/Friday-fun-how-create-tomorrow-s-green-cities-today-s-garbage/1050616/

Smart Cities Mission. (2019). *What is Smart City*. Available at: http://smartcities.gov.in/content/innerpage/what-is-smart-city.php

Smart Cities Team. (2018). *Smart City 3.0: Building Compelling Smart City Business Cases*. https://www.arcweb.com/blog/smart-city-30-building-compelling-smart-city-business-cases

Smartnet. (2019). *A Day in Data*. https://www.visualcapitalist.com/wp-content/uploads/2019/04/data-generated-each-day-wide.html

Smith, L. (2017). *Amsterdam Smart City: A World Leader in Smart City Development*. https://hub.beesmart.city/city-portraits/smart-city-portrait-amsterdam

Snort. (2020). https://www.snort.org/

Söderström, O., Paasche, T., & Klauser, F. (2014). Smart cities as corporate storytelling. *City*, *18*(3), 307–320. doi:10.1080/13604813.2014.906716

Soldatos, J., Kefalakis, N., Hauswirth, M., Serrano, M., Calbimonte, J.-P., Riahi, M., & … . (2015). OpenIoT: Open source internet-of-things in the cloud. In *Interoperability and open-source solutions for the internet of things* (pp. 13–25). Springer. doi:10.1007/978-3-319-16546-2_3

Somayya, M., & Ramaswamy, R. (2016), Amsterdam Smart City (ASC): fishing village to sustainable city. *Proceedings of the 11 International Conference on Urban Regeneration and Sustainability*, 831-*842*. 10.2495/SC160681

Soyata, T., Habibzadeh, H., Ekenna, C., Nussbaum, B., & Lozano, J. (2019). Smart City in Crisis: Technology and Policy Concerns. *Sustainable Cities and Society*, *50*, 101566. doi:10.1016/j.scs.2019.101566

Stacey, P. (2018). Urban development and emerging relations of informal property and land-based authority in Accra. *Africa*, *88*(1), 63–80. doi:10.1017/S0001972017000572

Steventon, A., & Wright, S. (Eds.). (2006). *Intelligent spaces: The application of pervasive ICT*. London: Springer. doi:10.1007/978-1-84628-429-8

Sticht, C. (2017, June 6). Smartphone als Segen für Sehbehinderte. *Frankfurter Rundschau*.

Stollenwerk, T. (2019). Quantum Annealing Applied to De-Conflicting Optimal Trajectories for Air Traffic Management. [*IEEE Transactions on Intelligent Transportation Systems*, 1–13.

Street-Savvy - Meeting the biggest challenges starts with the city. (2011). *Scientific American*, *305*(3), 27–29.

Streitz, N. A. (2011). Smart cities, ambient intelligence and universal access. In *International Conference on Universal Access in Human-Computer Interaction* (pp. 425-432). Springer.

Streitz, N., Charitos, D., Kaptein, M., & Böhlen, M. (2019). Grand challenges for ambient intelligence and implications for design contexts and smart societies. *Journal of Ambient Intelligence and Smart Environments*, *11*(1), 87–107.

Stytz, M. R., Lichtblau, D. E., & Banks, S. B. (2005). *Toward using intelligent agents to detect assess, and counter cyber-attacks in a network-centric environment*. Alexandria, VA: Belvoir Defense Technical Information Center. doi:10.21236/ADA464134

Sullins, J. P. (2008). Friends by design: A design philosophy for personal robotics technology. In P. E. Vermaas, P. Kroes, A. Light, & S. Moore (Eds.), *Philosophy and Design* (pp. 143–157). Springer Netherlands.

Sultan, B., Defrance, D., & Iizumi, T. (2019). Evidence of crop production losses in West Africa due to historical global warming in two crop models. *Scientific Reports*, *9*(1), 12834. doi:10.103841598-019-49167-0 PMID:31492929

Sundararajan, S. (2017). *Australian Government grants $8 million for blockchain energy pilot*. Retrieved from https://www.coindesk.com/australian-government-grants-8-million-for-blockchain-energy-pilot

Sun, K., & Bai, F. (2008). Mining weighted association rules without preassigned weights. *IEEE Transactions on Knowledge and Data Engineering*, *20*(4), 489–495. doi:10.1109/TKDE.2007.190723

Sweden. (2018). *Energy Use in Sweden*. https://sweden.se/society/energy-use-in-sweden/

Swift, D. (2019). Successful SIEM and Log. *SANS Institute - Information Security Reading Room, 11*(1), 1-40.

Syngas, G. (2019). *Waste to energy gasification.* Retrieved from https://www.globalsyngas.org/syngas-production/waste-to-energy-gasification/

Tahir, Z., & Mulek, J. A. (2016). Main criteria in the development of smart cities determinants using analytical method. *Journal of Malaysian Institute of Planners, 14*, 1–14. Retrieved from https://www.researchgate.net/publication/309958294_Main_criteria_in_the_development_of_smart_cities_determined_using_analytical_method

Tankard, C. (2011). Advanced Persistent threats and how to monitor and deter them. *Network Security, 2011*(8), 16–19. doi:10.1016/S1353-4858(11)70086-1

Tao, F., Murtagh, F., & Farid, M. (2003, August). Weighted association rule mining using weighted support and significance framework. In *Proceedings of the ninth ACM SIGKDD international conference on Knowledge discovery and data mining* (pp. 661-666). ACM. 10.1145/956750.956836

Tavallaee, M., Bagheri, E., Lu, W., & Ghorbani, A. A. (2009). A detailed analysis of the KDD CUP 99 data set. *2009 IEEE Symposium on Computational Intelligence for Security and Defense Applications*, 1-6. 10.1109/CISDA.2009.5356528

Technavio. (2019). *Top 10 Connected Car Companies Leading the Global Connected Car Market in 2019.* Available at: https://blog.technavio.com/blog/top-10-connected-car-companies-leading-global-connected-car-market

Techopedia. (2012). *Cloud Computing for Vehicles: Tomorrow's High-Tech Car.* Available at: https://www.techopedia.com/2/28137/trends/cloud-computing/cloud-computing-for-vehicles-tomorrows-high-tech-car

Thamilarasu, G., & Chawla, S. (2019). Towards Deep-Learning-Driven Intrusion Detection for the Internet of Things. *Sensors (Basel), 19*(09), 1977. doi:10.339019091977 PMID:31035611

The Future of Smart. (2016). Harnessing digital innovation to make London the best city in the world. Greater London Authority.

The Open Group. (2018). *The TOGAF Standard, Version 9.2 - Reference: C182*. Author.

The Rockefeller Foundation. (2019). *Accra*. Retrieved from https://nextcity.org/informalcity/city/accra

Thomas, C. (1999). *Female Forms: Experiencing and understanding disability*. Open University Press.

Thuzar, M. (2011). Urbanization in Southeast Asia: Developing smart cities for the future? *Regional Outlook*, 96.

Thuzar, M. (2011). Urbanization in SouthEast Asia: developing smart cities for the future? *Regional Outlook*, 96.

Together, S. L. (2018). The Mayor's roadmap to transform London into the smartest city in the world. Greater London Authority.

Torn, S. (2001). The social model of disability: An outdated ideology? In W. Nicholas, S. N. Barnartt, & B. M. Altman (Eds.), *Exploring Theories and Expanding Methodologies: Where we are and where we need to go* (Vol. 2, pp. 9–28). Emerald Group Publishing Limited. doi:10.1016/S1479-3547(01)80018-X

Toti, G., Vilalta, R., Lindner, P., Lefer, B., Macias, C., & Price, D. (2016). Analysis of correlation between pediatric asthma exacerbation and exposure to pollutant mixtures with association rule mining. *Artificial Intelligence in Medicine, 74*, 44–52. doi:10.1016/j.artmed.2016.11.003 PMID:27964802

Transparency International. (2019). *Ghana*. Retrieved from https://www.transparency.org/country/GHA

Tree Nation. (2019). *List of active projects*. Retrieved from www.tree.nation.com

Trendsetter. (2006). *Evaluation Report- Trendsetter*. Trendsetter Report No 2005:16, Trendsetter Deliverable No 4.7, Jun 2006.

Trifonov, R., Nakov, O., & Mladenov, V. (2017). *Artificial Intelligence in Cyber Threat Intelligence.* Sofia: Technical University of Sofia.

Turan, B., Demir, K. A., Soner, B., & Ergen, S. C. (2019). Visible Light Communications in Industrial Internet of Things (IIoT). In *The Internet of Things in the Industrial Sector* (pp. 163–191). Cham: Springer.

Tweneboah-Koduah, S., Skouby, K. E., & Tadayoni, R. (2017). Cyber Security Threats to IoT Applications and Service Domains. *Wireless Personal Communications*, *95*(1), 169–185. doi:10.100711277-017-4434-6

Twerefou, D. K. (2012). *Report on Sustainable Development Goals for the West Africa Subregion.* United Nations Economic and Social Council Economic Commission for Africa. Retrieved from https://www.uneca.org/sites/default/files/uploaded-documents/SDG/2013/sdg2013_draft-sdgs-report-west-africa_en.pdf

Tyugu, E. (2013). Artificial Intelligence in Cyber Defense. *International Conference on Cyber Conflict.*

UN Climate Change. (2016). *The Paris Agreement.* Retrieved from unfccc.int/process-and-meetings/the-paris-agreement/the-paris-agreement

UN ECOSOC (Economic and Social Council) Report. (2016). *Smart Cities and Infrastructure Report 26 February 2016.* UN.

UN SDGs. (2015). *Envision 2030: 17 goals to transform the world for persons with disabilities.* Retrieved from https://www.un.org/development/desa/disabilities/envision2030.html

UN. (2008). *United Smart Cities (USC).* Retrieved from https://sustainabledevelopment.un.org/partnership/?p=10009

UNE. (2017). *Sistemas Integrales de Gestión de la Ciudad Inteligente. Requisitos de interoperabilidad para una Plataforma de Ciudad Inteligente.* Madrid: UNE.

UNESCAP. (2019). *Water efficient infrastructure and water-smart cities (the case of Korea).* Retrieved from unescap.org/sites/default/files/session%205-Yongdeok%20Cho-AWC.pdf

UN-Habitat. (2013). *State of the World's cities 2012/2013 Prosperity of Cities.* UN-Habitat (United Nations Human Settlements Programme).

UN-Habitat. (2018). *SDG Indicator 11.1.1 Training Module: Adequate housing and slum upgrading.* United Nations Human Settlement Programme (UN-Habitat), Nairobi. Retrieved from https://unhabitat.org/wp-content/uploads/2019/04/Indicator-11.1.1-Training-Module.pdf

United Nations (UN). (2008). Benchmarking E-Government: A Global Perspective. Division of Public Economics and Public Administration Report United Nations, Department of Economic and Social Affairs, Population Division: World Urbanization Prospects, the 2011 Revision.

United Nations Department of Economic and Social Affairs. (2018). *68% of the world population projected to live in urban areas by 2050, says UN.* https://www.un.org/development/desa/en/news/population/2018-revision-of-world-urbanization-prospects.html

United Nations. (2014). *Cyber School Bus. Cities of Today, Cities of Tomorrow! Unit 1: What is a City?* Retrieved from https://www.un.org/cyberschoolbus/habitat/units/un01txt.asp

United Nations. (2019). *World Urbanization Prospects: The 2018 Revision (ST/ESA/SER.A/420).* Department of Economic and Social Affairs, Population Division. https://population.un.org/wup/Publications/Files/WUP2018-Report.pdf

United Nations. (2019a). *Sustainable development goals.* Retrieved from https://sustainabledevelopment.un.org/sdgs

United Nations. (2019b). *Sustainable Development Goal 11. Make cities and human settlements inclusive, safe, resilient and sustainable*. Retrieved from https://sustainabledevelopment.un.org/sdg11

United Nations. (2019c). *Growing at a slower pace, world population is expected to reach 9.7 billion in 2050 and could peak at nearly 11 billion around 2100*. Retrieved from https://www.un.org/development/desa/en/news/population/world-population-prospects-2019.html

USAID. (2019). *Agriculture and food security*. Retrieved from https://www.usaid.gov/west-africa-regional/agriculture-and-food-security

USDA. (2019). *Ghana Fish and Seafood Report*. Retrieved from https://apps.fas.usda.gov/newgainapi/api/report/downloadreportbyfilename?filename=Fish%20and%20Seafood%20Report_Accra_Ghana_3-8-2019.pdf

Vahlkamp, S. (2016). *de-klimaatstraat-duurzaam-ondernemen-en-wonen*. http://buurtwinkels.amsterdammuseum.nl/783/nl/de-klimaatstraat-duurzaam-ondernemen-en-wonen

van den Bosch, H. (2017). *Smart Beyond Technology Push*. https://smartcityhub.com/technology-innnovation/smart-beyond-technology-push/

Van Zoonen, L. (2016). Privacy concerns in smart cities. *Government Information Quarterly*, *33*(1), 472–480. doi:10.1016/j.giq.2016.06.004

Varmedja, D., Karanovic, S., Arsenovic, M., & Anderla, A. (2019). Credit Card Fraud Detection - Machine Learning methods. *2019 18th International Symposium INFOTEH-JAHORINA (INFOTEH)*.

Veruggio, G. (2007). *EURON Roboethics Roadmap. Release 1.2. January 2007*. Retrieved from http://www.roboethics.org/index_file/Roboethics%20Roadmap%20Rel.1.2.pdf

Veruggio, G., & Operto, F. (2008). Roboethics: Social and Ethical Implications. In B. Siciliano & O. Khatib (Eds.), *Springer Handbook of Robotics* (pp. 1499–1524). Springer International Publishing.

Verutes, G. M., Fiocco, M. B., Weeks, J. R., & Coulter, L. L. (2012). Health, poverty, and place in Accra, Ghana: Mapping Neighbourhoods. *Journal of Maps*, *8*(4), 369–373. doi:10.1080/17445647.2012.744365 PMID:23505395

Visser, H. J., & Vullers, R. J. (2013). RF energy harvesting and transport for wireless sensor network applications: Principles and requirements. *Proceedings of the IEEE*, *101*(6), 1410–1423.

Vitunskaite, M., He, Y., Brandstetter, T., & Janicke, H. (2019). Smart Cities and Cyber Security: Are We There yet? A Comparative Study on the Role of Standards, Third Party Risk Management and Security Ownership. *Computers & Security*, *83*, 313–331. doi:10.1016/j.cose.2019.02.009

Vuckovic, D. (2014). *Design and implementation of energy harvesting powered wireless sensor networks* (PhD Thesis). Technical University of Denmark (DTU).

Wahyudi, M. B., & Kalamullah, R. (2016). *Improving Intrusion Detection System Detection Accuracy and Reducing Learning Time by Combining Selected Features Selection and Parameters Optimization*. IEEE.

Walsh, N. P. (2019). *World's Largest 3D-Printed Concrete Pedestrian Bridge Completed in China*. https://www.archdaily.com/909534/worlds-largest-3d-printed-concrete-pedestrian-bridge-completed-in-china

Wang, B. Y., & Zhang, S. M. (2003, November). A mining algorithm for fuzzy weighted association rules. In *Proceedings of the 2003 International Conference on Machine Learning and Cybernetics (IEEE Cat. No. 03EX693)* (Vol. 4, pp. 2495-2499). IEEE. 10.1109/ICMLC.2003.1259932

Wang, F.-Y. (2010). The emergence of intelligent enterprises: From CPS to CPSS. *IEEE Intelligent Systems*, *25*(4), 85–88. doi:10.1109/MIS.2010.104

Wang, J., Li, H., Huang, J., & Su, C. (2016). Association rules mining based analysis of consequential alarm sequences in chemical processes. *Journal of Loss Prevention in the Process Industries*, *41*, 178–185. doi:10.1016/j.jlp.2016.03.022

Wang, W., Yang, J., & Philip, S. Y. (2000). *Efficient mining of weighted association rules (WAR). IBM Thomas J.* Watson Research Division.

Wang, Z., Tsonev, D., Videv, S., & Haas, H. (2015). On the design of a solar-panel receiver for optical wireless communications with simultaneous energy harvesting. *IEEE Journal on Selected Areas in Communications*, *33*(8), 1612–1623.

Wan, J., Zhang, D., Zhao, S., Yang, L., & Lloret, J. (2014). Context-aware vehicular cyber-physical systems with cloud support: Architecture, challenges, and solutions. *IEEE Communications Magazine*, *52*(8), 106–113. doi:10.1109/MCOM.2014.6871677

Warnick, J. (2013). *88 Acres How Microsoft Quietly Built the City of the Future*. https://www.microsoft.com/en-us/stories/88acres/

Water Technology. (2019). *Sorek Desalination Plant*. https://www.water-technology.net/projects/sorek-desalination-plant/

Weeks, S. (2019). *Whitepaper: Smart Cities - Beyond the hype*. Smart Cities World in Association with Interact.

Wein, C. (2019). *10 Millionen Fahrten bei Citybike Wien!* https://www.citybikewien.at/en/about-citybike/press/596-neuer-meilenstein-bei-citybike-wien-erreicht-2

Weiser, M. (1991). The computer for the 21st century. *Scientific American*, *265*(3), 94–104. PMID:1675486

Wen, J., Zhong, M., & Wang, Z. (2015). Activity recognition with weighted frequent patterns mining in smart environments. *Expert Systems with Applications*, *42*(17-18), 6423–6432. doi:10.1016/j.eswa.2015.04.020

Weyrich, M., & Ebert, C. (2016). Reference Architectures for the Internet of Things. *IEEE Software*, *33*(1), 112–116. doi:10.1109/MS.2016.20

Wharton. (2019). *What's Fueling the Smart City Backlash?* https://knowledge.wharton.upenn.edu/article/whats-behind-backlash-smart-cities/

Whitehead, D. E., Owens, K., Gammel, D., & Smith, J. (2017). *Ukraine Cyber-Indued Power Outage: Analysis and Practical Mitigation Strategies*. Houston: IEEE.

Whiting, N. (2018). Cyberspace Triggers a New Kind of Arms Race. *Signal*, *72*(6), 38–40.

WHO. (2019). *Health and sustainable development: Improved urban waste management*. Retrieved from who.int/sustainable-development-cities/strategies/urban-waste/en/

Whyte, C. (2019). *Google claims it has finally reached quantum supremacy*. https://www.newscientist.com/article/2217347-google-claims-it-has-finally-reached-quantum-supremacy/

Wiig, A. (2015). The empty rhetoric of the smart city: From digital inclusion to economic promotion in Philadelphia. *Urban Geography*, *37*(4), 535–553. doi:10.1080/02723638.2015.1065686

Williams, H. (2018). *What is a smart city? How to define a smart city*. Retrieved from computerworld.com/article/3427835/what-is-a-smart-city-how-to-define- a smart-city.html

Wisker, G. (2010). *The postgraduate research handbook: Succeed with your MA, MPhil, EdD and PhD*. Palgrave Macmillan.

Witte, A. D. (1996). Urban Crime: Issues and Policies. *Housing Policy Debate*, *7*(4), 731–749.

World Atlas. (2019). *What is the von Thunen model?* Retrieved from worldatlas.com-what-is-the von-thunen-model.html

World Bank. (2007). *Ghana country environmental analysis*. Retrieved from http://documents.worldbank.org/curated/en/719681468032166880/pdf/369850GH.pdf

World Bank. (2019). *Sustainable Energy for All (SE4ALL) database. Renewable energy consumption (% of total final energy consumption)*. https://data.worldbank.org/indicator/EG.FEC.RNEW.ZS?end=2015&start=1990,World

World Bank. (2019a). *Overview*. Retrieved from www.worldbank.org/en/country/ghana/overview

World Bank. (2019b). *Poverty overview*. Retrieved from https://www.worldbank.org/en/topic/poverty/overview

World Economic Forum. (2018). *Slums could inspire the cities of the future. Here's how*. Retrieved from https://www.weforum.org/agenda/2018/01/this-startup-is-turning-slums-into-microcities/

World Economic Forum. (2019). *What slums can teach us about building the cities of the future*. Retrieved from https://www.weforum.org/agenda/2019/03/why-slums-could-provide-the-housing-inspiration-of-the-future/

WorldData. (2019) *Climate for Greater Accra (Ghana)*. Retrieved from https://www.worlddata.info/africa/ghana/climate-greater-accra.php

Worth, N. (2013). Visual Impairment in the City: Young People's Social Strategies for Independent Mobility. *Urban Studies (Edinburgh, Scotland)*, *50*(3), 574–586. doi:10.1177/0042098012468898

Wu, W., Pirbhulal, S., Zhang, H., & Mukhopadhyay, S. C. (2019). Quantitative Assessment for Self-Tracking of Acute Stress Based on Triangulation Principle in a Wearable Sensor System. *IEEE Journal of Biomedical and Health Informatics, 23*(2), 703-713.

Wu, X., Wang, Q., Liu, Y., & Li, Y. (2017, December). Data-driven regional analysis of urban atmosphere pollution based on density clustering. In *2017 IEEE 2nd Information Technology, Networking, Electronic and Automation Control Conference (ITNEC)* (pp. 412-417). IEEE. 10.1109/ITNEC.2017.8284764

Xuan, S., Liu, G., Li, Z., Zheng, L., Wang, S., & Jiang, C. (2018). Random Forest for Credit Card Fraud Detection. *2018 IEEE 15th International Conference on Networking, Sensing and Control (ICNSC)*.

Xu, J., Zeng, Y., & Zhang, R. (2018). UAV-enabled wireless power transfer: Trajectory design and energy optimization. *IEEE Transactions on Wireless Communications*, *17*(8), 5092–5106.

Yang, C. H., Tsai, S. W., Chuang, L. Y., & Yang, A. C. H. (2012). An improved particle swarm optimization with double-bottom chaotic maps for numerical optimization. *Applied Mathematics and Computation*, *219*(1), 260–279. doi:10.1016/j.amc.2012.06.015

Yan, J., & He, H. (2017). Cyber-physical attacks and defences in the smart grid: A survey. *IET Cyber-Physical Systems: Theory & Applications*, *1*(1), 13–27.

Yaqoob, I., Ahmed, E., Rehman, M. H., Ahmed, A. I. A., Al-garadi, M. A., Imran, M., & Guizani, M. (2017). The Rise of Ransomware and Emerging Security Challenges in the Internet of Things. *Computer Networks*, *129*, 444–458. doi:10.1016/j.comnet.2017.09.003

Yavuz, F. Y. (2018). *Deep Learning in Cyber Security for Internet of Things*. Graduate School of Natural and Applied Sciences.

Yin, C., Xiong, Z., Chen, H., Wang, J., Cooper, D., & David, B. (2015). A literature survey on smart cities. *Science China. Information Sciences, 58*(10), 1–18. doi:10.100711432-015-5397-4

Yovanof, G. S., & Hazapis, G. N. (2009). An architectural framework and enabling wireless technologies for digital cities & intelligent urban environments. *Wireless Personal Communications, 49*(3), 445–463. doi:10.100711277-009-9693-4

Yue, J. S., Tsang, E., Yeung, D., & Shi, D. (2000, October). Mining fuzzy association rules with weighted items. In *Smc 2000 conference proceedings. 2000 ieee international conference on systems, man and cybernetics.'cybernetics evolving to systems, humans, organizations, and their complex interactions'(cat. no. 0)* (Vol. 3, pp. 1906-1911). IEEE.

Yun, U., & Leggett, J. J. (2005, April). WFIM: weighted frequent itemset mining with a weight range and a minimum weight. In *Proceedings of the 2005 SIAM international conference on data mining* (pp. 636-640). Society for Industrial and Applied Mathematics. 10.1137/1.9781611972757.76

Yun, U., & Leggett, J. J. (2006, April). WIP: mining Weighted Interesting Patterns with a strong weight and/or support affinity. In *Proceedings of the 2006 SIAM International Conference on Data Mining* (pp. 624-628). Society for Industrial and Applied Mathematics. 10.1137/1.9781611972764.73

Zanghi, B. (2017). *4 Key Principles for Smart City Solutions.* https://www.masabi.com/2017/01/17/4-key-principles-for-smart-city-solutions/

Zarpelao, B. B., Miani, R. S., Kawakani, C. T., & de Alvarenga, S. C. (2017). A survey of intrusion detection in Internet of Things. *ACM Digital Library, 84*(C), 25–35. doi:10.1016/j.jnca.2017.02.009

Zavratnik, V., Kos, A., & Duh, E. S. (2018). Smart Villages: Comprehensive Review of Initiatives and Practices. *Sustainability, 10*(7), 2559. doi:10.3390u10072559

Zeng, Y., & Xiang, K. (2017). Adaptive Sampling for Urban Air Quality through Participatory Sensing. *Sensors (Basel), 17*(11), 2531. doi:10.339017112531 PMID:29099766

Zeng, Y., Zhang, R., & Lim, T. J. (2016). Wireless communications with unmanned aerial vehicles: Opportunities and challenges. *IEEE Communications Magazine, 54*(5), 36–42.

Zhang, K., Ni, J., Yang, K., Liang, X., Ren, J., & Shen, S. (2017, January). Security and Privacy in Smart City Applications: Challenges and Solutions. *Enabling Mobile and Wireless Technologies for Smart Cities: Part 1*, 122-129.

Zhang, Y., Meratnia, N., & Havinga, P. J. (2013). Distributed online outlier detection in wireless sensor networks using ellipsoidal support vector machine. *Ad Hoc Networks, 11*(3), 1062–1074. doi:10.1016/j.adhoc.2012.11.001

Zhang, Z., Chen, E., Wang, J., & Wang, X. (2004, April). Enabling personalization recommendation with WeightedFP for text information retrieval based on user-focus. In *International Conference on Information Technology: Coding and Computing, 2004. Proceedings. ITCC 2004.* (Vol. 1, pp. 272-276). IEEE. 10.1109/ITCC.2004.1286465

Zimmermann, H. (1980). OSI reference model - the ISO model of architecture for open systems interconnection. *IEEE Transactions on Communications, 28*(4), 425–432. doi:10.1109/TCOM.1980.1094702

Zygiaris, S. (2012). Smart city reference model: Assisting planners to conceptualize the building of smart city innovation ecosystems. *Journal of the Knowledge Economy, 4*(2), 217–231. doi:10.100713132-012-0089-4

About the Contributors

Zaigham Mahmood is a published author/editor of thirty books on subjects including Electronic Government, Cloud Computing, Data Science, Big Data, Fog Computing, Internet of Things, Internet of Vehicles, Industrial IoT, Smart Cities, Ambient Intelligence, Project Management, and Software Engineering, including: Cloud Computing: Concepts, Technology & Architecture which is also published in Korean and Chinese languages. Additionally, he is developing two new books to appear later in the year. He has also published more than 150 articles and book chapters and organized numerous conference tracks and workshops. Professor Mahmood is Editor-in-Chief of Journal of E-Government Studies and Best Practices, Editor-in-Chief of the IGI book series on E-Government and Digital Divide, Technology Consultant at Debesis Education UK and Professor at the Shijiazhuang Tiedao University in Hebei, China. He further holds positions as Foreign Professor at NUST and IIU in Islamabad Pakistan. He has also served as a Reader (Associated Professor) at the University of Derby UK, and Professor Extraordinaire at the North West University South Africa. Professor Mahmood is a certified cloud computing instructor and a regular speaker at international conferences devoted to Distributed Computing and E-Government. Professor Mahmood's book publications can be viewed at: https://www.amazon.co.uk/Zaigham-Mahmood/e/B00B29OIK6.

* * *

Wissam Abbass received her M.S. degree from the University of Science of Marrakech in Information systems security. She is a Ph.D. at the National Institute of Posts and Telecommunications (INPT) in Rabat. Her research interests include Computer networking, Web development, Network and software security, risk management and A.I.

Amine Baina received the PhD in Computer Science from National Institute of Applied Sciences (INSA) in Toulouse, France, in 2009. He is an Assistant Professor in computer science at the National Institute of Posts and Telecommunications (INPT) in Rabat, Morocco since 2010. His research interests are in critical infrastructures protection, Electrical Grid, Information Security.

Mostafa Bellafkih received the PhD in Computer Science from University of Paris, France, in June 1994 and Doctorat Es Science in Computer Science from University of Mohammed V in Rabat, Morocco, in May 2001. He is a professor at the National Institute of Posts and Telecommunications (INPT) in Rabat, Morocco since 1995. His research interests include network management, Systems security, knowledge management, AI, Data mining and Software Engineering.

Derya Birant received her B.S., M.S. and Ph.D. degrees in Computer Engineering from Dokuz Eylul University, Turkey in 2000, 2002 and 2006 respectively. Since 2017, she has been an Associate Professor at the Computer Engineering Department of Dokuz Eylul University; and is now the vice-chair of the Computer Engineering Department. She was a Visiting Lecturer at the South East European University in 2006 and Ege University between 2010 and 2012. She is the author of 5 book chapters and more than 70 publications as journal articles and conference papers. Dr. Birant has also served as an organizing committee member for several conferences; and has been involved in more than 20 long-term interdisciplinary R&D projects. Dr. Birant has several Most Downloaded Article certifications and has graduation awards, ranking 2nd in the Faculty and the Department. She was also the recipient of the Outstanding Achievement Award in 2010.

Darrold L. Cordes holds a Master of Business Administration and a Master of Science in Management Information Systems. He is undertaking a Doctor in Sustainable Development degree by research at the Curtin University Sustainability Policy Institute, Perth, Australia while concurrently completing a Master of Science in Applied Economics and Data Intelligence at the University of Nevada, Las Vegas. His research interests are in the application of E-Commerce, Machine Learning, Data Analytics, Blockchain, Spatial Data Analysis, Design Science Research, and Agent-based Modelling to potential solutions for low-income community development in Sub-Saharan Africa.

Kadir Alpaslan Demir is an Associate Professor and program manager for a large-scale defense system and an associate professor of management of information systems." with "is a program manager for a large-scale defense system and an associate professor of management of information systems. Dr. Demir has an M.S. in Computer Science and another M.S. in Software Engineering, and a Ph.D. in Software Engineering from Naval Postgraduate School, USA. He also has an MBA from Gebze Technical University, Turkey. Dr. Demir served as a faculty member of the department of computer engineering at the Turkish Naval Academy. He also has significant experience in the development and management of large-scale defense and C4I systems. His research interests include defense systems development, software engineering, unmanned aerial vehicles, IT and software project management, information systems, human-robot co-working, and roboethics.

Francisco J. Duarte has a degree in Informatics Engineering, an MSc in Computer Science, and a PhD in Software Engineering, all from University of Minho, Portugal. Currently, he is responsible for the "Digital and Software" Technical Unit at Altran Portugal, managing technologies for Frontend, Backend, Testing, Verification and Validation in software development projects, and for the provision of digital services. Francisco has worked as a software developer, project manager, lean production coordinator, quality and customer assistance manager for automotive, and as information systems manager. He also created two startups related to software for libraries and interoperability of business processes. He was the coordinator of Industry 4.0 for Bosch entities in Portugal and Spain. He has more than 15 years experience as Invited Professor at Universidade do Minho. He was Vice-president of the DTx Digital Transformation Collaborative Laboratory. He is a researcher at Centro Algoritmi, University of Minho.

Péricles Varella Gomes is at Universidade Positivo in Brazil. He has a degree in Architecture and Urban Design from Universidade Federal do Paraná, a Master's Degree in Telecommunications and Multimidia from Michigan State University, and Ph.D in Educational Systems Development from Michigan

State University obtained in 1996. Gomes has experience in Usability, Virtual Reality and Gamification, with a focus on Information. He is a Senior Researcher at Positivo University Research Center, Curitiba, Brazil and a Visiting Professor at Michigan State University and at University Grenoble Alps France. Gomes plays the Cello and is also a Visual Artist. He has also been a researcher in Animal Welfare Indicators with the European Union from 2009 to 2016. Gomes is currently a Professor in Architecture for the Masters in Smart Cities at Positivo University in Brazil.

Fida Hussain is a full time PhD student at Canterbury Christ Church University, working on IoT to develop a security framework for Smart Home Automation. In 2018, Fida published a paper on smart home security entitled "Integrated Security Scheme for Smart Home" in International Conference on Natural Computation, Fuzzy Systems and Knowledge Discovery, Huangshan, China. From 2014 to 2017, Fida worked as IT Consultant for a hospitality company, after graduating with MSc in Computing from Canterbury Christ Church University in 2012. From 2012 to 2014 Fida also worked for two different charities as part time IT Support.

Abhaya Induruwa FIET, FIESL, CEng, FBCS, CITP is the inaugural endowed Professor of Computing at the University of Colombo School of Computing, Sri Lanka. He served as the Director of Cyber Innovation Hub at Canterbury Christ Church University, UK, and is now retired. He pioneered Computer Science & Engineering teaching in Sri Lanka at the University of Moratuwa in 1985, and in 1989, he pioneered data communication and Internet connectivity in Sri Lanka. He was inducted to the Internet Hall of Fame in 2014 and received the Internet Pioneer Award from ISOC-Sri Lanka in 2015. Sri Lanka Telecom conferred a Life Time Achievement Award in 2017 for his services to the digital arena. In 2019 he received the LEARN Founder Award for proposing the conceptual design in 1989, and implementing the national academic and research network (LEARN) in Sri Lanka which he connected to the Internet in 1995. He served on the Internet Hall of Fame Advisory Board from 2016 until 2018. He is a contributing author to the Asia Internet History project.

Rian Leevinson J. is a Senior Associate Analyst at Infosys Limited India with over two years of experience, working in the field of data science, machine learning, NLP, statistics, computer vision and predictive analytics. His research areas include applications of Industrial IoT, emerging technologies for smart cities, intelligent traffic management systems in smart cities, chatbot testing, quantum machine learning and emerging technologies. He has experience in building data driven solutions to industrial problems. He is skilled in Python, R, Rshiny, Azure ML Studio, Azure data services and Azure AutoML. His areas of interest include data science, advanced machine learning, quantum computing, intelligent autonomous systems and astronomy.

A. S. M. Kayes received his Ph.D. degree from Swinburne University of Technology, Australia, in 2014. He is currently a Lecturer in cybersecurity with the Department of Computer Science and Information Technology at La Trobe University, Australia. His research interests include information modelling, context-aware access control, big data integration, IoT, cloud and fog computing, advanced data analytics, fuzzy computation, and security and privacy protection.

Ricardo J. Machado is a full professor of Information Systems Engineering and Technology in the Dept. of Information Systems at the University of Minho (UMinho), School of Engineering. Within the Information Systems Engineering domain, his primary research interests are in modelling and requirements for systems analysis and design and in process and project management life-cycles. He has supervised 50 PhD and MSc theses in these areas. He has published over 150 scientific publications and has four industrial patents, and has also acted as coordinator and senior researcher of over 50 R&D projects. Currently, at UMinho, he is the vice-rector for institutional development, the director of the ALGORITMI Research Centre, the scientific coordinator of the EPMQ Laboratory at the CCG/ZGDV Institute, the director of the Doctoral Program in Advanced Engineering Systems for Industry, member of the Board of the UM-Cities Platform, and vice-president of the DTx Digital Transformation Collaborative Laboratory.

Omaru Maruatona is a Cyber security and Machine Learning practitioner with deep industry and research experience. Omaru has previously worked with a big Australian bank to investigate Machine Learning based approaches to Internet Banking fraud detection. He has also worked for a global financial services company as a Security Analyst and has consulted for PwC Australia in Cyber security architecture and strategy. Omaru is a recognised thought leader in the application of AI and smart cities and has published and spoken at various organisations and conferences in Australia.

Paula Monteiro is a researcher at CCG. She received the Lic. degree in Mathematics and Computer Science and the MSc in Informatics from the University of Minho; and has expertise in Distributed Systems, Computer Communications and Computer Architecture. She received the Ph.D. degree in Computer Science from University of Minho, University of Aveiro and University of Porto (joint Doctoral Programme). In the last 20 years, she has been involved in the development and implementation of several projects in the area of Software Engineering (model-driven development, requirements engineering, software quality). Paula is author of scientific publications in the area of Model-Driven Development and in the area of Software Process Improvement. She has already participated in European and national projects as a researcher.

Greg Morrison is the Director of the Curtin University Sustainability Policy Institute, Perth, Australia. He is the professor in Sustainable Cities. Professor Morrison has successfully initiated and run large-scale societal infrastructure related projects, with the most recent being the smart cities energy and water peer-to-peer trading projects in Fremantle, Australia.

Alex Ng holds a PhD degree in Computer Science, and a Master of Technology degree in Software Engineering both from Macquarie University. He is currently engaged as a Lecturer in Cybersecurity position at La Trobe University in Australia, and a Visiting Honorary Research Fellow position in the School of Medical Science at RMIT University. His research interests span the fields of Blockchain Security, Blockchain-based B2B Collaboration, AI-based Security Defence for Intelligent Cities, Malware Detection and Prevention Mechanisms and Chromatographic Fingerprinting techniques. He is a Certified Information Security Manager (CISM) and has been a senior member of the IEEE since 2018.

Jared Piconi is a cyber security professional with multiple theoretical and practical qualifications in the field. He graduated from La Trobe University with a Master of Cybersecurity (Computer Science) and has undertaken many additional industry leading certifications. Jared's previous degree and career in urban development provides him with a unique pairing of qualifications, giving him extensive understanding of the current needs of cyber security within smart cities and how to integrate technical solutions practically. Jared's research areas and specific interests include machine learning to aid cybersecurity, smart cities and their digital integration, wireless access point security, personal digital security, data handling and cybercrime incident handling.

Man Qi is a Senior Lecturer in Computing at Canterbury Christ Church University. Her research interests are in Cyber Security, Intelligent Computing and Applications, Internet of Things and HCI. She has been successfully funded by QR (Quality-related Research), the university research and internship scheme. Dr Qi has published over 70 research papers and is the editorial board member for 5 international journals. She has been a PhD external examiner for many Universities in Australia and UK. Dr Qi has served as chair/programme committee member for around 50 international conferences and been a long-term reviewer for many international journals. Dr Qi is a Fellow of British Computer Society (FBCS) and Fellow of Higher Education Academy (FHEA).

Verônica Isabela Quandt is at Universidade Positivo in Brazil. She earned her PhD in Biomedical Engineering from the Federal University of Technology in Paraná in 2015. She completed a master's degree in Aeronautical Engineering at the Technological Institute of Aeronautics in 2008 and a master's degree in IT from Mannheim University of Applied Sciences in Germany in 2004. Dr Quandt is currently an assistant professor at Universidade Positivo (UP) in Curitiba, Brazil, working on the courses of Electrical Engineering, Computer Engineering and Computer Science. She is a participant in the Research Group in Technology and Applied Science (CDTI) of Universidade Positivo (cdti.org.br), working in the research lines of Smart Cities and Services, and in the Well-Being and Digital Health line. She has experience in the areas of Cognitive Computing, AI, Signal Analysis and the Internet of Things.

Akanksha Rajendra Singh is a Consultant in Innovation Partner Program at Infosys Limited in India, with over 9 years of industry experience in digital and emerging technologies. She dons dual hats as Innovation Manager and Go-To-Market Consultant with elan. Her work involves consulting accounts on their innovation strategy, setting up innovation pipelines and defining Go-To-Market strategy for the Infosys Center of Emerging Technology Solutions. Apprising clients on the innovation culture at Infosys, spotting relevant mega trends, seeking their participation, and designing intellect property solutions for business needs while managing stakeholders, are a significant part of her work. Akanksha is passionate about working at the convergence point of business, technology, and solutions. She has a Financial Services background, and keeps a close tab on the technological advances in the BFSI sector.

Kwesi Atta Sakyi is a Lecturer at ZCAS University in Lusaka. He has teaching experience in secondary schools and colleges in Ghana, Nigeria, and Zambia. His teaching experience spans half a century. He has taught Economics and Business Management at the Int. School of Lusaka before moving on to Zambia Centre for Accountancy Studies to teach Strategic Management, Managing and Leading, and Research and Scholarship. Sakyi has a BA (Hons) degree in Economics from the University of Ghana, NDP and MPA from UNISA, and recently completed an MBA from UNZA/AstriaLearning. He has published two poetry books and has journal articles on Google Scholar. In July 2015, he became a member of the Oxford Round Table. He is an Executive member of the Economics Association of Zambia. His hobbies include photography and writing.

Dago Schelin is at Philipps University of Marburg, Germany, where he earned his PhD degree in Media Studies. His research focuses on the intersections of concepts of vision and blindness within media archeology both in theory and in practice. Dago Schelin also has a Masters in Media Production from the Hochschule Ostwestaflen-Lippe University of Applied Sciences, Germany, a B.A. in Popular Music from Faculdade de Artes do Paraná, Brazil, and a B.A. in Languages English/Portuguese from Pontifícia Universidade Católica do Paraná. His most recent book is titled Vision and Blindness in Film.

Nuno Soares is a University of Minho/CCG (Centro de Computação Gráfica) researcher and integrates the SEMAG (Software Engineering and Management Group) at ALGORITMI Research Centre. He holds a DEng degree in Engineering Systems and Computing and an MSc in Computing, specialization in Information Management, both from University of Minho (UMinho). He is currently pursuing a PhD in Advanced Engineering Systems for Industry (AESI@UMinho) with Bosch Car Multimedia Portugal as an industrial partner. He was an Associate Professor for 15 years, teaching Computing and Management in two private Universities and has been involved in research projects since 2014. His main areas of interest are Smart & Digital Cities and Digital Factories comprising their organisational, societal and technological derivations, Semantic Web, Linked Data, Web of Things, and other technical fields of semantic interoperability in the domains of IoT and Industry 4.0.

Buğra Turan is a principal researcher at the Vehicular Connectivity Research group of Wireless Networks Laboratory, Koc University. Mr. Turan received his M.S. degree from the University of Southern California, USA and is currently pursuing a Ph.D. at Koc University, Istanbul, Turkey, in electrical engineering. From 2010 to 2012, he was involved in Urban Mobility projects with Volkswagen Electronics Research Laboratory and the University of Southern California. From 2014 to 2018, he served as the Vehicular Connectivity Program coordinator for inter-vehicular and intra-vehicular connectivity projects between Koc University, Ozyegin University, Tofas Fiat and Ford OTOSAN. He also served as a Vehicular Connectivity Board Member in the high advisory council of TAYSAD, TOSB, 5G VATS, EU Instrument for Pre-Accession Assistance (IPA) Programme and Investment Office of the Presidency of Turkey. His research interests include Vehicular Communications, 5G, and Unmanned Air Vehicle Communications.

Goksu Tuysuzoglu received her BS in Information Systems Engineering in 2013 from Dogus University, Turkey. In the same year, she also received her double major in Industrial Engineering. Later, she received her MS in the department of computer engineering from Istanbul Technical University, Turkey in 2016. At the same time, she worked as a research and teaching assistant in there from 2014 to 2016. She is currently a PhD student in the department of computer engineering at Dokuz Eylul University, Turkey, and has worked as a research and teaching assistant in the same department since 2016. She has BS graduation awards with ranking 1st in the Department and ranking 3rd in the Faculty and University. Her research interests include data mining and machine learning.

Vijayaraghavan Varadharajan is a Principal Research Scientist at Infosys Limited in India, researching in the field of XAI, FoW, Security analytics, Cloud security, Security assessment, Authentication and Privacy protection. He has over 19 years of experience as a researcher in industry and academia. He also focuses on Emerging technology & Business opportunity identification, Incubation and Venture assessment for Investments. He has also led a security track in Sensor based Engagement for Improved

Health (SENSeI) project in COMMIT, a research organization in Netherlands. Dr. Vijay has six granted US patents and filed many US and Indian patents in key technology areas. He has published 50+ research publications in International journals & conferences and served as a Technical Reviewer, Program Committee member and Chair for many conferences around the globe.

Paul Watters began his first R&D role in security in 2002, joining the CSIRO's Networking Applications and Technologies (NAT) Group, and leading a programme in secure and distributed storage. In 2013, he took up a Professorship in IT at Massey University, New Zealand. In 2015, he became Adjunct Professor at Unitec Institute of Technology, the home of New Zealand's first Cyber Security Research Centre. He is currently a Professor and a Leading Expert in cyber security with the Department of Computer Science and Information Technology at La Trobe University in Australia. In recognition of his track record combating child abuse material online, he received an ARC Discovery grant in 2015.

Pornpit Wongthongtham is the recipient of a PhD in Information Systems, Master of Science in Computer Science, and Bachelor of Science in Mathematics. She is currently a research fellow at the Curtin University Sustainability Policy Institute, Perth, Australia. Her research interests include Data Analytics, Big Data, Artificial Intelligence, Blockchain Technology, and the like.

Index

K

L

M

N

O

P

Q

R

S

T

U

V

W

Z

www.ingramcontent.com/pod-product-compliance
Lightning Source LLC
LaVergne TN
LVHW060408300926
849162LV00006B/1369
* 9 7 8 1 7 9 9 8 5 0 6 2 5 *